TURBULENCE IN INTERNAL FLOWS

A PROJECT SQUID WORKSHOP

TURBULENCE IN INTERNAL FLOWS

Turbomachinery and Other Engineering Applications

EDITED BY

S. N. B. MURTHY

School of Mechanical Engineering
Purdue University
West Lafayette, Indiana

HEMISPHERE PUBLISHING CORPORATION
Washington London

Library of Congress Cataloging in Publication Data

Project Squid Workshop on Turbulence in Internal Flows: Turbomachinery and Other Applications, Airlie House, 1976.
Turbulence in internal flows.

At head of title: A Project Squid workshop.
Sponsored by the Naval Air Systems Command and the Office of Naval Research (Fluid Dynamics Program and Power Program).
Includes index.
1. Fluid dynamics–Congresses. 2. Turbulence–Congresses. 3. Turbomachines–Aerodynamics–Congresses. I. Murthy, S. N. B. II. Project Squid. III. United States. Naval Air Systems Command. IV. United States. Office of Naval Research. V. Title.
TA357.P75 1976 620.1'064 77-15615
ISBN 0-89116-073-6

Proceedings of the Project SQUID WORKSHOP ON TURBULENCE IN INTERNAL FLOWS: Turbomachinery and Other Applications, held at Airlie House, Warrenton, Va., June 14–15, 1976, and sponsored by the Naval Air Systems Command and the Office of Naval Research (Fluid Dynamics Program and Power Program).

1 2 3 4 5 6 7 8 9 0 DODO 7 8 3 2 1 0 9 8 7

CONTENTS

Preface vii

Thirty Years of Research ix
James R. Patton, Jr.

Welcoming Remarks xi
James R. Patton, Jr.

Introduction 1
S. N. B. Murthy

PART I
FUNDAMENTAL PROBLEMS

Preliminary Report on Sheared Cellular Motion as a Qualitative Model of Homogeneous Turbulent Shear Flow 11
Stanley Corrsin and *W. Kollmann*

Interacting Shear Layers in Turbomachines and Diffusers 35
Peter Bradshaw

Some Preliminary Observations on the Effect of Initial Conditions on the Structure of the Two-Dimensional Turbulent Mixing Layer 67
D. Oster, I. Wygnanski, and *H. Fiedler*

On the Developing Region of a Plane Mixing Layer 89
Stanley F. Birch

The Effects of an External Turbulent Uniform Shear Flow on a Turbulent Boundary Layer 101
Q. A. Ahmad, R. E. Luxton, and *R. A. Antonia*

The Effect of Swirl on the Turbulence Structure of Jets 113
R. E. Falco

Implications of the Structure of the Viscous Wall Layer 131
J. D. A. Walker and *D. E. Abbott*

Further Investigation of the Linear and Nonlinear Theory for Constant-Temperature Hot-Wire Anemometers 169
Peter Freymuth

PART II
MODELLING PROCEDURES

Results of a Two Equation Model for Turbulent Flows and Development of a Relaxation Stress Model for Application to Straining and Rotating Flows 191
P. G. Saffman

Application of the Turbulence-Model Transition-Prediction Method to Flight-Test Vehicles 233
T. L. Chambers and *D. C. Wilcox*

A Second Moment Turbulence Model Applied to Fully Separated Flows 249
M. Briggs, G. Mellor, and *T. Yamada*

The Modelling of a Turbulent Near Wake Using the Interactive Hypothesis 283
B. S. Ng and *G. David Huffman*

Some Important Physical Phenomena in Flows with Separated Turbulent Boundary Layers 311
Roger L. Simpson

Turbulence Velocity Scales for Swirling Flows 347
Ronald M. C. So

Effect of Freestream Turbulence and Initial Boundary Layers on the Development of Turbulent Mixing Layers 371
Otto Leuchter

PART III
TURBOMACHINERY APPLICATIONS

Measurements in Curved Flows 407
J. A. C. Humphrey and *J. H. Whitelaw*

Small Disturbances in a Compressor with Strong Swirl 439
Jack L. Kerrebrock

Wake Cutting Experiments 463
Leslie S. G. Kovasznay

Some Turbulence and Unsteadiness Effects in Turbomachinery 485
R. L. Evans

Visual Study of Oscillating Flow over a Stationary Airfoil 517
A. A. Fejer

Generation, Measurement and Suppression of Large Scale Vorticity in Internal Flows 521
R. A. Wigeland, M. Ahmed, and *H. M. Nagib*

PART IV
PANEL DISCUSSION AND SUMMARY

Panel Discussion 537

Summary Report 563
S. N. B. Murthy

Index 569

PREFACE

Turbulence is an important aspect of the fluid mechanics of turbomachinery including ducting and diffuser. Whether one is concerned with losses under steady state conditions or with flow distortion and nonsteady aerodynamic and aeroelastic forces, the presence of turbulence introduces many complexities. In general, turbulent shear flows in such machinery become complicated by distortion and interaction processes arising on account of geometry and flow field.

This volume is devoted to a consideration of such complex turbulent shear flows. It contains the contributions, papers, and discussions on them, made by a distinguished body of workers in the field at a Project SQUID Workshop. The contributions are divided into four groups. The first group is devoted to a discussion of some of the current ideas on turbulent structure in shear flows. In the second group, there are contributions dealing with analysis and measurement, including predictive techniques and numerical computation methods. Turbulent flow considerations in turbomachinery are discussed in the third group. Finally, a report of a panel discussion is presented that synthesizes the principal themes discussed at the workshop.

Project SQUID workshops are intended to establish directions in which further research will prove enlightening and useful in various areas. The outcome of this workshop is due, of course, entirely to the contributors to this volume. Professor Hans Liepmann should be mentioned specially for his active participation and perceptive contribution to the discussions.

The workshop was cosponsored by the Fluid Dynamics Program of the Office of Naval Research, the Naval Air Systems Command, and the Power Program of the Office of Naval Research. Dr. Ralph Cooper, Dr. Herb Mueller, and Mr. James R. Patton, Jr., supported the idea of the workshop and, in addition to providing valuable financial assistance, contributed in many ways in the formulation of the workshop program. Mr. Patton has very kindly written in the immediately following pages both about the association of this workshop with an anniversary of the ONR as well as about the background against which Project SQUID Workshops are organized.

Mrs. Barbara Pounds typed the entire manuscript, and this volume is a testimony to her careful and elegant workmanship. The artwork was done primarily by Mr. Kurt R. Sacksteder, and Mrs. Amanda Niemantsverdriet assisted in many ways in the preparation of the final manuscript. Each of these persons deserves the editor's appreciation.

Turbulent flows continue to present basic and far-reaching challenges. It is hoped that this volume will provide some inspiration to those engaged in understanding the nature of turbulent flows in the context of turbomachinery and in some respects in a wider context.

S. N. B. Murthy
Editor and Workshop Chairman

THIRTY YEARS OF RESEARCH

The main objective of the workshop on Turbulence in Internal Flows was to be a forum for assessing the direction of scientific and technological progress in this area. A secondary goal was to bring theoreticians and practical design groups together from industry, universities, and government, in an effort to use their combined talents. There is a close relation of turbulence problems in turbomachines with the disciplines covered in other SQUID workshops, such as measurement techniques and aerodynamic-structural integration (blade flutter), etc.

There has been a resurgence of interest in recent years in basic problems relating to air-breathing engines due to increasing demands for higher performance, for smaller and lighter-weight power plants, and for operation over a wider range of operating conditions. This interest is exemplified in the current need to develop deeper fundamental understanding of the physical phenomena involved in all aspects of engine design and development.

To satisfy this need, the Power Program of the Office of Naval Research, through Project SQUID, initiated a series of workshops to consider selected topics from the standpoint of (1) a critical evaluation of current efforts, (2) determining the extent of agreement in explaining various phenomena associated with the subject, and (3) discussion of possible new approaches to solution of problem areas. These workshops have been held:

- Research in Gas Dynamics of Jet Engines, ONR/Chicago, December 4-5, 1969; R. Goulard and M. L'Ecuyer, eds. Project SQUID Report.
- Fluid Dynamics of Unsteady 3-D Separated Flows, Georgia Institute of Technology, June 10-11, 1971; F. J. Marshall, ed. NTIS AD736248.
- The Use of the Laser Doppler Velocimeter for Flow Measurements, Purdue University, March 9-10, 1972; W. H. Stevenson and H. D. Thompson, eds. NTIS AD753243.
- Aeroelasticity in Turbomachines, Detroit Diesel Allison, June 1-2, 1972, S. Fleeter, ed. NTIS AD749680.
- Laser Raman Diagnostics, General Electric Research & Development Center, May 10-11, 1973; M. Lapp and C. M. Penney, eds. 1974. *Laser raman gas diagnostics*. New York: Plenum Press. Also NTIS AD782652.
- Second International Workshop on Laser Velocimetry, Purdue University, March 27-29, 1974; H. D. Thompson and W. H. Stevenson, eds. NTIS AD010223.
- Turbulent Mixing: Nonreactive and Reactive Flows, May 20-21, 1974; S. N. B. Murthy, ed. 1975. *Turbulent mixing in nonreactive and reactive flows*. New York: Plenum Press. Also NTIS AD006322.
- Unsteady Flows in Jet Engines, United Aircraft Research Laboratory, (UARL) now (UTRC), July 11-12, 1974; F. O. Carta, ed. NTIS AD003853.

- Combustion Measurements in Jet Propulsion Systems, Purdue University, May 22-23, 1975; R. Goulard, ed. *Combustion measurements: Modern techniques and instrumentation.* 1976. Washington, D.C.: Hemisphere Publishing Corporation.
- Transonic Flow Problems in Turbomachinery, Naval Postgraduate School, February 11-13, 1976; T. C. Adamson, Jr., and M. F. Platzer, eds. 1977. *Transonic flow problems in turbomachinery*. Washington, D.C.: Hemisphere Publishing Corporation.
- Turbulence in Internal Flows, Airlie House, Warrenton, Va., June 14-15, 1976 (*this volume*).

As the time of this workshop, the Office of Naval Research was actively planning its thirtieth anniversary for 1976. The year 1976 also marks the thirtieth anniversary of Project SQUID. This volume is part of a set of scientific publications released by ONR in recognition of, and on the occasion of, its thirtieth anniversary.

James R. Patton, Jr.
Power Program
Office of Naval Research
U.S. Department of the Navy

Anniversary Theme:

Exploring new horizons to protect our heritage

WELCOMING REMARKS

It is a great pleasure, and I am privileged, on behalf of the Office of Naval Research, to welcome this group to the workshop on Turbulence in Internal Flows. We are particularly happy to have in attendance what we consider to be a well-rounded group, with representation from the academic community, industry, and interested government agencies. We are also happy to join with the Fluid Dynamics Program of ONR and with the Naval Air Systems Command in sponsorship of the workshop. As I am sure you have already observed the Airlie House conference center is a delightful location to hold a meeting like this one. It offers great potential as a forum for an effective and productive meeting while being removed from urban distractions.

The year 1976 happens to be the thirtieth anniversary of the Office of Naval Research. In recognition of this anniversary, under Project SQUID, we planned this workshop plus another, which was held February 11-12, 1976, on the subject "Transonic Flow Problems in Turbomachinery" at the Naval Postgraduate School, Monterey, California. It is planned to publish the proceedings of both workshops in recognition of the thirtieth anniversary. These two workshops, in turn, are part of a series of workshops we have held starting in 1969, all on basic subjects relating to air-breathing engines for aircraft and missile applications.

Proper understanding of turbulence and mixing phenomena is an essential ingredient of jet engine design. This is an area in which we have been devoting considerable research effort for several years. It has close relevance to other areas of research in which we are involved, such as aerodynamics, measurement techniques, and chemical kinetics and combustion. Such research efforts, designed to develop accurate, detailed understanding of the internal flows and interactions within jet engines are considered to be just as vital and integral a part of an air-breathing propulsion system as is the development of the component parts of the engine. We feel that reviewing the status of research efforts in this field with selected investigators and program managers will have a strong impact on the direction of future research and on the activities of the R&D community. In this manner, we seek to provide more relevant long-range support to aeropropulsion programs as well as to provide improved predictive techniques and design information necessary for more efficient, lighter weight, and more reliable power plants capable of meeting more stringent operating requirements.

In addition to the invitee list for this workshop, there are a number of people in attendance representing the Propulsion and Energetics and the Fluid Dynamics Panels of AGARD (NATO). AGARD has cooperated in arranging for the workshop because of its interest in the subject. The two panels of AGARD are in the process of forming a working group on Turbulent Transport Phenomena and the results of this workshop will be used in the future deliberations of that group.

On behalf of the sponsors, I wish to thank the organizer, Dr. S. N. B. Murthy, for arranging and planning the workshop. We look forward to participation with this group in its deliberations, and we deeply appreciate the contribution of your expertise, knowledge, and time, so essential to the success of this workshop.

James R. Patton, Jr.
Power Program
Office of Naval Research
U.S. Department of the Navy

TURBULENCE IN INTERNAL FLOWS

Introduction

S. N. B. MURTHY
Purdue University
West Lafayette, Indiana (U.S.A.)

The last conference dealing specifically with "Turbulence in Internal Flows" was held in 1965 and the proceedings of that conference, edited by G. Sovran (Ref. 1), is a convenient starting point to examine the outcome of the current workshop.

In the 1965 conference, J. Ackeret referred to many implications of the presence of turbulence in turbomachinery flows and drew attention specifically to (a) secondary flow and induced losses in ducts, diffusers and compressors, (b) diffusers with various geometry and entry conditions under optimum, stalled and nonsteady state operating conditions, (c) elbows with separation and reattachment and (d) compressibility effects in cascade flows. In the past ten years there has been progress in understanding and calculating such flows but designers of turbomachinery and related flows continue to face difficulties in many flows of practical importance. It is of course clear that there is no conceivable way of predicting or measuring every practical flow in every detail.

In the 1965 conference, G. Sovran and F. D. Klomp presented very impressive correlations of diffuser data for two-dimensional, conical and axisymmetric annular diffusers based on area ratio, nondimensional wall length and blockage factor. The design of optimum diffusers for different inlet conditions was discussed. The passage of a nonuniformity or distortion through a diffuser is one aspect of that problem. It was shown that there are serious limitations to the application of both linearized theory (perturbed mean flow) and inviscid analysis (dominance of pressure forces) when the nonuniformity consists, for instance, in a relatively thick boundary layer surrounding a core flow in a diffuser. Diffusers in practice sometimes include geometrical three-dimensionality and curvature. Some experimental results were presented at the conference on such flows. On the other hand the analysis of such flows was shown by P. N. Joubert et al., E. A. Eichelbrenner and J. P. Johnston, among others, to be restricted to correlation studies.

J. C. Rotta gave a summary of calculation methods for turbulent boundary layers that were available in 1965. It was pointed out that the von Karman momentum integral equation approach with a first-order differential equation for the velocity profile shape parameter was especially unsatisfactory in the presence of pressure gradients. Another problem (related to which there was not adequate guidance from experiments) was the method of incorporating adequate upstream history of turbulent motion into the governing equations. Lastly, the concept of eddy viscosity, used in conjunction with a two-layer model, was discussed and it was pointed out that, in the absence of adequate physical information concerning production, considerable empiricism would be required in cases more general than F. Clauser's equilibrium boundary layers. Reference was made to the then-current ideas of G. L. Mellor and D. M. Gibson (1963), Mellor (1964, 1966) and P. Bradshaw and D. H. Ferris (1965).

Boundary layer separation is an important problem in turbomachinery and diffuser flows. Separation of skewed boundary layers was discussed by E. S. Taylor at the 1965 conference within the context of a definition for the process of separation in three-dimensional flows. It was clear that considerably more detailed measurements were needed in the "entire region of separation" before one could attempt to establish the kinematic features of separation to some degree of satisfaction.

Another example of practical importance discussed in the 1965 conference was the calculation of secondary flows (spiralling flows in ducts and blade passages) and of the development of such flows. In that connection, W. H. Hawthorne demonstrated the powerfulness and limitations of inviscid flow theory under various approximations. One of those, the large shear, small disturbance approximation, yields wavy flows, which is still a controversial phenomenon. Once an inviscid solution was available, a laminar or a turbulent boundary layer as the case may be was to be included based on a comparison of a relevant characteristic length of the boundary layer with a characteristic length associated with the geometry of the flow. No mention was made of the effect of freestream turbulence in that discussion.

Finally, among practical problems, there was discussion of jets and wakes in the presence of various types of initial conditions and pressure gradient along the flow. The limitations of calculation methods based on similarity assumptions were stressed by B. G. Newman not only in view of the uncertainty of the role of initial conditions in the development of flow but also in the far region where one ordinarily assumes self-similarity.

The 1965 conference also discussed in substantial detail two basic questions related to turbulent flows: (1) development of theory for turbulent flow (J. L. Lumley and the important discussion on the paper) and (2) S. J. Kline's observations on the structure of turbulent boundary layers including transition and relaminarization and the enlightening discussion on that paper. The two problems are of course related and continue to be a challenge in understanding turbulent flows that are even slightly more complicated than homogeneous and isotropic flows, with or without the influence of boundary walls, whether one pursues observation and measurements or one adopts experimentation with numerical-computation methods for postulated models. Although we are still far, to this date, from incorporating ideas on the detailed structure of turbulence in practical engineering model calculations, there are of course strong implications of advances in such fundamental ideas in the calculation of boundary layers and free shear layers, jets and wakes. As illustrations, one can cite questions pertaining to turbulence production processes and to the general relation between large scale eddies and transport processes. Regarding the latter, the nonlinear interactions between different ranges of frequencies in the turbulence spectrum is of central importance in understanding the interchange of energy between the large and small wave numbers, for example R. E. Kronauer illustrates a case where energy had been transferred to the dominant wave length of the pipe during turbulent flow in a pipe with an oscillating circular cylinder embedded in the flow.

A convenient entry point to the literature on turbulence in internal flows is Ref. 1. A second entry point is the interesting and lucidly written survey article by J. P. Johnston (Ref. 2) on "Internal Flows." The articles by P. Bradshaw, W. C. Reynolds and T. Cebeci and H.-H. Fernholz in the same volume as Ref. 2 complement the discussion on internal flows. Johnston's article is written with applications to ducts, diffusers and turbomachinery in mind. The duct flows include curvature, step and confined mixing effects.

Attention is also drawn to Refs. 3-52 that deal with various aspects of the flows under consideration in this volume. Those references are listed in alphabetical order and are indicative of (a) some of the problems in internal flows under investigation in the past few years and (b) some of the institutions, investigators and journals reporting on such research. References to those who have taken part in this workshop have generally been omitted.

ORGANIZATION OF PROCEEDINGS

The proceedings of the workshop have been divided into four parts.

Part I. Fundamental Problems: Corrsin, Bradshaw, Wygnanski, Birch, Antonia, Falco, Walker and Freymuth.

Part II. Modelling Procedures: Saffman, Wilcox, Mellor, Huffman, Simpson, So and Leuchter.

Part III. Turbomachinery Applications: Whitelaw, Kerrebrock, Kovasznay, Evans, Fejer and Nagib.

Part IV. Panel Discussion and Summary.

It is possible to read each of the first three parts separately. However, they are interrelated in the Panel Discussion and in the Summary Report.

REFERENCES

1. Sovran, G. (ed.) Fluid Mechanics of Internal Flow, Elsevier Publishing Company, New York, 1967.

2. Johnston, J. P., Internal flows in turbulence in Turbulence, edited by P. Bradshaw, Springer-Verlag, New York, 1976.

3. Blackwelder, R. F. and Kaplan, R. E., On the wall structure of turbulent boundary layer, Jour. Fluid Mech., Vol. 76, Pt. 1, 1976, p. 89.

4. Boldman, D. R., Brinich, P. F. and Goldstein, M. E., Vortex shedding from a blunt trailing edge with equal and unequal external mean velocities, Jour. Fluid Mech., 75, Pt. 4, 1976, p. 721.

5. Bradshaw, P., Complex turbulent flows in Theoretical and Applied Mechanics, edited by W. T. Koiter, North-Holland Publishing Co., Delft, 1976.

6. Brodkey, R. S., Nychas, S. G. and Taraba, J. L., Turbulent energy production, dissipation and transfer, The Physics of Fluids, Vol. 16, No. 1, 1973, p. 2010.

7. Browand, F. K. and Weidman, P. D., Large scales in the developing mixing layer, Jour. Fluid Mech., Vol. 76, Pt. 1, 1976, p. 127.

8. Comte-Bellot, Genevieve, Hot Wire Anemometry in Annual Review of Fluid Mechanics, Vol. 8, edited by Milton Van Dyke et al., Annual Reviews, Inc., Palo Alto, California, 1976.

9. Corrsin, S., Limitations of Gradient Transport Models in Random Walks and in Turbulence, in Turbulent Diffusion in Environmental Pollution edited by F. N. Frenkiel et al., Academic Press, New York, 1974.

10. Dean, Robert C. Jr., The fluid dynamic design of advanced centrifugal compressors, Tech. Note TN-185, Creare Inc., Hanover, New Hampshire.

11. Demetriades, A., Turbulence correlations in a compressible wake, Jour. Fluid Mech., 74, Pt. 2, 1976, p. 251.

12. Dimotakis, Paul E. and Brown, Garry L., The Mixing Layer at High Reynolds Number: Large-Structure Dynamics and Entrainment, Jour. Fluid Mech., 78, Pt. 3, 1976, p. 535.

13. Dzung, L. S. (ed.), Flow research on blading, Elsevier Publishing Co., New York, 1970.

14. Fannelop, T. K. and Krogstad, P. A., Three-dimensional turbulent boundary layers in external flows: a report on Euromech 60, J. Fluid Mech., Vol. 71, Pt. 4, 1975, p. 815.

15. Fillo, J. A., Transverse velocities in fully developed turbulent duct flows, The Physics of Fluids, Vol. 17, No. 8, Sept. 1974, p. 1778.

16. Galbraith, McD. R. A., Eddy viscosity and mixing length from measured boundary layer developments, Aeronautical Quarterly, May, 1975, p. 133.

17. Gessner, F. B., The origin of secondary flow in turbulent flow along a corner, J. Fluid Mech., Vol. 58, Pt. 1, 1973, p. 1.

18. Gordon, Cliff M., Period between bursts at high Reynolds number, The Physics of Fluids, Vol. 18, No. 2, 1975, p. 141.

19. Harlow, F. H., Turbulence transport modeling, AIAA Selected Reprint Series, Vol. XIV, 1973, American Institute of Aeronautics and Astronautics, New York.

20. Herring, J. R., Approach of axisymmetric turbulence to isotropy, The Physcis of Fluids, Vol. 17, No. 5, May, 1974, p. 859.

21. Kahawita, R. A. and Meroney, R. N., Longitudinal vortex instabilities in laminar boundary layers over curved heated surfaces, The Physics of Fluids, Vol. 17, No. 9, Sept. 1974, p. 1661.

22. Lakshminarayana, B., Britsch, W. R. and Gearhart, W. S. (eds.), Fluid Mechanics, Acoustics and Design of Turbomachinery, Pts. I and II, NASA SP-304, National Aeronautics and Space Administration, Washington D.C., 1974.

23. Lalas, D. P., The 'Richardson' criterion for compressible swishing flows, J. Fluid Mech., Vol. 69, Pt. 1, 1975, p. 65.

24. Launder, B. E., Reece, G. J. and Rodi, W., Progress in the development of a Reynolds-stress turbulence closure, J. Fluid Mech., Vol. 68, Pt. 3, 1975, p. 537.

25. Laufer, J., New trends in experimental turbulence research, in Annual Review of Fluid Mechanics, Vol. 7, edited by M. Van Dyke et al., Annual Reviews Inc., Palo Alto, California, 1975.

26. Margolis, D. P. and Lumley, J. L., Curved turbulent mixing layer, The Phys. of Fluids, Vol. 8, No. 10, Oct. 1965, p. 1775.

27. McCune, James E. and Hawthorne, William R., The effects of trailing vorticity on the flow through highly loaded cascades, Jour. Fluid Mech., 74, Pt. 4, 1976, p. 721.

28. Menkes, J., Sreedhar, R. and Hindman, C., Entrainment in turbulent wake flows, The Physics of Fluids, Vol. 17, No. 1, 1974, p. 37.

29. Mollo-Christensen, E., Physics of turbulent flow, AIAA Jour., 9, 1971, p. 1217.

30. Mulhearn, P. J., A note on several theories of turbulence in a uniform shear flow, ASME Jour. Fluids Engineering, June 1974, p. 181.

31. Munson, M. R., Experimental results for oscillating flow in a curved pipe, The Physics of Fluids, Vol. 18, No. 12, 1975, p. 1607.

32. Murata, S., Miyake, Y. and Inaba, T., Laminar flow in a curved pipe with varying curvature, J. Fluid Mech., Vol. 73, Pt. 4, 1975, p. 735.

33. Murthy, S. N. B., Survey of some aspects of swirling flows, Report No. ARL 71-0244, Air Force Systems Command, U.S. Air Force, Wright-Patterson A.F.B., Ohio, 1971.

34. O'Brien, W. F. and Moses, H. L., Instrumentation for flow measurements in turbomachine rotors, A.S.M.E. Paper No. 72-GT-55, 1973.

35. Parsons, D. J. and Hill, P. G., Effects of curvature on two-dimensional diffuser flow, A.S.M.E. Jour. Fluids Engineering, Sept. 1973, p. 349.

36. Patankar, S. V., Pratap, V. S. and Spalding, D. B., Prediction of turbulent flow in curved pipes, J. Fluid Mech., Vol. 67, Pt. 3, 1975, p. 583.

37. Pratap, V. S. and Spalding, D. S., Numerical computations of the flow in curved ducts, Aeronautical Quarterly, Aug. 1975, p. 219.

38. Raj, R., Form of the turbulence dissipation equation as applied to curved and rotating flows, The Physics of Fluids, Vol. 18, No. 10, 1975, p. 1241.

39. Rao, G. N. V. and Keshavan, N. R., Axisymmetric turbulent boundary layers in zero pressure-gradient flows, A.S.M.E. Jour. Applied Mech., March, 1972, p. 25.

40. Reynolds, W. C., Computation of Turbulent Flows in Annual Review of Fluid Mechanics, Vol. 8, edited by Milton Van Dyke et al., Annual Reviews, Inc., Palo Alto, California, 1976.

41. Sabot, Jean and Compte-Bellot, Genevieve, Intermittency of coherent structures in the core region of fully developed pipe flow, Jour. Fluid Mech., 74, Pt. 4, 1976, p. 767.

42. Sandborn, V. A., A review of turbulence measurements in compressible flow, NASA TMX-62,337, National Aeronautics and Space Administration, Washington D.C., 1974.

43. Smith, F. T., Upstream interactions in channel flows, Jour. Fluid Mech., 79, Pt. 4, 1977, p. 631.

44. Sobey, I. J., Inviscid secondary motions in a tube of slowly varying ellipticity, J. Fluid Mech., Vol. 73, Pt. 4, 1976, p. 621.

45. Stratford, B. S., The prevention of separation and flow reversal in the corner of compressor blade cascades, Aeronautical Jour., May, 1973, p. 249.

46. Townsend, A. A., The structure of turbulent shear flow, Second Edition, Cambridge University Press, Cambridge, 1976.

47. Ueda, H. and Hinze, J. O., Fine-structure turbulence in the wall region of a turbulent layer, J. Fluid Mech., Vol. 67, Pt. 1, 1975, p. 125.

48. Van Atta, Charles W., Sampling techniques in turbulence measurements, in Annual Review of Fluid Mechanics, Vol. 6, edited by M. Van Dyke et al., Annual Reviews Inc., Palo Alto, 1974.

49. Van Den Berg, B., A three-dimensional law of the wall for turbulent shear flows, J. Fluid Mech., Vol. 70, Pt. 2, 1975, p. 149.

50. Weyer, H. and Schodl, R., Methods for measurement of fluctuating quantities in turbomachinery, Report from DFVLR Institut fur Luftstrahlantriebe, Porz-Wahn, Germany, 1973.

51. Willmarth, W. W., Structure of Turbulence in Boundary Layers, in Advances in Applied Mechanics, Vol. 15, edited by Chia-Shun Yih, Academic Press, New York, 1975.

52. Yao, L.-S. and Berger, S. A., Entry flow in a curved pipe. J. Fluid Mech., Vol. 67, Pt. 1, 1975, p. 177.

PART I
FUNDAMENTAL PROBLEMS

Preliminary Report on Sheared Cellular Motion as a Qualitative Model of Homogeneous Turbulent Shear Flow

STANLEY CORRSIN
The Johns Hopkins University
Baltimore, Maryland (U.S.A.)

W. KOLLMANN
Technische Hochschule
Aachen, West Germany

ABSTRACT

In the hope of gaining some insight into homogeneous turbulent shear flow dynamics, the motion of a three-dimensional cellular velocity field under a constant mean strain-rate has been computed. Navier-Stokes velocity and pressure fields are calculated as power series in time with coefficients periodic in space. The results are used to compute various instantaneous fields, including vorticity, velocity products and pressure/strain-rate products. The cell-averaged results show rough agreement with statistical experiments on nearly homogeneous turbulent shear flow, although the total mean strain is limited by the computer time available to us.

The instantaneous fields allow a search for explicit hydrodynamic mechanisms of energy transfer, both from mean flow to "fluctuations," and (especially) between orthogonal velocity "fluctuation" components. The dominant "right way" intercomponent transfer paradigm is a kind of local stagnation point flow; local regions of "wrong way" transfer include reduced pressure swirls with particular relative orientations of local vorticity and strain-rate tensor. Although the effective Reynolds number is modest, most of the intercomponent transfer is contained in a rather small fraction of the cell.

* * * *

I am going to talk about relatively simple flows.

The attempt to understand the character of turbulent flows naturally tends to focus on the simpler ones, and we are still quite a way from what I consider satisfactory understanding of even the simplest kind: that is isotropic turbulence, as introduced many years ago by G. I. Taylor. Since this is a shear flow meeting, I am going to talk about some of our recent research on something which resembles the simplest type of turbulent shear flow; that is homogeneous shear flow. The sequence will be something like this: first, I shall show some of our earlier data on an attempt to generate an approximation to homogeneous shear flow in a wind tunnel. Our first paper on that was by Bill Rose (1966); then a few years later a somewhat improved version was done by Frank Champagne, Godfrey Harris, and myself (1970). What I am going to show you is still later development which was done quite a few years ago but has not been published yet (Graham, Harris, and Corrsin, 1970; Harris, Graham, and Corrsin, 1976) because we seem to procrastinate as well as we do research.

Then when Wolfgang Kollmann spent a year at Johns Hopkins on a visit from Aachen, we decided to try to generalize the Taylor-Green (1937) calculation of unsheared cellular motion. We decided it would be interesting to shear a cellular motion, finding out first whether it could in any sense be interpreted as a model of a homogeneous turbulent shear flow. Secondly, if it did indeed seem to give a rough approximation or even to show proper tendencies, we might follow a suggestion I made at the 1961 turbulence symposium in Marseille (1962), and use the computer as a "laboratory"; that is, not just compute average results, but rather use the instantaneous computer output field to look for fluid dynamic characteristics of local instantaneous flow phenomena, local in both space and time. This is the kind of information we'd like to get with probes, and the kind people have been getting with dye, hydrogen bubbles, smoke, and so on. The advantage of the computer is that it gives you a field full of numbers; the disadvantage is that you see your budget used up in a few minutes.

The results we have on the sheared cellular motion, which I am just going to outline briefly for you, are not too satisfactory. There were many other things that we would like to have done, but we ran out of computing time very quickly.

The first slide shows simply the ideal of homogeneous turbulent shear flow, together with its simplest moment equations. This is an idealized case that we wondered about once upon a time, i.e. whether it would be possible to have a turbulent flow with a uniform shear and, if so, whether it would be stationary in time. We found out that it couldn't be. As indicated, the mean momentum equation for an ideal homogeneous turbulent shear flow becomes trivial. The equation for kinetic energy of the mean flow becomes trivial. The

Homogeneous Turbulent Shear Flow
Momentum, Kinetic Energy and
Shear Stress Equations

Mean Mom.: $\frac{\partial \overline{U}_1}{\partial t} = 0$

K.E. of mean flow: $\frac{\partial}{\partial t}\left(\frac{1}{2}\,\overline{U}_1^2\right) = 0$

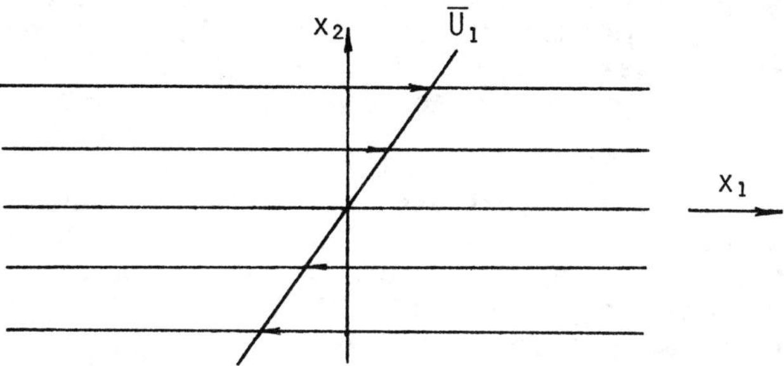

K.E. of turbulent motion:

$$\frac{d}{dt}\left(\frac{1}{2}\,\overline{u_k u_k}\right) = -\overline{u_1 u_2}\,\frac{d\overline{U}_1}{dx_2} - \nu\,\overline{\frac{\partial u_i}{\partial x_j}\,\frac{\partial u_i}{\partial x_j}}$$

$$\begin{pmatrix}\text{time rate of}\\ \text{increase}\end{pmatrix} = \begin{pmatrix}\text{production}\\ \text{rate}\end{pmatrix} - \begin{pmatrix}\text{dissipation}\\ \text{rate}\end{pmatrix}$$

Slide No. 1

kinetic of the turbulent motion has the three familiar attributes; this is homogeneous in space. We have the time rate of increase of kinetic energy equal to the difference between the rate of production and the rate of viscous dissipation.

On the second slide are the averaged component energy equations, which are familiar to most of you. The downstream velocity component of the turbulence receives all the energy, which it kindly doles out to the other two components through the pressure/strain-rate covariances; the viscous dissipation rate terms are also there. The turbulent shear stress balance equation for this example is written below. The production rate is proportional to the second moment of velocity component in the gradient direction times the (fixed) mean velocity gradient. The viscous terms are presumably not important at large Reynolds numbers, because of the tendency toward "local isotropy"; in isotropic turbulence they are zero. The destruction of the Reynolds stress comes about due to the off-diagonal term of the pressure/strain-rate covariance, and we have determined the components of that covariance tensor in our laboratory approximation to this idealized flow.

The wind tunnel versions of this that we have created (Rose, 1966; Champagne et al., 1970) involve steady flow through some kind of turbulent shear flow generator, with the hope that far downstream we might get an approximation to homogeneity. It turns out that the asymptotic state is only approximately homogeneous. As we showed at an American Physical Society meeting about half a dozen years ago (Graham, Harris, and Corrsin, 1970),* it has constantly increasing scale and constantly increasing turbulent energy.

Our present shear flow generator amounts to a series of parallel jets which are individually controlled by screens that throttle the individual channel flows by different amounts at the upstream ends. If we are lucky, we get a uniform velocity gradient far downstream, outside of the wind tunnel boundary layers. We were able to get a moderately homogeneous flow field. I am not going to show you the degree, but over a distance of several integral scales transversely we were able to get uniformity within a few percent.

Slide No. 3 shows the downstream development of the three Cartesian velocity component energies (Harris, Graham, and Corrsin, 1976). We are interested in whether we get a well-defined asymptotic state, stationary or not. The experiment reported in the paper by Champagne, Harris, and myself (1970) had a much smaller mean velocity gradient at the same mean velocity. The velocity gradient (hence the

---

* See Harris, Graham, and Corrsin (1976).

K.E. of Cartesian Turbulent Components:

$$\frac{d}{dt}\left(\frac{\overline{u_1^2}}{2}\right) = -\overline{u_1 u_2}\,\frac{d\overline{U}_1}{dx_2} + \frac{1}{\rho}\,\overline{p\,\frac{\partial u_1}{\partial x_1}} - \nu\,\overline{\frac{\partial u_1}{\partial x_j}\,\frac{\partial u_1}{\partial x_j}}$$

$$\frac{d}{dt}\left(\frac{\overline{u_2^2}}{2}\right) = \frac{1}{\rho}\,\overline{p\,\frac{\partial u_2}{\partial x_2}} - \nu\,\overline{\frac{\partial u_2}{x_i}\,\frac{\partial u_2}{x_i}}$$

$$\frac{d}{dt}\left(\frac{\overline{u_3^2}}{2}\right) = \frac{1}{\rho}\,\overline{p\,\frac{\partial u_3}{\partial x_3}} - \nu\,\overline{\frac{\partial u_3}{\partial x_k}\,\frac{\partial u_3}{\partial x_k}}$$

intercomponent
exchange

Turbulent Shear Stress Balance:

$$\frac{d}{dt}\left(\overline{u_1 u_2}\right) = -\overline{u_2^2}\,\frac{d\overline{U}_1}{dx_2} + \frac{1}{\rho}\,\overline{p\left(\frac{\partial u_1}{\partial x_2} + \frac{\partial u_2}{\partial x_1}\right)} - 2\nu\,\frac{\partial u_1}{\partial x_j}\,\frac{\partial u_2}{\partial x_j}$$

production destruction

* * * * * *

At large enough Reynolds number, we expect "local isotropy": then,

$$\nu\,\overline{\frac{\partial u_1}{\partial x_i}\,\frac{\partial u_1}{\partial x_i}} \approx \nu\,\overline{\frac{\partial u_2}{\partial x_j}\,\frac{\partial u_2}{\partial x_j}} \approx \nu\,\overline{\frac{\partial u_3}{\partial x_k}\,\frac{\partial u_3}{\partial x_k}} \approx \frac{1}{3}\,\varepsilon$$

and

$$2\nu\left|\frac{\partial u_1}{\partial x_i}\,\frac{\partial u_2}{\partial x_i}\right| \ll \frac{1}{\rho}\left|p\left(\frac{\partial u_1}{\partial x_2} + \frac{\partial u_2}{\partial x_1}\right)\right| .$$

Slide No. 2

$\frac{\overline{u_i^2}}{\overline{U}_c^2} \times 10^4$ (not summed)
$\frac{u_1'^2}{\overline{U}_c^2}$
$\frac{u_2'^2}{\overline{U}_c^2}$
$\frac{u_3'^2}{\overline{U}_c^2}$
$-\frac{\overline{u_1 u_2}}{u_1' u_2'}$
Harris, Graham & Corrsin
(J. Fl. Mech., in press)
downstream distance →
x_1/h
homogeneous shear flow: TURBULENT ENERGY

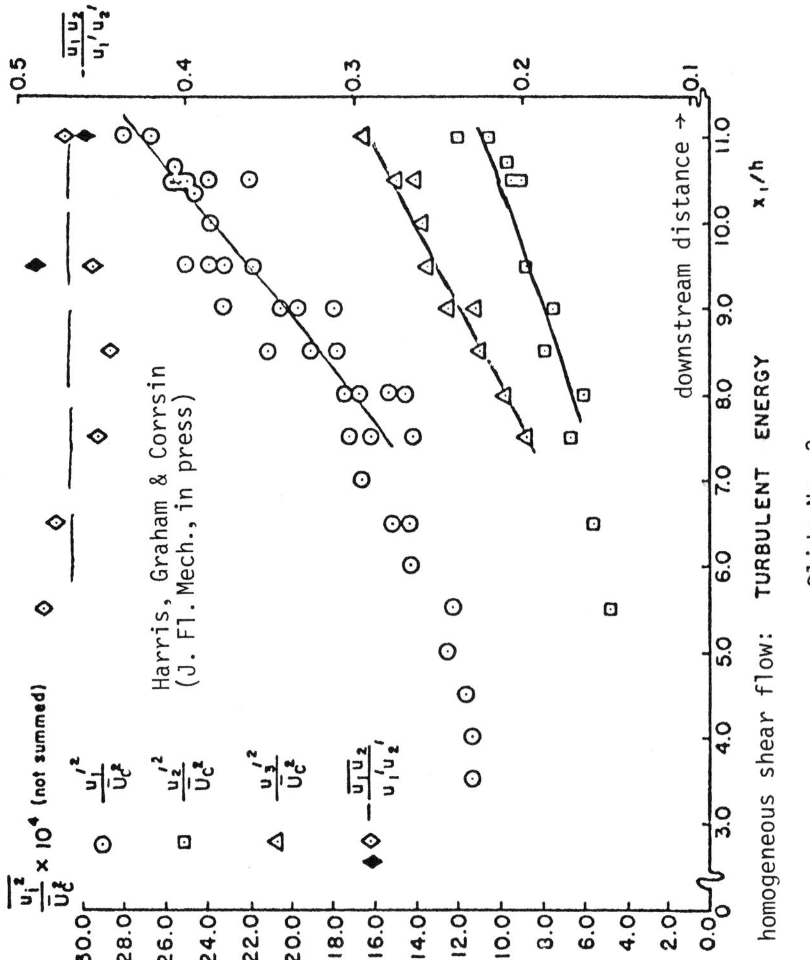

Slide No. 3

mean strain-rate) was a factor of 3 smaller than the case shown here. Therefore, the total strain reached in the equal test section length was one-third of that reached here. By coincidence, the total strain, $(\overline{U}_1/x_1)(d\overline{U}_1/dx_2)$ for the flow shown here is very nearly equal to the normalized downstream distance x_1/h. h is tunnel height, an irrelevant length.

The earlier experiment ran out of test section at a total mean strain of 3.5, just about when turbulent energy leveled off. We realized that things couldn't remain that way, however, as we remarked in our paper. Sure enough, if we go to a total strain of about 9 or 10, things have pretty well settled to an asymptotic growth state. The Reynolds shear stress has long ago reached a fairly typical value of about -0.5 or a little bit less in magnitude. So, we think that maybe we are pretty well in a nonequilibrium asymptotic state.

Now I'll tabulate some of the properties which seem to be typical for turbulent shear flows. Some of them are things we shall look for in our sheared cellular flow. One of the things that is fairly typical for nearly homogeneous turbulent shear flow, and for almost all shear flows away from boundaries, is the turbulent shear stress correlation coefficient magnitude tending to be about 0.5. Another result very familiar to those of you who do experiments is that, perhaps because the energy is fed into the downstream turbulent velocity component from the mean flow, that component tends to be the most energetic. It is hard to have any intuition about the reason for the next property, found in boundary layers, wakes, pipes, and channels: the component along the gradient direction, u_2 here, tends to be the least energetic of the three. The fact that the two transverse component energies, $\overline{u_2^2}$ and $\overline{u_3^2}$, are unequal makes it a little bit doubtful that a linear intercomponent energy exchange hypothesis (Rotta, 1951) will turn out to be correct. In summary, $\overline{u_1^2} > \overline{u_3^2} > \overline{u_2^2}$ in nearly rectilinear shear flows, if x_1 is the dominant flow direction and x_2 is the velocity gradient direction.

Another thing that we have found clearly is that with the uniform shear, the turbulent structures are convected with the mean flow. This makes "wave models" of this turbulent shear flow seem (to put it euphemistically) inappropriate, because nothing propagates relative to the flow.

Slide No. 4 has to do with intercomponent energy transfer, the particular physical phenomenon I want to emphasize in this talk. About 25 years ago, J. C. Rotta (1951) proposed that the tendency toward equipartition which comes from the pressure/strain-rate covariance, as shown in the equations, at the top of Slide No. 2, could be approximated as a simple linear approach to equipartition; that is, the pressure/strain-rate products could be proportional to

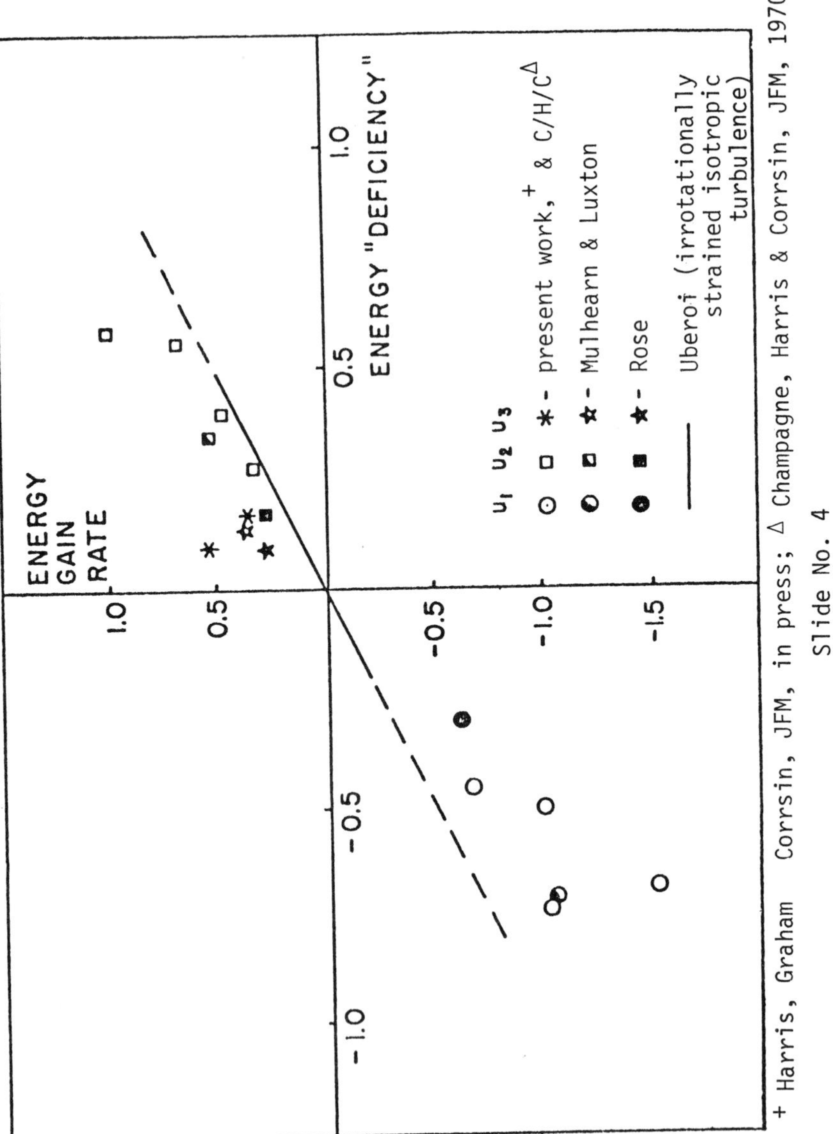

+ Harris, Graham Corrsin, JFM, in press; Δ Champagne, Harris & Corrsin, JFM, 1970

Slide No. 4

the departure of the kinetic energy of a particular component from the mean energy in the three velocity fluctuation components. Some people tell me that he didn't propose that the three "constants" of proportionality should be equal, but a rereading of his paper still seems to suggest that assumption. In any case, our experiments show that where they are roughly constant they are unequal--and that they are not necessarily constant. They remain fairly uniform until we reach the asymptotic region, where one of them changes drastically. Slide No. 4 shows the kinetic energy gain rate of each component plotted against its "deficiency" compared with the component average. The solid line is taken from the contraction-strained turbulence data of Uberoi (1957) as computed by Rotta (1962).

The circles correspond to the downstream component, $\overline{u_1^2}$, which receives energy directly from the mean flow, so it has an energy excess, and is in the third quadrant. The squares correspond to the component $\overline{u_2^2}$ along the gradient direction. It has the lowest level. Those two components don't seem to follow a linear intercomponent transfer relation, but their behavior could be easily approximated by adding a parabola to Rotta's linear hypothesis. This would be even with the same linear term constant. However, the component $\overline{u_3^2}$, normal to both the mean velocity and the gradient, exhibits relatively "pathological" behavior and, although our data are too slight to be certain, it even seems to slope the "wrong" way. It doesn't look as though it is "trying" for equipartition.

The intercomponent exchange assumptions of simple linear, or at least monotonic, dependence on departure from equipartition may work sometimes, but nobody knows when. One of the things that we hoped to learn from our sheared cellular flow computations was information about this particular attribute.

Slide No. 5 illustrates our sheared generalization of the Taylor-Green type calculation, schematically indicated. We start with a simple cellular motion, with only two elements indicated by the trigonometric expression at the top. We take constant mean shear, $d\overline{U}_1/dx_2$ = constant.

We all know very well that turbulence is not a cellular motion, so we wanted to find out whether the cell-averaged values of physical variables would roughly approximate, or at least qualitative resemble, the experimental values in the actual homogeneous turbulent shear flow. A brute force time series computation was done with machine time granted by the National Center for Atmospheric Research. The power series in time was truncated at fourth power. Convergence was good up to dimensionless time of 0.2, which corresponded to tilting an originally rectangular three-dimensional cell by roughly the amount sketched in Slide No. 5, about 1/3 of the wavelength.

Uniformly Sheared Cellular Motion

(A model of homogeneous turbulent shear flow?)

e.g.

$$u_i(x,0)=\begin{pmatrix}\frac{d\overline{U}_1}{dx_2}x_2\\ 0\\ 0\end{pmatrix}+\begin{pmatrix}\alpha^{(1)}\cos \ell^{(1)}x_1\sin m^{(1)}x_2\sin n^{(1)}x_3+\beta^{(1)}\sin q^{(1)}x_1\sin r^{(1)}x_2\sin s^{(1)}x_3\\ \alpha^{(2)}\sin \ell^{(2)}x_1\cos m^{(2)}x_2\sin n^{(2)}x_3+\beta^{(2)}\cos q^{(2)}x_1\cos r^{(2)}x_2\sin s^{(2)}x_3\\ \alpha^{(3)}\sin \ell^{(3)}x_1\sin m^{(3)}x_2\cos n^{(3)}x_3+\beta^{(3)}\cos q^{(3)}x_1\sin r^{(3)}x_2\cos s^{(3)}x_3\end{pmatrix}$$

Initial condition for Navier-Stokes calculation in time series.

Extent of computation: 4 terms in Taylor series → 48 Fourier modes and dimensionless time $t = 0.2$

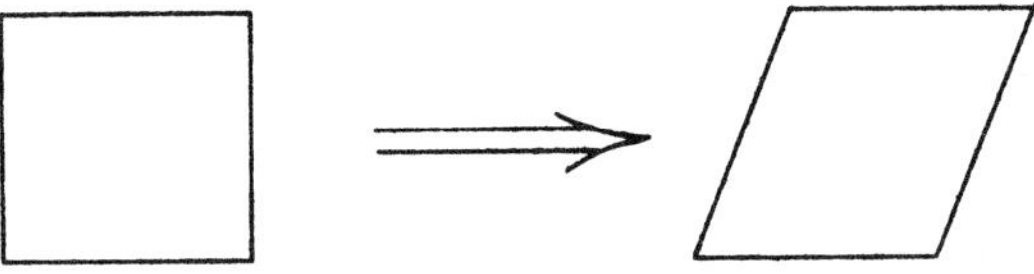

→ Compare cell-averaged quantities with turbulent shear flow experiment

→ Explore instantaneous fields for possible dominant flow configurations
- in turbulent energy production
- in intercomponent transfer
- in dissipation

etc.

Slide No. 5

Slide No. 6 shows the evolution of cell-averaged quantities. The initial state was chosen with $\overline{u_2^2}$, $\overline{u_3^2} < \overline{u_1^2}$ in order to hasten the approach to a condition resembling the turbulence. However, we decided to start with $\overline{u_2^2} = \overline{u_3^2}$, in the hope that the ubiquitous inequality $\overline{u_3^2} > \overline{u_2^2}$ would emerge. Also we decided to start without turbulent shear stress ($\overline{u_1 u_2} = 0$), in the hope that this would be created before our eyes. Slide No. 6 shows that both of the latter events occurred as we had hoped.*

A principal shortcoming of the calculation is that we were unable to push it to t values large enough for $\overline{u_2^2}$ and $\overline{u_3^2}$ to increase to be large fractions of $\overline{u_1^2}$. Preferable initial energy ratios would be $\overline{u_2^2}/\overline{u_1^2} = \overline{u_3^2}/\overline{u_1^2} = 0.5$, but the initial wave-number choices plus mass conservation limited the choices of Fourier coefficients.

In any case, this configuration can be said to have averaged behavior with a qualitative resemblance to the turbulent shear flow, so we decided to look at detailed instantaneous flow configurations, to see how the energy and shear are produced, in general what kinds of paradigms of hydrodynamic structures we could find in the regions where various key events took place. My presentation is on one process, the intercomponent energy transfer, the pressure/strain-rate covariance.

Slide No. 7 shows freehand sketches of the evolution of the pressure field and of the leading diagonal component of the strain-rate tensor, $\partial u_1/\partial x_1$. For each variable we see the initial field at the left, then the two alternative evolved fields, the top one of each pair being with zero mean shear, the bottom one with positive shear, $dU_1/dx_2 > 0$. The sheared case is sketched in terms of unsheared coordinates, to make comparison easier.

The initial pressure field is symmetric about a line parallel to x_1, and the $\partial u_1/\partial x_1$ strain-rate field is anti-symmetric. Therefore, their product $p(\partial u_1/\partial x_1)$, the initial local rate of gain of u_1^2 energy from u_2^2 and u_3^2 is anti-symmetric. Therefore its cell-averaged initial mean value is zero, even though $\overline{u_1^2}$ has the most energy and should be losing it to $\overline{u_2^2}$ and $\overline{u_3^2}$. In each sketch, negative regions are indicated by "—," positive regions by "+."** The dashed lines are zero lines. At time t = 0.2, when we look at the contours of constant pressure, we see relatively modest changes if there is no shear. In contrast, the sheared case shows intensification of some

---

* For somewhat larger t, where convergence is less certain, $\overline{u_1 u_2}/(\overline{u_1^2}\,\overline{u_2^2})^{\frac{1}{2}}$ increases back toward zero.

** Color coding made this clearer in the actual slides.

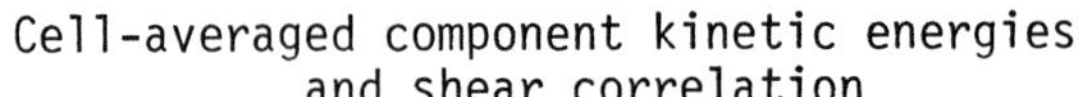
Cell-averaged component kinetic energies and shear correlation

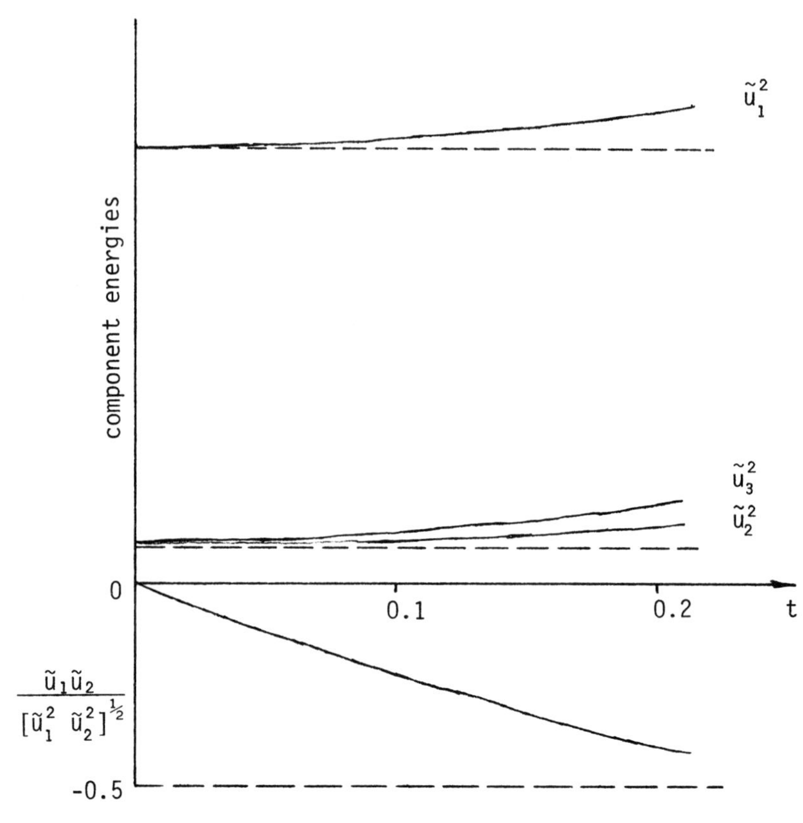

Slide No. 6

Typical fields: (slide at x_3 = constant)
contours of p = const.
t = 0
x_2
x_1
$\left\{ \begin{array}{l} p > 0 \\ p < 0 \end{array} \right\} \begin{array}{l} (+) \\ (-) \end{array}$
t = 0.2
$\frac{d\bar{U}_1}{dx_2} = 0$
$\frac{d\bar{U}_1}{dx_2} > 0$
(strained field is plotted in mean original coordinates)
The product $p\left(\frac{\partial u_i}{\partial x_k} + \frac{\partial u_k}{\partial x_i}\right)$ is of special importance--
contours of $\left(\frac{\partial u_i}{\partial x_k} + \frac{\partial u_k}{\partial x_i}\right)_{i=k=1}$ = const.
t = 0.2
t = 0
x_2
x_1
$\frac{d\bar{U}_1}{dx_2} = 0$
$\frac{d\bar{U}_1}{dx_2} > 0$
$\left\{ \begin{array}{l} \frac{\partial u_1}{\partial x_1} > 0 \\ \frac{\partial u_1}{\partial x_1} < 0 \end{array} \right\} \begin{array}{l} (+) \\ (-) \end{array}$

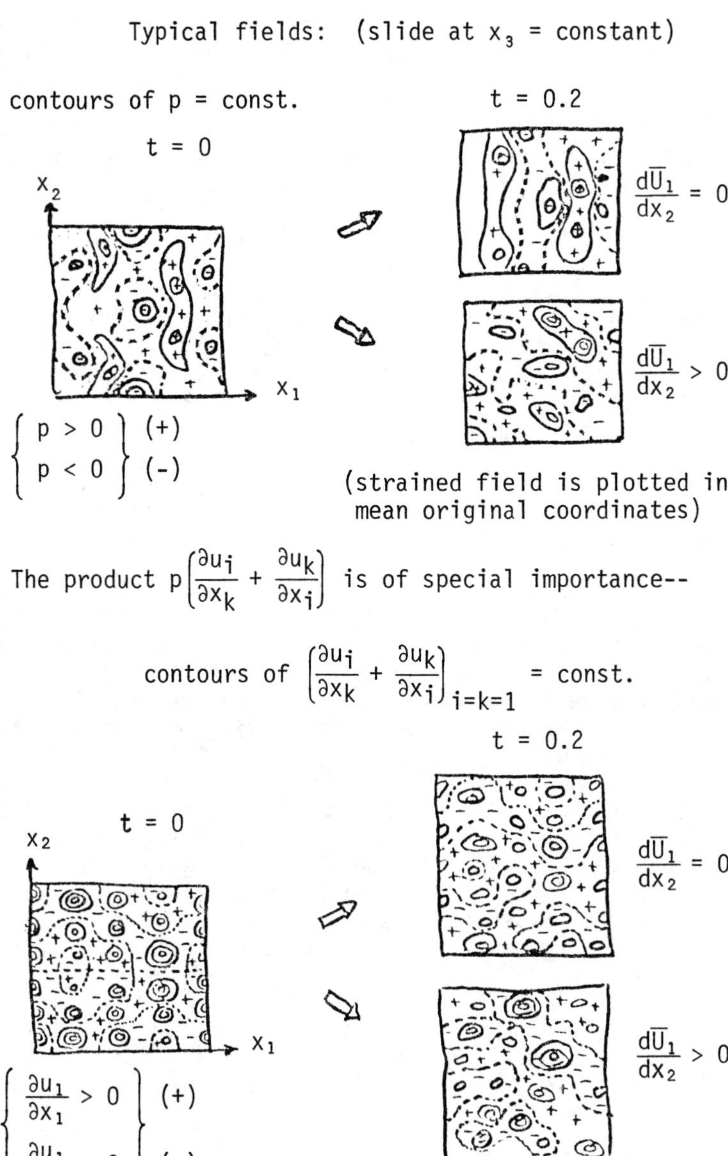

Slide No. 7

pressure peaks and destruction of others. Correspondingly the strain-rate field evolution without shear is modest; with shear it is considerable. But the mere fact that addition of a rather strong mean shear causes big changes is to be expected. The most important observation is the nature of those changes. A look at the $p(\partial u_1/\partial x_1)$ product field (which we see only for $t = 0.2$ and with shear, on Slide No. 8) shows that the principal effect is to amplify the local flow structures which transfer energy the "right" way (away from the most energetic component), and to suppress those which transfer energy the "wrong" way (into the most energetic component)!

The obvious thing to do then is to look more closely at both kinds of regions, in order to identify the flow configurations, hoping eventually to find out why the wrong-way transfer configuration got suppressed, and why the right-way one was amplified. Evidently the mean shear favors certain types of structures, which may initially occur by chance, but then are amplified, and presumably it quickly destroys flow configurations which locally transfer energy in the wrong direction.

The energy exchange regions are relatively localized, although the Reynolds number is not very large. The velocity derivative field had no particular appearance of intermittency, however.

A detailed look at the velocity fields showed us that the two strong negative regions in $p(\partial u_1/\partial x_1)$ (identified as 1 and 5 in Slide No. 8) are roughly axisymmetric "squashing"-type stagnation point flows, i.e. with one negative principal strain-rate and two positive ones. The negative one is essentially in the x_1 direction. Obviously such a flow structure transfers kinetic energy from u_1^2 to u_2^2 and u_3^2, the "right way."

The high pressure region located at a point symmetrical to region 1 (top left of Slide No. 7) is also a stagnation region, but a reversed one (a "stretching"-type); it has $\partial u_1/\partial x_1 > 0$, and $\partial u_2/\partial x_3$, $\partial u_3/\partial x_3 < 0$, so at $t = 0$ it was transferring energy into u at exactly the same rate region 1 was transferring it out. The remains of this "wrong way" structure sits at location 4 in Slide No. 8!*

The extreme numerical values (in arbitrary units) associated with the five locations tagged in Slide No. 8 are tabulated in Slide No. 9. We see, for example, that in the relatively short time of

---

* We may recall that for isotropic turbulence Betchov (1956) suggested a predominance of squashing over stretching stagnation regions, based on measured values of the skewness factor of the velocity spatial derivative.

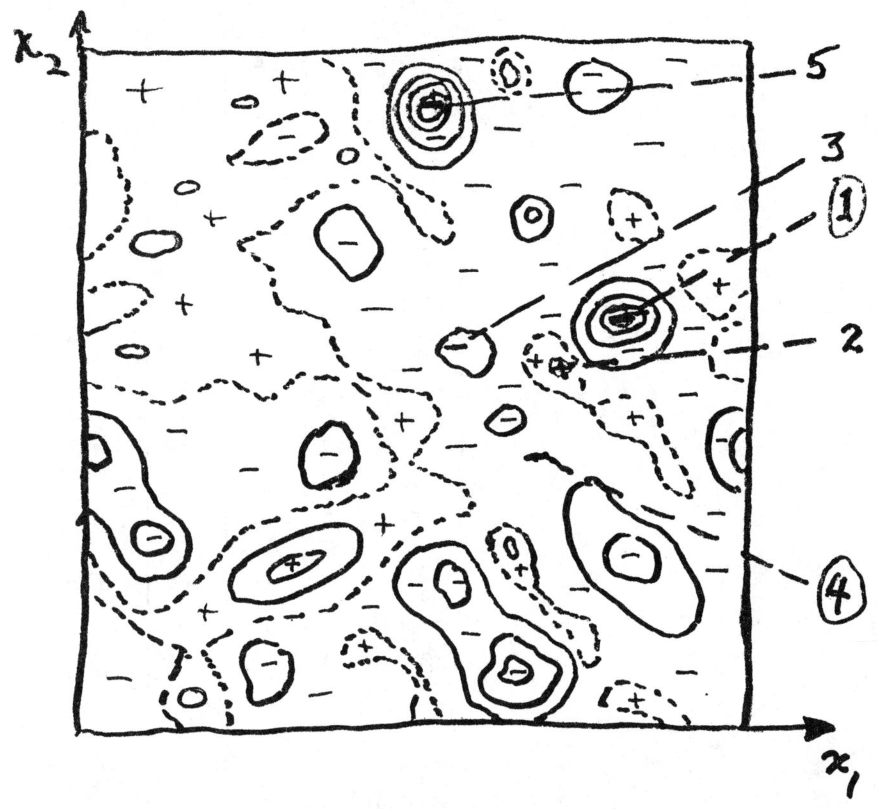

$P\frac{\partial u_1}{\partial x_1}$

IN PLANE $X_3 = 0.5$

TIME $t = 0.2$

MEAN SHEAR $\frac{d\bar{U}_1}{dx_2} = 3.5$

Reynolds no. based on cell size and initial cell flow speed $R = 500$

"Points" 1 and 5 contribute most of the "Correct" intercomponent energy transfer

Slide No. 8

at $t = 0$

field \ point	1	2
$\sim p$	.42	.42
$\sim p \frac{\partial u_1}{\partial x_1}$	-.90	.90
$\sim p \frac{\partial u_2}{\partial x_2}$	.45	-.45
$\sim p \frac{\partial u_3}{\partial x_3}$	.45	-.45

at $t = 0.2$

field \ point	1	4	2	3
$\sim p$	.54	-.05	-.20	-.24
$\sim p \frac{\partial u_1}{\partial x_1}$	-1.4	-.1	.12	-.24
$\sim p \frac{\partial u_2}{\partial x_2}$	.66	.04	-.06	.07
$\sim p \frac{\partial u_3}{\partial x_3}$	.74	.05	-.06	.17

Initially,

Region #1 is an ordinary stagnation with symmetry axis in mean flow direction,

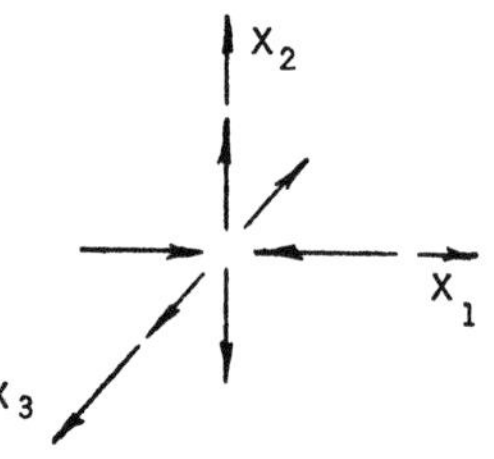

and

Region #4 is a "radial" stagnation with symmetry axis in mean flow direction.

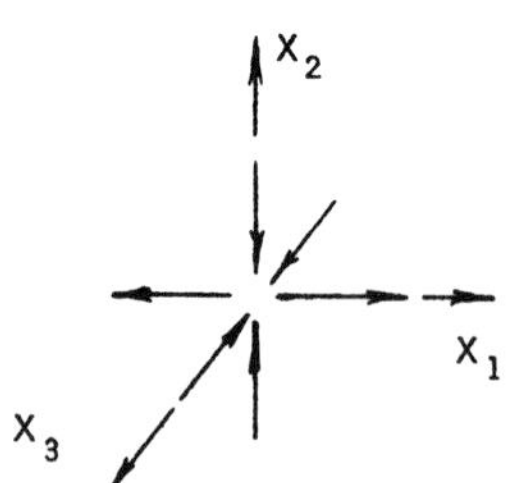

Slide No. 9

t = 0.2, region 4 has changed from "wrong way" exchange to "right way" exchange.

At the bottom of Slide No. 9, qualitative sketches symbolize the contrapuntal initial structures of regions 1 and 4.

There is at least one other region of peculiar interest: region 2 is transferring energy the wrong way even at time t = 0.2. If we scrutinize the velocity field in that neighborhood, we find large vorticity and small strain-rate; it is nearly in local rigid rotation.

We have done some theoretical analysis based on the differential equations for the strain-rate tensor, the vorticity tensor, and the pressure fluctuations, looking to see which terms are the ones that contribute to this rapid destruction of the "wrong way" structure at point #4, for example. It turns out that the pressure fluctuation fields play a major role. The mean vorticity rotates the principal axes of the strain-rate tensors a bit differently at #1 and #4, but the change in the p field has the greater effect on the change in the pressure/strain-rate products.

We have also looked at other local events important to the energy and shear stress balances, but time is too short to allow me to describe them.

* * * * *

This research has been supported generally by the Office of Naval Research, Fluid Dynamics Program; the computation was done at the National Center for Atmospheric Research, which is supported by the National Science Foundation. W. Kollmann was supported by a postdoctoral fellowship from the Deutschen Furschungsgemeinschaft (Federal Republic of Germany).

REFERENCES

1. Betchov, R. 1956, Jour. Fluid Mech., 1.

2. Champagne, F. H., Harris, V. G. and Corrsin, S. 1970, Jour. Fluid Mech. 41.

3. Corrsin, S. 1962, in Mecanique de la Turbulence, Editions du C. N. R. S., Paris (in U.S., Gordon & Breach, New York).

4. Graham, J. A. H., Harris, V. G. and Corrsin, S. 1970, Bull. Am. Phys. Soc. 15 (11) p. 1544 (abstract only)

5. Harris, V. G., Graham, J. A. H. and Corrsin, S. 1976, Jour. Fluid Mech., in press.

6. Rose, W. G. 1966, Jour. Fluid Mech., 25.

7. Rotta, J. C. 1951, Zeits f. Physik, 129 and 131.

8. Rotta, J. C. 1962, in "Progress in Aeronautical Sciences," 2, Pergamon, Oxford.

9. Taylor, G. I. and Green, A. E. 1937, Proc. Roy. Soc. a, 158.

10. Uberoi, M. S. 1957, Jour. Appl. Phys., 28.

DISCUSSION

BRADSHAW: (Imperial College)

I would like to start by saying that that was one of the most fascinating papers I have heard for many years. We all very much look forward to seeing it in black and white instead of red and green.

You said that intercomponent exchange gets rather spotty in space. So, of course, do the u_1, u_2 Reynolds stresses. But would you say that the pressure-strain correlation should really be represented in a turbulence model by scales of the small scale motion rather than the energy containing range?

CORRSIN:

I haven't looked at it in detail. First of all let me say that since we have a rather small time computation with only about 48 modes, all the real fine scale hasn't had time to develop yet. But in looking at the pressure-strain rate covariances, we have one element of the product which is characterized by very large scale, namely the static pressure which tends to be very large-structured, and the spatial derivative of velocity which emphasizes the fine structure. So, if you have a product of a large structure and a fine structure component you get (kind of) a modulated picture and it is not easy to give a simple answer to that. I intuitively tend to think of the smaller structure as dominating in that case but that may be because of some other application that I was once involved in. I cannot psychoanalyze myself on the spur of the moment.

BRADSHAW:

The really difficult part about the velocity gradient terms is whether the part of the velocity gradient that correlates with the pressure comes from the large scale motion or the fine scale motion. I am not trying to set you up, I just wonder if you had any real thoughts about it.

CORRSIN:

Generally speaking what we've seen, I think, is that the regions which do the right kind of "right-direction" energy transfer do tend to be stagnation points of rather classical types with a modest degree of axial symmetry. The "wrong-way" regions which of course are very much in the minority, tend to be vortical regions. I am not sure that the characteristic lengths associated with vorticity and with strain-rate fluctuations are any different from each other, since they are both first-order velocity derivatives. For instance, in isotropic turbulence their spectra are identical.

BRADSHAW:

One last point. Perhaps you may have been a little bit unfair to Julius Rotta who, in his 1951 paper,* did point out what we all promptly forgot for 15 years, that the pressure-strain term had two parts: the turbulence-turbulence part, which he suggested could be represented as a return to isotropy, and of course a part that depends on the mean strain rate. So, maybe, he would not be too surprised at the constant going off to infinity.

CORRSIN:

I suspect that in our computations of the inter-component transfer data, and we have also done that with some of the Tucker and Reynolds data, which also gives us a point in the wrong quadrant, that may be so in part, because our assumption of homogeneity for their experimental configuration within their straining region is not valid; we assumed homogeneous equations for interpreting their data.

KLEBANOFF:

I have a follow up comment. It might be well to note that in a turbulent boundary layer, if one looks at the spectrum of the transverse fluctuation relative to that for the longitudinal fluctuation, one finds that the latter has a -5/3 range and the former does not. Perhaps the reason for this may be the effect of a pressure-strain

---

* Rotta, J. C., "Statistische Theore Nichtomogener Turbulenz," Z. für Physik, Vol. 129, p. 547 and Vol. 131, p. 51 (1951).

correlation which is represented by scales different from those for the Reynolds stress.

CORRSIN:

That could be. I am trying to remember as you say that, how we were spatially oriented. I don't remember now, for instance, whether the negative uv correlation had its most active regions in the same parts of the cell as the stagnation point. Maybe Dr. Kollmann remembers. (To Dr. Kollman) Do you remember whether the instantaneous uv product had its extreme values in the same region as the pressure-strain rate had its extreme value?

KOLLMANN: (Lehrstuhl f. Techn. Thermodynamik)

They tended to be more or less in the same parts of the cell.

MELLOR: (Princeton University)

As you went down your wind tunnel this time your length scale never increased. Did it ever get large compared to the tunnel dimensions?

CORRSIN:

As far down as we went, the region of pretty good homogeneity was still about five times the integral length scale, maybe four times, outside the boundary layers. The boundary layers were getting pretty thick, of course, and they do induce fluctuations on the axis and all that, so we weren't perfectly free of boundary interference.

MELLOR:

I don't understand my question now!

Your previous published data in the Journal of Fluid Mechanics tend asymptotically to a point where the reduction in spatial is fairly well balanced but the advection term was not. If we go further now, how does the advection term compare?

CORRSIN:

The advection term is still negligible and the triple correlation term, the turbulent transport of turbulent energy, which we checked, also is still negligible, but the homogeneity is not good if we use convected coordinates.

KOVASZNAY: (ONR Tokyo)

Did you give much thought to Rotta's later formulation when he splits the term and only one part of it goes to this mechanism?

CORRSIN:

No, I hadn't given much thought to it. That was what Peter Bradshaw just said.

LIEPMANN: (California Institute of Technology)

I think that we should avoid using pressure as a variable in incompressible turbulence. The crucial variable is vorticity.

CORRSIN:

Yes, we really do. We really use the velocity integral and all that sort of stuff with the solution to the Poisson equation for pressure.

It turns out that vorticity is not the be-all and end-all. It turns out that working with the strain-rate conservation equation is just as important as working with the vorticity conservation equation.

It is true that we do get rid of the pressure in making the calculations. But I find that, in thinking about mechanisms locally, I can (maybe) fool myself, but I think I can think a little bit better about pressure gradients than I can think about infinite integrals over all space of this complicated integrand with the kernel being the sum of the strain rate and the vorticity tensors divided by the radius and all that sort of stuff. I find that hard to think about intuitively.

I am motivated to try again to measure static pressure fluctuations in the middle of the flow and one of the nicest ways to do it is with a computer, of course.

FALCO: (Cambridge University)

I agree with you. I am very excited about what you have just presented and I just want to make a comment that might help clear the picture a little bit. It is certainly my experience in observing the instantaneous motions in turbulent boundary layers that the Reynolds stress producing motion decrease in scale as Reynolds number increases, but my recent work suggests that large scale motions may produce important large scale (i.e., Reynolds number independent pressure gradients). At low Reynolds numbers, the features which produce large Reynolds stress contributions are large scale, so that it will not be easy to differentiate between scales which produce the large scale pressure fluctuations and those which produce the Reynolds stress, if such a differentiation is warranted. I am not sure how much computer time you need to increase your effective Reynolds number, but if your present Reynolds number is low, maybe a factor of two will be sufficient to show a difference.

CORRSIN:

A Reynolds number based on cell size and initial cellular velocity amplitude is about 500, which probably means R may be on the order of 50 or less.

FALCO:

I don't think that there is an effective differentiation at Reynolds numbers that low.

LUXTON: (University of Adelaide)

The fact that you can stack up your low strain rate results and your high strain rate results suggests very much that the turbulence does indeed forget its origin. Would you like to comment on that?

CORRSIN:

I started doing turbulent shear flow in round turbulent jets many years ago and in those days I tended to think of turbulence as forgetting its origin, but John Laufer was convincing us last week that if the jet blows rings maybe it does not forget its rings; I don't know. I think in the case of the moderate homogeneous shear flow it may. In fact, in grid-generated nearly isotropic turbulence, we found very little difference between turbulence generated by "silver dollars" and by round rods and square rods and stuff like that. We didn't look in very sophisticated detail at condition-sampling data: maybe the rings behind our silver dollar grid were hidden conditionally.

LAUFER: (University of Southern California)

I think you are misquoting me; we were looking at the early development of a shear flow and jet flow and in those flows there seems to be some indication that initial conditions are important, but we certainly don't want to imply that further downstream the "memory" still exists.

CORRSIN:

Actually, although it also hasn't been published, I had a Dr. Anatoly Utionkov in my lab for a year, some years ago (he was from the Moscow Steel and Alloy Institute, later from the University of Cairo, now back at the Moscow Steel and Alloy Institute), and he did a series of experiments of the early neighborhood of jet development in round jets with tailored velocity profiles at the start. He was interested in furnace jets. He made velocity distributions ranging from rectangular to conical by using variable solidity screens, and the difference was detectable in the turbulence level, for instance, only out to about 20 diameters.

BRADSHAW:

Since silver dollars have been mentioned I wonder if we could ask how much the calculations with what one might describe as your ravioli-cellular model cost per round?

CORRSIN:

It is $2000 per hour and it takes 8 minutes so that the cost is approximately $300.

BRADSHAW:

So it is equal to the cost of about 3 or 4 cross-wire probes?

CORRSIN:

Who makes your cross-wire probes?!

Interacting Shear Layers in Turbomachines and Diffusers

PETER BRADSHAW
Imperial College
London, England

ABSTRACT

Three types of interaction between a shear layer and another turbulence field are discussed. They are (i) the effect of freestream turbulence on a boundary layer or mixing layer, (ii) the merging of two boundary layers to form an aerofoil or blade wake, (iii) the interaction of two perpendicular boundary layers in a wing-body or blade-hub junction. A brief description is given of recent experimental work on these problems, including flow visualization and temperature-conditioned sampling.

1. INTRODUCTION

The two main phenomena that distinguish real-life "complex" flows (e.g. Bradshaw, 1975, 1976; Bradshaw in Murthy, 1974, p. 243) from simple shear layers, are distortion of a shear layer by extra rates of strain and the interaction between a shear layer and another turbulence field. The second turbulence field can be unsheared "freestream" turbulence; or a shear layer with the same direction of shear as the first one, as in the internal boundary layer that grows behind a change in surface roughness; or a shear layer with the opposite direction of shear, as in the merging of two boundary layers to form an aerofoil wake; or a shear layer whose plane of mean shear is inclined to the first, as in the flow along a wing-body junction.

In the present state of our empirical knowledge of turbulence these have to be treated as separate problems, and experiments should precede calculation methods. Experimental work on all the above types of interaction, except the internal boundary layer which has been investigated in depth by Antonia and Luxton (1971, 1972, 1974) and others, is in progress at Imperial College. All the experiments are being done in low-speed wind tunnels and in all cases detailed measurements of turbulence structure are being made, with application to calculation methods based on the Reynolds stress transport equations. "Turbulence structure" means dimensionless quantities like anisotropy ratios, spectrum shapes, or even the empirical constants that appear in transport-equation turbulence models. Frequent use has been made of flow visualization, or its quantitative analogue "temperature-conditioned sampling" (in which heat is introduced instead of smoke and a fast-response resistance thermometer generates an on-off conditioning function so that velocity-fluctuation statistics can be accumulated for the "hot" regions only). Paradoxes appear: the interaction of two opposed plane shear layers is much easier to predict than the effect of unsheared freestream turbulence, because the changes in turbulence structure in the former case are quite small. Again, the dominant effect in a wing-body junction is not the secondary flow induced by Reynolds-stress gradients but the axial vortex produced by quasi-inviscid skewing of the body boundary layer around the wing leading edge.

This paper is a progress report on the Imperial College work; more detailed accounts have been, or are being, prepared by my associates named below, and I am grateful to them for permission to reproduce their unpublished work.

2. FREESTREAM TURBULENCE

Especially in turbomachine applications a distinction must be made between "ordered unsteadiness" with wavelengths many times the thickness of a shear layer, and small-scale turbulence with wavelengths of the order of the shear-layer thickness. The response of a shear layer to ordered unsteadiness could be quite well predicted by ignoring effects on the turbulence structure--that is, by using the same turbulence model as for statistically-steady flow but performing a time-dependent calculation for the "mean" flow. For example, the left-hand side of the transport equation for any quantity Q, whether velocity or Reynolds stress, can be written as DQ/Dt; the operator D/Dt is equal to $\partial/\partial t + U_\ell \, \partial/\partial x_\ell$ but except for numerical difficulties it is quite immaterial whether the substantial derivative is wholly temporal, wholly spatial, or mixed. Roughly speaking, calculations for a shear layer with ordered unsteadiness in freestream will break down only when D(shear stress)/Dt exceeds the value at which the transport-equation model becomes unsatisfactory in <u>steady</u> flow. A pictorial way of making the same point is given in Figure 2 (p. 41) showing the respective effects of small-scale and large-scale turbulence on a jet in a moving stream. Ordered unsteadiness or large-scale turbulence just causes the jet to flap from side to side: the mean rate of spread, and the apparent Reynolds shear stress $-\overline{uv}$, are increased but the intrinsic turbulence of the jet is little affected.

Note that, near a solid surface, even near-isotropic freestream turbulence may count as ordered unsteadiness if its length scale, say the longitudinal integral scale $L_{x,e}$ where suffix e denotes the external-stream value, is large enough. Since the normal-component fluctuation v is reduced to zero on a solid surface, the continuity equation $\partial v/\partial y = -(\partial u/\partial x + \partial w/\partial z)$ implies $\overline{v^2} \propto \overline{u^2}\, y^2/L_x^2$ for y rather smaller than L_x: the constant of proportionality should be of the order of unity, so that a boundary layer of thickness δ much smaller than L_x will see a freestream turbulence field consisting mainly of large-scale "sloshing" in the x and z directions, with little y-component motion. The measurements of Karlsson (1959) suggest that the effect of pure u-component oscillations on the mean motion is very small.

Most of the available data for the effect of freestream turbulence on shear layers have--very sensibly--been obtained for the simplest case, a flat-plate boundary layer below a grid-turbulence field with L_x/δ not much larger than unity. The data are summarized by Bradshaw (1974). Consider the rough rules

$$\delta \approx 0.015\,(x - x_\ell) \tag{1a}$$

$$L_{x,e} \approx 0.09\,\sqrt{(x - x_g)M} \tag{1b}$$

$$\frac{\sqrt{\overline{u_e^2}}}{U_e} \approx 0.1\left(\frac{x - x_g}{M} - 10\right)^{-\frac{1}{2}} \tag{1c}$$

where suffix e denotes the external stream, x_L and x_g are the positions of the leading edge of the flat plate and of the grid, respectively, and the ratio of grid mesh size M to bar diameter d is about 5 (a typical value, giving a pressure-drop coefficient of about unity). At $(x - x_g)/M = 20$, about the smallest distance at which the turbulence is acceptably homogeneous, we have $\sqrt{\overline{u_e^2}}/U_e \approx 0.03$ and $L_{x,e} \approx 0.02\,(x - x_g)$, so that even if the leading edge is close to the grid $L_{x,e}/\delta$ is somewhat larger than unity. Smaller values of $L_{x,e}/\delta$ can be achieved only at larger downstream distances where $\overline{u^2}$ is smaller. Any freestream turbulence field that roughly resembles grid turbulence will maintain a high turbulence level down the length of a boundary layer only if the length scale is rather large compared to δ. An alternative way of seeing this--and an alternative way of measuring a length scale--is to consider the decay equation for isotropic turbulence

$$U_e \frac{d}{dx}\left(\frac{3}{2}\overline{u_e^2}\right) = -\varepsilon \equiv -\frac{(\overline{u_e^2})^{3/2}}{L_\varepsilon} \quad (2)$$

where the dissipation length parameter L_ε is about 0.9 L_x and can be deduced directly from decay measurements. If $L_\varepsilon = \delta$ and $\sqrt{\overline{u_e^2}}/U_e = 0.05$, the local decay rate of $\sqrt{\overline{u^2}}/U_e$ is 0.05 per 60δ, indicating a large fractional decay in the distance taken for δ to double. If the decay rate is as fast as this it is meaningless to quote shear-layer parameters and give the local value of $\overline{u_e^2}$ alone. Equation (2) implies, however, that the decay rate can be specified if the length scale is known. There are therefore two possible reasons why freestream turbulence effects can depend on the length scale, as well as the intensity, of the freestream turbulence; the ratio of freestream eddy size to boundary layer eddy size, represented by L_x/δ or L_ε/δ, is presumably important in its own right, and the dimensionless decay rate $(\delta/U_e^2)\ d\,\overline{u_e^2}/dx$ also depends on L_ε/δ via (2). It is not clear how much of the scatter in existing plots of--say--the skin-friction ratio $c_f/c_{f,o}$ against $\sqrt{\overline{u_e^2}}/U_e$ for L_x/δ near unity is attributable to neglected effects of length scale. Our own explorations suggest that one of the most important causes of scatter is the effect of three-dimensionality of the boundary layer caused by persistence of the wakes of the grid bars. This is most noticeable if one attempts to obtain high intensities by putting the plate leading edge near the grid.

The existing results for the effect of freestream turbulence on boundary layers all agree that the logarithmic "law of the wall" is not significantly affected by freestream turbulence: this is consistent with the continued validity of the logarithmic law in boundary layers which are themselves highly turbulent. It follows that the increase in c_f due to freestream turbulence is the result of changes in outer-layer structure, leading to a decrease in the "wake parameter" Π. Simple extension of the conventional outer-layer similarity analysis, based on the assumption that the freestream turbulence is adequately described by $\overline{u_e^2}$ and L_x, gives

$$\frac{U_e - U}{u_\tau} = f\left(\frac{y}{\delta}, \frac{\sqrt{\overline{u_e^2}}}{u_\tau}, \frac{L_x}{\delta}\right) \quad (3)$$

so

$$\Pi - \Pi_0 = f\left(\frac{\sqrt{\overline{u_e^2}}}{u_\tau}, \frac{L_x}{\delta}\right) \quad (4)$$

If the law of the wall remains unchanged, the "overlap" skin friction formula is unchanged except for the value of Π, and Bradshaw (1974) shows that, for small changes in Π,

$$\frac{c_f}{c_{f,o}} \approx 1 - \frac{4}{K}(\Pi - \Pi_0)\sqrt{\frac{c_{f,o}}{2}} \quad (5)$$

Experiments show that, for $\sqrt{\overline{u_e^2}}/U_e < 0.05$ at least, $\Pi - \Pi_0$ is directly proportional to $\sqrt{\overline{u_e^2}}/u_\tau$ and so (5) gives

$$\frac{c_f}{c_{f,o}} \approx 1 + \text{constant} \times \frac{\sqrt{\overline{u_e^2}}}{U_e} \quad (6)$$

rather than the dependence on $\sqrt{\overline{u_e^2}}/u_\tau$ which one might have expected a priori. The form (6) has been used, without justification, by most experimenters. The "constant" is nominally a function of L_x/δ but most of the data imply values of the constant lying between 3 and 5.

The physics of freestream turbulence effects are not well understood. Flow-visualization pictures like those of Figure 1 (unpublished work at Imperial College by P. E. Hancock) show that the intermittency interface between (smoke-filled) boundary layer fluid and (smoke-free) freestream fluid becomes much more ragged when freestream turbulence is present, with thin streamers of boundary layer fluid being dragged out into the freestream. The result is that the profiles of shear stress and velocity defect are flatter near the outer edge (corresponding to reduced Π) but although this seems obvious it becomes less obvious on further thought, because there is no fundamental reason why the interaction of unsheared freestream turbulence with the delicately-structured eddies of a shear layer should increase the shear stress, rather than reducing it by interfering with the existing organization. Charnay (1974) has made some conditionally-sampled measurements in which the boundary-layer fluid was heated to distinguish it from freestream fluid, analogous to the qualitative use of smoke in the work shown in Figure 1. Unfortunately Charnay's system did not pick up the fine detail of the interface structure: further work using temperature-conditioned sampling is in progress at Imperial College.

A contaminant plume in an isotropic turbulence field will spread out with a standard deviation of $\sqrt{\overline{u^2}}x/U$ and a half-thickness from center-line to edge of about three times this, or 0.15x if $\sqrt{\overline{u^2}}/U$ = 0.05. Now in a boundary layer with $\sqrt{\overline{u_e^2}}/U_e$ = 0.05 the growth rate is about 0.04, i.e., $\delta \approx 0.04\,x$, this is less than one-third of the growth rate expected if boundary layer fluid were passively convected by the freestream turbulence. Part of the reason may be that the

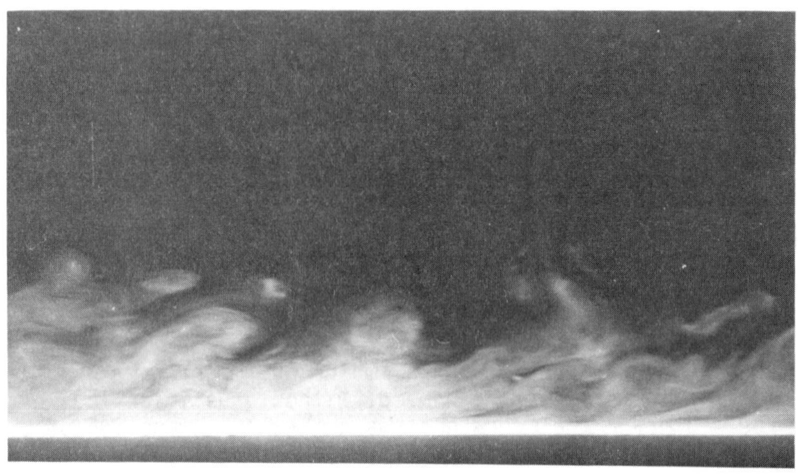

(a) Freestream Turbulence ≈ 0.2 percent.

(b) Freestream Turbulence ≈ 3 percent, $L_x/\delta \approx 0.5$.

Figure 1. Instantaneous Edge of Smoke-Filled Boundary Layer at $U_e\ \theta/\nu \approx 700$.

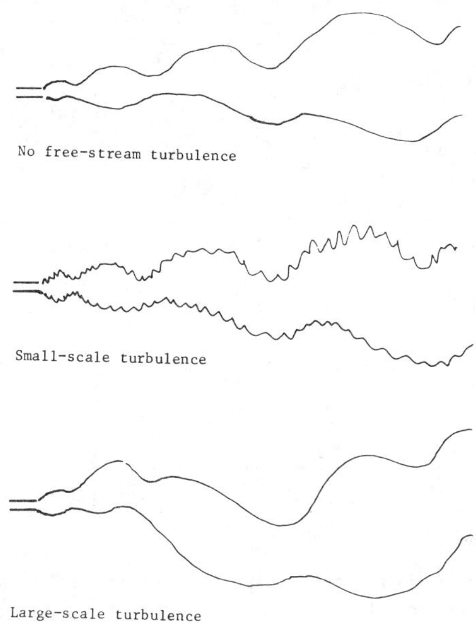

Figure 2. Effect of Freestream Turbulence on Jets.

interaction decreases the efficiency of the shear-stress-producing processes, as hinted above: but a simpler reason is the reduction of $\overline{v^2}$ near a solid surface, also mentioned above.

P. E. Hancock and N. H. Thomas (to be published) have investigated the effect of the inviscid "no-transpiration" boundary condition, v = 0, and the viscous "no-slip" boundary conditions u = 0, w = 0, on grid turbulence near a wall. Ideally the experiment should be done near the floor of a box filled with decaying turbulence, but the absence of a mean translational velocity makes hot-wire measurements difficult; Hancock and Thomas have done a steady-state experiment, downstream of a grid and close to an endless belt which forms the floor of the wind tunnel and moves at the speed of the freestream. Intensity measurements for one of the configurations tested are shown in Figure 3, and it is seen that the inhomogeneity layer extends for a distance of nearly $2L_x$ above the surface. The increase of $\overline{u^2}$, which would be accompanied by an equal increase in $\overline{w^2}$ if the turbulence were truly isotropic, is the natural consequence of the decrease in $\overline{v^2}$: eddies "splash" on the floor. At the surface, of course, u = w = 0, but the thickness of the layer dominated by the viscous boundary condition was too small to resolve in the present experiment. The previous experiment of Uzkan and

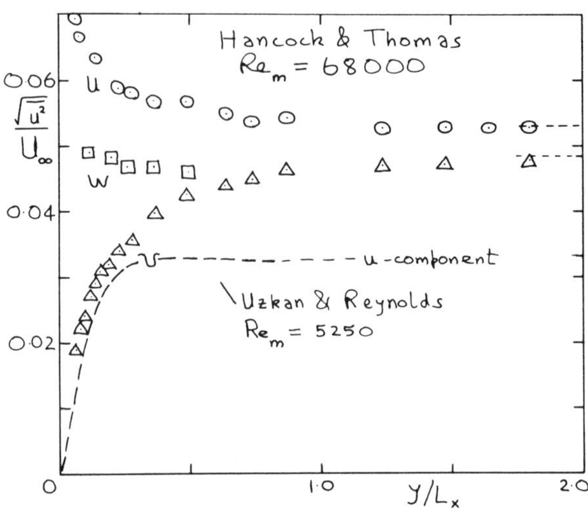

Figure 3. Turbulence Intensity in the Unsheared "Inhomogeneity Layer" near a Moving Wall: Unpublished Imperial College Compared with Data of Uzkan and Reynolds (1967).

Reynolds (1967), at a mesh Reynolds number UM/ν less than a tenth of the present one, was dominated by the viscous boundary condition. They stated, correctly, that the viscosity-dominated region is roughly the same thickness as a laminar boundary layer, $5\sqrt{\nu x/U_e}$, but in their experiment L_x was almost the same as this. The inviscid "splashing" effect was therefore not detectable. Both sets of experiments agree well with an analysis by J. C. R. Hunt and J. M. R. Graham (to be published) using rapid-distortion theory for the inviscid effect.

As well as affecting fully-developed turbulent shear layers, freestream turbulence can have a large effect on transition. Our measurements (Chandrsuda et al., 1976) strongly suggest that this is the main reason why the two-dimensional orderly structure found in the two-stream mixing layer of Brown and Roshko (1974: see also Roshko in Murthy, 1974, p. 295) was not found in previous work on mixing layers in still air. We set up a two-stream mixing layer in the 4 ft. × 2 ft. (120 cm. × 60 cm.) smoke tunnel at Imperial College, with the configuration shown in Figure 4. When the turbulence in the two streams was low (the tunnel turbulence level without the splitter plate is about 0.2 percent) the two-dimensional "spanwise vortices" that arise from instability of the laminar mixing layer grew to a large amplitude, pairing in an approximately self-preserving fashion and gradually acquiring some small-scale three-dimensional

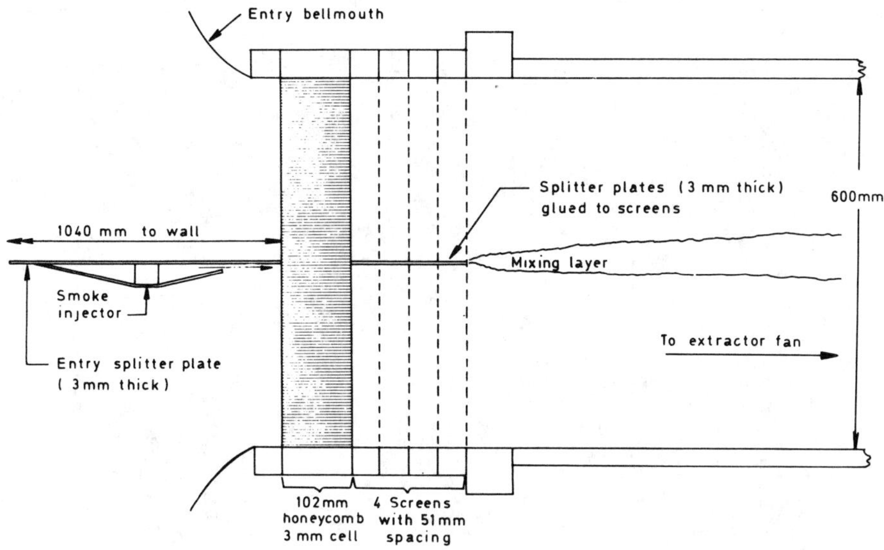

Figure 4. Conversion of Smoke Tunnel for Study of Two-Stream Mixing Layer.

turbulence (Figure 5a). Our photographs, taken with smoke injection, look somewhat different from the shadowgraphs of Brown and Roshko but evidently show the same phenomenon. The intensity of the two-dimensional structure increased as the stream speed ratio U_2/U_1 decreased towards zero. When the turbulence level in one stream was high, breakdown to fully-three-dimensional turbulence occurred after only one or two stages of pairing. We introduced turbulence by sticking a grid of masking tape to the last screen, but as this also caused strong three-dimensionality of the mean flow near the grid it could be criticized as an unrealistic configuration. Figure 5b, however, was taken with one stream completely shut off so that the flow approximated that over a backward-facing step--typical of many industrial examples of mixing-layer flows. Turbulence in the secondary "stream" was supplied by the unsteady recirculation from the reattachment point. Figure 5b clearly shows conventional (three-dimensional) turbulence in the shear layer. Because of the rapid diffusion of the smoke it was difficult to observe the flow much further downstream than in Figure 5, but even with low free-stream turbulence the flow pattern seemed to become increasingly three-dimensional with increasing distance downstream.

The smoke-tunnel experiment was in fact the last in a series of flow-visualization tests whose early stages were reported in

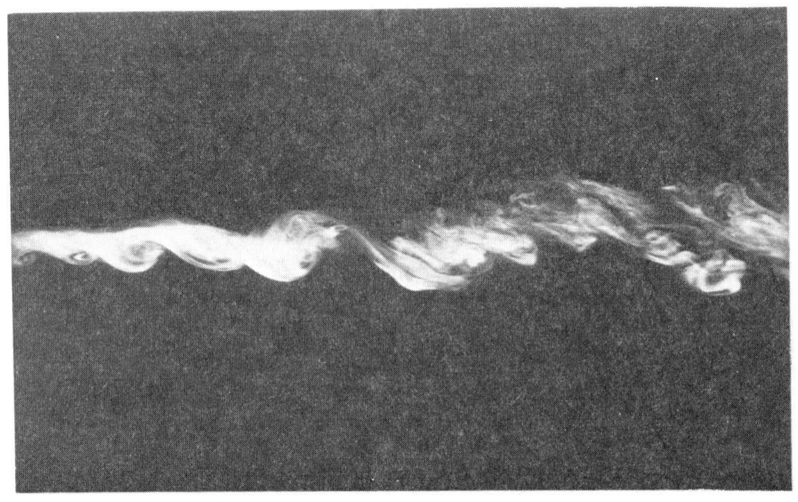

(a) $U_2/U_1 \sim 0.03$ (low freestream turbulence).

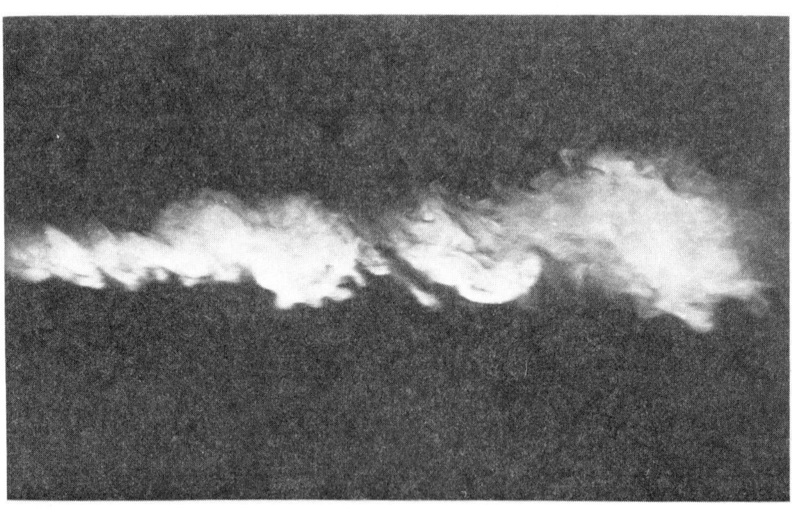

(b) $U_2/U_1 = 0$ (high turbulence in recirculating flow below mixing layer).

Figure 5. Two-Stream Mixing Layer in Smoke Tunnel, High-Speed Stream Below. Picture width about 60000 ν/U_1.

Murthy (1974, p. 243). As mentioned there, studies of a smoke-filled plane mixing layer in "still air" showed that two adjacent quasi-laminar vortices often became significantly non-parallel so that when pairing occurred the vortices merged into a double helix, as seen in the plan view of Figure 6, and breakdown to three-dimensionality ensued. It is now clear that "still air" in a laboratory can in fact be sufficiently highly turbulent to introduce three-dimensionality into the transition pattern, because the laboratory air is stirred up by the jet flow and then re-entrained into it. Our jet nozzle, 75 × 12.5 cm., was near the middle of a large and otherwise undisturbed laboratory (about 10 m. × 7 m. × 5 m.) about 1.5 m. from the floor, which suggests that almost any mixing layer not formed between two low-turbulence streams will become conventionally turbulent quite close to the nozzle rather than developing the two-dimensional large structure of Brown and Roshko. However, observation does suggest that axisymmetric disturbances in circular jets in "still air" are rather more persistent than the corresponding two-dimensional flow, probably because axisymmetric pressure fluctuations help to synchronize the transitional disturbances and prevent helical pairing.

Our experiments do not prove that the two-dimensional large structure in Brown and Roshko's experiment would eventually have distintegrated into the conventional turbulence structure found when two-dimensionality of the transitional disturbances is destroyed by freestream turbulence, but it seems very likely that this is the case. With hindsight and the aid of Figure 7 it can now be seen that the measurements of Bradshaw, Ferriss and Johnson (1964) were made in "conventional" mixing-layer turbulence. Figure 7 is a photograph taken by them at a speed between the two cases shown in their paper and may be compared with the latter. They also found very distinct large structures in the flow but correlation measurements showed that these structures were three-dimensional, with spanwise scales much less than the longitudinal scales, and appeared to be roughly the same as the "mixing jets" found in other shear layers (Grant, 1958). Townsend (1976) shows that the "mixing jet" approximates to an inclined horseshoe vortex (pointing downstream) and its induced velocity pattern. Now the inclined horseshoe vortex is exactly the instability mode of an isolated spanwise vortex in the presence of a mean shear once it is perturbed by a spanwise-periodic disturbance (Figure 8): Hama (1963) shows that an isolated vortex is unstable, in a horseshoe mode, because of its self-induced velocity. Figure 7 suggests that vortex-ring disturbances in a circular jet may break down directly into a ring of horseshoe vortices without helical vortex pairing. The circumferential wavelength of the secondary instability seen in Figure 7 is roughly equal to the primary wavelength (vortex spacing) at the position where the secondary instability is first noticeable, and therefore varies with jet speed; it does not scale on jet diameter. Similar secondary

(a) Side View. Picture width about 10000 ν/U_1.

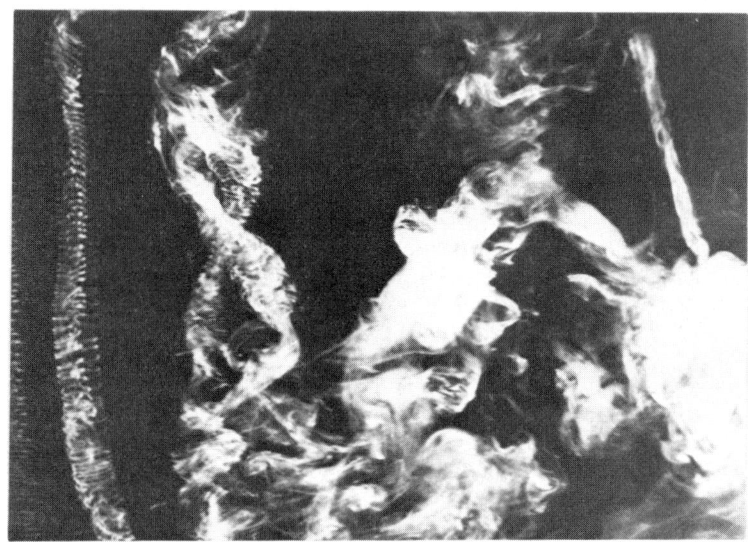

(b) Plan View. Picture width about 40000 ν/U_1: each array of streaks is a vortex core (streaks arise from smoke injection through discrete holes).

Figure 6. Smoke Pictures of Mixing Layer in "Still Air," High-Speed Stream Below.

Figure 7. Schlieren Picture of Circular Jet (with Small Concentration of CO_2) in "Still Air." Picture width about 400000 ν/U_1.

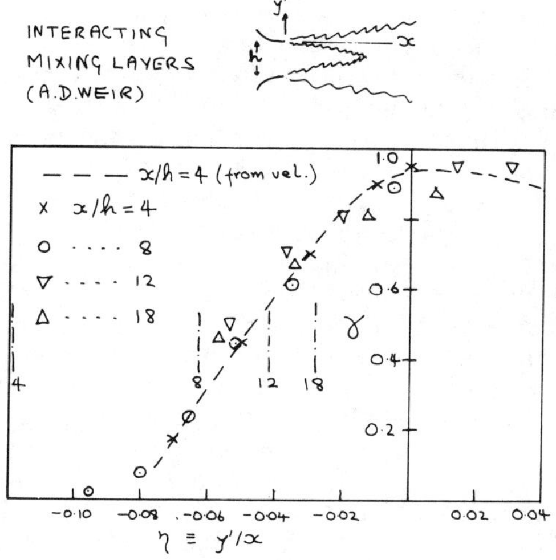

Figure 8. Intermittency of Temperature Fluctuations in a Plane Jet Plotted in Similarity Coordinates for the Heated (Upper) Mixing Layer.

instabilities are seen in Brown and Roshko's Figure 8 but apparently do not lead to full three-dimensionality whereas in the circular jet they do, the longitudinal streaks in Figure 7 being obliterated at about one diameter from the exit. The above evidence strongly suggests that the life history of a mixing layer developing from laminar flow at the nozzle lip is

(i) instability of the laminar mixing layer in a spanwise-vortex mode,
(ii) quasi-self-preserving "pairing" of the spanwise vortices,
(iii) development of secondary instabilities, either (a) small-scale local ones or (b) the large-scale phenomenon of helical pairing, the probability being that (b) precedes (a) if the freestream turbulence is large enough for vortices to become significantly non-parallel between stages of pairing,
(iv) onset of fully-three-dimensional turbulence, at a much slower rate following (iiia) than following (iiib),
(v) breakdown of any quasi-two-dimensional large structure into horseshoe-vortex/mixing-jet large eddies,
(vi) a final self-preserving state.

Clearly option (iiia) allows the spanwise vortices (the inviscid instability mode) to persist until a fluctuation with large spanwise scale somehow arises and precipitates helical pairing. This appears to be the course of the Brown-Roshko experiment and of the low-turbulence case in our own work (Figure 5a). Option (iiib), which can occur when fluctuations with large spanwise scale are imposed externally, by freestream turbulence (Figure 5b) or otherwise, seems to be typical of experiments on mixing layers in still air. Judging by the work of Hill (1976) the two-dimensional (or axisymmetric) large structure is not found at all in plane or circular jets which originate from fully turbulent boundary layers, and it seems likely that the tripped boundary layer in Brown and Roshko's experiment did not become three-dimensionally turbulent before reaching the nozzle exit. There is no reason to suppose that the final states are different in case (iiia), case (iiib) and the case of an initially turbulent boundary layer; but in case (iiia) it may not be reached before the end of the test rig. It must be recalled that even in case (iiib) self-preservation is not reached for a distance of order 1000 times the initial momentum thickness, while Dimotakis and Brown (1975) have pointed out that pairing is an infrequent process, so that breakdown depending on gradual loss of organization will take a very long streamwise distance indeed, perhaps almost as long as the decay of the Karman vortex street behind a two-dimensional bluff body. Once a fully-three-dimensional state is established it seems improbable that correlations over large spanwise distances can recur, even though the inviscid instability mode is two-dimensional: the medium-scale turbulence will ensure that breakdown to horseshoe vortices occurs. While transient disturbances coherent over a spanwise distance of a few shear-layer thicknesses might well occur, they are evidently rare enough not to have much effect on spatial correlations (a two-dimensional large structure would be shown up by conventional correlation measurements, as an unusually large spanwise integral scale, despite the suggestions to the contrary in Brown and Roshko (1974)).

The characteristic Reynolds number of a mixing layer is that based on, say, the momentum thickness at the nozzle lip: the Reynolds number based on the length of the apparatus is not relevant as such, although the ratio of apparatus length to initial momentum thickness is important since it determines the maximum number of pairings that can occur. Even the initial momentum-thickness Reynolds number does not play a very important part because the basic instability of a laminar free shear layer is inviscid, and the growth rates are practically independent of Reynolds number for $U_e x/\nu > 10^4$, as can be deduced from Fiture IX.22 of Rosenhead (1963) where $R = (U_e x/\nu)^{\frac{1}{2}}$. Therefore the demonstration by Dimotakis and Brown (1975) that large-scale structures occur at higher Reynolds numbers than in Brown and Roshko's experiment does not prove that the large-scale structures are not of transitional origin. In any case, Dimotakis and Brown apparently did not check the two-dimensionality of the large structures they observed, and the presence or absence of two-dimensional large structures is the central point of the controversy: Bradshaw et al. (1964) showed that the mixing layer is certainly dominated by large structures, but in that case they were three-dimensional large eddies.

3. THE NEAR-WAKE OF AN AEROFOIL

In earlier work on interacting shear layers at Imperial College, briefly described by Bradshaw (1974) (see also Bradshaw, Dean & McEligot (1973) and for an account of experimental work, see Dean & Bradshaw (1976)) we found that the turbulence structure in the core of a plane duct flow was close to that of two boundary layers whose intermittent regions "time-shared" near the duct center-line. In Reynolds averaging, time-sharing is indistinguishable from superposition, and good results were obtained from a calculation method treating duct flows as superposed boundary layers. Superposition as such is not possible in a non-linear system, but the rms turbulence intensity on the center-line of a duct is only about 3 percent of the mean velocity so that the duct is not a very severe test case. Work on the merging of two mixing layers at the end of the potential core of a plane jet is now being written up by Weir and Bradshaw. As in Dean's work, one shear layer was heated so that its fluid could be distinguished from that in the other layer even after the two started to merge. Figure 8 shows the temperature intermittency profile (nominally identical, in an isolated mixing layer, to the usual velocity-intermittency profile) plotted against the similarity coordinate for an isolated mixing layer, y/x, with origin at the start of the "hot" layer. Even at 18 nozzle heights downstream three or four times the distance at which the two mixing layers start to overlap, the collapse is still good. This is not a very severe test of the time-sharing or superposition hypothesis but more detailed results suggest that the latter is a good approximation.

J. Andreopoulos and the author have recently started experimental work in the wake of a flat plate, one of whose boundary layers is heated to permit study of the near-wake interaction in the same way as in the duct or jet interactions. We do not expect the wake to behave like two superposed boundary layers; it is probably closer to the superposition of two mixing layers with very thick initial boundary layers. A plate 2.5 cm. thick, 2.5 m. long was used, to provide sufficient thermal insulation. The trailing-edge angle is about 3 degrees, but the center-line velocity in the wake seems to be nearly the same, in inner-layer similarity coordinates, as in the measurements of Chevray and Kovasznay (1969) on a thin flat plate (Figure 9). Inner-layer scaling, using the friction velocity u_τ at the trailing edge as a velocity scale, is appropriate for a flat-plate wake close to the trailing edge. According to our hyperbolic turbulence model (see for instance Bradshaw & Unsworth (1974)) a small disturbance at the surface below a boundary layer propagates outwards at an angle close to $0.55\ u_\tau/U$, about 0.03 if we take $U/u_\tau = 15\text{-}20$, and therefore reaches the outer edge of the inner layer, $y = 0.2\delta$ say, at a distance of about 7δ downstream of the disturbance. Roughly the same limit should apply to the strong disturbance constituted by the termination of the solid surface at the trailing edge. The general result for the mean velocity profile in the region $x < 7\delta$, $-0.2\delta < y < 0.2\delta$, where x and y are measured from the trailing edge, is

$$\frac{U}{u_\tau} = f_1\left(\frac{u_\tau x}{\nu}, \frac{u_\tau y}{\nu}\right) \quad \text{or} \quad f_2\left(\frac{u_\tau x}{\nu}, \frac{y}{x}\right) \tag{7}$$

of which the scaling in Figure 9 is a special case. Another useful special case is

$$V_{edge} \approx V_{y=0.2\delta} = u_\tau\ f_3\ (u_\tau\ x/\nu) \tag{8}$$

where suffix "edge" denotes the local edge of the "inner wake" and the approximation sign implies that $\partial U/\partial x$ is small in the region between the edge of the inner wake and the point $y = 0.2\delta$. Equation (8) provides a universal solution for the displacement thickness of the first 7δ of a flat-plate wake. Since for $x < 7\delta$ outer-layer inviscid scaling should apply if the viscous sublayer at the trailing edge is negligibly thin, we also have the following.

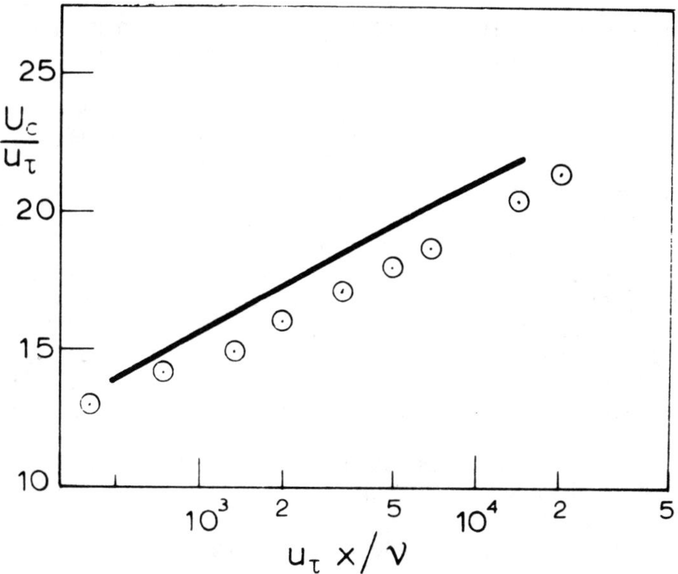

Figure 9. Center-Line Velocity in the Wake of a Flat Plate, Plotted in Inner-Layer Similarity Coordinates with u_τ as Trailing Edge Value. ———, Chevray & Kovasznay (1969). ⊙ J. Andreopoulos (unpublished).

$$\frac{U_e - U}{u_\tau} = f_4\left(\frac{x}{\delta}, \frac{y}{x}\right) \tag{9}$$

and

$$\frac{U_e - U_{CL}}{u_\tau} = f_5\left(\frac{x}{\delta}\right) \tag{10}$$

Extensions to the above scaling can be made for a thin flat plate with different values of u_τ on each side, inserting the friction-velocity ratio as an extra parameter. It is of course invalidated by pressure gradient.

So far our data analysis of turbulence recordings has been confined to the region outside the near wake: unconditional-average intensities and the intermittency (deduced from the temperature signals) agree well with previous data. Analysis of the data in the near wake is about to begin, with the object of improving our turbulence model, referred to above and under investigation by Huffman (as reported elsewhere in these Proceedings).

4. FLOW IN A SIMULATED WING-BODY JUNCTION

In turbulent flow in a long, straight non-circular duct, cross-stream gradients of Reynolds stress set up circulations in the cross-stream plane (i.e. streamwise vorticity) called secondary flow of Prandtl's second kind, or "stress-induced" secondary flow (see Johnston, 1976). Even in inviscid flow, lateral skewing of the streamlines of a shear flow can induce streamwise voriticty via the $(\Omega\cdot\nabla)$ U term in the vorticity equation, and this is secondary flow of Prandtl's second kind, or "skew-induced" secondary flow. We have made measurements in a simplified wing-body or blade-hub junction consisting of an "aerofoil" of constant thickness following a semi-elliptical nose, mounted on a flat tunnel wall simulating the body. The body boundary layer thickness at the leading edge of the wing was about half the wing thickness. Figure 10 (Shabaka, 1975) shows the longitudinal-component velocity at a distance measured

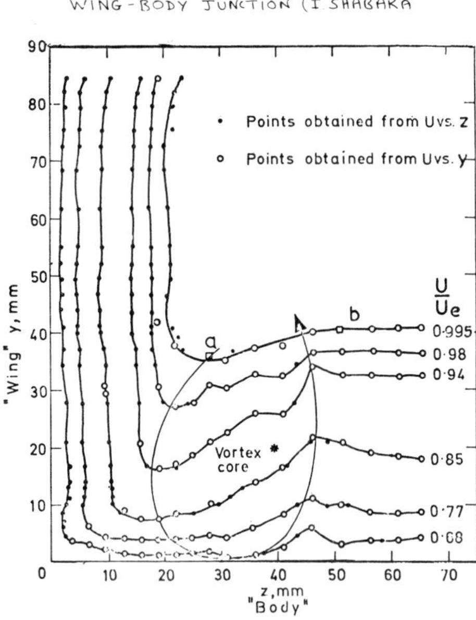

Figure 10. Longitudinal-Velocity Contours in Idealized Wing-Body Junction, 26 Wing Thicknesses Downstream of Leading Edge. Wing Thickness 50 mm, Initial Boundary-Layer Thickness on Body 25 mm (Shabaka, 1975).

from the leading edge equal to 26 wing thicknesses (the trailing edge of an aerofoil 4 percent thick). The secondary flow of the first kind generated by the skewing of flow at the leading edge (the horseshoe vortex), resulting in a trailing vortex in the corner, has completely overwhelmed the stress-induced secondary flow. This suggests that the rather large amount of work done on stress-induced secondary flow is not likely to be relevant to many streamwise corner flows, because there can be few practical cases in which non-circular ducts are straight for a long enough distance for skew-induced secondary flow to decay and stress-induced secondary flow to take over. In Figure 10 the position of the horseshoe vortex is conjectural: V and W component mean-velocity measurements are about to begin. Figure 11 shows the u-component intensity measurements at the same station as Figure 10, and the vortex position shown is at least plausible. The vortex can still be detected, by the eye of faith, in surface oil-flow patterns at this station. Of course, the circulation around the vortex is gradually decaying, under the action of Reynolds-stress gradients--thus the latter, far from generating secondary flow, act to reduce it. It seems rather unlikely that the same turbulence model will suffice to predict both the single-vortex pattern of decaying skew-induced secondary flow and the double-vortex pattern of stress-induced secondary flow.

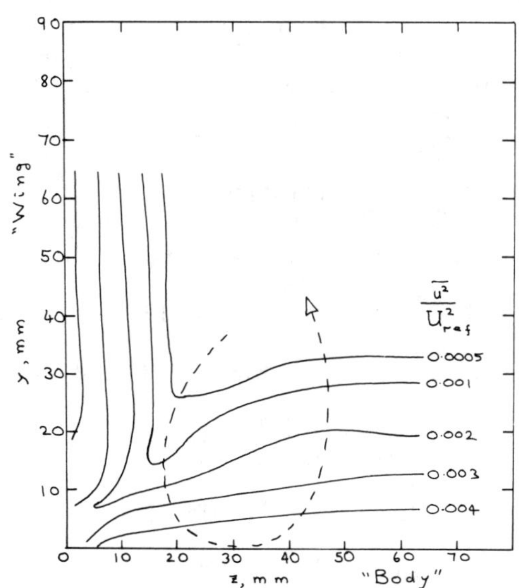

Figure 11. Longitudinal-Intensity Contours in Wing-Body Junction. Conditions as in Figure 10 (Shabaka, unpublished).

5. CONCLUSIONS

The experiments described above relate to highly idealized situations, which were chosen so as to show up the main phenomenon being studied. The data are being, or will be, used in calculation methods, and this writer believes that such basic experiments are a necessary preliminary to the development of reliable calculation methods. The assumption commonly used is that the effects of two or more basic phenomena on the empirical constants or functions of a turbulence model are independent and therefore additive (or at least expressible as separable-variable functions). The success of this assumption depends, of course, on choosing the phenomena carefully: almost certainly, the turbulence structure of the near-wake of an unstalled aerofoil is not affected by freestream turbulence (except that the problem may become an unsteady one), so that "near wakes" and "freestream turbulence" are independent phenomena in the present sense; but the horseshoe trailing vortex in a wing-body junction interacts so strongly with the two boundary layers that the phenomenon "wing-body junction flow" cannot be broken down into the phenomena "isolated trailing vortex" and "stress-induced secondary flow in streamwise corner." We now have enough information on complex flow phenomena for the temptation to assemble practical flows from several known phenomena to be a strong one. However it must be remembered that the complex-flow concept is based on perturbations of simple shear layers, by distortion or by interaction with another turbulence field. Further perturbations by interaction of two or more complex-flow phenomena need treating with respect!

REFERENCES

1. Antonia, R. A. and Luxton, R. E. 1971 J. Fluid Mech. 48, 721
2. Antonia, R. A. and Luxton, R. E. 1972 J. Fluid Mech. 53, 737
3. Antonia, R. A. and Luxton, R. E. 1974 Adv. Geophys. 18A, 263
4. Bradshaw, P. 1974 Imperial College Aero Rept 74-10, ARC paper 35648, STAR index no. N75-20669
5. Bradshaw, P. 1975 ASME J. Fluids Engg. 97, 146
6. Bradshaw, P. 1976 (in) Proceedings, 14th Int. Congr. Appl. Mech., Delft (North Holland Publishing Co., in press)
7. Bradshaw, P. and Unsworth, K. 1974 Imperial College Aero Rept 74-02, STAR index no. N74-33802

8. Bradshaw, P., Dean, R. B. and McEligot, D. M. 1973 ASME J. Fluids Engg. 95, 214

9. Bradshaw, P., Ferriss, D. H. and Johnston, R. F., 1964 J. Fluid Mech. 19, 591

10. Brown, G. L. and Roshko, A. 1974 J. Fluid Mech. 64, 775

11. Chandrsuda, C., Mehta, R. D., Weir, A. D. and Bradshaw, P. 1976 "Effect of Free-Stream Turbulence on Large Structure in Turbulent Mixing Layers" submitted to J. Fluid Mech.

12. Charnay, J. 1974 PhD Thesis, Université de Lyon

13. Chevray, R. and Kovasznay, L. S. G. 1969 AIAA J. 7, 1641

14. Dean, R. B. and Bradshaw, P. 1976 "Measurements of Interacting Turbulent Shear Layers in a Duct" submitted to J. Fluid Mech.

15. Dimotakis, P. and Brown, G. L. 1975 Project SQUID Rept. TR-CIT-7-PU

16. Grant, H. L. 1958 J. Fluid Mech. 4, 149

17. Hama, F. R. 1963 Phys. Fluids 6, 526

18. Hill, W. G. 1976 Grumman Aerospace Co. Rept. RM-612, AD-A020509/6GA

19. Johnston, J. P. 1976 (in) Topics in Applied Physics; Vol. 12 - Turbulence (P. Bradshaw, Ed.) Springer, Heidelberg

20. Karlsson, S. K. F. 1959 J. Fluid Mech. 5, 622

21. Murthy, S. N. B. (Ed.) 1975 Turbulent Mixing in Non-reactive and Reactive Flows. Plenum, New York

22. Rosenhead, L. (Ed.) 1963 Laminar Boundary Layers. Clarendon, Oxford

23. Shabaka, I. M. M. A. 1975 Imperial College Aero Rept. 75-05 STAR index no. N76-13028

24. Townsend, A. A. 1976 The Structure of Turbulent Shear Flow (2nd ed.) University Press, Cambridge

25. Uzkan, T. and Reynolds, W. C. 1967 J. Fluid Mech. 28, 803

DISCUSSION

SAFFMAN: (California Institute of Technology)

I want to make a comment on the effect of freestream turbulence on the large structures in mixing layers. It is to point out, as has recently been shown, that arrays of vortices are, in fact, subject to a three-dimensional instability, the axial wavelength of the instability being about the diameter of the vortex core. I just wonder whether the freestream turbulence may not be exciting this instability, which may be a partial reason for the ability of freestream turbulence to destroy the organized structures.

BRADSHAW:

This is at least possible. I wasn't aware of this work. Is this an extension of the work that Jimenez* did where he proved stability of a co-rotating vortex pair in the linear case?

SAFFMAN:

No, Jimenez uses the cutoff approximation for the vortices. He ignores the internal structure. The instability that I am referring to is an effect of the internal structure. The calculation has only been done, so far for the inviscid vortices with uniform cores and no turbulence. A reference is D. W. Moore and P. G. Saffman: Proc. Roy. Soc. A346, 413 (1975).

LIEPMANN: (California Institute of Technology)

At the IUTAM meeting last week Garry Brown** discussed experiments by Konrad† demonstrating the development of three-dimensional structures from two-dimensional ones. The most exciting result obtained was the reappearance of the two-dimensional structures further downstream. This fact is not too difficult to understand in mixing layers because large scale two-dimensional vortices are being produced all the time, hence the overall tendency to two-dimensional flow. Turbulence in the freestream should definitely

---

* Jimenez, J. "Stability of a pair of co-rotating vortices," Phys. Fluids 18, 1580 (1975).

** Proceedings of IUTAM Symposium on Structure of Turbulence and Drag Reduction, Washington, D.C., June 7-12, 1976, to appear in Physics of Fluids.

† Konrad, J. H., Caltech, Pasadena.

affect the phases of the two-dimensional vortices coming off and this fact is, after all, already clear from the Brown and Roshko experiments. A stationary hot wire exposed to these vortices with random phases should show a perfectly normal turbulence spectrum. Furthermore, it is possible that a slight modulation spanwise of the initial phase could produce a more rapid transition to three dimensionality.

FALCO: (Cambridge University)

I have a comment and a question. The first point is, that the first slide (Figure 1 of your paper) you showed appeared to me to be a truly very low Reynolds number turbulent boundary layer with R_θ on the order of 500 rather than 5000 as you indicated. That is probably worth looking at again. The second is that I tried a year and a half ago to look at smoke filled turbulent boundary layers that were generated downstream of the grid, in this case a bi-plane grid (M/D = 4) with nine meshes across my tunnel. But, I couldn't in fact come up with an edge to the boundary layer. What I did was to put the grid in a boundary which already was turbulent and smoke filled and looked downstream of the grid. The angle between the smoke and the wall was truly phenomenal and smoke got right across to the other side of my tunnel in about eight feet. I had a four inch thick boundary layer before it hit the grid. At that point I just gave up. So I am quite interested to know if you started out with a turbulent boundary layer or if your grid was placed upstream of the tripping point in the boundary layer?

BRADSHAW:

We have looked both at the tunnel floor boundary layer and at the boundary layer on a flat plate with its leading edge some distance downstream of the grid. I strongly recommend anyone who wants to do experiments on freestream turbulence not to use the wind tunnel floor, because all sorts of funnies happen around the bottom of the grid. The main purpose of the flow visualization experiments which I showed was to sort out this matter for ourselves. There are certain conceptual advantages in using the tunnel floor because you then don't have to worry about the possibility of transient separations, pressure gradients and odd behavior near the leading edge of a flat plate in freestream turbulence. But in fact the hassles that you get into with the flow around the bottom half of the grid are quite considerable whether the incoming boundary layers are laminar or turbulent. Things are worse if they are laminar, but in any case you are liable to find boundary fluid being drawn up behind the vertical bars of the grid.

The two pictures in Figure 1 of my paper were at about the same R_θ, roughly 750 and for the purposes of this lecture I am merely saying that there was a fairly spectacular difference between them.

KLINE: (Stanford University)

I want to say first that I always appreciate hearing what is new, different and well-put, which, as usual, characterizes Peter Bradshaw's presentation. I want to ask a question about the comparison between the Reynolds-Acharya* experiment and your data. You explained it, but can you expand on that Reynolds number effects a little bit, why you think you should get different results in the two cases?

BRADSHAW:

I think I owe Bill Reynolds an apology for showing Figure 3 of my paper in public before having a chance to show it to him privately, but it was made very shortly before I left. The basic point is that the effect of the no-slip condition spreads out like a laminar boundary layer. This is what the Uzkan-Reynolds scaling shows. If the scale of the freestream turbulence at given X-position is large compared with the thickness a laminar boundary layer would have at the same X-position, then it is the normal component constraint, the "splat effect," that matters and you get a high Reynolds number behavior. If however the experiment is done at so low a Reynolds number that far downstream the thickness of a laminar boundary layer would be about the same as the integral scale of the freestream turbulence (as was the case in the Uzkan-Reynolds experiment) then you get a monotonic decrease in u-component turbulence to the wall.

KLINE:

So you are visualizing the effect of the incoming disturbance acting on the wall as if it were then some superposed laminar boundary layer. Is that the concept you are using?

BRADSHAW:

Just for purposes of exposition, yes. I am not claiming that there is anything one would strictly call a laminar boundary layer, simply that in the case of the no-slip condition, as distinct from the no-transpiration condition, the effects propagate outwards as viscous diffusion.

---

* Acharya, M. and Reynolds, W. C. "Measurements and Predictions of a Fully Developed Turbulent Channel Flow with Imposed Controlled Oscillations." Stanford M.E. Department Report TF-8, May 1975.

CORRSIN: (Johns Hopkins University)

You said that you were embarrassed by the anisotropy of the turbulence coming off your freestream turbulence generating grid but that is the way that almost everyone finds it, so why be embarrassed?

BRADSHAW:

I was embarrassed because we didn't have the time to put in a contraction downstream of the grid to produce a better approximation to isotropy.

LAKSHMINARAYANA: (Pennsylvania State University)

The difference between your picture and the Brown and Roshko picture is obviously the way you visualize the flow. You introduce smoke for visualization. Was the smoke injected uniformly?

BRADSHAW:

The smoke was injected as uniformly as we could manage. We did in fact try some shadowgraph pictures in the single stream mixing layer (which was the one that I showed first), and the results certainly did not show the Brown and Roshko type of structure. I am not, I think, prepared to go very much further than that, because we had certain difficulties with the shadowgraph installation. I grant that there is a difference between the two techniques: the shadowgraph technique is much more likely to show up (and indeed to emphasize) two-dimensionality. I think that I can claim that the two-stream mixing layer pictures without freestream turbulence (Figure 5 of the paper) did in fact have a pretty strongly two-dimensional structure. It really is very spectacular to see these things. Most of you have seen them in two-dimensional movies, but seeing them in a three-dimensional flow is spectacular. The Brown and Roshko structures undoubtedly exist and there are undoubtedly cases where they are important but those cases do not include mixing layers in "still air" or, I suspect, most industrial type mixing layers in which at least one of the streams will be either recirculating or, for other reasons, will have a fairly high turbulence level.

LAKSHMINARAYANA:

Can you elaborate on the secondary flow you mentioned in connection with wing-body combination flows?

BRADSHAW:

The present case is the sort of secondary flow that occurs in a long straight duct, the secondary flow of Prandtl's second kind, which I prefer to call stress-induced secondary flow. On the other hand, there is secondary flow of Prandtl's "first kind," the quasi-inviscid secondary flow type which turbomachinery people are accustomed to, and that is what one would expect to get (admittedly in modest quantities) in horseshoe vortices around the front of the thin wing. If we had used a thicker wing the results would have been more spectacular. The point is that even this comparatively weak skew-induced secondary flow completely obliterates the stress-induced (long duct) type of secondary flow.

COMTE-BELLOT: (University of Grenoble)

I would like to come back to the channel flow. I think that you said that you have investigated the merging of the two lateral boundary layers with heat as a tracer. Could you describe how the merging takes place? For example, what does happen to the superlayers?

BRADSHAW:

I can't go into too much detail. Consider the boundary layer that grows on the upper wall, and the secondary boundary layer that grows on the lower wall of a duct. One of them is heated, the other is not and one can detect the statistics of the interface between the heated and unheated fluid. Is that the detail that you wanted or did you want some more?

COMTE-BELLOT:

I would like to know more precisely how the two boundary layer edges, which may be quite contorted, can fit together.

BRADSHAW:

The large eddies do seem to fit together, in fact they seem to time-share. It is not really superposition. What in fact seems to happen is a sort of "meshing" like gear wheels. In other words, the large eddies from one side and the large eddies from the other side reach a sort of détenté and establish their own spheres of influence. The superlayers do not fit in detail. There seems to be a thin region of low-intensity turbulence separating the large eddies. This is a crude way of looking at it!

WALLACE: (University of Maryland)

Could you tell me how far downstream, in terms of channel thicknesses, this persists, this clear delineation of the outer layers?

BRADSHAW:

We were able to detect it in our experiments to be about twice the distance that the boundary layers took to meet.

WALLACE:

There does it die out or you can't detect it?

BRADSHAW:

No, it doesn't die out. I think it is true that, far enough downstream in our experiment, small-scale mixing would give uniform temperature across the flow, but large-eddy time sharing can be regarded as happening indefinitely further downstream. The practical use of the superposition idea is that you can take a boundary layer calculation method and by calculating both a "hot" shear stress and a "cold" shear stress, you can calculate a duct flow and we succeeded in getting very good results for fully developed duct flow. So we regarded that as being a sort of post-diction if you like, showing that the superposition hypothesis was not too far out even indefinitely further downstream.

WALLACE:

If I understand correctly then, you see this sort of delineation right to the end of your channel?

BRADSHAW:

I think that if we introduced a new source of heat on one side only, far downstream, then we would again see the same sort of interface between one shear layer's fluid and the other shear layer's fluid.

WALLACE:

What would then be the relationship between the traditional parameters for determining fully developed channel flow and the distance that this interface persists?

BRADSHAW:

I think that one would simply define full development in the conventional way. I think the interface persists indefinitely, there are always two "boundary layers" on the two sides of the duct, they've always got large eddies, the large eddies always interact in a certain manner.

MILLER: (Naval Postgraduate School)

I was happy with this experiment until you drew it for us on the board! Was it heated above or below and what was roughly the Grashof number?

BRADSHAW:

We convinced ourselves that the buoyancy effect was negligible. The flow speed was about 30 meters a second, the duct height was 5 centimeters, the typical temperature difference was on the order of a couple of degrees Centigrade, the lower side being the hotter. We did in fact do some checks: things like the unconditioned intensity profiles were as symmetrical across the duct as one could expect in this imperfect world.

NAGIB: (Illinois Institute of Technology)

As far as the case of the turbulence in the free shear layer, did you see any pairing? We saw just a still picture and you did not comment on the pairs you get.

BRADSHAW:

What seems to happen is that the usual transitional oscillations appear, but after, at the most, one or two stages of pairing the freestream turbulence stirred the flow up to such strong three dimensionality that any further vortex pairing resulted in the double-helix type of pairing and then virtually an explosion into turbulent flow. After the double-helix pairing occurs there is a very rapid development of the small scale structure because there are all sorts of horrible tertiary instabilities that occur.

CORRSIN:

I was going to comment on the doctoral dissertations of Don Johnson (at Johns Hopkins) and Chris Nicholl (at Cambridge) back in the 50's, where they ran turbulent boundary layers over small step functions in temperature in fully developed turbulence.*

The point is that the sharpening of the temperature jump interface within the fully turbulent fluid takes place by mechanisms which are closely analogous to the one that Kistler and I proposed for vorticity interface at freestream with turbulence (NACA Report 1244, 1955). The only difference is that this one depends on the surface stretch rate of the fluid elements as contrasted with the line stretching rate. Except for this counterpoint, they are analogous.

BRADSHAW:

If I can chip in myself I find one of the most interesting things in all of this is that although the interface between a boundary layer and unsheared freestream turbulence is all mixed up and raggedy, as I showed in Figure 1, yet the time-sharing-cum-superposition of two shear layers still seems to work rather well. I am almost prepared to say that unsheared freestream turbulence is likely to be <u>more</u> difficult than at least the simpler cases of interacting opposed shear layers.

WILLMARTH: (University of Michigan)

Does the contaminant (smoke or heat) always follow the boundary between the unsheared freestream turbulence and the sheared and stretched turbulence at the edge that Corrsin is talking about?

BRADSHAW:

The honest answer is that we don't know because we haven't checked it. The slightly more helpful answer is that although the Schmidt number for smoke is of order 30,000 the Reynolds number in the practical cases (though not necessarily in our flow visualization experiments) is probably high enough that the thickness of the viscous superlayer is probably fairly small. Essentially if the Schmidt number is large, the smoke is going to extend out to the inside of the viscous superlayer, whereas, the vorticity fluctuations are going to extend to the outside of the viscous superlayer. This is a crude way of putting it but I think it is the most helpful.

---

* D. S. Johnson, Jour. Appl. Mech. <u>24</u> (Trans. A.S.M.E., Vol. <u>79</u>) pp. 2-8, 1957.

D. S. Johnson, Jour. Appl. Mech. <u>26</u> (Trans. A.S.M.E., Vol. <u>81</u>) pp. 325-326, 1959.

C. I. H. Nicholl, Jour. Fluid Mech., <u>40</u>, pp. 361-384, 1970.

WALLACE:

Would you say, based on this picture then, that a fully developed channel flow could be treated as a boundary layer with a very high freestream turbulence?

BRADSHAW:

No, one couldn't. This is what I was commenting on a little while ago. I think that in order for the superposition ideas to work reasonably well one has got to have two interacting flows which are roughly similar. I think one could probably stand a factor of four or five on length scale, so maybe this idea would still work for, say, a wall jet in which you have a fairly small scale boundary layer interacting with a jet-like layer. But certainly when the turbulence structure of one layer is totally foreign to that of the other they are not really likely to reach what I call the détanté very well.

NAGIB:

As far as the interaction between the base and the wall, we made some measurements around a bluff body in a thick turbulent boundary layer.* While trying to trace the horseshoe vortices around the body we found out that the vertical velocity away from the wall, and its fluctuations, and the cross correlation in the Reynolds stress, between the vertical and the streamwise velocity, are much more sensitive in detecting this kind of large scale vortical motions compared to the streamwise velocity. I just wondered if you wanted to comment.

BRADSHAW:

That is a very helpful comment. It will give us a little more hope in making turbulence measurements. There are very few worthwhile measurements of the vw Reynolds stress in this business so far, but we are just hoping to be able to make some. I think that we may decide to go a little further upstream than the plane which I showed where the secondary flow is really comparatively quick.

SOVRAN: (General Motors Research Laboratories)

How would you visualize this merging of boundary layers occurring for an axisymmetric flow like that in a pipe?

---

* Corke, T. C. and Nagib, H. M., "Sensitivity of Flow Around and Pressures on a Building Model to Changes in Simulated Atmospheric Surface Layer Characteristics," <u>IIT Fluids and Heat Transfer Report</u> R76-1, 1976.

BRADSHAW:

I think one probably has 360 boundary layers all merging, so if 35 people from the audience (with 10 fingers each) would like to come up and help me I will demonstrate. To give a serious answer, I very much doubt if the superposition idea would be worth applying in that case. I think probably the better way of dealing with a pipe flow would be to try and regard the core part of the flow as a turbulence module in its own right fed by energy from the surrounding highly sheared region. Also I doubt if it is worthwhile trying to use the superposition idea for dealing with wing-body junctions, because everything is dominated by the secondary flow (whether of the first or the second kind).

Some Preliminary Observations on the Effect of Initial Conditions on the Structure of the Two-Dimensional Turbulent Mixing Layer

D. OSTER, I. WYGNANSKI,* AND H. FIEDLER†
Tel-Aviv University
Tel-Aviv, Israel

ABSTRACT

The effect of a trip wire on the large structure in the mixing layer is examined at moderately high Reynolds numbers. It is observed that the trip wire may either enhance or inhibit the spreading rate of the mixing layer depending on the velocity ratio between the two streams. The effect of the trip wire extends downstream beyond the 1000 initial momentum thicknesses and is felt in a region which is supposed to be self-preserving.

Correlation and spectral results indicate that the large structures in the mixing layer consist of quasi two-dimensional row of vortices which are convected approximately at the average velocity of the two streams. The abrupt change in the predominant wave length of the correlations with downstream distance suggest that vortices interact and possibly pair while being convected.

1. INTRODUCTION

The mechanism of growth of the two-dimensional turbulent mixing region is not completely understood in spite of the fact that the flow has been continuously investigated during the last decade. A detailed discussion of the views held by many researchers on turbulent mixing is presented in a volume edited by Murthy (1975) to which the reader may refer.

* Present address Department of Aerospace Engineering, University of Southern California, Los Angeles, California.

† Hermann Föttinger Institut für Thermal und Fluiddynamik Tecknische Universität, Berlin.

It is becoming increasingly evident, however, that the flow in the turbulent mixing layer is dominated by a row of large vortices which retain their basic structure over long distances. The individual vortices in the row grow as they proceed downstream, but the overall spread of the mixing layer is attributed largely to the interaction between neighboring vortices which roll around each other to form larger coherent structures. Since the initial observations on vortex amalgamations were made at low Reynolds numbers (Winant and Browand, 1974, reported Re $\approx 10^3$) it is important to establish that the mechanism persists at Reynolds numbers for which viscous effects are considered to be negligible. Dimotakis and Brown (1975) showed that the fundamental periodic structure of the mixing layer is retained at Re = 3.10^6 but the observed moderate pairing process of adjacent vortices is more often than not replaced by a more violent interaction causing a momentary disappearance of identifiable order.

In contrast to many other turbulent shear flows the mean properties of the mixing layer are not universally determined. Spreading rates which are summarized in Figure 10 of Brown and Roshko (1974) or in Table 1 of Champagne, et al. (1976) vary by as much as 30%. Variations of this magnitude cannot be easily attributed to errors of measurement, so there is reason to believe that the effects of initial conditions persist over long distances downstream in spite of the fact that the flow passes the general criteria of self preservation (Townsend, 1956). Unfortunately, the effects of the initial conditions were seldom examined whenever similarity of velocity and intensity profiles was established in a given flow, because theoretical consideration of such similarity imply that the flow is independent of the initial conditions. Although Bradshaw (1966) concluded that as many as 1,000 initial momentum thicknesses may be required for the flow to become independent of its origin, effects of initial conditions on spreading rate of the mixing layer were not seriously considered nor documented.

When Wygnanski and Fiedler (1970) reported that their mixing layer was spreading more rapidly than previously observed by Liepman and Laufer (1947) they alluded to the possibility that a trip wire which they had placed on the splitter plate was responsible for the difference but it was Batt, et al. (1970) who actually proved that the trip wire enhanced the spreading of the mixing layer between a uniform stream and a quiescent surrounding fluid.

The ability to control the spreading rate of the mixing layer by a simple device like a trip wire is interesting because it may have numerous technological applications, furthermore a careful assessment of the effects of the initial conditions on the growth of the mixing layer may contribute to our understanding of the complicated process at hand.

2. EXPERIMENTAL APPARATUS

The apparatus which was constructed by Dov Oster consists of two independent cascade blower tunnels discharging into a common test section (Figure 1). The two streams are separated initially by a splitter plate which extends upstream through the contraction section and into the settling chamber. The splitter plate ends 20 cms downstream of the contraction allowing the two streams to become parallel before mixing. The spatial velocity variations did not exceed 1% outside the boundary layers and the turbulence level in that region is approximately 0.2%. The trailing edge of the splitter plate was milled out of an aluminum plate at an included angle of 3°. The test section is 50 cm high, 60 cm wide and 150 cm long. For the series of tests reported here the velocity of the lower stream was maintained at 15 m/sec while the velocity of the upper stream was varied from 0 to 15 m/sec. Pressure gradients at various velocity ratios were eliminated by adjusting the top and bottom walls of the test section which are mounted on small screw jacks. To test the effect of changing the initial conditions a trip wire was placed on the surface of the splitter plate which faces the high velocity stream, 10 mm upstream of the trailing edge. The wire was 1.6 mm in diameter giving a Re based on the freestream velocity and the diameter of the trip of $\approx 1.6 \cdot 10^3$.

In future experiments an orderly disturbance will be introduced into the mixing layer by replacing the trip wire with a vibrating ribbon or an oscillating small flap.

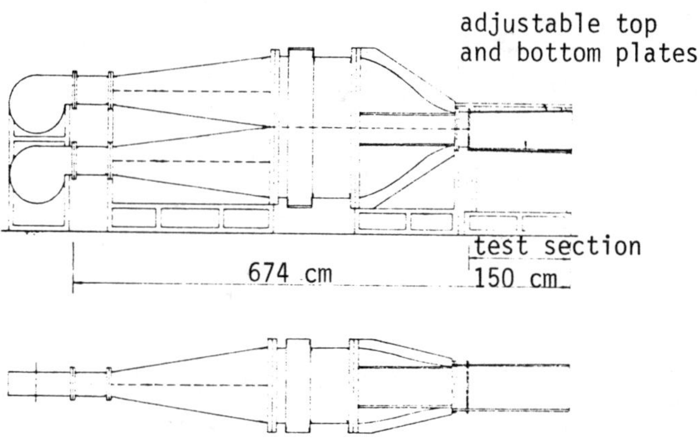

Figure 1. The Wind Tunnel.

3. RESULTS AND DISCUSSION

Four velocity profiles plotted in the usual similarity coordinates are shown in Figures 2a, b and 3a, b. ($\eta = (y - y_{0.5})/(x - x_0)$ where $y_{0.5}$ gives the y coordinate at which $(\overline{U}(y) - U_1)/(U_2 - U_1) = 0.5$ and x_0 presents the location of the virtual origin from the trailing edge of the splitter plate). In Figure 2 the mixing occurs between a single stream and a quiescent surrounding fluid ($U_1 = 0$) while in Figure 3 $U_1/U_2 = 0.4$. Part a of each figure represents the result without a trip wire while in part b, the trip was attached to the splitter plate.

In all cases the velocity profile becomes self similar at distances larger than 500 mm from the discontinuity, and, yet, the spreading rate of the mixing layer is affected significantly by the

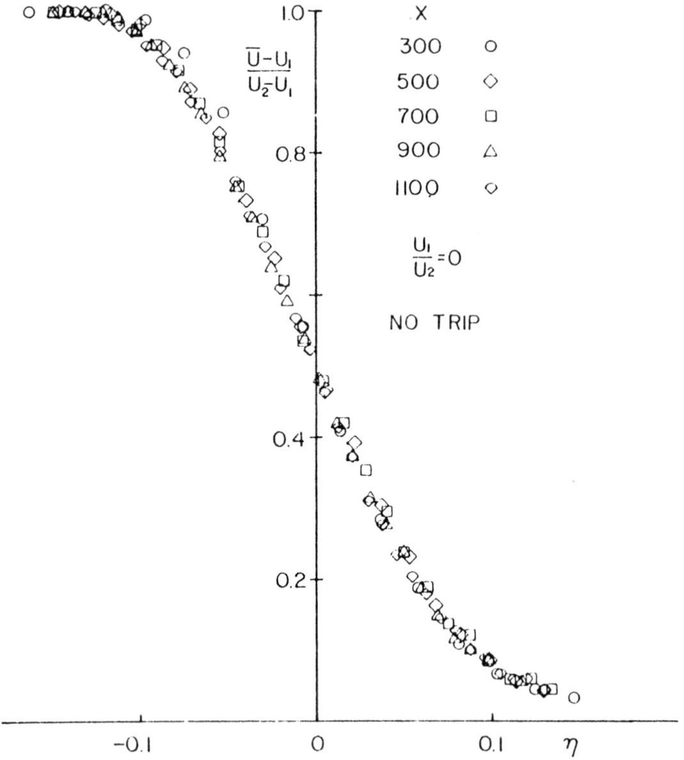

Figure 2a. The Mean Velocity Profile in Similarity Coordinates. $U_1/U_2 = 0$, no trip.

presence of the trip wire. The results are summarized in Table 1 where $db/dx = (y_{0.1} - y_{0.95})/(x - x_0)$ for which the subscripts 0.1 and 0.95 represent the elevations at which $U - U_1/U_2 - U_1 = 0.1$ and 0.95 respectively and * represents the presence of a trip wire. Note that the trip wire caused an increase in db/dx when $U_1/U_2 = 0$. These results are in good agreement with the previous findings of Liepmann and Laufer (1947), Wygnanski and Fiedler (1970) and Batt, et al. (1970), thus reducing significantly the possibility that the large discrepancies in the reported spreading rates are due to experimental difficulties. More surprising however, is the apparent reduction in db/dx due to the presence of the trip at $U_1/U_2 = 0.4$ and 0.6.

When the results are plotted on Figure 10 of Brown and Roshko (using the appropriate variables) the Abramovich-Sabin relation

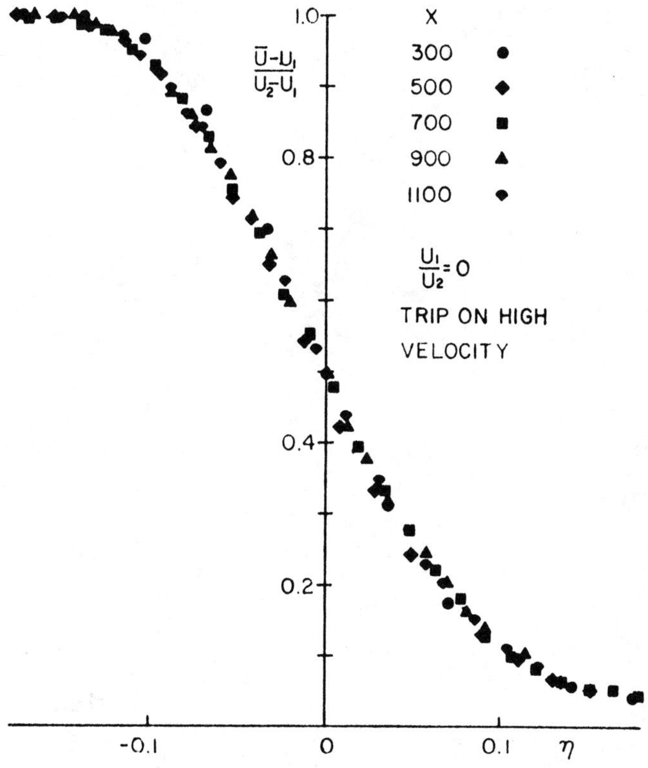

Figure 2b. The Mean Velocity Profile in Similarity Coordinates. $U_1/U_2 = 0$, with trip wire.

$$b = 0.18 \frac{U_1 - U_2}{U_1 + U_2} = 0.18\,\lambda$$

fits the untripped data fairly well; on the other hand, the tripped results define a curve of b vs. λ which is concave upwards and, thus, in contraction with models suggesting that it should be concave downwards (see Brown & Roshko, 1974, pp. 787-789).

Preliminary measurements of the longitudinal component of the turbulent intensity were made using a DISA 55D35 rms meter. Although these results should be viewed with caution they indicate the effect of the trip wire on the fluctuation intensity u'. Thus, while the trip was responsible for an increase in u' at $U_1/U_2 = 0$ (see also Figure 5 of Champagne, et al., 1976) it had the opposite effect at

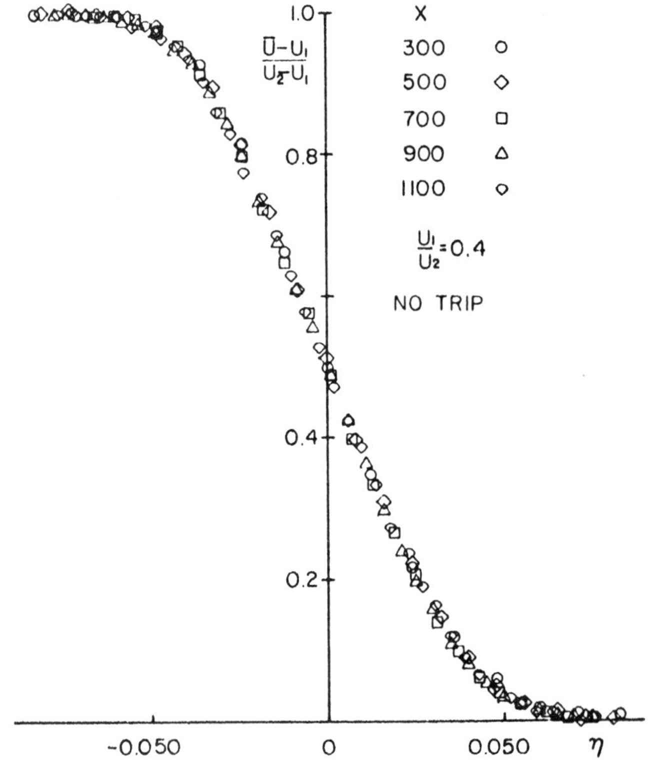

Figure 3a. The Mean Velocity Profile in Similarity Coordinates. $U_1/U_2 = 0.4$, no trip.

U_1/U_2 = 0.4 and 0.6. In general, $u'/(U_2 - U_1)$ attains a constant level independent of x (i.e., indicating self preservation) except in the two cases marked by + for which a very slow dependence on x was detected. It is possible that for U_1/U_2 = 0.6 the tripped mixing layer, which develops rather slowly does not attain its asymptotic similarity within the test section, but the relatively low level of u' at U_1/U_2 = 0 is attributed to the lack of low frequency response of the rms meter.

The initial momentum thickness θ_i which was calculated from measurements at x = 10 mm is shown in Table 1 for all cases under consideration. It indicates that the length of the test section is adequate to attain self preservation according to Bradshaw's (1966) criterion. A difficulty arises when one considers the two limiting cases U_1/U_2 = 0 and $U_1/U_2 \rightarrow 0$. In the first case only the boundary

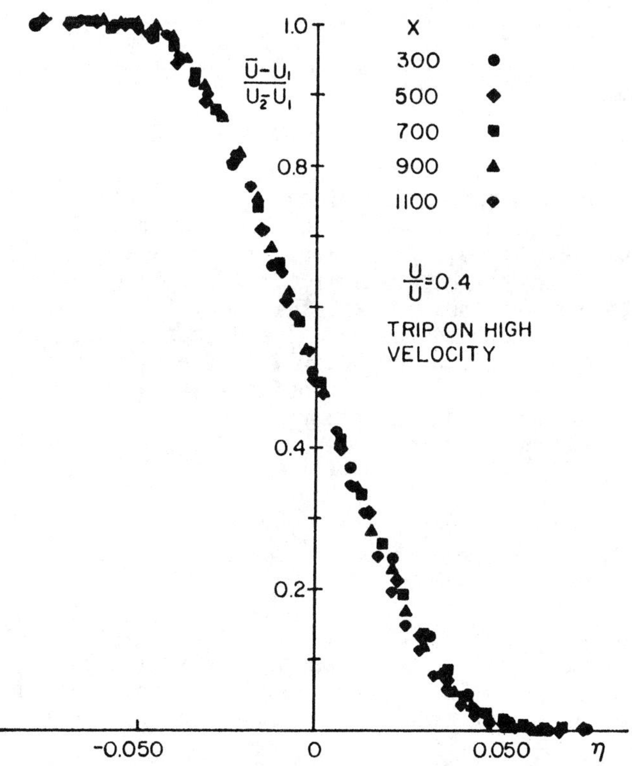

Figure 3b. The Mean Velocity Profile in Similarity Coordinates. U_1/U_2 = 0.4, with trip wire.

Table 1

U_1/U_2	θ_i mm	x_0 cm	$\frac{y_{0.5} - y_0}{x - x_0}$	$\frac{db}{dx}$	$\left(\frac{u'}{U_2 - U_1}\right)$
0	0.46	-15	0.049	0.178	0.140
0*	0.62	+1.5	0.058	0.215	0.146†
0.4	1.13	-12	0.021	0.080	0.166
0.4*	1.35	-14	0.019	0.073	0.142
0.6	1.01	-18.5	0.013	0.050	0.166
0.6*	1.11	-28.7	0.014	0.043	0.13†

† The asymptotic condition which is independent of x was not attained.

layer on the high speed side of the splitter plate contributes to θ_i, in the second the momentum thickness of the low speed boundary layer has to be added. Thus, according to Bradshaw's criterion the mixing layer may not become self similar when $U_1/U_2 \to 0$ at any reasonable distance downstream of the splitter plate, for the fixed geometries under consideration. The discontinuous jump in θ_i when the second stream was added underlines the difficulty in applying Bradshaw's criterion. If one considers an initial vorticity thickness in a manner analogous to Birch and Egger's (1972) for the fully developed mixing layer (see also Brown & Roshko, 1974), then

$$\delta\omega_i = \frac{U_2 - U_1}{\left[\frac{\partial U_2}{\partial y} - \frac{\partial U_1}{\partial y}\right]_{max}}$$

where $(\partial U/\partial y)_{max}$ represents the slope of the velocity profile at the surface of the splitter plate near the trailing edge. If the two boundary layers develop in the same manner on both sides of the plate and are both laminar, they may be expressed by Blasius' solution:

$$\delta\omega_i = \frac{(U_2^{\frac{1}{2}} - U_1^{\frac{1}{2}})x^{3/2}}{\nu^{\frac{1}{2}} * 0.332}$$

This length scale remains continuous as the low speed stream U_1 is turned off, however, it presents a problem when $U_2 \to U_1$ because the initial wake component associated with θ_i is neglected. Thus, the attempt to scale the effect of initial conditions by a single length scale may be an oversimplification, in particular, when $U_1/U_2 \neq 0$.

The initial wake regime generated by the two boundary layers on opposite sides of the splitter plate is associated with a wake characterized by discrete vorticity of opposite signs, and the decay of the wake-like component of the initial mixing layer profile is characterized by the cancellation of these vortices. This process occurs at a different rate than the growth of the mixing layer vortex street, which is essentially characterized by discrete vortices of the same sign.

Preliminary measurements of power spectra and auto-correlations near the trailing edge of the splitter plate indicate that:

1. A strong peak in the spectrum at f = 225 Hz disappeared after the trip wire was introduced at $U_1/U_2 = 0$.

2. The dominant peak in the power spectrum (at f = 460) was not destroyed by the introduction of the trip wire at $U_1/U_2 = 0.4$.

Additional correlations and power spectra of the fluctuations existing outside the turbulent interface of the shear layer were computed. Dimotakis and Brown (1975) argue that the surprisingly high intensity of these fluctuations can only be associated with the large organized structure within the shear layer. They also show that the fundamental periodicity of the auto-correlation function is consistent with the similarity scale (i.e., $0.4 < (\tau_0 U_C/x - x_0) < 0.5$ where τ_0 is defined as twice the time lag to the first minimum of the auto-correlation function and U_C is the convection velocity). Consequently, the fundamental periodicity should be altered by the introduction of the trip wire. We, thus, expected $\tau_0^* > \tau_0$ for $U_1/U_2 = 0$ and $\tau_0^* < \tau_0$ for $U_1/U_2 = 0.4$ provided the convection velocity remained unchanged by introducing the trip. The results for $U_1/U_2 = 0.4$ are shown in Figures 4a, b, c and d. The fundamental periodicity as defined above, indeed, increased with x but the expected relationship between the tripped and untripped case did not materialize. On the contrary, the untripped time scale is always smaller for $U_1/U_2 = 0.4$ than the tripped time scale. Comparison of Figure 4d with 4b indicates that the fundamental time scale made a quantum jump of a factor of 4 which may be interpreted as indicating that on the average two amalgamations take place between x = 700 mm and x = 1100 mm. However, the power spectra shown in Figure 4 indicate that the above definition of the fundamental periodicity may be misleading because a fundamental frequency of f_0 = 11.5 Hz which

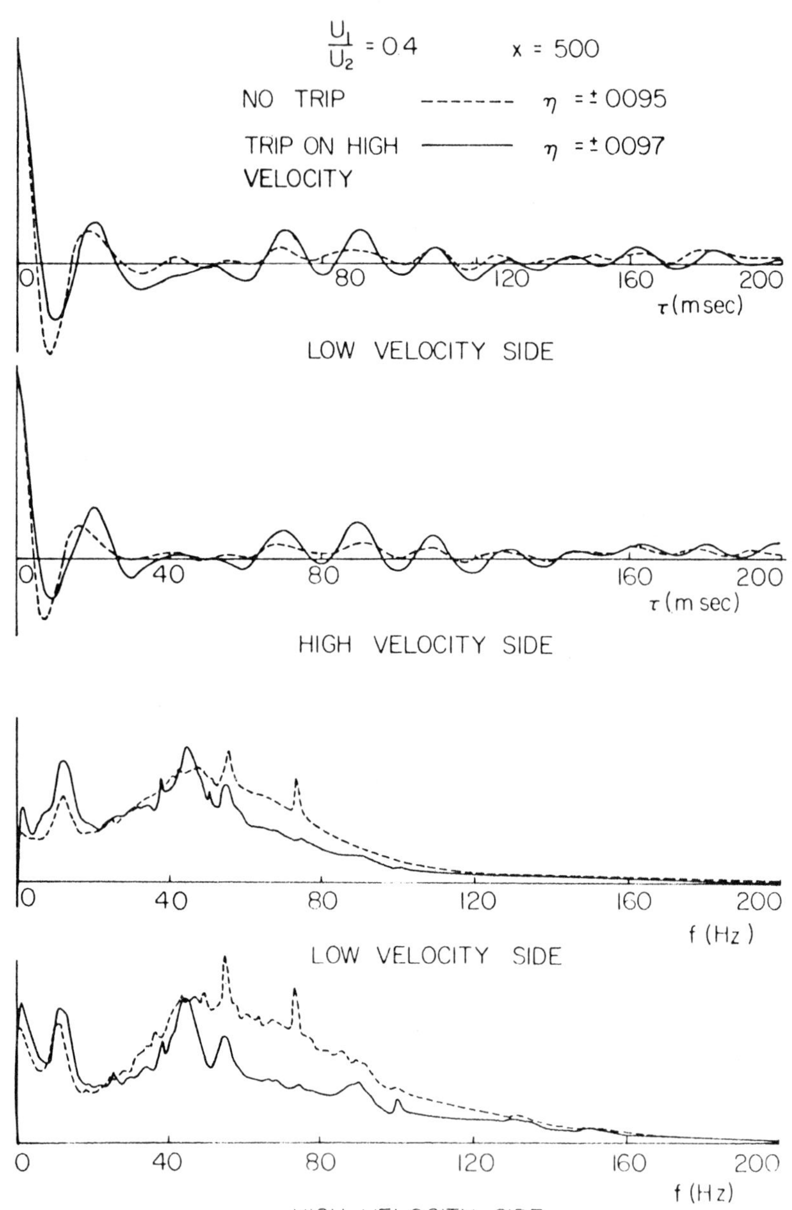

Figure 4a. Auto-correlation and Frequency Spectra. $U_1/U_2 = 0.4$, $x = 500$.

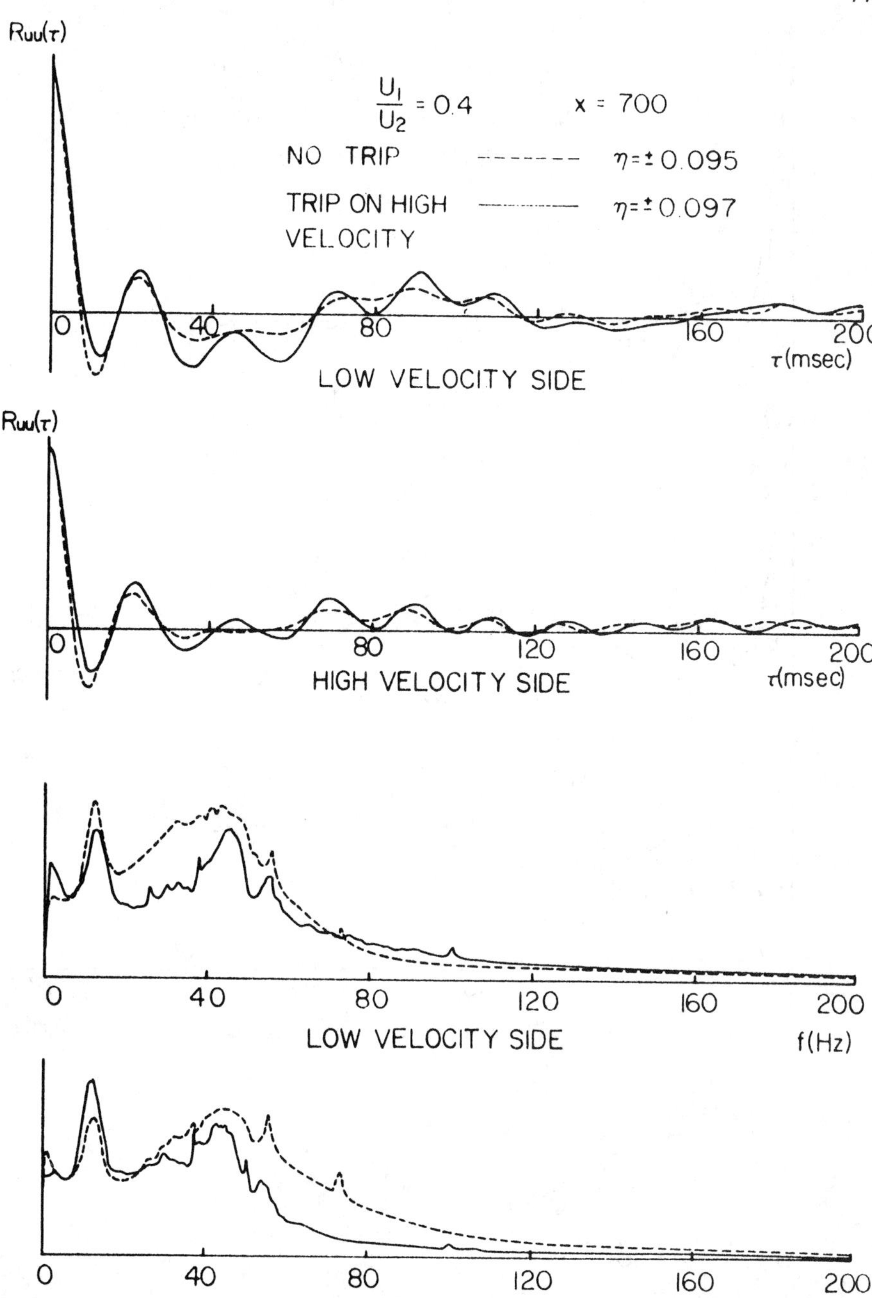

Figure 4b. Auto-correlation and Frequency Spectra. $U_1/U_2 = 0.4$, x = 700.

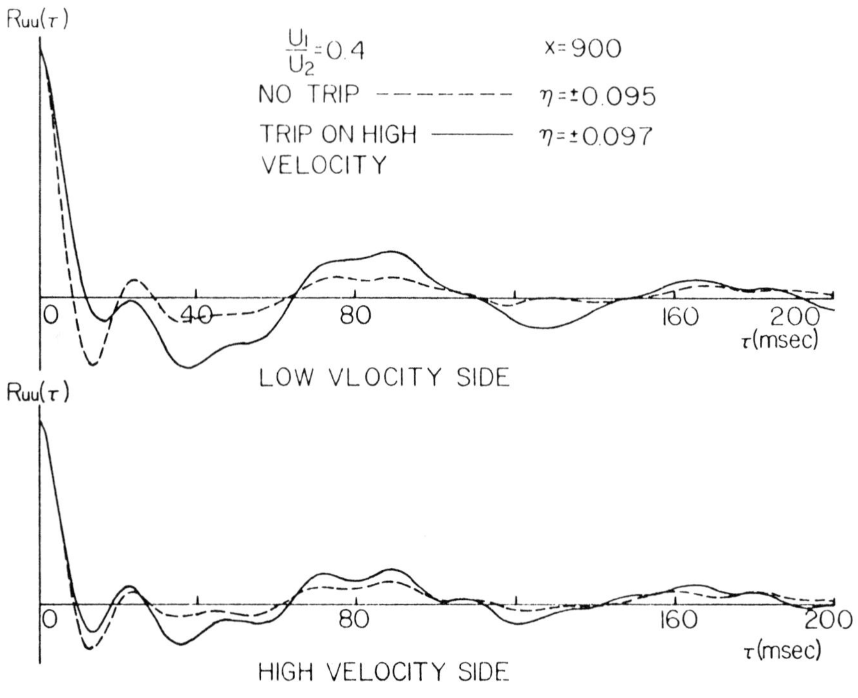

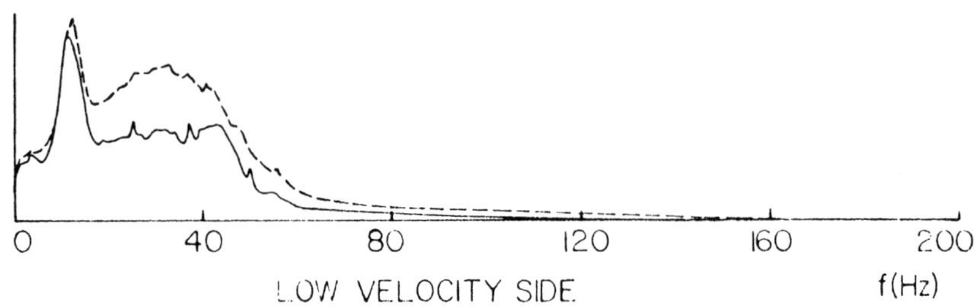

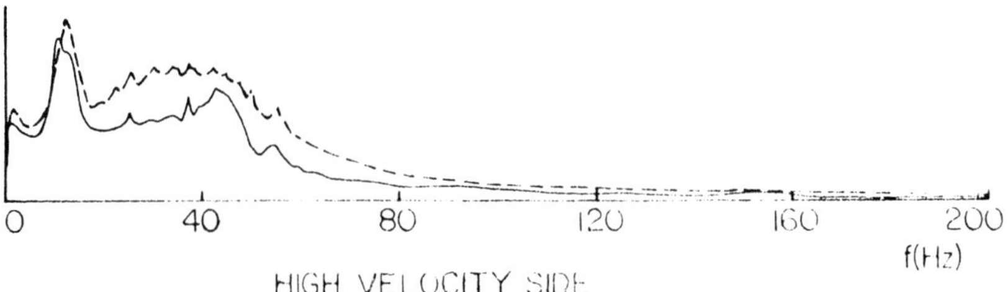

Figure 4c. Auto-correlation and Frequency Spectra. $U_1/U_2 = 0.4$, $x = 900$.

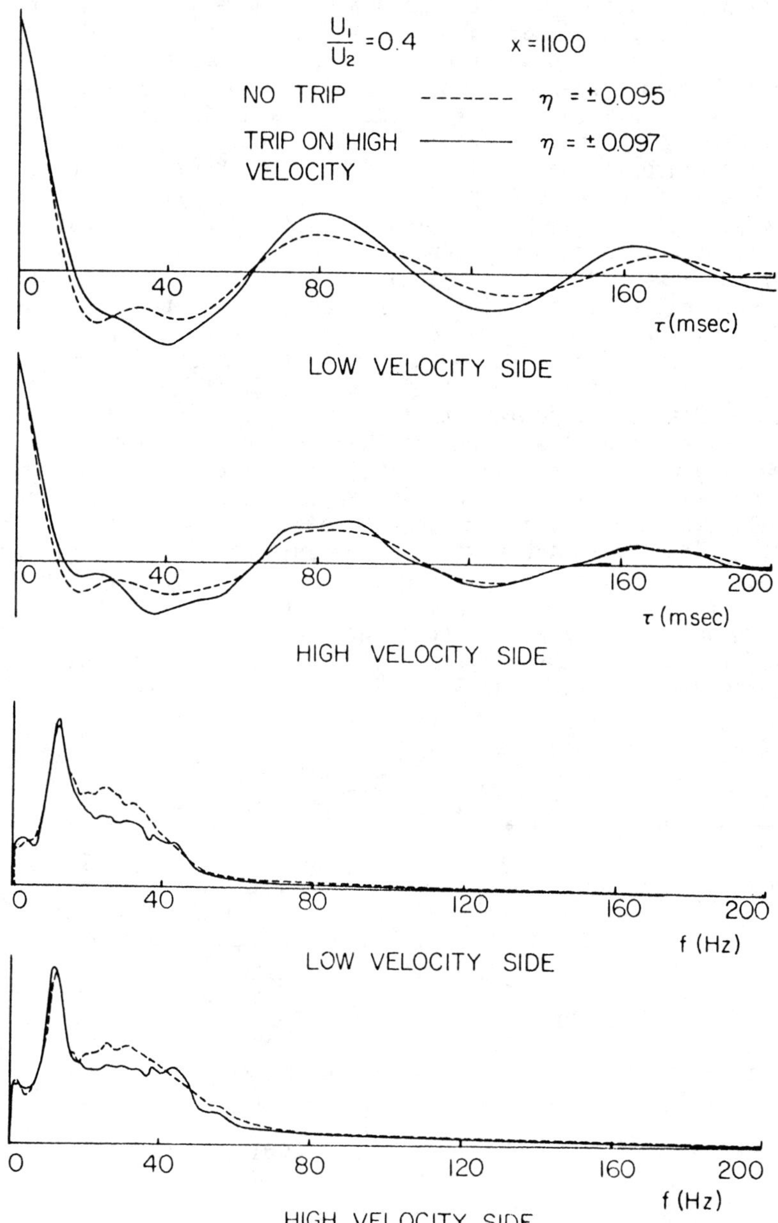

Figure 4d. Auto-correlation and Frequency Spectra. U_1/U_2 = 0.4, x = 1100.

corresponds to the largest time scale (Figure 4d) exists at all values of x. It is more apparent far downstream because the high frequencies in the power spectra disappear and, in particular, the fourth harmonic frequency which is so dominant at x = 500 mm and 700 mm. The existence of the strong peak in the spectrum at f for all values of x shown in Figure 4 may render the similarity scaling invalid since it implies that a feedback mechanism associated with the scale of the entire apparatus may exist ($U_C/f_0 \approx 1$ meter). Because the measurements in Figure 4 were taken at almost identical values of η in both the tripped and untripped case (i.e., the co-ordinate y was adjusted for a given x) the stronger intensity of the spectral peak at f_0 in the tripped case may be interpreted as a larger degree of order which is introduced by the trip at U_1/U_2 = 0.4. This effect is consistent with the measured correlations in Figure 4 and is in agreement with the findings near the splitter plate. One may conjecture that the more orderly structure inhibits the occurrence amalgamations of the large eddies, thus, reducing the spreading of the shear layer. The auto-correlations on both sides of the mixing layer are very similar in spite of the fact that the local velocities differ by a factor of 2.5. The most plausible explanation for this behavior is the existence of a single row of large eddies moving with a unique convection velocity. Space-time correlation measurements on the same side of the shear layer outside the turbulent interface at constant η enabled us to calculate the velocity at which the large structures are being convected. The results are summarized in Table 2.

Table 2

U /U	X mm	$2U_C/U_1 + U_2)$ High vel. side	$2U_C/(U_1 + U_2)$ Low vel. side
0.4	300	1.087	0.935
0.4	1100	1.059	0.955
0.4*	300	1.087	0.916
0.4*	1100	1.051	0.966
0	300	1.077	-
0	900	1.077	-

The convection velocity on the high speed side of the layer is between 10% - 15% higher than on the low velocity side; it also decreases slightly with increasing x while the convection velocity on the low speed side increases with x. A single row of eddies is, thus, being slowly stretched and tilted while proceeding downstream. This process may reflect itself in the continuous growth of the fundamental time scale τ_0 which was discussed before. The trip wire has no affect on these convection velocities. Cross correlation measurements between two probes positioned on opposite sides of the shear layer at the same x and z locations are shown in Figure 5. The negative correlation at $\tau = 0$ indicates antisymmetric oscillations in the potential flow which are consistent with the shadowgraph pictures of Brown and Roshko (1974). Finally, we addressed ourselves to the question of the two-dimensionality of these large eddies by measuring two point correlations of space and time in the z direction on the same side of the shear layer (Figure 6a, b). Since the shadowgraph technique used by Brown and Roshko averages in the spanwise direction a possibility of helical motion and local interaction may exist, but the results in Figure 6 indicate an incredible persistence of quasi two-dimensional structures in the z direction. The correlation at a given z actually increased at larger distances from the splitter plate.

4. CONCLUSIONS

The spreading rate of the mixing layer is affected by the initial conditions at the discontinuity in spite of the fact that the velocity and intensity profiles appear to follow the similarity scaling. A trip wire placed on the surface of a splitter plate enhances the spread of the mixing layer into quiescent fluid but inhibits the spread of the mixing layer between two parallel streams of different velocity.

Preliminary correlation and spectral data indicate that the large structures consist of quasi two-dimensional row of vortices which are convected at approximately the average velocity of the two streams. The trip wire has some effect on the degree of order existing in those large structures.

REFERENCES

1. Batt, R. G., Kubota, T. and Laufer, J. 1970 AIAA Paper 70-721.

2. Birch, S. F. and Eggers, J. M. 1972 NASA SP-321.

3. Bradshaw, P. 1966 JFM 26 275-236.

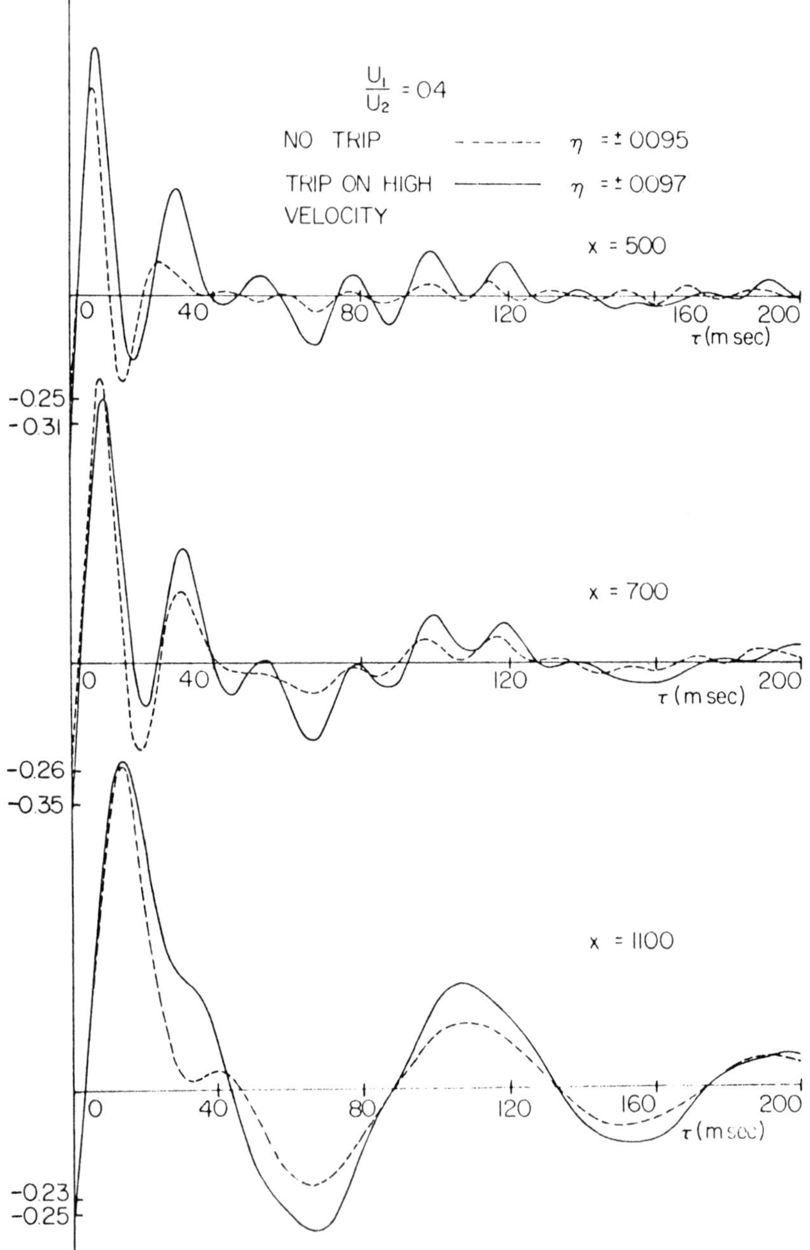

Figure 5. Two Point Cross Correlation Between both Sides of the Mixing Layer, U_1/U_2 = 0.4, x = 500, 700, 1100.

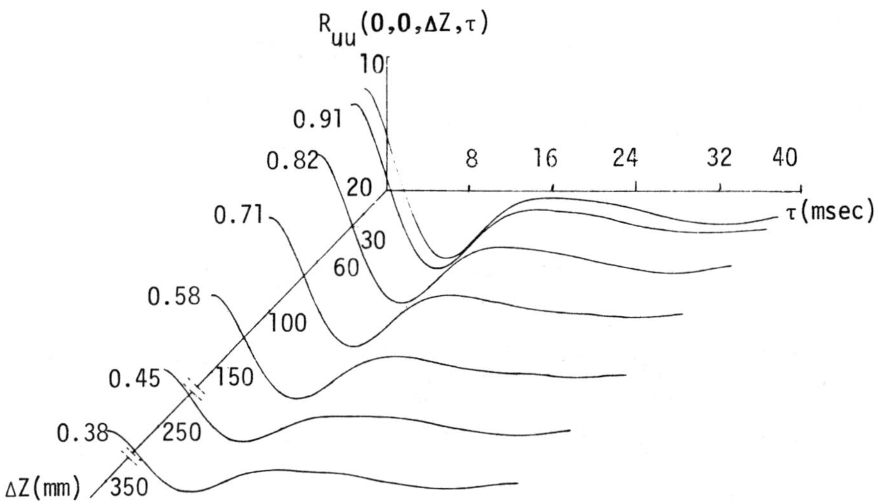

Figure 6a. Space-Time Correlations in the Spanwise Direction. $U_1/U_2 = 0.4$, no trip. x = 500, low velocity side.

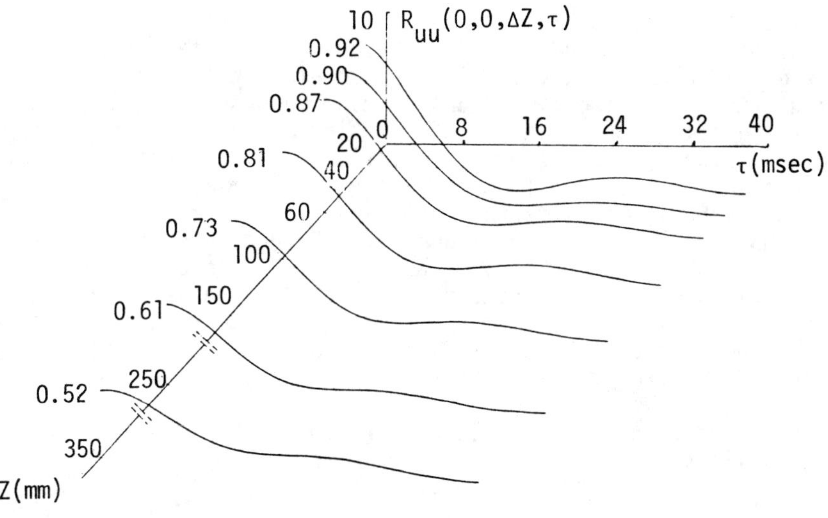

Figure 6b. Space-Time Correlation in the Spanwise Direction. $U_1/U_2 = 0.4$, no trip. x = 1100, high velocity side.

4. Brown, G. L. and Roshko, A. 1974 JFM 64 775-816.

5. Champagne, F. H., Pao, Y. H. and Wygnanski, I. 1976 JFM 74 209-250.

6. Dimotakis, P. E. and Brown, G. L. 1975 Project Squid Technical Report CIT-7-PU.

7. Liepmann, H. W. and Laufer, J. 1947 NACB TN-1257.

8. Murthy, S. N. B. 1975 (Editor) Turbulent mixing in non-reactive and reactive flows, Plenum Press, New York.

9. Townsend, A. A. 1956 The structure of turbulent shear flow, Cambridge University Press, London.

10. Winant, C. B. and Browand, F. K. 1974 JFM 63 237-255.

11. Wygnanski, I. and Fiedler, H. E. 1970 JFM 41 327-361.

DISCUSSION

GODLSCHMIDT: (Purdue University)

Just two curves which relate to your comments, although not directly to the topic of this symposium. In an earlier publication (Goldschmidt, V. W. & Bradshaw, P., "Flapping of a Plane Jet," Physics of Fluids 16, 354 (1973)) we documented what we called a flapping motion in a two-dimensional jet. Two probes on opposite sides of the jet showed a large scale motion which could be interpreted as a sideways "flapping" like motion. The frequency of that motion was easily measured from time correlation measurements. In his doctoral thesis J. Cervantes ("An Experimental Study of the Flapping Motion of a Turbulent Plane Jet," Purdue University, 1976) shows a curve which we now present as Figure 1. The ordinate is the Strouhal number of this sideways flapping motion in terms of the exit slot width and velocity. The frequency decreases gradually, and is independent of the lateral location of the probes. There may be an interesting relationship between this and your results.

The second curve is a result of work done a few years ago but not yet published. This relates to the effect of upstream disturbances on the widening rate. By placing screens of different types before the final contracting nozzle (near the exit of the jet exit) we were able to vary the turbulence intensity at the mouth. The widening rate was then measured and compared with the turbulence intensity. Figure 2 summarizes the results, taken in collaboration with Peter Bradshaw at Imperial College. Now I'm not sure how this

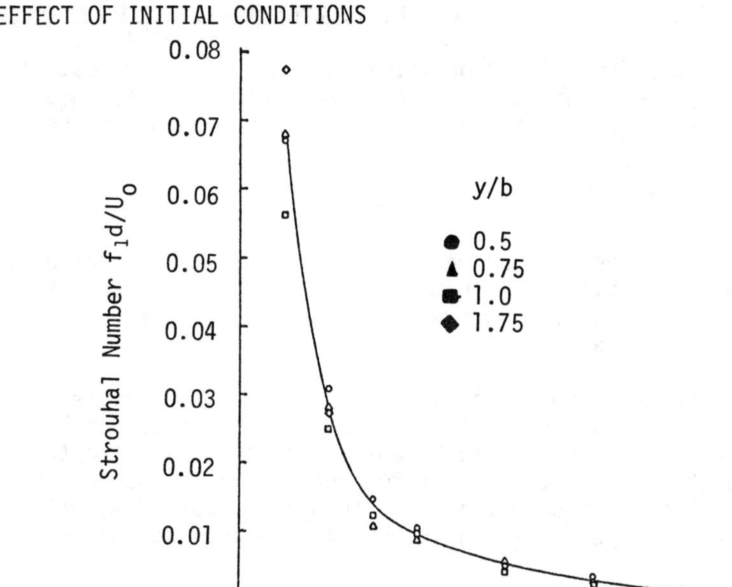

Figure 1. Strouhal Number f_1d/U_0 vs. x/d

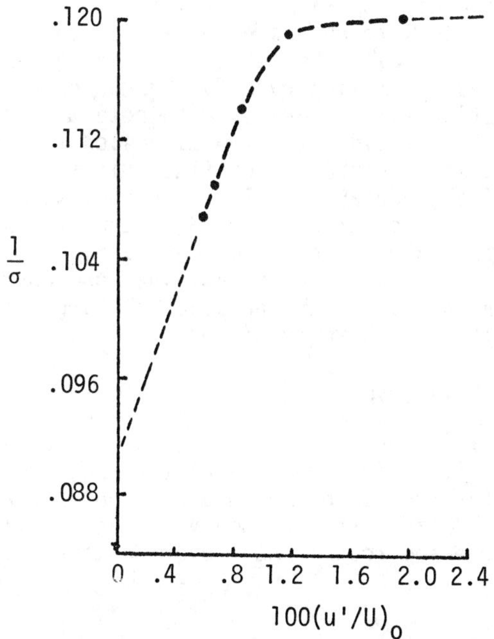

Figure 2. Upstream Effects; Turbulence Intensity at Plane Jet Nozzle Versus Widening Rate

applies to flows being tripped on the side but it does indicate that maybe we can relate things through the turbulence intensity near the mouth, and furthermore, that there may be an upper bound to the widening rate.

WYGNANSKI:

The trip wire was located near the discontinuity and introduced quasi-sinusoidal oscillations in the flow. It did not produce a fully developed turbulent boundary layer. The amplitude of the oscillation was very small.

Introducing two trips on each side of the splitter plate did not alter the resulting spread of the shear layer significantly. I do not feel that the level of turbulent intensity is so important. Patel showed that the intensity has to exceed 0.6% before any effect became noticeable, so probably one can change the spreading rate by introducing a small perturbation at the right place and the right frequency.

LAUFER: (University of Southern California)

Just a short remark in connection with the last two presentations. Fred Browand at USC is looking into the question of the effect of the initial conditions in the geometry of the two dimensional mixing layer and came up with exactly the same conclusion although his measurements were not as extensive as those presented by Dr. Wygnanski. Namely, with the two stream mixing condition the trip actually decreases the mixing rate further downstream. It might also be interesting for Mr. Bradshaw to note that for the case of no fluid velocity on one side, the two dimensionality of these structures is in fact still present. We find this to be the case in the initial region of a circular jet also. In fact we made extensive measurements of the type indicated by Dr. Goldschmidt. We found strong correlation, provided one does the correlation conditionally, past the potential core (in a circular jet) when one correlated across the jet, again indicating the presence of large structures that have certain azimuthal phase relations.

BRADSHAW: (Imperial College)

I will believe almost any ill of a circular jet because there are such good opportunities for pressure disturbances to correlate things around circumference at the mouth of the jet. I wonder if I could ask John Laufer whether the frequency of the oscillations which he observed decreased roughly as 1 over distance downstream?

LAUFER:

Yes, it did decrease approximately in that fashion.

BRADSHAW:

So that was more or less genuine turbulence as opposed to just an oscillation of the jet as a column.

LAUFER:

We measured it only to four diameters, and at that station it was definitely turbulent.

COLES: (California Institute of Technology)

Tell me if I am correct in my impressions. You measured the flow 120 centimeters downstream, at which station I would expect your mixing layer to be 25 to 30 centimeters wide to the point of zero intermittency. This is done in a channel 60 centimeters wide. I have a reservation, and I hope that you will be able to remove it; but this is not what most of us think of as a well-constructed two-dimensional experiment. Are you absolutely certain that there is no effect of the side walls on the whole problem of the variation of spreading rate with tripping device?

WYGNANSKI:

In the two stream case ($U_1/U_2 = 0.4$, say), I could say with very little reservation that I am certain, because in this case the mixing layer is much more narrow. Typically, 1 meter downstream, it is only 10 cms wide; yet the effect of the trip is noticeable. In the one stream case ($U_1/U_2 = 0$) the size of the test section becomes marginal, but if the flow is already self preserving 50 cms downstream, when the width of the mixing layer is still small (i.e. comparable to the width of the two stream mixing layer 120 cms downstream), one may conclude that the effect of the side wall is not important.

COLES:

But you made no measurements off the lateral plane of symmetry?

WYGNANSKI:

No, no surveys.

On the Developing Region of a Plane Mixing Layer

STANLEY F. BIRCH
The Boeing Company
Seattle, Washington (U.S.A.)

ABSTRACT

Some preliminary results are presented from an experimental study of the effects of initial conditions on the development of a plane mixing layer. It is concluded that, for sufficiently high Reynolds numbers, the spreading rate does become independent of initial conditions. When the initial boundary layer is turbulent this may require a Reynolds number of about 2×10^6. The spreading rate and the turbulence structure in a helium/air mixing layer is also briefly discussed.

1. INTRODUCTION

The plane turbulent mixing layer formed between two uniform and initially parallel streams is a flow of considerable importance, and it has been extensively studied. Most of the work reported in the literature, however, concerns the fully developed, or what is purported to be the fully developed region of the flow. There is very limited detailed information available on the initial developing region, or on how the extent of this region depends on the properties of the flow at the separation point.

Much of the available information on the effects of initial conditions is contained in a paper by Bradshaw (1966). Using the near field of a 2.0 in. jet, Bradshaw showed that when the initial boundary layer is laminar, the Reynolds number, based on the distance required for the flow to become fully developed, is approximately constant and equal to about 4×10^5. When the initial boundary layer was turbulent, however, the flow did not appear to become fully developed within the quasi-plane region of the mixing layer. Therefore, the conditions required to achieve a fully developed mixing layer, in the practically important case of a mixing layer developing from an initially turbulent wall boundary layer, remained to be determined.

More recent work by Wygnanski and Fiedler (1970) and by Batt (1975) shows that the spreading rate of a mixing layer appears to increase by about 20% when the initial boundary layer is tripped. In view of Bradshaw's earlier results one might suspect that these flows were not fully developed. Yet, a detailed study of the turbulence data is inconclusive. In any case, it is clear that the spreading rate of a mixing layer is not independent of initial conditions for the Reynolds numbers generally used in those studies. Champagne, Pao and Wygnanski (1976) speculate that a truly universal self-preserving flow may not be achievable in practice. They conclude, however, that based on the available experimental data no definitive resolution of the problem is possible. This is similar to the conclusion reached by Birch and Eggers (1973). Nevertheless, the spreading rate obtained by Liepmann and Laufer (1947) in their classic study is still widely regarded as being the most reliable estimate for the spreading rate of a fully developed mixing layer. This spreading rate, which corresponds to a spreading parameter σ of 11.0, is also in good agreement with the results obtained by Bradshaw (1966).

In an effort to resolve some of the problems discussed above, measurements were taken in a plane mixing layer developing from an initially turbulent wall boundary layer. The work reported here is a preliminary discussion of some of those results.

2. EXPERIMENTAL APPARATUS

To minimize spurious Reynolds number effects, which are believed to have plagued earlier work, the apparatus was designed to run at as high a Reynolds number as possible. The apparatus is shown diagrammatically in Figure 1. The test section was 2.0 ft long and the primary stream at its entrance was 8.0 in. wide by 3.0 in. high. Air entrained by the mixing layer was supplied through a porous wall at the bottom of the test section. To form a heterogeneous mixing layer helium could be substituted for air as the entrainment fluid. This flow was adjusted to give a constant static pressure within the test section. For the experiments to be discussed here there was no flow in the secondary stream.

At a freestream velocity of 300 ft/sec the Reynolds number, based on the test chamber length, was about 4×10^6. Air was

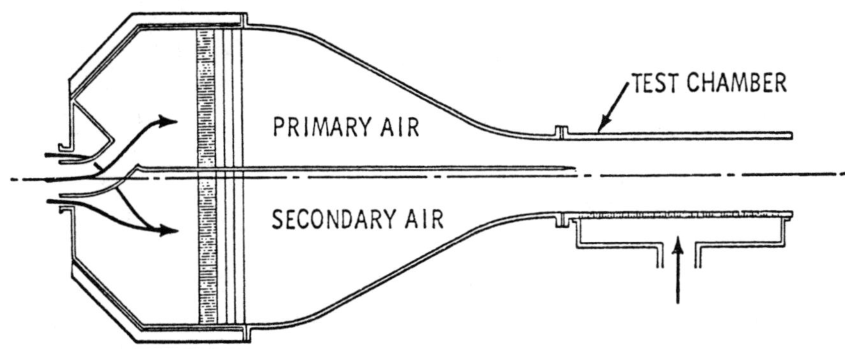

Figure 1. Schematic diagram of apparatus.

supplied from a bank of high pressure storage bottles and the available flow was sufficient to nearly double the Reynolds quoted above.

3. EFFECTS OF INITIAL CONDITIONS

For the first series of experiments a trip wire, 0.020 in. in diameter, was placed 3.0 in. upstream of the end of the splitter plate (Figure 2A). The boundary layer at the separation point was turbulent and was approximately 0.25 in. thick. A number of mean velocity profiles were taken using a total pressure probe, a single hot film probe, and a split film probe. The nominal freestream velocity U_1 was 300 ft/sec.

The results, shown in Figure 3, indicate an asymptotic spreading rate, dL/dX of 0.119. The width of the mixing layer L is defined here as the distance between the points at which the velocity U is given by $(U/U_1)^2 = 0.9$ and $(U/U_1)^2 = 0.1$.

When the velocity was reduced to 150 ft/sec, the spreading rate, both in the developing region and in the fully developed region, was essentially unchanged. Normalized mean velocity profiles, taken 18 in. downstream of the end of the splitter, for the two freestream velocities, are shown in Figure 4. This suggests that the flow under these conditions is largely Reynolds number independent.

The asymptotic spreading rate quoted above corresponds to a spreading parameter σ of approximately 11.0. Although this

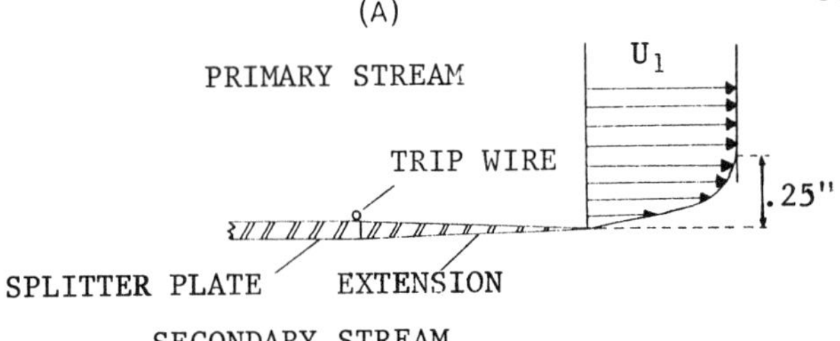

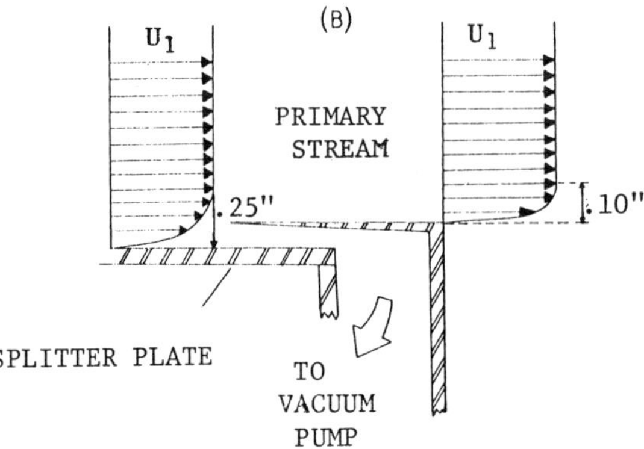

Figure 2. Splitter Plate Configurations

spreading rate agrees well with that obtained by Liepmann and Laufer (1947), it is about 20% lower than previous measurement for mixing layers developing from initially turbulent boundary layers. Note, however, that even at the lower velocity used here, the Reynolds number is still substantially higher than that generally used for mixing layer studies: approximately four times that used by Batt (1975) and by Wygnanski and Fiedler (1970). If, in the present experiments, data for only the first 10 in. were considered, the width of the mixing layer would still appear to be increasing linearly. But the spreading rate would be about 20% higher. This suggests that the high spreading rates, found previously for flows developing from initially turbulent boundary layers, were probably caused by the use of too low a Reynolds number; and that the mixing layer does achieve an asymptotic spreading rate, independent of initial conditions, if the Reynolds number is sufficiently high.

To investigate the effect of initial boundary layer thickness, a boundary layer scoop was fitted to the end of the splitter plate (Figure 2B). At a freestream velocity of 300 ft/sec, this reduced

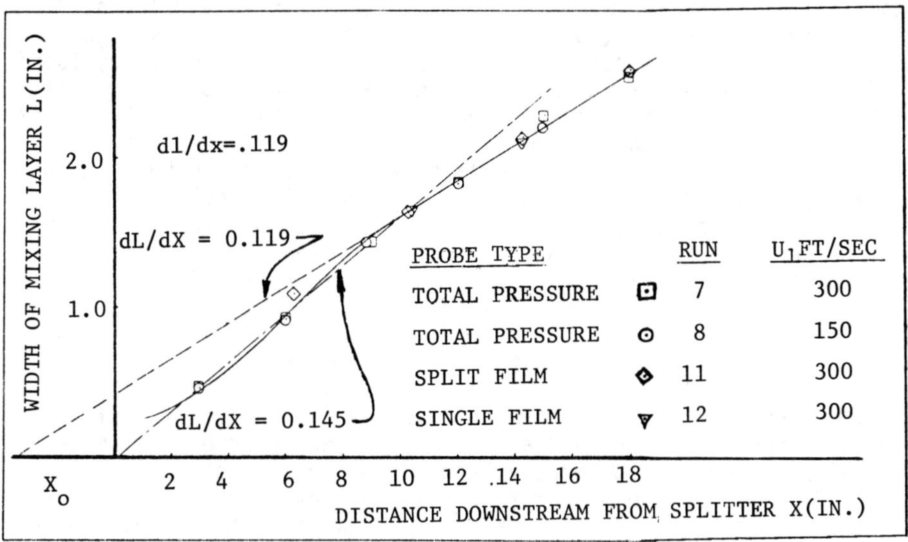

Figure 3. Spreading rate for mean velocity profiles; initial boundary layer thickness 0.25 in.

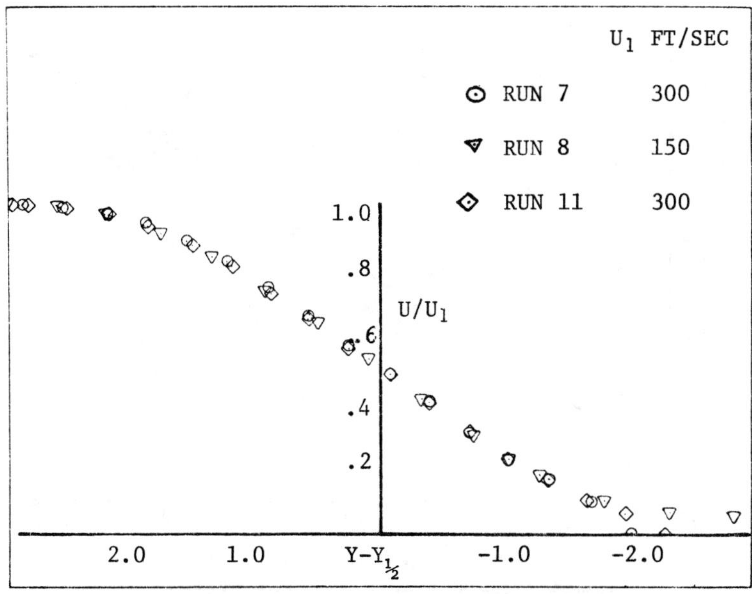

Figure 4. Mean velocity profiles 18 in. downstream from end of splitter plate.

the initial boundary layer thickness to approximately 0.10 in. The boundary layer at the separation was still turbulent. Again, the flow appeared to require 10 to 12 in. to become fully developed (Figure 5). A similarity plot of the mean velocity profiles in the fully developed region of the flow is shown in Figure 6. Similarity plots of the turbulence components are shown in Figures 7 and 8. Both sets of turbulence data collapse well to single curves and show very little scatter. These data were taken with a split film probe and were taken primarily to evaluate the probe for high intensity turbulence measurements. As such, their reliability cannot be guaranteed at present, although they do appear to be within about 10% of hot-wire data taken previously in similar flows.

In contrast to the results found by Bradshaw (1966) for flows developing from laminar initial boundary layers, the present results seem to indicate that the developing region is independent of both Reynolds number and of initial boundary layer thickness. Since turbulent flows tend to become Reynolds number independent at high Reynolds numbers, the former result is not too surprising. The reason for the absence of a change in the length of the developing region with a change in the initial boundary layer thickness is less obvious. In a fully turbulent flow the developing region

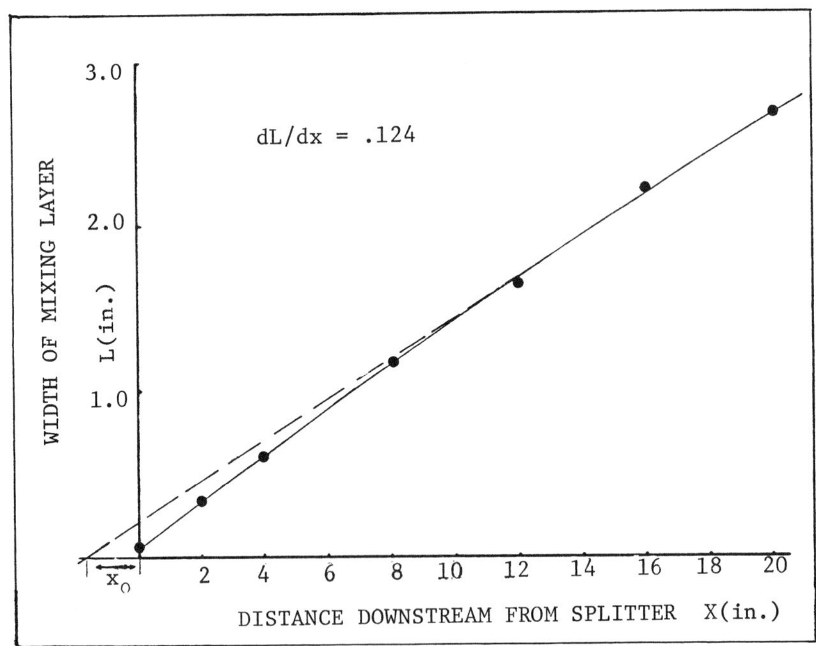

Figure 5. Spreading rate for mean velocity profiles, initial boundary layer thickness 0.10 in.

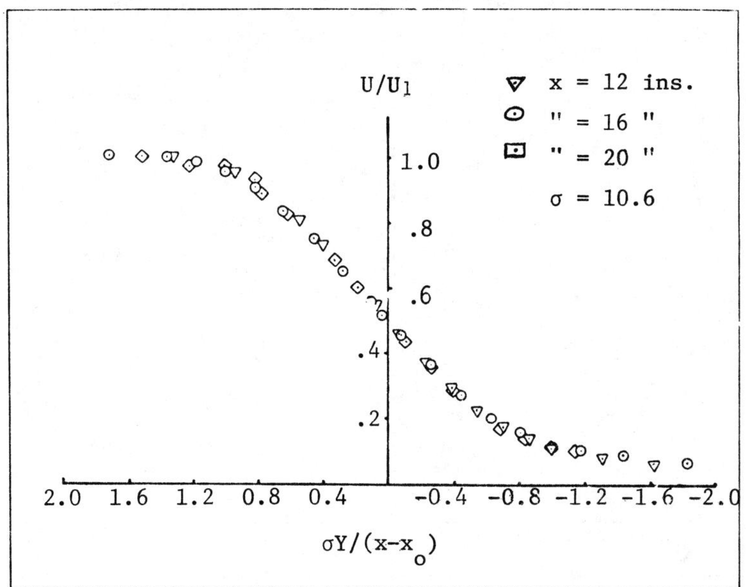

Figure 6. Similarity plot of mean velocity profiles.

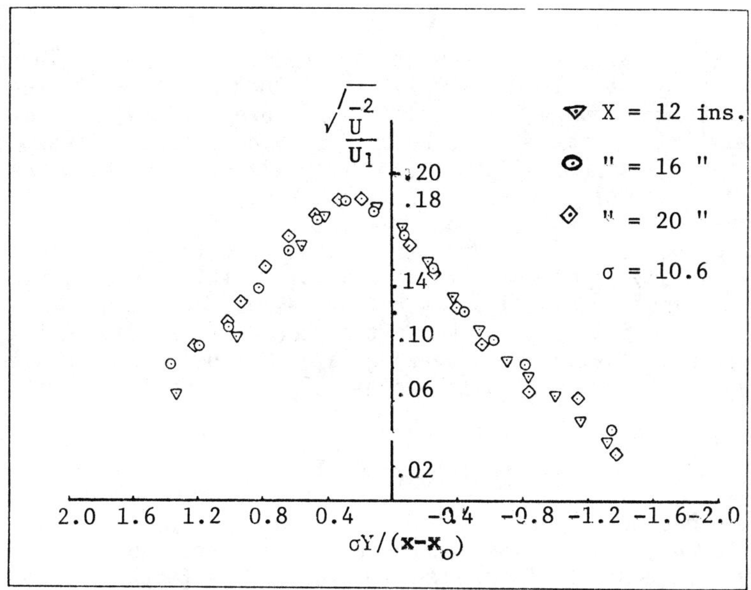

Figure 7. Similarity plot of turbulence component.

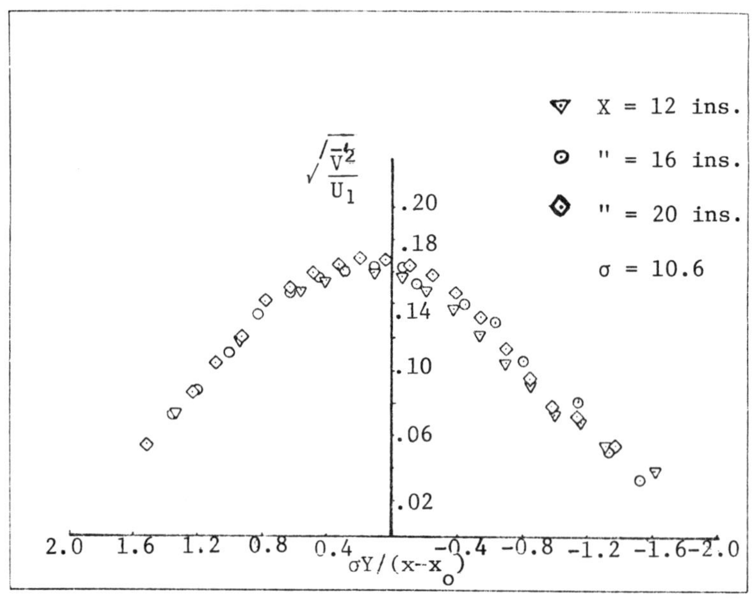

Figure 8. Similarity plot of turbulence component.

should scale locally with the width of the mixing region. Therefore, close to the end of the splitter plate the flow must depend on the initial boundary layer thickness. However, the total distance required for the flow to become fully developed will depend on some average of the width of the mixing layer over the developing region. If the initial boundary layer is thin compared with this average, the total distance required for the flow to become fully developed does not appear to be sensitive to changes in the initial boundary layer thickness. Obviously, this can only be true if the initial boundary layer is relatively thin, probably not more than 0.5 in. When the initial boundary layer is thick we must expect that it will have a greater influence on the distance required for the flow to become fully developed.

4. HELIUM/AIR MIXING

Some shadowgraphs of a mixing layer in the same apparatus, when helium was used as the entrainment fluid, are shown in Figure 9. The objective here was to study the effect of large density differences on the spreading rate of a plane mixing layer at Reynolds numbers higher than those used in previous studies.

A

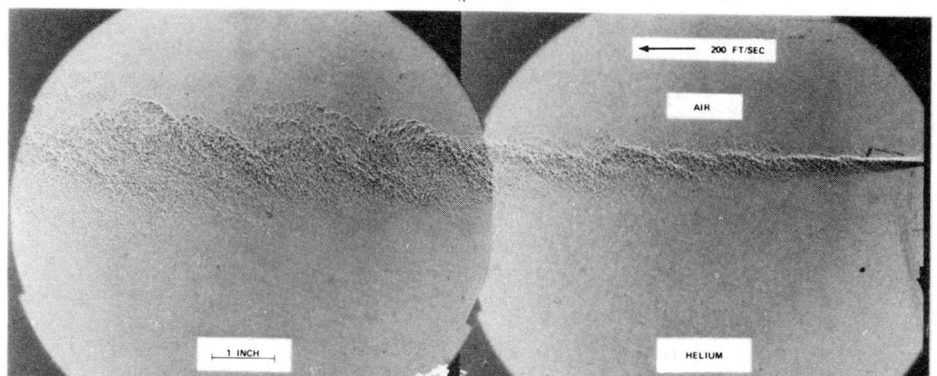

B

C

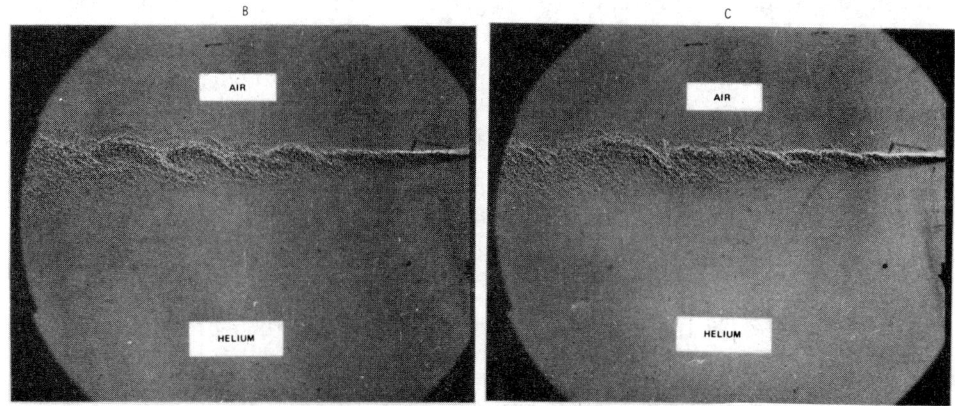

D

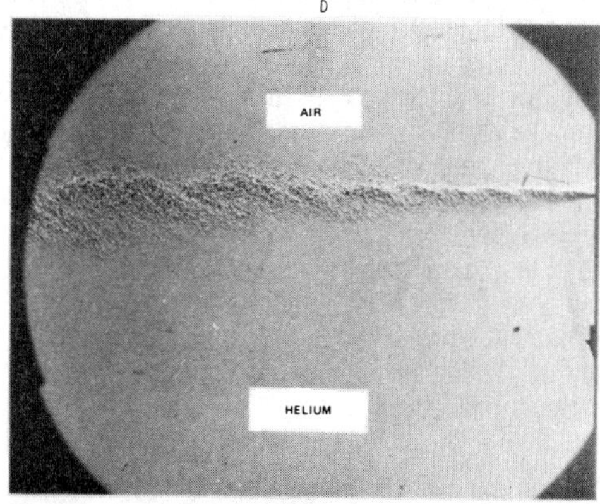

Figure 9. Shadowgraphs of air/helium mixing layer with velocity of air stream = 200 ft/sec (nominal), velocity ratio $U_2/U_1 = 0$.

A secondary objective was to obtain additional information on the large eddy structure in such flows. Note that the two halves of the photograph shown in Figure 9A were not taken at the same time.

The results obtained to date support the conclusion, reached by Brown and Roshko (1974), that large density gradients do not have a large effect on the spreading rate of a free mixing layer. Certainly it seems clear that the large changes in spreading rate noted at high Mach numbers (Morrisette, Birch and Wagner, 1973) cannot be attributed simply to the associated density changes across the mixing layer.

The developing region of the flow, the first 10 in. or so, shows a well defined large eddy structure which appears to be similar to that observed by Brown and Roshko (1976). There is, however, significant variation between the shadowgraphs shown in Figure 9, all taken under the same flow conditions, and this suggests that the structure is quite unsteady and intermittent. As the flow develops the structure appears to become weaker or more three-dimensional. Remember that a shadowgraph integrates over the width of the apparatus, in this case 8 in., so that a well defined large eddy structure will not be visible unless it has significant two-dimensionality. A superposition of two or three of these shadowgraphs shows little evidence of a large eddy structure. Brown and Roshko (1974) illustrate a similar effect. But, since the aspect ratios of these eddies decrease as we move downstream (the diameters of the eddies increase as the mixing layer spreads while their lengths are constrained by width of the apparatus), the shadowgraphs should show an increasingly stronger and better defined structure as the distance from the splitter increases, if the relative strengths of the eddies and their degree of two-dimensionality remained the same. Clearly this does not happen. Yet, the shadowgraphs did occasionally show strongly two-dimensional disturbance in the fully developed region of the flow.

A full discussion of the evidence for a well defined large eddy structure in mixing layers is beyond the scope of the present paper. This has, in any case, been tackled recently by Chandrsuda, Mehta, Weir and Bradshaw (1976). Suffice it to say that the present results suggest that a large eddy structure does persist in the fully developed region of the flow and that it can at times show considerable two-dimensionality. But this structure occurs intermittently and, in the present author's opinion, is not necessarily inconsistent with the rather small lateral correlations generally obtained from long time average measurements in such flows.

5. CONCLUSIONS

The development of a free mixing layer from an initially turbulent wall boundary layer appears to resemble a mixing layer developing from a laminar boundary layer in that the spreading rate, and presumably also the shear stress, initially overshoot their equilibrium values and then slowly relax to their asymptotic state farther downstream. A more detailed discussion of the relaxation of a perturbed mixing layer is given by Castro and Bradshaw (1976). When the boundary layer is turbulent, the initial increase in spreading rate is fairly slow, so that the total distance required for the flow to become fully developed is greater than if the boundary layer was initially laminar. Turbulent initial boundary layers should, therefore, be avoided if the objective is to generate a fully developed mixing layer.

Based on the results at present available, which are admittedly limited, it appears that, for an initially thin turbulent boundary layer and at a sufficiently high Reynolds number, the distance required for the flow to become fully developed is largely independent of both the initial boundary layer thickness and of the flow Reynolds number. As the initial boundary layer becomes thicker it is expected that it will have a more pronounced effect on the developing region. The exact nature of this dependence is at present unclear.

Shadowgraphs of the mixing layer formed between air and helium streams indicate that the effect of large density gradients is insufficient to explain the reduced spreading rates observed in high Mach number mixing layers. A large eddy structure similar to that observed by Brown and Roshko (1974) is observed in the developing region of the flow. A strongly intermittent and more three-dimensional structure was also observed in the fully developed region. This structure does not appear to lead to spreading rates which differ from those expected for a fully developed mixing layer (assuming that a similar structure exists in the homogeneous mixing layer), nor does it seem to be inconsistent with available hot wire data.

REFERENCES

1. Batt, R. G., 1975, A.I.A.A. J., 13, 245.

2. Birch, S. F., and Eggers, J. E., 1973, NASA SP-321, 11

3. Bradshaw, P., 1966, J. Fluid Mech., 26, 225.

4. Brown, G. L., and Roshko, A., 1974, J. Fluid Mech., 64, 775.

5. Champagne, F. H., Pao, Y. H., and Wygnanski, I. J., 1976, J. Fluid Mech., 74, 209.

6. Chandrsuda, C., Mehta, R. D., Weir, A. D., and Bradshaw, P., 1976, submitted to J. Fluid Mech.

7. Liepmann, H. W., and Laufer, J., 1947, NASA Tech. Note No. 1257.

8. Morrisette, E. L., Birch, S. F., and Wagner, R. D., 1973, Paper presented at ASME applied Mechanics/Fluids Engineering Conference, Atlanta, Georgia.

The Effects of an External Turbulent Uniform Shear Flow on a Turbulent Boundary Layer

Q. A. AHMAD, R. E. LUXTON, AND R. A. ANTONIA
University of Sydney
Sydney, Australia

ABSTRACT

A summary is presented of experimental results on the effect of an external uniform shear flow on a turbulent boundary layer. Both positively and negatively sheared external streams are considered and their influence on the boundary layer is compared with that of an external stream with zero shear but with a turbulence level comparable to that for the sheared freestream.

1. INTRODUCTION

The effects of freestream turbulence on the structure and development of turbulent shear flows (the turbulent boundary layers in particular) have been discussed in the literature and in papers that have been presented at this Workshop. In this paper, we investigate the effect of an external turbulent uniform shear flow on the growth of a turbulent boundary layer. The main motivation for adding the extra complication of a non-zero shear to that of a non-zero external turbulence level stems mainly from a hope of gaining further insight into the turbulence structure of a turbulent boundary layer by applying yet another type of perturbation at its boundary. The details of the present investigation have been presented elsewhere (Ahmad et al., 1976). Here, we will summarize briefly the main findings of our experimental work and present some new results that are relevant to the region of interaction between the external shear flow and the turbulent boundary layer.

2. EXPERIMENTAL CONDITIONS

Only a brief description of experimental conditions is given here as details of the wind tunnel and uniform shear flow generator have been reported elsewhere (Mulhearn and Luxton, 1975 and Ahmad et al., 1976). The shear flow generator consists of a grid of non-uniformly spaced steel rods (3.2 mm square cross section) situated at the entrance of the working section of the tunnel. At 76.2 cm downstream of the grid a 6.4 mm hexagonal cell honeycomb, of 23 cm depth, is inserted to impose a uniform length scale on the shear flow. Two different grids are used to obtain reasonable approximations to constant velocity gradients λ ($\equiv dU_1/dy$) of $+6s^{-1}$ and $-6s^{-1}$ respectively.

The characteristics of the positively sheared flow have been studied in detail (Mulhearn, 1971) and found to be in good agreement with those obtained by Champagne et al. (1970) and those reported by Corrsin (1976) at this Workshop. The boundary layer investigation is made at a sufficiently large distance downstream of the honeycomb* over a range of x for which the external uniform shear flow may be considered to be quasi-homogeneous. Turbulent energy budget measurements in the uniform shear flow indicate that the diffusion term, inferred by difference, is negligible. For $\lambda = +6s^{-1}$, the advection is of the same sign but of smaller magnitude than the dissipation term. For $\lambda = -6s^{-1}$, the advection is of the same sign and of somewhat larger magnitude than the production term. When λ is equal to zero (i.e. when the grid is removed), the advection is in close balance with the dissipation term.

For all three values of λ considered here, the freestream turbulence level T_u ($\equiv \overline{u_1^2}^{\frac{1}{2}}/U_W$, where $\overline{u_1^2}^{\frac{1}{2}}$ is the rms level of freestream u fluctuations and U_W is the value of the linear velocity profile extrapolated to the wall) is about 1.5%. For $\lambda = \pm 6s^{-1}$, the freestream shear stress $-(uv)_1$, is equal to about one-tenth the wall shear stress τ_W, while the magnitude of λ is of the same order as $\tau_W^{\frac{1}{2}}/\delta$, δ being the nominal boundary layer thickness. For $\lambda = 0$, $+6s^{-1}$ and $-6s^{-1}$, the measured turbulence integral length scale is approximately equal to 20%, 60% and 80% respectively of δ.

3. EXPERIMENTAL RESULTS AND DISCUSSION

Mean velocity profiles for $\lambda = \pm 6s^{-1}$ are shown in schematic form in Figure 1. A noticeable feature of the $\lambda = +6s^{-1}$ profile

---

* The flow immediately downstream of the honeycomb is largely dominated by the decay of turbulence generated by the honeycomb.

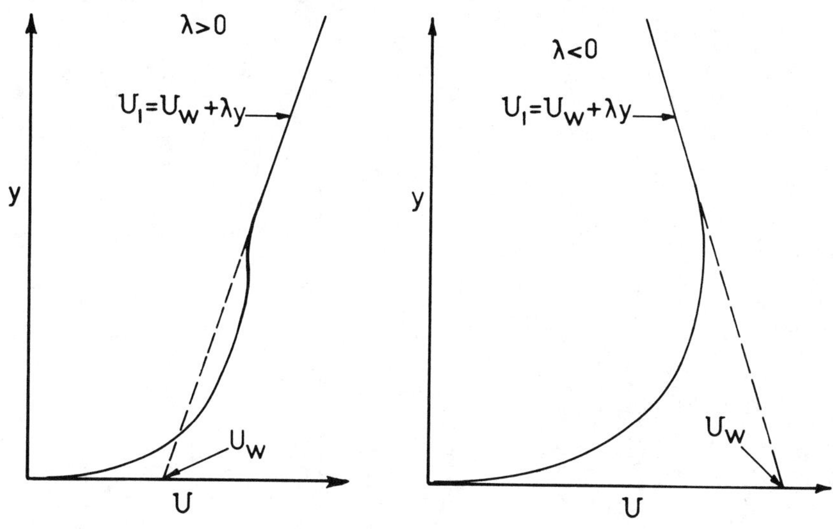

Figure 1. Schematic velocity profiles for $\lambda > 0$ and $\lambda < 0$.

is the appearance of a region of constant mean velocity U in the outer part of the layer, prior to the merging of boundary layer and external shear flow profiles. In the inner region of the layer, the universal viscous sublayer and logarithmic profiles are unaffected by the presence of non-zero Tu or λ. Although the reason for the constant U region in the outer layer when $\lambda = +6s^{-1}$ is not well understood, the plateau in U is consistent with the observed zero shear stress (Figure 2). In Figure 2, the difference between the local and external Reynolds shear stress is normalized by τ_W and plotted against y/Δ, where Δ is the Clauser thickness parameter, defined as $\Delta = \int_0^\infty (U_1 - U)\, dy/\tau_W^{1/2}$. For a given value of λ, the Reynolds shear stress defect in the layer may be considered to be self-preserving in that profiles shown in Figure 2 do not depend on the actual value of x, at least over the range of streamwise distances considered in the experiments. For the positive shear flow, Figure 2 shows that $\overline{uv}$ first goes to zero in the outer part of the layer and then increases to reach the constant $-(\overline{uv})_1$ value of the external stream. In the "wake" part of the layer, the shear stress is significantly larger than in the corresponding part of the layer with no external shear flow. On the other hand, the Reynolds shear stress distribution in the "wake" region of the negatively sheared

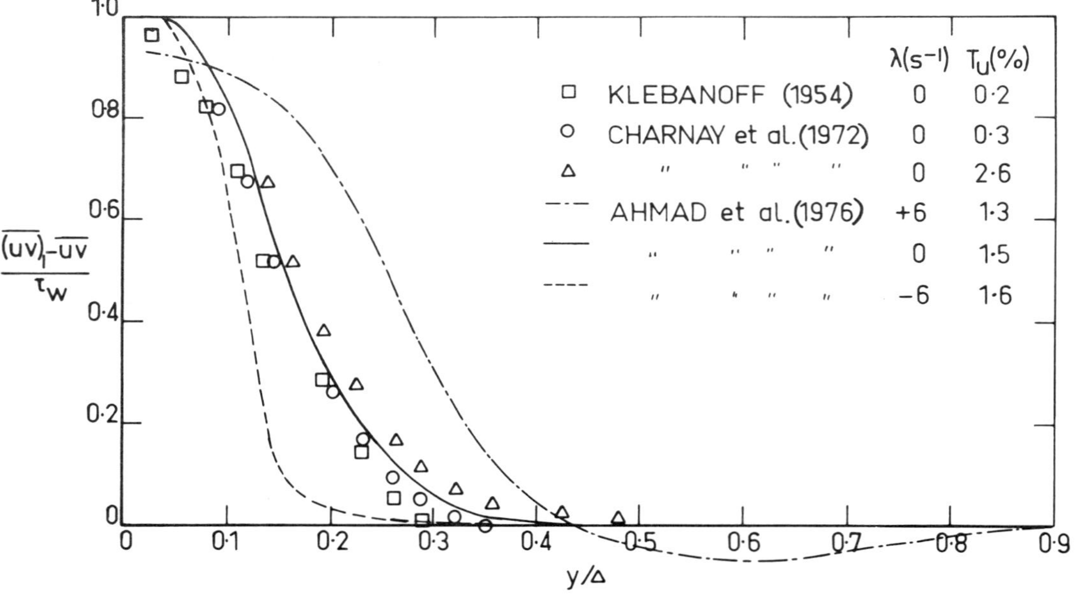

Figure 2. Reynolds shear stress distributions in the boundary layer.

layer is smaller than in the zero shear case. R.M.S. u and v intensities (not shown here) are in qualitative agreement with the trends exhibited by the Reynolds shear stress in the "wake" region of the boundary layer for the three values of λ. When λ is zero, the position at which the local shear stress goes to zero tends to increase, proportionately to Δ as Tu increases. The trend of the present results is consistent with that shown by distributions measured by Kelbanoff (1954) and Charnay et al. (1972a). It should also be noted that values of c_f (= τ_w/U_w^2, the skin friction coefficient) obtained for $\lambda = 0$ are in good agreement with Bradshaw's (1974) result that $(c_f - c_{f_0})/c_{f_0} \propto Tu$, c_{f_0} being the skin friction coefficient appropriate to the case Tu = 0. Positive and negative values of λ lead to respective increases and decreases in $(c_f - c_{f_0})/c_{f_0}$ in relation to the $\lambda = 0$ case. This trend in c_f appears to be consistent with the results of Figure 2.

A useful study of the interaction between a turbulent but shearless freestream and the boundary layer was made by Charnay et al. (1972b, 1974) who slightly heated the plate on which the layer developed and used the passive temperature contaminant to distinguish between the rotational velocity fluctuations in the boundary layer and those associated with freestream turbulence. Charnay et al. observed that the standard deviation of the turbulent/turbulent "interface" at the edge of the layer increased significantly as the freestream turbulence level was increased.* This result is consistent with Bradshaw's (1976) flow-visualization photograph of a highly re-entrant interface of a mixing layer with an effectively high freestream turbulence. It is also consistent with measurements by Antonia et al. (1975) of an internal thermal layer, completely immersed within a fully turbulent boundary layer, that is subjected to a step change in surface heat flux. The standard deviation of the thermal interface is as high as 50% of the thermal layer thickness at small distances from the step, where the external local turbulence intensity of the boundary layer is quite high.

In the present investigation, the boundary layer fluid was not tagged by heat but some information about the interaction between the boundary layer and the external shear flow has been obtained with measurements of the skewness and flatness factors of velocity fluctuations. Figure 3 shows the flatness factor $F_u \equiv (\overline{u^4}/\overline{u^2}^2)$ of u fluctuations at the same value of x but for the the three values of λ considered here. The distribution of F_u for $\lambda = 0$ shows that the maximum value of F_u is considerably reduced

---

* The boundary layer growth was correspondingly increased as Tu increased. Wigeland and Nagib (1974) found that the growth of a cylinder wake was also increased as Tu was increased.

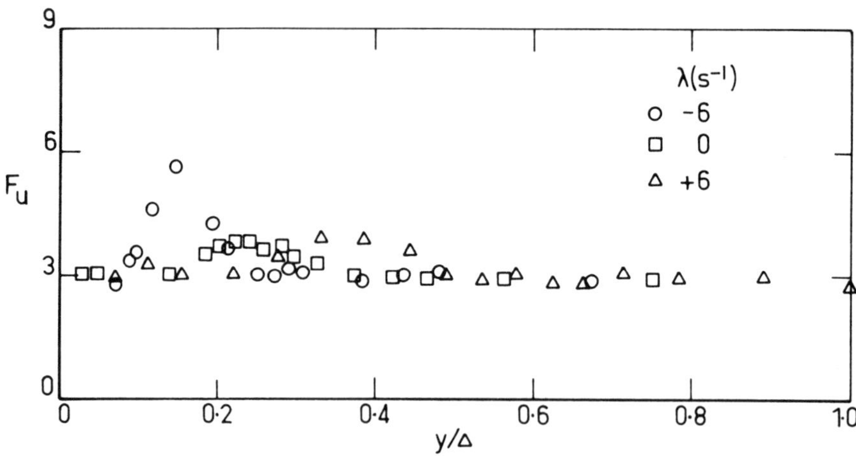

Figure 3. Flatness factor of u fluctuations.

by the presence of a relatively small freestream turbulence level. For $\lambda \neq 0$, the maximum value of F_u is somewhat larger, especially when $\lambda < 0$. At large values of y/Δ, F_u and F_v (Figure 4) are close to the Gaussian value of 3. In Figure 5, S_u and S_v are found to approach the Gaussian value of zero for $y/\Delta > 1$. The maximum value of F_v for $\lambda < 0$ is significantly larger than the maximum value for $\lambda > 0$, which is itself larger than that corresponding to $\lambda = 0$. This trend is also supported by results of the skewness ($\equiv \overline{v^3}/\overline{v^2}^{3/2}$) of v, shown in Figure 5. The results of Figures 3 to 5 appear to suggest that differences between turbulence characteristics of the external shear and boundary layer flows are more marked when $\lambda < 0$ than for the other two cases. This is not unreasonable in that the negative sign of the external shear should effectively inhibit if not destroy any boundary layer "eddies" with opposite mean shear. Differences between the maximum values of F_u and F_v for the cases $\lambda > 0$ and $\lambda = 0$ seem in qualitative agreement with the maximum values of F_u and F_v reported by Antonia and Luxton (1974) at the edge of an internal layer that propagates inside a turbulent boundary layer downstream of a step change in surface roughness.

ACKNOWLEDGEMENT

This work was supported by the Australian Research Grants Committee and the Australian Institute of Nuclear Science and Engineering.

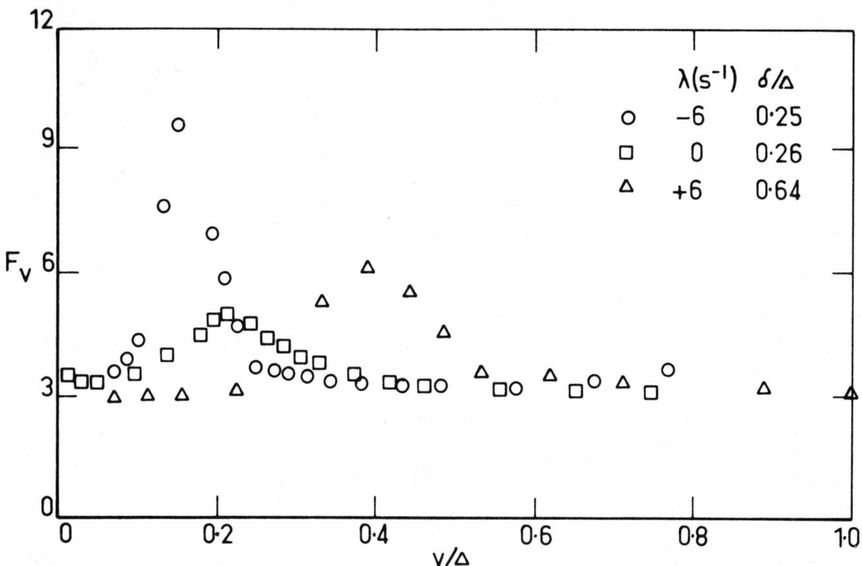

Figure 4. Flatness factor of v fluctuations.

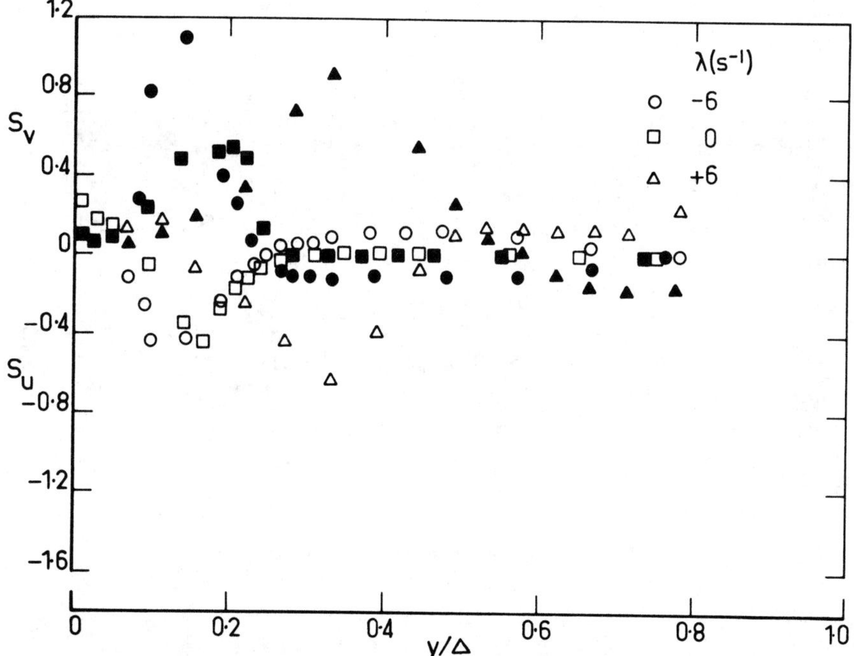

Figure 5. Skewness coefficients of u and v.

REFERENCES

1. Ahmad, Q. A., Luxton, R. E. & Antonia, R. A., 1976. Characteristics of a turbulent boundary layer with an external turbulent uniform shear flow, J. Fluid Mech. (to appear).

2. Antonia, R. A., Danh, H. Q. & Prabhu, A., 1975. Response of a turbulent boundary layer to a step change in surface heat flux. Charles Kolling Res. Lab., Dept. Mech. Engng., University of Sydney, Tech. Note F-74.

3. Antonia, R. A. & Luxton, R. E., 1974. Characteristics of turbulence within an internal boundary layer. Advances in Geophysics 18A, 263.

4. Bradshaw, P., 1974. Effect of free stream turbulence on turbulent shear layers, I.C. Aero Report 74-10, Dept. of Aeronautics, Imperial College of Science and Technology, London.

5. Bradshaw, P., 1976. Interacting shear layers in turbomachines, in this volume.

6. Champagne, F. H., Harris, V. G. & Corrsin, S., 1970. Experiments on nearly homogeneous turbulent shear flow, J. Fluid Mech. 41, 81.

7. Charnay, G., Comte-Bellot, G. & Mathieu, J., 1972a. Etat des contraintes et des fluctuations de vitesse dans une couche limite perturbée, C. R. Acad. Sc. Paris, Ser. A, 274, 1843.

8. Charnay, G., Comte-Bellot, G. & Mathieu, J., 1972b. Etude de la frontière libre d'une couche limite perturbée, C. R. Acad. Sc. Paris, Ser. A., 275, 615.

9. Charnay, G., Gence, J. N., Comte-Bellot, G. & Mathieu, J., 1974. Mesures échantillonnées des coefficients de dissymétrie et d'aplatissement de la fluctuation longitudinale de vitesse dans la zone intermittente d'une couche limite perturbée, C R. Acad. Sc. Paris, Ser. B, 278, 441.

10. Corrsin, S., 1976. Sheared cellular motion and homogeneous turbulent shear flow, in this volume.

11. Klebanoff, P. S., 1955. Characteristics of turbulence in a boundary layer with zero pressure gradient, N.A.C.A. Rept. 1247.

12. Mulhearn, P. J. & Luxton, R. E., 1975. The development of turbulence structure in a uniform shear flow, J. Fluid Mech. 68, 577.

13. Wigeland, R. A. & Nagib, H. M., 1975. Diffusion from a periodically heated line-source segment and its application to measurements in turbulent flows, in Proceedings of SQUID Workshop on "Turbulent Mixing in Non-Reactive and Reactive Flows" (ed. S. N. B. Murthy), p. 417, Plenum Press.

DISCUSSION

NAGIB: (Illinois Institute of Technology)

On the length scale graph, I couldn't see the left-hand side. Was it normalized with some parameter or was it actually in physical size?

ANTONIA:

It should have been normalized by δ. I should say also that I think the external length scale of Charnay et al. (1972) was in the range 1 to 4 boundary layer thicknesses. In fact, the external length scale range for all available measurements in the literature is about 0.25δ to 4δ.

FALCO: (Cambridge University)

I was interested in your flatness factors which in normal shear layers tend to get very large in the outer intermittent part of the layers. Would you comment on the fact that yours did not seem to increase at all as you got near the outer edge?

ANTONIA:

That is the effect of freestream turbulence.

FALCO:

It certainly confuses the boundary of the shear layer, and suggests that the flatness factor cannot be used to distinguish between grid turbulence and shear flow turbulence, an unexpected result.

ANTONIA:

You can see this clearly in Charnay et al.'s (1974) results.

BUSHNELL: (NASA Langley)

Do you have any feel for what the effect of a more narrow band disturbance would be? This is fairly wide band stuff that you are putting on, if I understand correctly.

ANTONIA:

We certainly have not done anything on that. I am not sure who in fact has looked at this. Karlsson (J. Fluid Mech., Vol. 5, p. 622, 1959) looked at a boundary layer with a sinusoidal fluctuation in freestream velocity. Does Luxton know?

LUXTON: (University of Adelaide)

I can't think of any off hand. I think that Stan Corrsin, or was it Clauser, put a cylinder into a boundary layer so that it had a periodic wake. Perhaps the work at Illinois Institute of Technology by Morkovin's group would be relevant? I am sorry that I can't shed more light on this.

CORRSIN: (Johns Hopkins University)

I know of none. Professor P. T. Fink of the University of New South Wales may know if anything was written up.

BRADSHAW: (Imperial College)

Some recent work was done under A. D. Young's supervision* on the turbulent boundary layer below an oscillating freestream. This was a traveling wave type of oscillation, as opposed to Karlsson's standing-wave oscillation but again the effect on the turbulence structure was pretty small. Wavelengths were many times the boundary layer thickness which is why the effects were small.

ANTONIA:

I think that in Karlsson's case, the length scale was about 5 or 6 times the shear layer thickness.

FALCO: (University of Cambridge)

In one of the slides with laminar positive you showed us a time averaged Reynolds stress which went negative for some region near the outer part of the layer. Is that correct?

---

* Patel, M. H. Proc. Roy. Soc. A (in press, 1976)

ANTONIA:

Yes, we subtracted $-\overline{(uv)}_1$, the constant Reynolds shear stress of the external stream, from the local Reynolds shear stress.

The Effect of Swirl on the Turbulence Structure of Jets

R. E. FALCO
Cambridge University
Cambridge, England

ABSTRACT

A volume flow visualization technique from which information in selected planes can be obtained is used to study the difference in the instantaneous turbulence structure of jets with and without swirl. This technique exhibits large scale motions as well as microscale motions. It was found that the large scale vorticity of the swirl resulted in azimuthal asymmetries in the large scale motions, which are associated with the formation of more Reynolds number dependent microscale size eddies in which there is a high concentration of vorticity. These eddies look like vortex rings. A model of the structural differences is given in analogy with the flow in the wake region of turbulent boundary layers.

INTRODUCTION

This paper discusses the differences between the instantaneous turbulent structure found in a swirled turbulent jet from that found in an un-swirled turbulent jet. The observations are for low Reynolds number jets. We are beginning to understand some aspects of the instantaneous turbulent structure in unswirled jets, but practically nothing is known about instantaneous turbulent structure changes that result when a jet is swirled. Since the Reynolds stress for moderate swirl (i.e., swirl with no mean recirculation region) can be three times the value found in unswirled jets, and it can be ten times the unswirled jet values for strongly swirled jets, it was expected that visual observations would reveal clearly defined changes in jet structure. A model which emphasizes the changes is presented. Swirl is usually used as an aid to mixing, and thus we will consider both large scale and microscale eddies. Since microscale eddies are Reynolds number dependent they will be a larger fraction of the overall shear layer width in low Reynolds number flows, and thus more easily examined. At the lower end of the Reynolds number range covered in these experiments $5{,}000 < R_D < 20{,}000$ the microscale eddies were a large fraction of the swirled shear layer width.

RESULTS

The basic result is that the large scale vorticity of the swirl results in azimuthal asymmetries in the large scale motions, which are associated with the formation of more Reynolds number dependent microscale eddies which look like vortex rings. We call these Typical Eddies.

Typical Eddies are a fundamental feature of low and moderate Reynolds number turbulent boundary layers, and have been extensively studied by Falco (1974, 1976a,b). They form as a result of a re-organization of vorticity which has been diffused or convected away from the wall. In boundary layer flows they are responsible for the intermittent peaks of instantaneous Reynolds stress observed in the outer part of the layer, and in fact they are responsible for a large fraction of the Reynolds stress in the outer part of low Reynolds number boundary layers. Our observations of similar data in swirled and unswirled jets suggest that a similar reorganization takes place of the additional vorticity introduced by the swirl, which results in more of the vortex ring like Typical Eddies.

By using a fog of tiny oil droplets, and illuminating only a slice of the visualized shear layer, we can see the internal structure of swirled and unswirled jets.

Figure 1 shows photographs of jets issuing into still air taken with the slit light plane technique. Both jets have the same reservoir pressure. The top photo is of an unswirled jet, and the lower photo shows a swirled jet with swirl angle about 45 degrees (i.e., S = 1). Swirl was generated by a "vortex whistle" with 4 tangential inlets similar to that used by Chanaud (1965). The vertical lines in the top view are spurious optical effects due to the use of a poor quality lens, and don't represent flow structure. Reviewing the structure of the unswirled jet, it emerges laminar, and undergoes the now well documented "transition" process to a turbulent jet. The slit lighting technique would reveal smoke free regions or regions of low smoke concentration that mark the boundaries of large scale motions if they extended into the center of the jet. We can see these gradients beginning to form by about five diameters. Furthermore, examining one of the bulges we see

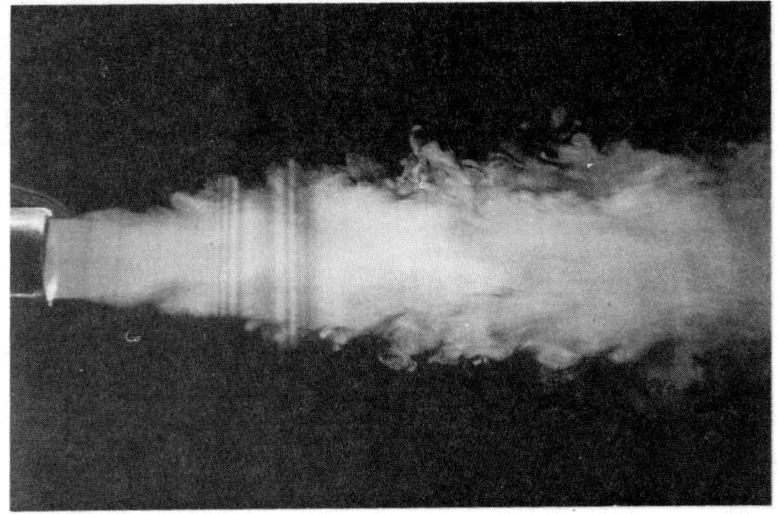

a

b

Figure 1. Photographs of swirled and unswirled jet. Exit diameter and tank pressure are the same.

that it contains much finer scale structure. Thus the large scale motions don't strongly influence the fluid near the center of this jet until about 5 diameters (the end of the "potential core"). However, even after the end of the potential core, we see that the flow still appears laminar like along the axis. Looking at the swirled jet we are immediately impressed by the much larger spread, and looking at the internal structure, we see that it has become turbulent at the jet exit. By 5 diameters we see extensive smoke free or low concentration regions in the smoke. Thus ambient fluid has been entrained into the center of the swirled jet. If we look closely at the regions of high smoke concentration we can see that they look like cuts through laminar vortex rings, which are of course highly distorted. These vortex ring like eddies, or Typical Eddies, are seen to exist across the whole of the swirled jet, and it is clear from the shapes of the regions of smoke free ambient fluid which have penetrated deep into the jet that the entrainment of this fluid is closely connected with the Typical Eddies. These Typical Eddies are larger than those found in the unswirled jet.

We are going to concentrate on these Typical Eddies. The reason for this is that it appears that the mixing phenomena may be predictable for scales of motion equal to and smaller than Typical Eddies, which have been identified with the Taylor microscale in wakes and boundary layers (see Falco, 1974, 1976b). If Typical Eddies are highly distorted vortex rings, then equations for vortex ring growth, and understanding of their internal diffusion processes combined with knowledge of their frequency of occurrence, should establish the limiting rates of mixing. This is true because the rate of molecular diffusion which occurs at Typical Eddy/vortex ring boundaries is the highest in the flow field, simply because the internal rotation of a vortex ring keeps the gradients of vorticity at the ring boundaries sharp and therefore results in a high rate of diffusion. This model of a vortex ring has been developed by Maxworthy (1972).

For the Typical Eddy, this enhanced diffusion rate occurs only during and shortly after its formation. Once the Typical Eddy has formed, and this may take one or two shear layer thicknesses, we observed it to be highly stable, and in boundary layers for example, it can be followed for five to ten shear layer thicknesses. Thus the volume of fluid concerned, after the early stages of the event, will essentially be isolated from further enhancement of diffusion, until the coherency of the Typical Eddy is destroyed. The rate of mixing will be increased only by producing more Typical Eddies/vortex rings (in this model), and preferably decreasing the scale of these Typical Eddies. We will show the photo indicates that the swirled jet develops more Typical Eddies more quickly, which should enhance mixing rates, however, the scale of the Typical Eddies is larger than in an unswirled jet. Thus after a few diameters, we should find a decrease in the mixing rate with respect to conditions in an unswirled jet at the same location.

There is evidence that a cross-over point in the magnitudes of the turbulent energy and Reynolds stress occurs in swirled jets. Figure 2 shows $\overline{uv}$ measurements in a jet as the swirl is increased. These are measurements of Allen (1970). We see that $\overline{uv}$ is increased by a factor of 3 near the jet exit as the swirl parameter S is increased to .6. S = .6 corresponds to the magnitude of swirl necessary to induce flow reversal near the jet exit. This increase in $\overline{uv}$ for S = .6 (a moderate level of swirl), is much larger than the increases obtainable by pulsing a jet or by perturbing a flow with a turbulence producing grid. However, as indicated in the figure, after about three diameters, the Reynolds stress of the swirled jets is less than that of the unswirled jet, it being lowest for the most strongly swirled jet. Allen (1970) found similar cross overs occurred in $\overline{u^2}$, $\overline{v^2}$ and $\overline{w^2}$. Syred, Bėer and Chiger (1971) have measured the $\overline{uw}$ distribution for a strongly swirled jet, S = 2.2. This showed very high values near the jet exit, which decreased by a factor of ten after one diameter.

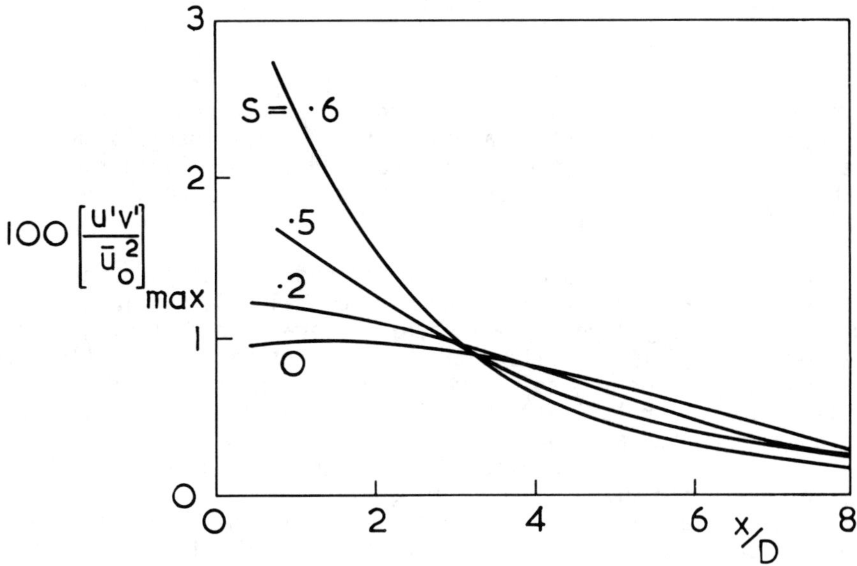

Figure 2. Increases in Reynolds stress induced by swirl. From Allen (1970).

Flow in turbulent boundary layers contributes the following information on Typical Eddies and their relationship to large scales:

1) The Typical Eddy scales are strongly Reynolds number dependent, the streamwise scale varies as $R_\theta^{-.8}$ (Falco, 1974).
2) Typical Eddies are on average formed on the backs (upstream side) of large scale bulges of the boundary layer.
3) They produce the intermittent bursts of Reynolds stress, and account for almost all of the Reynolds stress in the outer part of low Reynolds numbers layers.

It is clear that Typical Eddy formation involves a redistribution of vorticity by a large scale perturbation. If Typical Eddies were vortex rings then a force is necessary for their formation.

Returning to the swirled jet, swirl introduces large scale distributed vorticity in the streamwise direction. Our observations of the cross-stream structure of a swirled jet show that an asymmetry develops in the large scale motions, which resembles that found in the streamwise slices of large scale motions. Figure 3, again taken with the light plane technique, shows this asymmetry in the large scale motions as seen in the y-z plane. This photograph is of a higher Reynolds number swirled jet, which shows the large scale motions more clearly (the Typical Eddies are a smaller fraction of the shear layer thickness in this case). Asymmetries in the smoke marked shape of large scale motions in turbulent boundary layer flows is closely associated with the production of Typical Eddies. We will now discuss a model of the relationship between the large scale motions and the Typical Eddies, developed for the turbulent boundary layer which helps to explain why we see more Typical Eddies in swirled jets.

Figure 4 shows a model of the relationship between outer layer bursts which are Typical Eddies and large scale motions in the turbulent boundary layer. We see that Typical Eddies form on the upstream side of large scale motions, which characteristically have the asymmetry shown in the figure. The downstream side of large scale motions is relatively uneventful in boundary layers, when we are away from the wall region. An analogous picture in a non-swirled jet is shown in Figure 5. This model is suggesting that most of the rapid mixing is occurring at the downstream sides of the large scale motions, i.e., in the Typical Eddy generation region, and that the regions interior to the large scale motions are regions of lower velocity gradients and relatively low mixing rates.

Our picture of a swirled jet looked different however. Remember that we saw Typical Eddies apparently much more uniformly distributed. A reason for the large number of Typical Eddies is

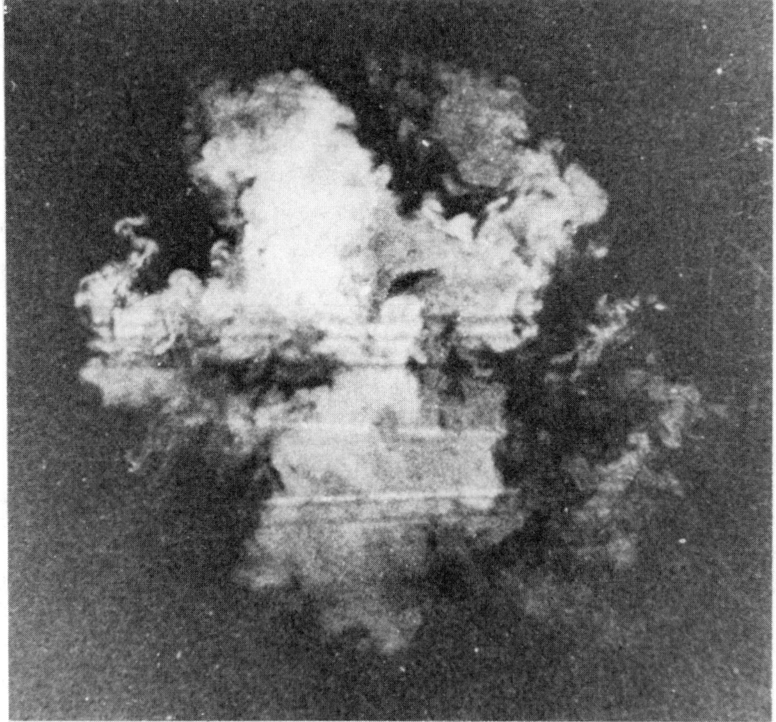

Figure 3. Photographs of cross-stream flow of a swirled jet through a plane at 6 diameters downstream. This jet is at a higher Reynolds number than the swirled jet in Figure 1, and emphasizes the large scale structure which develops. Five skewed large lobes are apparent in this photo. The jet is rotating clockwise.

suggested from observations of the cross-stream photos. These show that the large scale motions have additional Typical Eddies which are seen on the inclined portions of the large scale motions in the y-z plane (see Figure 6). The picture of the position of Typical Eddies on the azimuthally upstream side of the large scale motions is remarkably similar to the picture found in streamwise light planes. A sketch is shown in Figure 7. The Typical Eddies which form in the y-z plane can be thought of as being a direct re-distribution of the large scale vorticity introduced by swirling the jet. Because of these additional Typical Eddies, effectively on the sides of the large scale motions as seen in streamwise planes, we see a larger number of them in streamwise planes.

We have performed an independent experiment to confirm the observation that more Typical Eddies appear when additional vorticity is present, and present in large scale distributed form.

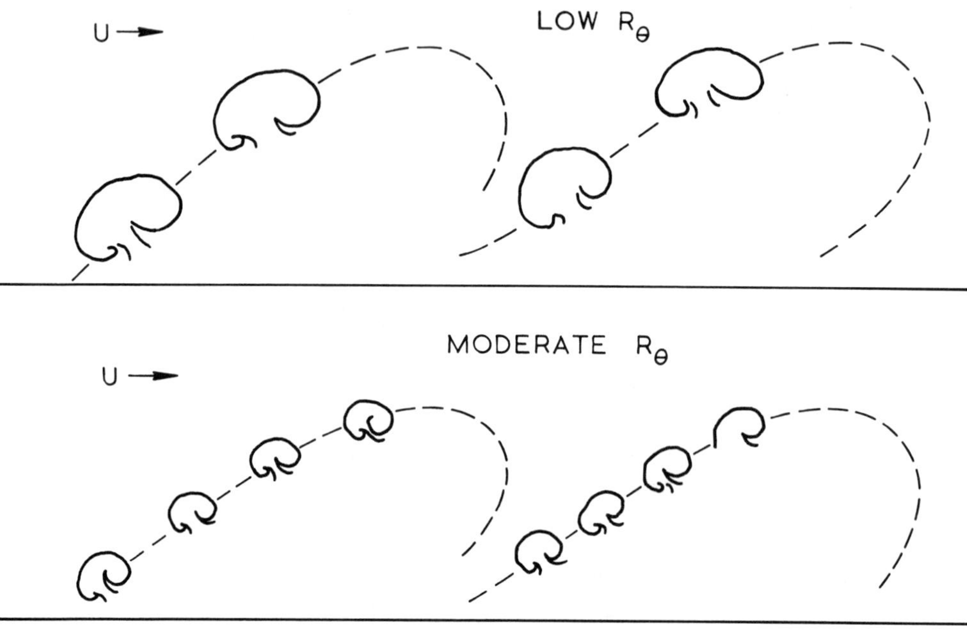

Figure 4. Model of the relationship between Typical Eddies and large scale motions in the turbulent boundary layer at low and moderate Reynolds numbers.

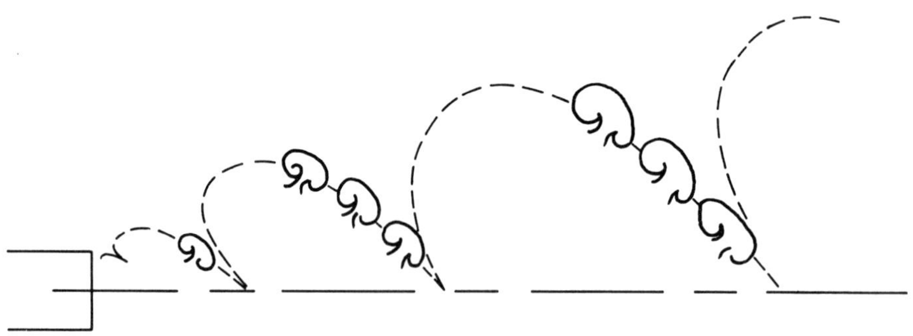

Figure 5. Model of the large eddy and Typical Eddy relationship in an unswirled jet of low Reynolds number.

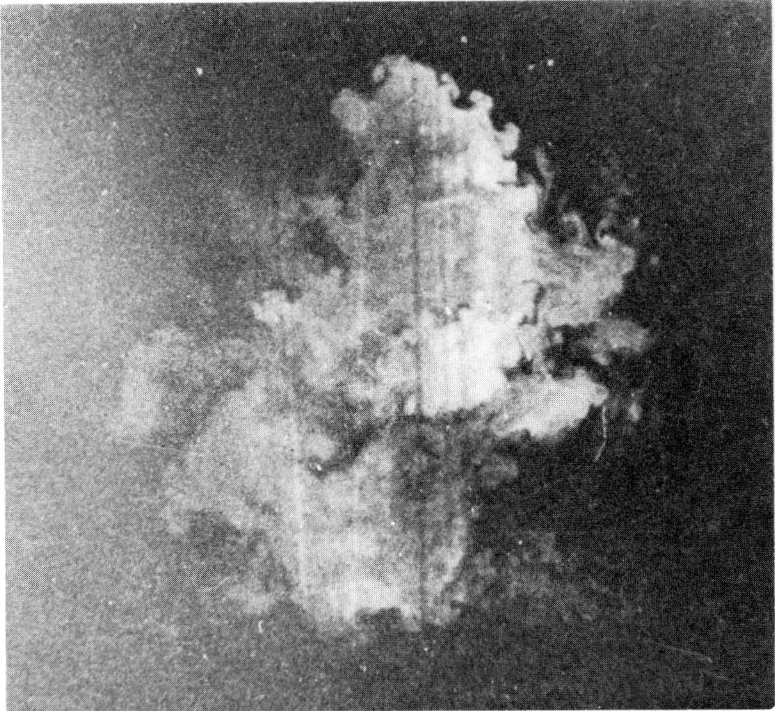

Figure 6. Cross-stream flow of a swirled jet through a plane at 6 diameters downstream. Conditions are the same as for the jet in Figure 3, however this photo shows only two large skewed lobes, with Typical Eddies clearly apparent on one of them (jet is rotating clockwise).

Figure 8 shows a turbulent boundary layer developing in a large scale laminar shear flow produced by a shear gauze. Observations of several pictures such as this indicated that the formation of more Typical Eddies at the edge of the boundary layer occurred than found for a layer developing into a uniform stream.

Now the re-distribution of vorticity into more Typical Eddies, as indicated previously, is a mixed blessing. This is due to the relatively long life of Typical Eddies/vortex rings, which are highly stable. Thus if we distribute the available vorticity into a couple of large Typical Eddies, then, after the initial period of development, mixing of the fluid contained in them will be un-enhanced until the vortex ring has itself grown to a scale at which it is unstable and its vorticity is again re-distributed by non-viscous processes. We have seen that the large scale vorticity introduced into the flow by swirl gets partially re-distributed

Figure 7. Model of the cross-stream large eddy structure in a swirled jet, showing the asymmetry and the production of additional Typical Eddies as a result of the swirl (clockwise rotation).

into Typical Eddies. However, it appears that the scale of the Typical Eddies, both those formed in the cross-stream planes, and the streamwise planes is considerably larger than the scale found in unswirled jets. Thus, although jet fluid may be convected considerably further out into the ambient by these motions, after the initial formation period, it appears that mixing may be slower. As shown in Figure 2 after three diameters, the Reynolds stress of the swirled jet is lower than that of the unswirled jet. The reason for the larger scale of the Typical Eddies is only partly understood. As indicated, Typical Eddies are strongly Reynolds number dependent, and are larger in lower Reynolds number flows. Now, it is certainly the case that the effective Reynolds number in the cross-stream direction, the azimuthal Reynolds number, rapidly increases as the jet spreads, and this is also true for the streamwise Reynolds number. Thus, the Typical Eddies which result will be larger than those found in an unswirled jet of similar exit Reynolds number at a given distance downstream. However, Typical Eddy scales may also be strongly affected by the mean

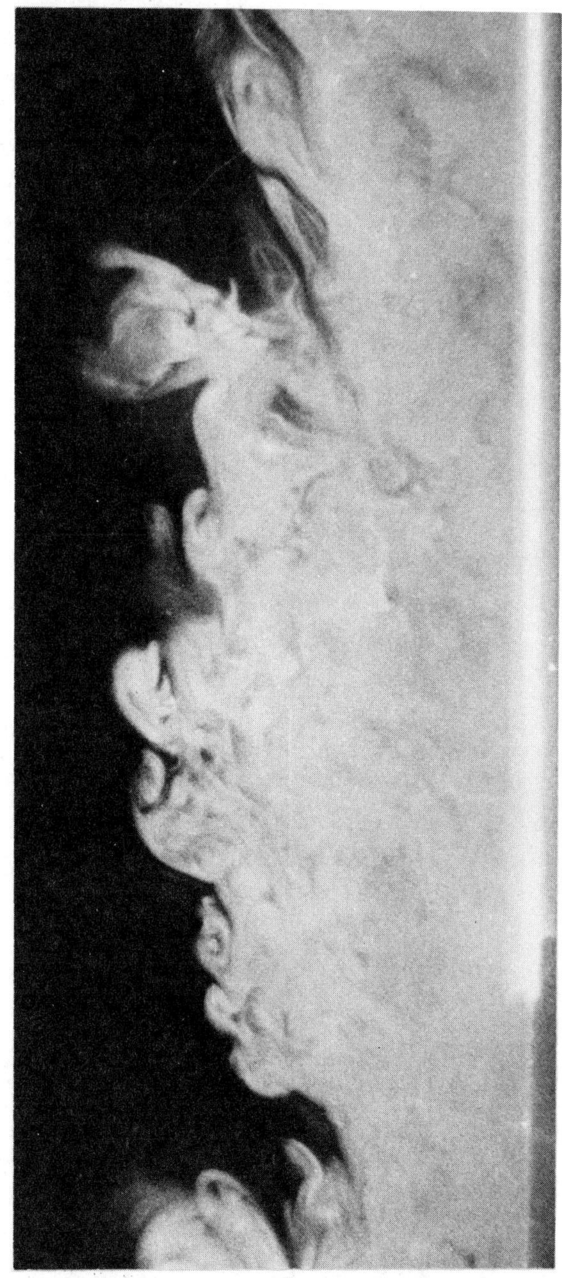

Figure 8. Photograph of a turbulent boundary layer growing in a large scale laminar shear flow. More Typical Eddies are observed to form in the outer part of the layer.

flow strain, which can produce vortex stretching and re-orientation. The flow emerging from a swirled jet is straining rapidly.

In conclusion, it appears that although a strongly swirled jet spreads more rapidly, the Typical Eddy dominated internal structure suggests that after the first few diameters, the mixing process may actually be hampered.

ACKNOWLEDGEMENTS

The author gratefully acknowledges the helpful discussions with Professor J. E. Ffowcs Williams. Financial support was provided by the Scientific Research Council.

REFERENCES

1. Allen, R. A. 1970. Aerodynamics and interaction of single and multiple jets with rotation. Ph.D. thesis, University of Sheffield.

2. Chanaud, R. C. 1965. Observations of oscillatory motion in certain swirling flows. J. Fluid Mech. 21, pp. 111-127.

3. Falco, R. E. 1974. Some comments on turbulent boundary layer structure inferred from the movements of a passive contaminant. AIAA Paper No. 74-99.

4. Falco, R. E. 1976a. Coherent motions in the outer region of turbulent boundary layers; the relationship between Reynolds number dependent and independent coherent motions. To be published in Physics of Fluids.

5. Falco, R. E. 1976b. An experimental study of Reynolds stress producing motions in the outer part of turbulent boundary layers. Part 1; the shape, scale, evolution and Reynolds number dependence. Submitted to J. Fluid Mech.

6. Maxworthy, T. 1972. The structure and stability of vortex rings. J. Fluid Mech., 51, 15.

7. Syred, N., Bėer, J. M. and Chigier, N. A. 1971. Turbulence measurements in swirling recirculating flows. Proc. Internal Flows Symp., Salford Univ., pp. B27-B36.

DISCUSSION

BRADSHAW: (Imperial College)

Do you believe that the spreading rate of the jet (or therefore the Reynolds stresses) depends on Reynolds numbers?

FALCO:

The experiment certainly shows that it doesn't.

BRADSHAW:

Does it not follow that your Reynolds number dependent vortex-ring structures are not part of the Reynolds-stress-producing mechanism in the jet?

FALCO:

The answer to that question eludes me. I was saying to Stan Corrsin early today that in the turbulent boundary layer (and in fact you may have made this point to me as well), at R_θ of the order of 5,000 or 6,000 the law of the wake holds. This is certainly not directly Reynolds numbers dependent. However, if you take a hot wire and measure the Reynolds stress characteristics in a boundary layer at this Reynolds number, as Bill Willmarth* has done, you find that a very large contribution to the Reynolds stress in the outer part of that layer is at scales that scale very nicely with my typical eddy scales, which are Reynolds number dependent.

KOVASZNAY: (Johns Hopkins University)

The evidence is that there is Reynolds stress dependence on that scale.

FALCO:

The evidence that I presented last week at the IUTAM Conference† is that if you take the uv signal and you look at the scale of the peaks (you have to choose some discriminating level), then

---

* Lu, S. S. and Willmarth, W. W., 1973, J. Fluid Mech., 60, 481.

† IUTAM Symposium on Structure of Turbulence and Drag Reduction, 7-12 June 1976, Washington, D.C. To be published in Physics of Fluids.

you find that scale is the Typical Eddy scale.* That typical eddy scale is very strongly Reynolds number dependent.

I don't have an answer to Mr. Bradshaw's question but that is the evidence I have.

WYGNANSKI: (Tel Aviv University)

You presented uv as evidence to the effect of swirl in the jet. Can you show the effect of uw and W so that we can see what happens. Is there any transfer from one component to the other?

FALCO:

Those were not my measurements. They were measurements of Allen (reference in text) at the University of Sheffield. Yes, I think he has measured uw and so has Syred, et al.,† but I don't know of any $\overline{uw}$ measurements in non-swirled jets.

WALLACE: (The University of Maryland)

In both the jet and the boundary layer, how did you determine that the smoke only marks the rotational flow and how do you demonstrate that this is Reynolds number independent?

FALCO:

This question was essentially asked before. The way I look at it is the following. The oil droplet air mixture has a very high Schmidt number, it is about 38,000 in fact. On the whole you would expect that a given isovorticity line will certainly be, if you pick one of arbitrarily small value to mark the edge of the boundary layer at any instant, further out than my isoconcentration "edge" and that undoubtedly is true. However, the point that makes this technique work, as far as I am concerned, is that the part of the turbulent motion that contributes most of the Reynolds stress has a great deal of normal motion to it, and the convective time scale of this normal motion is considerably shorter than the diffusion time scale. Thus the Schmidt number argument as far as the Reynolds stress producing motion is concerned is not a limiting argument. It is important, however, as far as the remaining part of the interface is concerned and that needs further investigation.

---

* Falco, R. E., 1974, AIAA Paper 74-99.

† Syred, N., Beér, J. M. and Chigier, N. A., 1971, Symposium on Internal Flows, Univ. of Salford, Paper 13, B27-36.

KERREBROCK: (M.I.T.)

Do I understand correctly that only the jet had swirl? That is, is a swirling jet entering into a stagnant atmosphere?

FALCO:

Yes, that is right.

KERREBROCK:

Then I have a question. Perhaps I just do not understand. It seems to me that the jet is unstable because of the fact that the circulation decreases outward and what you are seeing is just a large ordered motion due to that instability. If that is the case I wonder why you characterize it as turbulence?

FALCO:

The way I look at the problem in general is that I am applying a large scale vorticity field to an initially turbulent flow and examining the resulting changes in turbulent structure. The additional steady vorticity vector is initially in the axial direction, and I have shown that there appears in the flow field a skewing of the large scale motions in the cross stream direction and additional small scale regions of high vorticity concentration that would result in a fluctuating vorticity signal if we placed a fixed vorticity measuring probe in the flow. These additional features appear similar to the turbulence structure found in shear layers developing into a uniform flow when observations normal to the mean vorticity vector are made.

BRADSHAW:

I think I agree with Kerrebrock that it is just the effect of centrifugal instability on an existing turbulence field. It is an extreme case of the usual effects of streamline curvature on turbulence. I did some experiments on swirling jets myself more than 10 years ago* and again we got very large increases in spreading rate. And again, we observed a notable tendency for the mean profiles in the jet to become non-axisymmetric. At the time I attributed this just to the effect of the nozzle we were using. But I am inclined to think that there is a tendency even for mean nonuniformities in the nozzle to get rather amplified as you go downstream.

---

* Bradshaw, P. "Preliminary note on a mixing-nozzle ejector-shroud combination for jet noise reduction," NPLACO Report 1116 (1964).

NAGIB: (Illinois Institute of Technology)

Something bothers me, it goes back to the data from Allen and I think that the swirl rate as you call it indicates that you are getting only a substantial effect as you reach about 0.2 or 0.3, which is when the recirculation sets on. Does that mean most of the extra spreading and extra Reynolds stress is governed by this mean flow three-dimensionality? I mean that most of the effects show only when your swirl rate is large enough to get recirculation, right?

FALCO:

That is not true. Allen's curve (reference in text) shows that there is a cross-over effect even for very low swirl levels. Even for S (the ratio of angular to axial momentum times radius) of 0.2 S equal to 0.6 is roughly the swirl level needed for onset of reverse flow near the orifice.

GOLDSCHMIDT: (Purdue University)

I guess the answer you gave Jim Wallace bothered me a little bit. What is your particle response time? Do you have a feeling for that?

FALCO:

I have an indirect feeling from a paper of Jim Whitelaw's group* where he was doing laser doppler anemometry with particles and suggested that for roughly the same size particles one should be able to pick up completely any frequency under 700 Hz with an accuracy of 1%. At the Reynolds number I am dealing with I don't think that there is any problem at all.

MOREL: (General Motors Research Laboratory)

I noticed that several times you came back to Figure 2 showing curves of shear stress and wonder if they are divided by the initial jet velocity?

FALCO:

Yes, in that case.

---

* Baker, R. J., Bourke, P. J. and Whitelaw, J. H., 1973, Fourteenth Symposium on Combustion, pp. 699-706.

MOREL:

If you would divide the shear stress by the local jet velocity, would there still be this rapid decrease in the dimensionless shear stress?

FALCO:

Well, certainly the jet velocity goes down very quickly.

MOREL:

That might be the reason why uv divided by the initial velocity drops down so rapidly with χ/D giving the impression that the shear stress is being strongly suppressed by the swirl. However, should one use a more appropriate scale, such as the local jet velocity, then perhaps it would turn out that the suppression is much smaller or that it does not occur at all.

FALCO:

That is possible, although the choice of a local velocity scale would be complicated when reversed flow occurred.

UBEROI: (University of Colorado)

This business of swirl with axial velocity is rather complicated. You can show by using the linear and nonlinear stability theory that you get propagating waves as well as amplifying and decaying waves and the point at which these waves appear depends on the swirl ratio. In addition we find that just taking pictures is not really enough, you have to make careful measurements. There are many other points, but I would frankly say that these are merely preliminary and somewhat speculative measurements.

FALCO:

Well, I am flattered. I think that is absolutely right. Yes, they are preliminary and they are put up with the intention of initiating discussion on the modification of instantaneous turbulence structure by large scale vorticity.

KLINE: (Stanford University)

I just wanted to comment on Mr. Bradshaw's question. It doesn't seem obvious to me that because one observes the defect law, or Coles law, in the outer portion of the boundary layer, that you can't have the kind of scaling that you are suggesting. You are talking about the size of eddy that carries the Reynolds stress along, not the mean profile scales. Perhaps there is an inconsistency in the differences of scale, but there is also another possibility, namely, that there is some scaling effect that makes the two quantitatively consistent. I don't see that consistency is ruled out. I don't know the answer at this moment, but there does seem to be another possibility so that both scalings could be correct without inconsistency.

COLES: (California Institute of Technology)

When you showed the slide with the three lobes and what you called the Reynolds-stress producing elements on it, I thought, as I have before, of a drop of milk in a glass of water. There is an instability that as far as I know nobody really understands that causes a cascading of vortices from the original large vortex to smaller and smaller ones. I think that is what you are looking at. I think you are describing the cascade process.

FALCO:

Undoubtedly I am describing a cascade process. However, it is probably worth pointing out that there is quite a gap between large scales and typical eddy scales particularly at high Reynolds numbers.

Implications of the Structure of the Viscous Wall Layer

J. D. A. WALKER AND D. E. ABBOTT
Purdue University
West Lafayette, Indiana (U.S.A.)

ABSTRACT

The wall layer of a turbulent boundary layer, defined here as the region between the wall and the log law region, has been extensively investigated by experiment over the past fifteen years. These experiments reveal a remarkably well ordered time-dependent flow within the wall layer. In particular two features are known to dominate the character of the flow, namely (a) the longitudinal streak structure which is present and relatively stable during the quiescent period (as defined by Kline et al., 1967) and (b) the localized breakdown of the wall layer known as the bursting process. Neither feature is well understood on a theoretical basis and are the subject of the present paper. First, the governing equations during the quiescent period are investigated in the rational mathematical context of the Reynolds number becoming large. No complete solution for the wall layer flow is to be expected but representative solutions may be obtained. The object is to obtain realistic expressions for the mean profile and Reynolds stress terms within the wall layer. Secondly, the laminar boundary layer created by a rectilinear vortex in motion above an infinite plane wall is considered. The goal here is to investigate what type of boundary layer motion is to be expected when concentrated vorticity is near a wall. The results show that a boundary-layer eruption is to be expected and it is suggested that a possible physical mechanism for the burst is a moving outflow stagnation point.

1. INTRODUCTION

Experimental observations over the past fifteen years have demonstrated that the wall region of a turbulent boundary layer, which is two-dimensional and steady in the time-mean sense, possesses a remarkable degree of coherent structure. The experiments of Kline et al. (1967) were the first indication of this and a recent review article by Willmarth (1975) details the numerous experimental contributions that have been made since then. It is fair to conclude at this stage that there are many unanswered questions in regard to the causes and effects of what is being observed and it is clear that the dynamics of the turbulent boundary layer are not well understood; on the other hand, what is gradually emerging from experiments is an indication, that while the observed flow patterns are complex, there may be very little that is truly random about boundary-layer turbulence. This is an exciting prospect since, if it is eventually borne out, a rational theory of turbulence can be expected at some point in the future. In this connection, the two-dimensional time-mean boundary layer represents the most fundamental problem since it is the simplest type of boundary layer and a good theoretical understanding of this case seems essential to provide a firm basis for further research into the more complex types of turbulent boundary layers encountered in engineering practice.

The two-dimensional time-mean boundary layer is known to be a double structured layer consisting of an inner wall layer and a much thicker outer inviscid layer. Little is known about the dynamics of the outer layer except that it appears to be dominated by the presence of large scale vortex structures, usually referred to as eddies. Insofar as the wall layer is concerned there are two features which are known to characterize the time-dependent flow, namely the burst and the streak structure. The burst is an event which is observed to initiate deep within the wall layer and is characterized in its latter stages by a rapid and violent ejection of fluid from the wall layer. The bursting fluid appears to leave the wall layer as a discrete entity suggestive of a three-dimensional vortex structure (Offen & Kline, 1975) and as it leaves it is undercut by fluid rushing in from the outer layer. This latter event has been referred to as the sweep by Corino and Brodkey (1969) and it is during the latter stages of the sweep that the second feature of the wall layer flow appears; this is the streak structure which persists for a relatively long period of time, known as the quiescent period. During the quiescent period, the streaks are observed to be relatively stable and the wall layer maintains its integrity. At a later stage another burst occurs and with this the streaky patterns are obliterated. The burst may be viewed as a localized breakdown of the wall layer flow which results in an inviscid-viscous interaction between the inner and outer layers. The bursting repeats cyclically but not periodically at what appear to be random streamwise and spanwise locations; further because of the observed localized nature of this breakdown, it would seem appropriate to characterize the time T between bursts as a burst rate per unit area. This is essentially what is measured with the visual counting method used by Kline et al. (1967); it is not clear what is being measured in this respect with instrumented burst detection schemes (Offen & Kline, 1973).

Despite a wealth of experimentation on the observed effects, the causes of these phenomena are not known and these are fundamental questions which must be answered if the dynamics of the turbulent boundary layer are to be understood. Furthermore until the dynamics are understood one cannot hope to do anything but make guesses for the unknown terms which appear in the time-mean equations. At present there is no adequate theory which remotely explains what causes the streaks or the burst; what can be said with a degree of certainty is that (1) the streaks are present during the quiescent period and (2) the burst is very localized and seems to be connected in some way with the passage of a large scale vortex structure in the outer layer (Nychas et al., 1973). At the current level of technology, experiments alone cannot be expected to resolve all of the questions surrounding turbulent boundary layers; a major limitation of present experimental methods is that velocity measurements must be taken at isolated points and inferences in regard to the complete flow field are extrapolations. Further the meaning of current conditional averaging procedures in a flow which is cyclic but not periodic is unclear and assumptions which are sometimes invoked involving complete reproducibility seem unsatisfactory. Flow visualization is a useful technique but if it is indeed the motion of large scale vortex structures within the outer layer which lead to the streaks and bursting, visualization in the laboratory frame may be misleading since what is relevant is the flow relative to the vortex itself. As experimental techniques improve, an increase in understanding is to be expected but at the same time, a theoretical approach to the problem is essential if the observed phenomena are to be translated into improved and realistic prediction methods.

The general problem of time-dependent flow in a turbulent boundary layer is a complex problem and involves many aspects; the present paper focuses on two of these. In §3, the governing equations during the quiescent period are investigated in the asymptotic limit of the Reynolds number, $Re \to \infty$ and it is demonstrated that the leading order terms for all three velocity components in the wall layer satisfy linear equations of the heat conduction type during the quiescent period. In order to set the stage for the analysis in §3, it is worthwhile to briefly summarize some recent theoretical analyses concerning the form of the leading order equations (as $Re \to \infty$) that govern the time-mean flow and this is done in §2. No complete solution for the wall layer flow is to be expected but representative motions may be considered. From these solutions, an expression for the mean profile may be obtained and direct comparison with experimental data indicates that this profile represents a significant improvement over the Van Driest (1956) profile and its various modifications.

The second feature considered here (in §4) is the physical mechanism for the burst itself and to this end and as a preliminary step, the behavior of the unsteady laminar boundary layer due to the motion of a rectilinear vortex above a wall is discussed. The object is to understand how a boundary layer may be expected to behave when concentrated vorticity in an inviscid flow is present near a wall. It is emphasized at the outset that no attempt is being made here to model the flow in the wake layer of a turbulent boundary with a two-dimensional vortex. Heretofore it has been assumed by a number of authors that the burst is a manifestation of an instability of the wall layer; this view is purely conjectural and here an alternative mechanism is investigated. The results indicate that shortly after the creation of a rectilinear line vortex, separation occurs in the boundary layer and it is argued that an explosive eruption of the boundary layer along with a major modification of the inviscid flow is to be expected. This eruption takes place in a frame of reference which moves with the vortex and bears a resemblance to the observed bursting phenomenon.

2. THE ASYMPTOTIC STRUCTURE OF THE TIME-MEAN FLOW

In considering the possible time-dependent motions within the wall layer, there are a number of constraints in the form of a large body of experimental data that establish certain well accepted results about the time-mean quantities, such as the mean profile, turbulence intensities and the $\overline{u'v'}$ product. Clearly any theory which is incompatible with these results is not acceptable; at the same time, the measured quantities also serve to rule out various types of motion within the wall layer. If one accepts the Reynolds principle of similarity then all length and time scales for the turbulent boundary layer must depend on the Reynolds number in some way as $Re \to \infty$ and in a rational theory it is crucial to discern precisely what this Reynolds number dependence is. There have been a number of attempts in recent times to apply modern asymptotic analysis to turbulent boundary layers and of particular interest here is the study by Fendell (1972). The important assumptions in Fendell's (1972) analysis are as follows: The turbulence may be modeled as an additive stress σ, so that if (x,y) measure distance in the streamwise direction and normal to the wall, respectively, with corresponding mean velocity components $(\overline{u},\overline{v})$, the momentum equation is

$$\bar{u}\frac{\partial \bar{u}}{\partial x} + \bar{v}\frac{\partial \bar{u}}{\partial y} = U_\infty \frac{dU_\infty}{dx} + \frac{\partial \sigma}{\partial y} + \frac{1}{Re}\frac{\partial^2 \bar{u}}{\partial y^2} \tag{1}$$

Here σ may be related to the $\overline{u'v'}$ product and all lengths and velocities have been made dimensionless with respect to a representative length L and velocity U_0, respectively; the Reynolds number is defined as $Re = U_0L/\nu$, where ν is the kinematic viscosity. In addition it is assumed that there are two layers only and that in the outer layer the tangential velocity is to leading order independent of y (the defect law). An important parameter which emerges from the analysis is $u^* = u_\tau/U_\infty(x)$, where u_τ is the dimensionless friction velocity and $U_\infty(x)$ is the local dimensionless mainstream velocity; to leading order it may be shown that

$$u^* \sim \frac{\kappa}{log Re_\delta} \quad \text{as} \quad Re_\delta \to \infty, \tag{2}$$

where κ is the von Kármán constant and Re_δ is the Reynolds number based on the local boundary layer thickness. The outer region is found to have a thickness $O(u^*)$ and to leading order

$$\bar{u} = U_\infty(x)\left\{1 + u^* \frac{\partial F}{\partial \eta}(\eta,x) + \cdots\right\}, \quad \sigma = u^{*2}\Sigma_1(\eta,x) + \cdots \tag{3}$$

where $\eta = y/u^*$ is the scaled outer variable. The inner region has a thickness $O([Re\, u^*]^{-1})$ and here to leading order

$$\bar{u} = u^* \frac{\partial f}{\partial y^+}(y^+,x) + \cdots, \quad \sigma = u^{*2}\sigma_1(y^+,x) + \cdots \tag{4}$$

where $y^+ = Re\, u^* y$ is the scaled inner variable. Matching takes place in the limits $\eta \to 0$, $y^+ \to \infty$ and on the basis of an argument due to Millikan (1932), it emerges that

$$\frac{\partial F}{\partial \eta} \sim \alpha(x) log \eta + \beta_0(x), \quad \frac{\partial f}{\partial y^+} \sim \alpha(x) log y^+ + \beta_i(x) \tag{5}$$

in the region (fully turbulent zone) where the match takes place. Experimental data appear to justify the assumption that α is constant and equal to κ^{-1} but there seems to be little justification for also assuming β_i is constant.

A number of points deserve attention here. The first of these is that an additional assumption appears to be involved in obtaining equations (5) in that Millikan's argument is essentially an assumption of the experimentally observed law of the wall. The main idea involved is to match the quantity $y(\partial u/\partial y)$; the rationale is that this quantity is independent of the length scales of the inner and outer layer and hence that a separation-of-variables argument may be used. However this is true for any composite layer and if such a matching principle was used in general, one would be led to incorrect results when the leading terms for u from either side are independent of y. At the same time if the leading terms for u behaved algebraically in the matching zone, matching $y(\partial u/\partial y)$ would be an acceptable matching principle; however by assuming that the limit is nonzero as the matching region is approached, the "law of the wall" result is in reality assumed rather than derived. However, this in no way obviates the results of the paper; it simply means that an additional assumption appears to be required. Since the function σ in equation (1) is not known it is clearly unreasonable to expect that the asymptotic structure may be deduced without recourse to <u>any</u> assumptions. The assumptions in Fendell's (1972) paper all seem reasonable and substantiated by experiment; the work is important because it puts the equations in the framework of the rational mathematical context of the limit as $Re \to \infty$.

The second point is that there are few new results in Fendell's paper but that in the proper asymptotic context, the "classical" results of turbulence which are overwhelmingly suggested by experiment are confirmed in the limit $Re \to \infty$; indeed this is a stated purpose of the paper. An important feature is that the Reynolds number dependence of the inner and outer length scales and of u_τ is systematically obtained. Equation (2) is compared directly with zero pressure gradient data in Table I and the favorable comparison gives confidence in the result. This information is important insofar as the present analysis is concerned since it gives some guidance as to which types of time-dependent flows are possible and which are not.

Finally it should be mentioned that in the Fendell (1972) analysis and in other studies which have followed there is no attempt to resolve the closure problem. Although the order of magnitude of σ is fixed in each layer, the functions σ_1 and Σ_1 are unknown and to make progress here one must either guess at the functional form or attempt to understand the dynamics of the turbulent boundary layer. It is the second alternative which we begin to explore in the present paper.

3. THE QUIESCENT PERIOD

There have been a number of attempts to consider the time-dependent motion in the wall layer, most notably those by Einstein and Li (1956) and Black (1968). To some extent these have been motivated by the relatively ordered flow observed within the wall layer; there are a number of obvious criticisms that may be levelled at these studies and since this has occurred so many times in the literature this will not be done here. Nevertheless, there are a number of appealing features to these analyses which make them difficult to dismiss out of hand. Perhaps the most frequent objections that are made concern, first, the neglect of the convective terms and, second, failure to consider the three-dimensional motion that is experimentally observed. A prime consideration in the present paper, then, concerns the form of the equations governing the leading order flow.

An important first question is related to the Reynolds number dependence of the mean streak spacing in the wall layer. Most measurements of λ have been taken in a limited Reynolds number range and although there is a fair amount of scatter in the results, there is some support for the view that $\lambda^+ = \lambda u_\tau/\nu$ is a number on the order of 100. The main difficulty is that the physical mechanism which causes the streaks is not known. It is clear that while the wall layer is in the quiescent state (when the streaks are observed), only events which are occurring in the outer layer above can have an influence. Since the wake layer is dominated by large scale vortex structures, it seems reasonable to assume that the streaks are in some way connected with these eddies. Because the vertical dimension of the eddies must be comparable to δ, and hence $O(u^*)$, the spanwise and streamwise dimensions must be of $O(u^*)$ as well; otherwise as $Re \to \infty$, the streamwise and spanwise dimensions of an eddy would become either small or large with respect to the boundary layer thickness and neither of these alternatives seems reasonable. In Table I, zero pressure gradient data from Gupta et al. (1971) and

Table I. Zero Pressure Gradient Experimental Data for Increasing Re_δ

	Re_δ	u_τ/U_∞	$\kappa/\mathit{log}Re_\delta$	λ^+	λ/δ
Kline et al. (1967)	9,282	0.0424	0.0449	91	0.231
	11,936	0.0420	0.0437	106	0.211
Gupta et al. (1971)	21,129	0.0407	0.0412	97.5	0.112
	31,725	0.0394	0.0396	85.0	0.071
	43,750	0.0384	0.0384	109.7	0.065
	57,604	0.0375	0.0374	151.2	0.070

Kline et al. (1967) are summarized and arranged in order of increasing Re_δ. In the second and third columns the measured values of $u^* = u_\tau/U_\infty$ are compared with the asymptotic result (2) and agreement is increasingly satisfactory with increasing Re_δ. In addition, there appears to be a trend for λ^+ to increase but λ/δ to decrease to an approximately constant value. If $\lambda = O(u^*)$, then λ/δ should approach a constant but $\lambda^+ \to \infty$ as $Re_\delta \to \infty$. Although this trend appears to be supported in Table I, there is not enough data to be conclusive and moreover there is as yet no independent verification of the high Reynolds number results of Gupta et al. (1971). In any case, the only other alternative compatible with experiment is that λ^+ approaches a large positive constant as $Re \to \infty$.

If attention is now focused on an area of the plate, it is clear from observations of Kline et al. (1967) and Kim et al. (1971) that there is a period of time, during which the streaks are in place and no bursting occurs. This quiescent period of time is referred to as T_q; the period of time during which an actual burst and sweep takes place is referred to as T_B and so the period between bursts is $T = T_q + T_B$. During the quiescent period the integrity of the wall layer is intact and the normal length scale is therefore $O([Re\, u^*]^{-1})$; it follows from the momentum equations that T_q is $O([Re\, u^{*2}]^{-1})$, which is the time scale of the inner layer. The Reynolds number dependence of T_B is not known and this can only be answered by discovery of the precise mechanism that causes the burst. However, two possibilities exist; either T_B/T_q approaches zero or a small constant as $Re \to \infty$. Which limit is appropriate is not known but the visual observations of Kline et al. (1967) do reveal that $T_B \ll T_q$.

Turbulence intensity measurements indicate that all three intensities are of the same order of magnitude, and since the quiescent period comprises a majority of the total time within the wall layer, it would be incorrect to make an order of magnitude distinction at least between u and w. Consequently if (x,y,z) are cartesian coordinates measuring dimensionless distance in the streamwise, normal and spanwise directions, respectively, with corresponding dimensionless velocity components (u,v,w), then scaled velocities and coordinates are defined according to

$$u^+ = \frac{u}{u^*}, \quad v^+ = \frac{v}{u^*}, \quad w^+ = \frac{w}{u^*}, \tag{6}$$

$$x^+ = \frac{x}{\lambda^*}, \quad y^+ = Re\, u^* y, \quad z^+ = \frac{z}{\lambda^*}. \tag{7}$$

Here $\lambda^* = \lambda/L$ is the dimensionless mean streak spacing. The scale on the x-coordinate is motivated by substitution of (6) and (7) in the continuity equation

$$\frac{1}{\lambda^*}\frac{\partial u^+}{\partial x^+} + u^*\mathrm{Re}\,\frac{\partial v^+}{\partial y^+} + \frac{1}{\lambda^*}\frac{\partial w^+}{\partial z^+} = 0 \tag{8}$$

and is chosen so that the $\partial u/\partial x$ term balances the last term in equation (8) so that in what follows v is dependent on both u and w. The other possibility is that the length scale in the x-direction is much larger than λ^* in which case it emerges that v is dependent on w only. In either situation it will be found the dependence on x is parametric and enters only from the boundary conditions. It now follows from (8) that the three velocity components must be written to leading order during the quiescent period as

$$u^+ = u_0(x^+,y^+,z^+,t^+) + \cdots, \quad v^+ = \frac{1}{\mathrm{Re}\,u^*\lambda^*}\,v_0(x^+,y^+,z^+,t^+) + \cdots$$

$$w^+ = w_0(x^+,y^+,z^+,t^+) + \cdots \tag{9}$$

Substitution of (6), (7), and (9) into the Navier-Stokes equations leads to

$$\frac{\partial u_0}{\partial t^+} + \frac{1}{\mathrm{Re}\,u^*\lambda^*}\left\{u_0\frac{\partial u_0}{\partial x^+} + v_0\frac{\partial u_0}{\partial y^+} + w_0\frac{\partial u_0}{\partial z^+}\right\} = -\frac{1}{\mathrm{Re}\,u^{*3}\lambda^*}\frac{\partial p}{\partial x^+}$$
$$+ \frac{\partial^2 u_0}{\partial y^{+2}} + \frac{1}{\mathrm{Re}^2 u^{*2}\lambda^{*2}}\left\{\frac{\partial^2 u_0}{\partial x^{+2}} + \frac{\partial^2 u_0}{\partial z^{+2}}\right\} \tag{10}$$

$$\frac{\partial v_0}{\partial t^+} + \frac{1}{\mathrm{Re}\,u^*\lambda^*}\left\{u_0\frac{\partial v_0}{\partial x^+} + v_0\frac{\partial v_0}{\partial y^+} + w_0\frac{\partial v_0}{\partial z^+}\right\} = -\frac{\mathrm{Re}\,\lambda^*}{u^*}\frac{\partial p}{\partial y^+}$$
$$+ \frac{\partial^2 v_0}{\partial y^{+2}} + \frac{1}{\mathrm{Re}^2 u^{*2}\lambda^{*2}}\left\{\frac{\partial^2 v_0}{\partial x^{+2}} + \frac{\partial^2 v_0}{\partial z^{+2}}\right\} \tag{11}$$

$$\frac{\partial w_0}{\partial t^+} + \frac{1}{\mathrm{Re}\,u^*\lambda^*}\left\{u_0\frac{\partial w_0}{\partial x^+} + v_0\frac{\partial w_0}{\partial y^+} + w_0\frac{\partial w_0}{\partial z^+}\right\} = -\frac{1}{\mathrm{Re}\,u^{*3}\lambda^*}\frac{\partial p}{\partial z^+}$$
$$+ \frac{\partial^2 w_0}{\partial y^{+2}} + \frac{1}{\mathrm{Re}^2 u^{*2}\lambda^{*2}}\left\{\frac{\partial^2 w_0}{\partial x^{+2}} + \frac{\partial^2 w_0}{\partial z^{+2}}\right\} \tag{12}$$

Here $t^+ = u^{*2}\mathrm{Re}\,t$ is the scaled time whose form follows by insisting that the unsteady terms in the momentum equations balance the normal viscous stress (any other choice for t^+ leads to a contradiction). Note that the quantity $Re\, u^*\lambda^*$ is the conventionally defined local value of λ^+.

It now follows from (11) that the normal pressure gradient is

$$\frac{\partial p}{\partial y^+} = O(u^*/Re\lambda^*) = O(u^{*2}/\lambda^+) \qquad (13)$$

and as $Re \to \infty$, there is no variation in pressure across the wall layer to leading order during the quiescent period. Consequently the dimensionless pressure within the wall layer may be written as:

$$p = p_\infty(x) + p'(x^+,\infty,z^+,t^+;Re) + \frac{u^*}{Re\,\lambda^*}\, p_0(x^+,y^+,z^+,t^+) + \cdots$$

Here p_∞ is the mainstream pressure, p' is the fluctuating pressure due to unorganized or random motion at the edge of the wall layer, and p_0 is a lower order term due to the organized motion within the wall layer. It is clear from (10) and (12) that the lower order term p_0 will not enter the x or z momentum equations to leading order as $Re \to \infty$. Consideration of the fluctuating pressure term p' in (10) and (12) is somewhat more difficult. The data of Emmerling (1973) provide conclusive evidence of intense localized centers of pressure within the turbulent boundary layer and large fluctuating pressure gradients may not be ruled out entirely. At the same time if such a large pressure fluctuation is convected into the region under consideration, above the wall layer, a breakdown or burst might be expected. On the other hand, lower order pressure gradient fluctuations are possible if they are truly random with a zero time-mean average. Furthermore, it may be concluded from the observed well ordered motion in the wall layer during the quiescent period that it is unreasonable to expect that small scale unorganized pressure fluctuations to dominate the wall layer flow. Note, however, that this view is contrary to the premise underlying the Van Driest (1956) model, although it should be remembered that in the derivation of this model a troublesome cosine term is neglected in an ad hoc manner. It may be shown (Walker et al., 1976) that subsequent modifications of the Van Driest model to include pressure gradient can lead to erroneous formulations that are contrary to experimental evidence.

To leading order, then, the form of the Navier-Stokes equations becomes:

$$\frac{\partial u_0}{\partial t^+} = -p^+ + \frac{\partial^2 u_0}{\partial y^{+2}}, \quad \frac{\partial v_0}{\partial t^+} = -\frac{\partial p_0}{\partial y^+} + \frac{\partial^2 v_0}{\partial y^{+2}}, \quad \frac{\partial w_0}{\partial t^+} = \frac{\partial^2 w_0}{\partial y^{+2}}$$

$$\frac{\partial u_0}{\partial x^+} + \frac{\partial v_0}{\partial y^+} + \frac{\partial w_0}{\partial z^+} = 0 \tag{14}$$

Here, the mainstream pressure gradient term has been defined in the usual way

$$p^+ = \frac{1}{u^{*3}Re}\frac{dp_\infty}{dx}$$

and the p' terms have been omitted for the reasons just discussed. Note that in most situations p^+ is negligibly small, becoming O(1) only when $dp_\infty/dx = O(u^{*3}Re)$; with Re fixed and large, this can occur as $u^* \to 0$ or specifically when $u^* = O(Re^{-1/3})$. This latter case is only possible in an adverse pressure gradient and as a flow approaches separation. In the following analysis, it is only important to recognize that p^+ is at most O(1) and it may be retained without complication.

With attention now focused on equations (14) the question that immediately arises is, which possible solutions of these equations are compatible with theory and experiment. Since the dominant term in the time-mean profile behaves like $\log y^+$ as $y^+ \to \infty$, all solutions for u_0 and w_0 which behave like $y^{+\alpha}$ with $\alpha > 0$ must be discounted; the reason for this is associated with well verified measurements for the turbulence intensities $\overline{u'^2}$ and $\overline{w'^2}$ which show these quantities approaching a constant at the edge of the wall layer. A time-dependent solution for u_0 which has $u_0 \sim y^{+\alpha}$ as $y^+ \to \infty$, for $\alpha > 0$, would result in a calculated time average of u'^2 over the quiescent period which would be algebraically increasing at the wall layer edge and this is incompatible with experiment.

During the quiescent period, the streaks are observed to be in place and relatively stable; the consequence of this is that there are space curves in the x^+z^+ plane across which there is no significant motion. Since the wall layer is passive during the quiescent period, the streaks are a manifestation of lines of stagnation imposed by events which occurred or are occurring in the outer flow. The cause of the streaks is difficult to explain because the precise nature of these events in the outer layer is not known. However, it has been suggested by a number of authors that the wake layer is dominated by the motion of large scale eddies which bear a resemblance to the turbulent spots observed in transition. Since streaks have been observed under transition spots this seems a reasonable point of view. Further, since such outer flow structures or eddies are observed to move within the wake layer at essentially constant speed, it would seem reasonable to expect that movement of the streaks within the wall layer will be related to the large eddy motion. With this hypothesis, consider the possible induced motion of the streaks during the quiescent period. Suppose that the velocity of the eddy is AU_∞ where A is O(1). Then on the time scale of the wall layer, which is $O(1/u^{*2}Re)$, the eddy will move a streamwise distance $D_S = O(U_\infty/u^{*2}Re)$. Since the normal and spanwise components of velocity are at most $O(u^*)$ within the wake layer, then on the time scale of the wall layer, fluid particles in the wake layer will move a lateral distance $D_L = O(1/Re\, u^*)$. Thus D_S/λ^* and D_L/λ^* become small as $Re \to \infty$ and on the time scale of the wall layer the streaks are frozen in place.

The simplest type of structure is that where the streaks are straight lines over an area of the plate having spanwise and streamwise dimensions $O(\lambda^*)$; this assumption about the streaks is not completely compatible with experiment and is not really necessary. However it is instructive to consider this case first and the discussion of curved streaks will be given elsewhere. Suppose that the dimensionless spanwise distance between the i and i + 1 streak is λ_i and let z^+ be measured from the left edge of the λ_0 streak. Then during a quiescent period the solution in the λ_0 streak may be written as a Fourier series

$$\begin{aligned} u_0 &= \sum_1^\infty h_n(x^+,y^+,t^+)\cos n\pi z^+/\lambda_0 + h_0(y^+,t^+) \\ v_0 &= \sum_1^\infty g_n(x^+,y^+,t^+)\cos n\pi z^+/\lambda_0 \\ w_0 &= \sum_1^\infty f_n(x^+,y^+,t^+)\sin n\pi z^+/\lambda_0 \end{aligned} \tag{15}$$

With corresponding expansions within adjacent streaks, such a formulation ensures continuity of velocity components and their z^+ derivatives across each streak and allows a different flow development within each pair of streaks as the quiescent period progresses. In writing down equations (15), terms arising from small scale random fluctuations have been omitted on the grounds that if such motion really is present it cannot be a dominant feature because of the relatively ordered flow observed within the wall layer; in any case such motion will not produce any contribution to the mean profile.

Substitution of equations (15) into (14) leads to heat conduction type equations for all functional coefficients of the Fourier series in (15) and we seek solutions of these which lead to an orderly flow development during the quiescent period. A complete discussion of such solutions will appear elsewhere and only the results will be summarized here. Briefly the solutions for f_n and h_n are of the form

$$b_\eta(x^+)F(\eta,t^+) \tag{16}$$

where $\eta = y^+/2(t^+ + t_0^+)^{\frac{1}{2}}$ and t_0^+ is a parameter corresponding to an uncertainty in the time origin; because of the form of equations (14) the variable x^+ appears parametrically in (16). There are four possible types of solutions which vanish at $y^+ = 0$ and which have

(i) $F \sim \mathit{log} y^+$, (ii) $F = 0(1)$

(iii) F exponentially small, (iv) $F = 0(y^{+^{-\alpha}})$, $\alpha > 0$

as $y^+ \to \infty$. Of these possible solutions, for $\eta \geq 1$ turbulence intensity measurements preclude a solution of type (i) for f_n and h_n. It may be shown that for $n \geq 1$ solutions of type (ii) are of the form

$$b_\eta(x^+)\mathit{erf}\,\eta \tag{17}$$

This type of three-dimensional motion is fixed by a knowledge of the three-dimensional motion at the outer edge of the wall layer and, as the measurements of Kline and Schraub (1965) show, can have an amplitude up to 50% of the mean profile at any y^+ station. The solutuions indicated by types (iii) and (iv) represent eigenfunctions whose strength decays with time and for which the $b_\eta(x^+)$ in (16) may be chosen to represent the initial profiles at $t^+ = 0$, the beginning of the quiescent period. The details of these solutions have been analyzed but because of their complexity will be reported elsewhere.

Turning now to the mean velocity profile, the contribution from the instantaneous velocity during the quiescent period may be obtained according to

$$u^+ = \frac{\overline{u}}{u_\tau} = \frac{1}{T_q}\int_0^{T_q} u^+ dt \tag{18}$$

However, the complete period between bursts in a given area of the x^+z^+ plane is $T = T_q + T_B$. Since experimental evidence suggests that $T_B << T_q$, then (18) represents an approximation to the mean streamwise profile in the wall layer. Also since the streaks are observed to rearrange themselves after each burst in what appear to be random locations over long periods of time, the z-dependent motion in (15) will not contribute to the mean profile in (18) and the total contribution must come from the term $h_0(y^+,t^+)$. Furthermore, there are three conditions which the mean profile must satisfy:

$$U^+ \sim \frac{1}{\kappa}\, log y^+ + C(x^+) \qquad \text{as} \quad y^+ \to \infty, \tag{19}$$

$$\frac{\partial U^+}{\partial y^+} = 1, \quad \frac{\partial^3 U^+}{\partial y^{+3}} = 0 \qquad \text{at} \quad y^+ = 0, \tag{20}$$

and to satisfy (19) motion of type (i) must be included. The complete results will not be given here for either the time-dependent motion or the mean profile, but they are very similar to those reported by Walker et al. (1976). The time-dependent development of the instantaneous streamwise velocity function $h_0(y^+,t^+)$ for a zero pressure gradient flow is given in Figure 1; it should be stressed that during a quiescent period, in addition to the motion

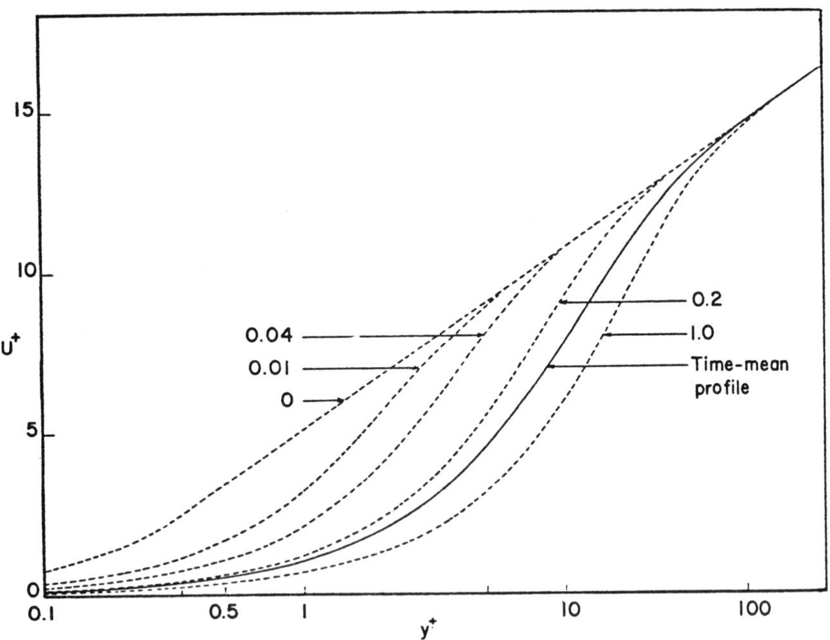

Figure 1. Instantaneous Streamwise Profiles During Quiescent Period (Labels correspond to different values of t/T).

in Figure 1, motion of type (ii) given in (17) as well as the eigenfunction solutions of types (iii) and (iv) will be superimposed and this can represent excursions away from the instantaneous profiles of Figure 1 which may be up to 50% of the mean. The type of relaxing flow illustrated in Figure 1 is the same type of relaxing flow observed by Kim et al. (1968) and Blackwelder and Kaplan (1972) between bursts. No attempt has been made here to compare instantaneous profiles with Blackwelder and Kaplan's (1972) data since it has been conditionally averaged and the interpretation of such results, in a flow which is cyclic but not periodic, is unclear.

The mean profile which is also given in Figure 1 is in general a function of H and p^+, where $H = y^+/2S$ and $S = u_\tau\sqrt{T_q/\nu}$. Note also that as $y^+ \to \infty$,

$$U^+ \sim \frac{1}{\kappa}\, log y^+ + C(p^+, S) \tag{21}$$

where the log law "constant" is not, in general, a constant. Thus if we set $T_q = T$ as an approximation, the mean profile is a function of the two parameters u_τ and T. The difficulty with comparing the mean profile is that, while both parameters are physical features of the flow, T is in general not known and there are no direct measurements of u_τ. Consequently, with the view of verifying whether the present mean profile is capable of representing turbulent boundary layer data, these parameters were optimized to obtain the best fit in the least squares sense to experimental profile data. In Figure 2, the mean velocity profile is used in combination with the Coles' law of the wake to define a composite profile; this is a three parameter fit (u_τ, T, δ) to the zero pressure gradient data of Anderson et al. (1972) and represents an extremely close fit having an average root mean square error of 0.006.

A question which naturally arises is whether any profile which has the asymptotic form given by (21) would be just as good. The profile given by Van Driest (1956) falls into this latter class and since the Andersen et al. (1972) data contain a considerable number of wall layer points, it provides a convenient data set to answer the question. Detailed comparisons of both the present and the Van Driest inner region profiles are given in Scharnhorst et al. (1976) and for all cases the present profile gives a better fit to the data; in most cases the difference in the root mean square error is substantial, being typically twice as large with the Van Driest profile. A number of other data sets with pressure gradients are considered by Scharnhorst et al. (1976) and the conclusions there are the same.

There are certain objections to employing Coles' law of the wake, namely that while it represents a useful device for curve fitting zero and adverse pressure gradient data; (i) it contains the ill-defined

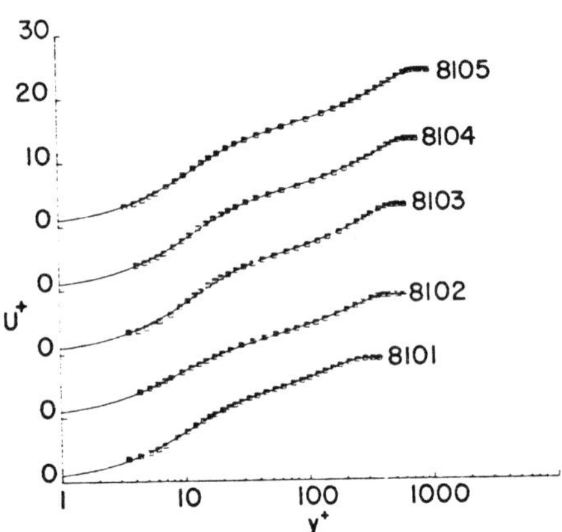

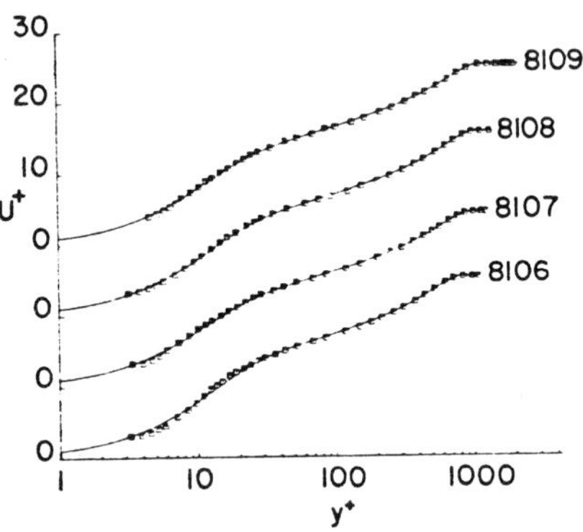

Figure 2. Comparison of Composite Profile Consisting of Present Inner Region Profile and the Law of the Wake with Data of Andersen et al. (1972), (data stations are labeled in increasing order in the streamwise direction).

boundary layer thickness δ, (ii) it fails to represent data in moderately accelerated flows, and (iii) the particular functional form assumed appears to be incompatible with the time-mean equations in the outer layer (see also the discussion by Fendell, 1972). An alternative outer region profile may be obtained from the work of Mellor and Gibson (1966); this is developed by assuming an eddy viscosity in the outer layer proportional to the displacement thickness δ^*. The constant of proportionality is denoted by K and taken equal to 0.0168 as in the Cebeci-Smith (1974) model. Upon assuming similarity, an ordinary differential equation may be obtained for the outer region profile (see Fendell, 1972) and when this is used with the present inner region profile, a one parameter (u_τ) profile results. This is compared with the Andersen data in Figure 3, where u_τ has been varied to obtain the optimum fit and where the variations from the experimentally quoted values of u_τ are slight. Such a procedure only makes sense in a truly similar flow, and since similarity may only be achieved in an equilibrium boundary layer at large distances downstream, the trend in Figure 3 is encouraging. In Figure 4, the constant of proportionality K is varied to obtain the optimal fit and this composite profile fits the data almost as well as the law of the wake, but with one less parameter.† Moreover, this type of profile represents accelerated flow data much more closely than the law of the wake, and in cases where the law of the wake fails (Scharnhorst et al., 1976).

An independent check on the inner region profile is also provided in the following way. The Fendell (1972) analysis shows that to leading order the convective terms are negligible in the time-mean equations and consequently a knowledge of the mean profile implies a knowledge of the $\overline{u'v'}$ term and vice versa, so that

$$0 = -p^+ + \frac{\partial^2 U^+}{\partial y^{+2}} - \frac{\partial}{\partial y^+}\left\{\frac{\overline{u'v'}}{u^{*2}}\right\} \tag{22}$$

Equation (22) represents the leading order time averaged streamwise momentum equation in the wall layer and it is of interest to examine where the contribution to the $\overline{u'v'}$ product comes from. Since $\overline{u'v'}$ must be $O(u^{*2})$, it is clear from equation (9) that during the quiescent period the v component of velocity is too small to contribute to leading order. Thus the major contribution must come from the bursting process, a result which is compatible with experiment (Nychas et al., 1973). At the same time it has been argued that the major contribution to the mean profile comes during the quiescent period and consequently an approximation

---

† The displacement thickness δ^* is taken from experiment.

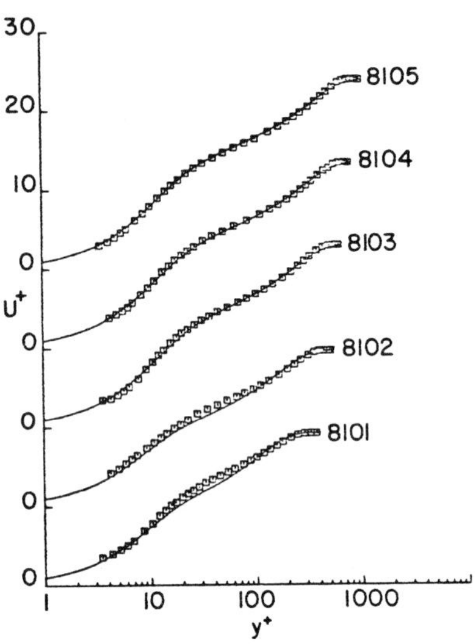

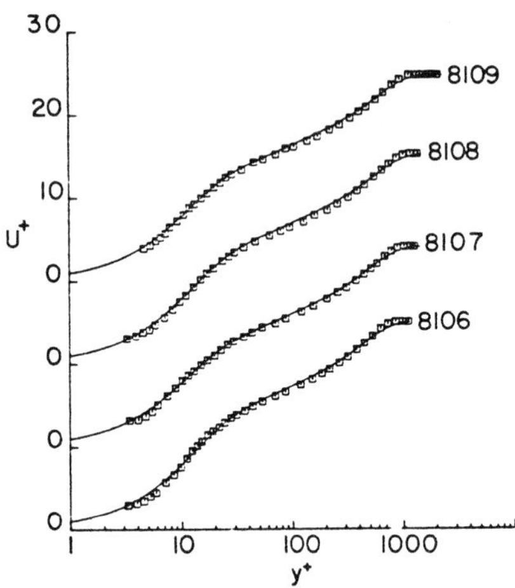

Figure 3. Comparison of Composite Profile Consisting of Present Inner Region Profile and Outer Profile of Mellor and Gibson (1966); K = 0.0168 and labels are as in Figure 2.

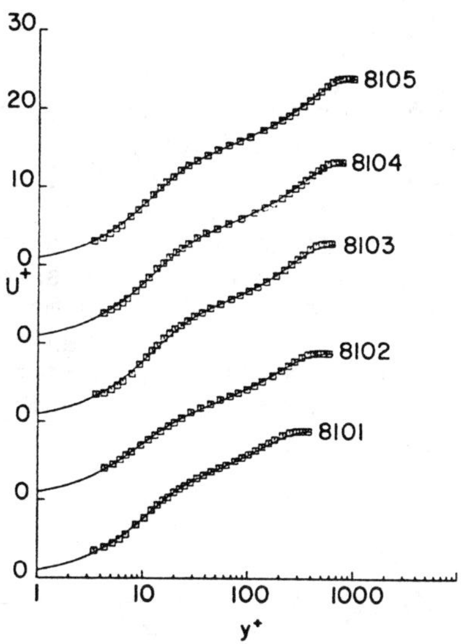

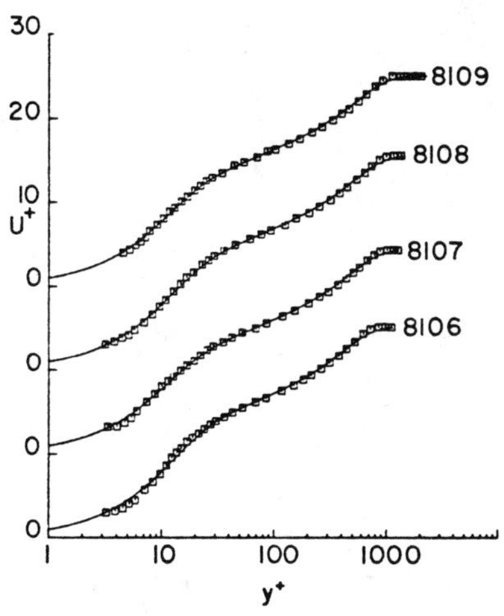

Figure 4. Comparison of Composite Profile Consisting of Present Inner Region Profile and Outer Profile of Mellor and Gibson (1966) with K Varied.

$$-\frac{\overline{u'v'}}{u^{*2}} = 1 + p^+y^+ - \frac{\partial U^+}{\partial y^+} \tag{23}$$

to the $\overline{u'v'}$ product may be obtained by integrating equation (22). Equation (23) is plotted in Figure 5 for $p^+ = 0$; the right side of (23) is a function only of S and included in Figure 5 are plots for S = 8, 12 and 10.67. These values of S are the smallest and largest values of S obtained in fitting zero pressure gradient data (Scharnhorst et al., 1976) and the value which gives a log law constant of 5.1, respectively. The value of S = 10.67 is most typical of the values obtained in the fitting. The data given in Figure 5 were obtained by Schubauer (1954) and although there is some scatter in the data, agreement with the theory is encouraging.

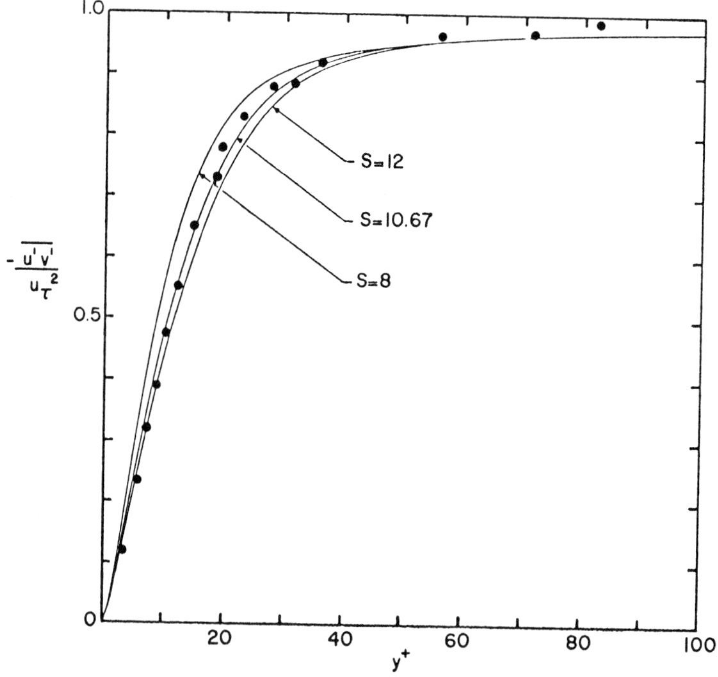

Figure 5. Comparison of Present Theory with Data of Schubauer (1954).

4. A POSSIBLE MECHANISM FOR THE BURST

Although it was possible to make some progress in the latter section for the wall layer flow, the theoretical ideas are highly motivated by experimental observation and many questions remain. In particular it is not possible to treat the burst itself without knowing what physical mechanism is responsible and a major difficulty is that very little is known about the dynamics of the outer layer. Although the outer layer is inviscid in character, experiments reveal that a substantial amount of vorticity is present principally in the form of three-dimensional eddy structures. At least part of the vorticity in the outer layer appears to have its genesis in the bursting phenomenon but although the study by Nychas et al. (1973) suggests that the eruptions of the wall layer are in some way associated with the passage of a large scale structure in the outer layer, the entire regenerative process is not understood.

It has been suggested by some authors that the bursting phenomenon is a hydrodynamic instability of the Kelvin-Helmholtz type (see Bark, 1975); that is, if it is assumed that on either side of a streak there are streamwise vortices which are counter-rotating during the quiescent period, then on the upflow plane (streak) it is hypothesized there is a shear layer which eventually develops an instability. Although this is an interesting suggestion, there is little evidence to support it and the view that this is the dominant mechanism for the burst must be regarded as conjectural. Another possibility has been recently suggested by Coles and Barker (1975). Based upon speculation by a number of authors that the motion in the wake layer is dominated by large U-shaped vortex structures similar to turbulent transition spots, Coles and Barker (1975) created artificial spots through the use of intermittent jets emanating from the wall below a laminar boundary layer. Upon assuming complete reproducibility, measurements were taken at various points and the outline of the artificial spot appears to be a U-shaped structure. Because of limitations of the measuring apparatus, detailed instantaneous streamline patterns within the spot cannot be obtained and the internal motion is conjectured to be that of a simple U-shaped vortex. There is some uncertainty as to the nature of the flow near the trailing legs of the vortex. Certainly, if the vortex is imagined to be made up of vortex tubes, these cannot close on the wall as is sometimes suggested (in view of the no-slip condition). The difficulty here is that the motion of such vortex structures is not understood. It is clear, however, that fluid will be forced up between the observed trailing arms. Coles and Barker (1975) suggest that this upwelling behind the leading edge of the vortex is the bursting. This type of continuous process does not seem compatible with a considerable body of experimental evidence which strongly suggests that the burst is a highly localized breakdown of the wall layer flow.

The problem of concentrated vorticity and how it interacts with a solid boundary is an important subject in its own right, about which relatively little is known on a theoretical basis. As a first step the problem investigated here represents the simplest situation, namely a rectilinear line filament in motion above a plane wall. It is emphasized at the outset that no attempt is being made here to model the turbulent boundary layer with line vortices. The treatment was motivated in part by a model discussed by Willmarth (1975), which appeared to give a qualitative correspondence with wall pressure fluctuations in a turbulent boundary layer, and also by the view that, before the boundary layer motion created by complex three-dimensional structures could be understood, it was important to understand the simplest possible case.

Consider the situation illustrated in Figure 6 where a rectilinear vortex of positive rotation κ is located a distance, a, above an infinite plane wall. The inviscid theory for this configuration is discussed by Milne-Thomson (1962, p. 359) and predicts that the vortex will move, under the influence of the image vortex below the plate, to the right with constant speed $V = \kappa/2a$. The inviscid solution is not uniformly valid since as $y^* \to 0$, the tangential velocity is given by

$$u^* \to \frac{4a^2V}{(x^* - Vt^*)^2 + a^2}$$

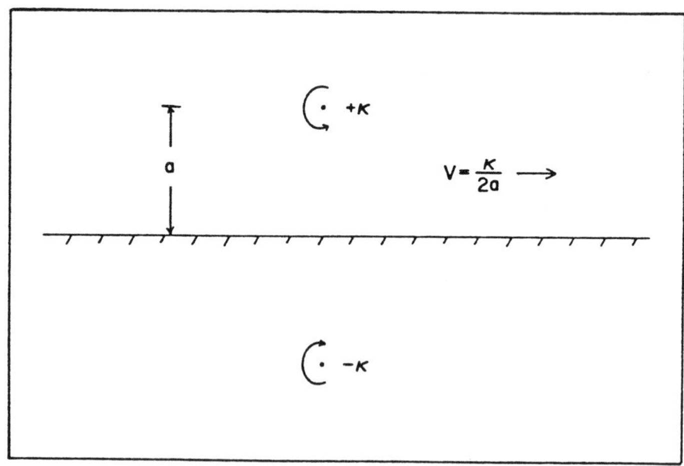

Figure 6. Geometrical Configuration for a Rectilinear Vortex in Motion above a Plane Wall.

where the vortex is located at $x^* = 0$, $y^* = a$ at time $t^* = 0$. Consequently a thin unsteady boundary layer near the plate is required in order to reduce the flow to relative rest. In the laboratory frame of reference the inviscid motion is unsteady but in a frame of reference which moves uniformly with the vortex, the inviscid motion is steady and the relative flow pattern is sketched in Figure 7. Redefining (x^*,y^*) as cartesian coordinates relative to the vortex, with corresponding velocity components (u^*,v^*), it may be observed in Figure 7 that there are two stagnation points in the relative inviscid flow. The first of these at $(\sqrt{3}\,a,0)$ will be referred to as the leading relative stagnation point and is characterized by flow away from the wall; the trailing relative stagnation point at $(-\sqrt{3}\,a,0)$ is characterized by flow toward the wall.

Dimensionless boundary layer variables are now defined according to

$$x = \frac{x^*}{a}, \qquad y = \frac{y^* Re^{\frac{1}{2}}}{a}, \qquad t = \frac{Vt^*}{a}$$

$$u = \frac{u^*}{V}, \qquad v = \frac{v^* Re^{\frac{1}{2}}}{V},$$

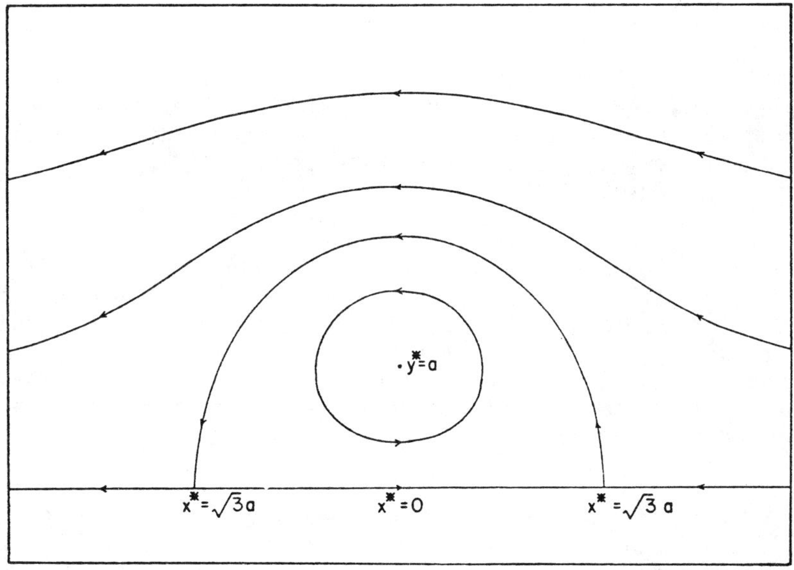

Figure 7. Inviscid Flow Relative to the Vortex (in this moving coordinate system the wall moves to the left with speed V).

where Re is the Reynolds number defined as Re = Va/ν and ν is the kinematic viscosity. The laminar boundary-layer equations are

$$\frac{\partial u}{\partial t} + u\frac{\partial u}{\partial x} + v\frac{\partial u}{\partial y} = U_\infty\frac{dU_\infty}{dx} + \frac{\partial^2 u}{\partial y^2} ,$$

$$\frac{\partial u}{\partial x} + \frac{\partial v}{\partial y} = 0 , \qquad (24)$$

where

$$U_\infty(x) = -1 + \frac{4}{1 + x^2}$$

and the boundary conditions associated with (24) are

$$u = -1 , \quad v = 0 \quad \text{at} \quad y = 0 , \quad u \to U_\infty(x) \quad \text{as} \quad y \to \infty . \qquad (25)$$

The solution of the boundary-layer problem is discussed in detail by Walker (1976) and only a brief summary of the results will be given here.

Because the inviscid motion in the moving reference frame is steady, it is natural to seek a steady boundary-layer solution to the system (24). Surprisingly, however, it has been demonstrated by Walker (1976) that there is no steady solution to the problem. Consequently the flow in the boundary layer is unsteady, even in the moving reference frame. In order to understand the nature of this resulting flow, the following problem is formulated: for $t < 0$ a vortex pair moves to the right with speed V in an unbounded fluid and at $t = 0$ an infinite plate is suddenly inserted on the symmetry plane of the two vortices. Although the inviscid flow is not affected by the sudden insertion of the plate, the inviscid flow is no longer uniformly valid for $t > 0$ and a thin unsteady boundary layer will develop on the plate in order that the no-slip condition be satisfied. An alternative physical situation, which may seem less artificial in the experimental context and which leads to the same boundary-layer development, is that where a rectilinear vortex is imagined to be created abruptly at $t^* = 0$ at a distance a above a plane wall.

Rather than proceed with the coordinate x, it is convenient to introduce a new streamwise coordinate ξ, defined by the Görtler type transformation

$$\xi = \frac{1}{2\pi}\int_x^\infty \frac{4dx}{1 + x^2} = 1 - \frac{2}{\pi}\tan^{-1}x \qquad (26)$$

whereupon the mainstream velocity becomes

$$U(\xi) = -1 + U_e(\xi), \qquad U_e(\xi) = 2(1 - \cos\pi\xi).$$

This transformation is one-to-one and compresses the doubly infinite range of x to the finite range [0,2] for ξ; the effect of this transformation is summarized in Table II for some of the critical x-locations in the flow.

An unsteady stream function $\psi(\xi,y,t)$ is now defined by

$$u = 1 + U_e(\xi)\,\frac{\partial\psi}{\partial y}, \qquad v = \frac{1}{2\pi}\,U_e(\xi)\,\frac{\partial}{\partial\xi}\,[U_e(\xi)\psi]$$

and Rayleigh variables are introduced according to

$$\eta = y/2\sqrt{t}\,, \qquad \psi = 2\sqrt{t}\,\Psi(\xi,\eta,t).$$

The boundary layer equations become,

$$\frac{\partial^3\Psi}{\partial\eta^3} + 2\eta\,\frac{\partial^2\Psi}{\partial\eta^2} - 4t\,\frac{\partial^2\Psi}{\partial\eta\partial t} = 4t\left[\sin\pi\xi(1 - 2\cos\pi\xi) + \frac{1}{2\pi}\,\frac{\partial}{\partial\xi}\left[U_e\,\frac{\partial\Psi}{\partial\eta}\right]\right.$$
$$\left. + \frac{1}{2\pi}\,U_e^2\left[\frac{\partial\Psi}{\partial\xi}\,\frac{\partial^2\Psi}{\partial\eta^2} - \frac{\partial\Psi}{\partial\eta}\,\frac{\partial^2\Psi}{\partial\xi\partial\eta}\right] + \frac{1}{2\pi}\,U_eU_e'\left[\Psi\,\frac{\partial^2\Psi}{\partial\eta^2} - \left(\frac{\partial\Psi}{\partial\eta}\right)^2\right]\right] \tag{27}$$

The boundary conditions are

$$\Psi(\xi,0,t) = \frac{\partial\Psi}{\partial\eta}\,(\xi,0,t) = 0, \qquad \frac{\partial\Psi}{\partial\eta}\,(\xi,\eta,t) \to 1 \text{ as } \eta \to \infty \tag{28}$$

Table II. Effect of Transformation (26)

x	ξ	Streamwise Location
$+\infty$	0	upstream infinity
$\sqrt{3}$	$1/3$	leading relative stagnation point
0	1	vortex center
$-\sqrt{3}$	$5/3$	trailing relative stagnation point
$-\infty$	2	downstream infinity

and since the right side of equation (27) vanishes as $\xi \to 0,2$ the boundary condition at $\xi = 0,2$ compatible with (28), is $\partial\Psi/\partial\eta = \mathit{erf}\eta$. The initial condition at $t = 0$ for all ξ and η follows by letting $t \to 0$ in (27) and is also $\partial\Psi/\partial\eta = \mathit{erf}\eta$ for all ξ. The object now is to trace the boundary-layer development forward in time.

An exact solution, in the form of a power series in time, may be found with a procedure similar to that used by Goldstein and Rosenhead (1936) to calculate the initial flow past an impulsively started circular cylinder; details are given by Walker (1976). This time series solution is only valid for small time and to extend the solution to higher times equation (27) was integrated numerically; this is discussed by Walker (1976). In order to plot the instantaneous streamlines a stream function $\bar{\psi}$ is defined by $u = \partial\bar{\psi}/\partial y$ where $\bar{\psi}$ is related to Ψ by

$$\bar{\psi} = 2\sqrt{t}\, F(\xi,\eta,t); \qquad F = -\eta + U_e(\xi)\Psi(\xi,\eta,t)$$

and consequently a plot of lines of constant F gives the instantaneous streamline patterns. It is worthwhile to point out that in the following figures, the streamline patterns are relative to the vortex. Thus the vortex is towing the developing boundary-layer patterns with it as it moves along the plate.

In Figure 8 the instantaneous streamline patterns are plotted at $t = 0.2$ where the direction of flow is indicated by arrows. For $0 \leq \xi \leq 0.33$, inflow from upstream infinity and upflow toward the leading relative stagnation point may be observed, while for $1.67 \leq \xi \leq 2$, the motion is downward and toward downstream infinity. Between the relative stagnation points the pattern is characterized by an inflow-outflow behavior.

As the solution proceeds in time, an interesting feature develops. Specifically, at $t_s = 0.278$ and $\xi_s = 0.592$, $y_s = 0.655$, flow separation occurs, where separation is defined here to imply the presence of a closed recirculating eddy in the flow field. At a later stage the wall shear vanishes ($t_r = 0.324$), and subsequently becomes negative in the region under the developing eddy. The development of the resulting eddy is typical of that observed in the classical separation behind bluff bodies (see, for example, Dennis & Walker, 1971, 1972); once the eddy first appears, it develops rapidly in the direction tangential (parallel) to the wall. During the early period of time, there is some development in the direction normal to the wall. However, once the rate of tangential development begins to slow, there is a noticeable acceleration in the rate of growth of the normal dimension of the eddy. The situation is illustrated in Figure 9 where the instantaneous streamlines at $t = 0.6$ are plotted. The eddy lies beneath and to the left of

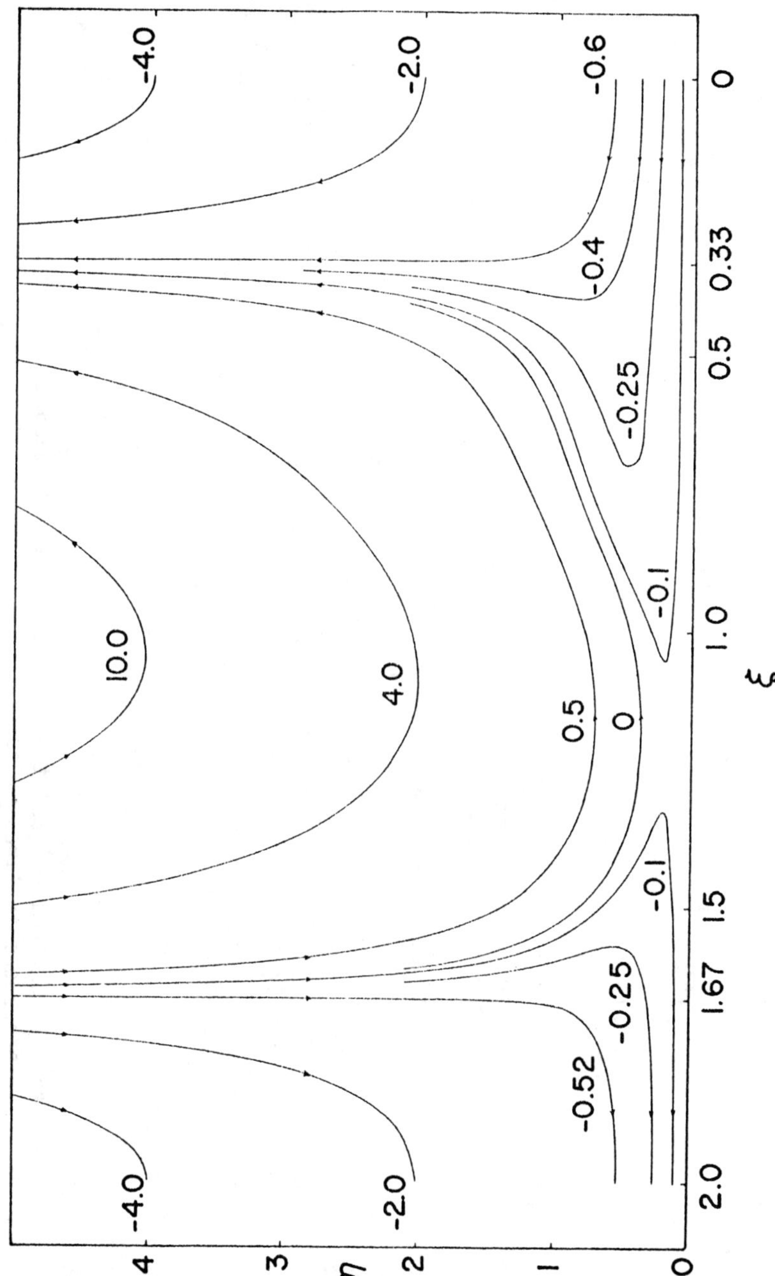

Figure 8. Instantaneous Streamlines Relative to the Vortex in the Boundary Layer at t = 0.2 (Labels correspond to lines of constant F).

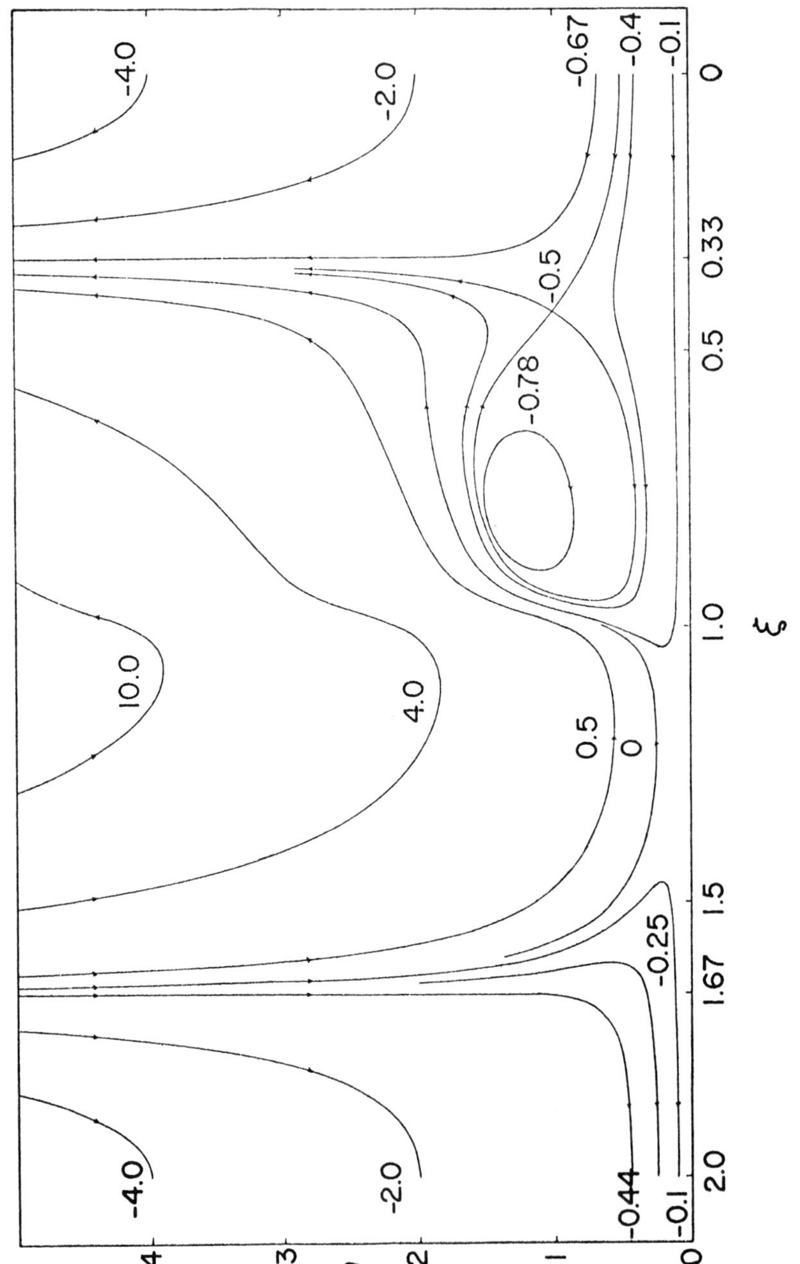

Figure 9. Instantaneous Streamlines Relative to the Vortex in the Boundary Layer at $t = 0.6$ (Labels correspond to lines of constant F).

the leading relative stagnation point. There are two stagnation points associated with the eddy: one at the eddy center and one on the limiting streamline which defines the eddy edge. At this stage of its development, the rate of increase of the eddy's streamwise dimension is decaying but the eddy is beginning to grow substantially in the direction normal to the wall, and at the same time it is lifting off the wall.

Finally, it was found that at approximately t = 0.675, the numerical integration failed to converge despite varied attempts to continue it further. A similar situation occurs with the numerical solution of boundary-layer flow past impulsively started bluff bodies (Dennis & Walker, 1972; Collins & Dennis, 1973) and this is discussed in detail by Walker (1976). The underlying reasons for this behavior in the bluff body case is thought to be associated with the form of the large time solution in the boundary layer near the rear stagnation point. The work of Proudman and Johnson (1962) and Robins and Howarth (1972) shows that for large time, the structure of the boundary-layer is such that to leading order there is (i) a thin inner viscous layer near the wall and (ii) an outer layer which is inviscid in character and which thickens exponentially with time. In this connection, Riley (1975) argues that as $t \to \infty$ ($t = O(\log \mathrm{Re})$) an eruption of the boundary-layer is to be expected in the vicinity of the rear stagnation point, along with a major modification of the external inviscid flow.

The Robins and Howarth (1972) analysis has been extended by Walker (1976) to consider the case of a stagnation point in motion with constant velocity above a plane wall. For a stagnation point where the inviscid flow is away from the wall it is found that the boundary-layer structure is very similar to that for a two-dimensional bluff body; that is for large time and in a frame of reference moving with the stagnation point, the boundary layer assumes a double structure. In particular, there is an outer region of the boundary-layer which to leading order is inviscid in character and which thickens exponentially with time. Presumably this is the reason for the difficulty experienced with the present numerical solutions at about t = 0.675; the flow is entering a phase where the exponential thickening of the boundary-layer is commencing and the outer boundary (where the mainstream velocity is imposed in the numerical scheme as an approximation) cannot be moved out quickly enough to take into account the exponential growth.

Although the details of the present large time solution for the vortex flow are not known, it can be said with a degree of confidence that an eruption of the boundary layer in the moving frame of reference is to be expected for large time and it is tempting to speculate on what should eventually transpire. Since the eddy is increasing rapidly in its normal dimensions and moving away from the wall, it appears plausible that for large time the eddy will emerge intact from the boundary-layer. Further, since the eddy is rotating in the opposite direction to the vortex and since the eddy is not symmetrical with respect to the vortex, it might be expected that one modification to the inviscid flow may be that the vortex itself will be driven away from the wall.

The rapid occurrence of this phenomenon, and the explosive nature of the boundary-layer flow discussed here, bear a striking resemblance to the bursting phenomenon observed in turbulent boundary layers. At this stage it would be premature to claim anything more than this; however there are a number of points worth considering. First, this example shows for the first time how concentrated vorticity can result in a breakdown, and upward eruption, of boundary-layer flow. In addition, it is important to properly visualize such a flow such that the experimental observations should be done in a frame of reference moving with the vortex structure. A plot of the instantaneous streamlines in the laboratory reference frame at t = 0.6 is given in Walker (1976) and the streamline patterns appear quite different and tend to mask the true physical features of what is transpiring within the boundary-layer. In addition, it is suggested that a problem which may be more relevant to the turbulent boundary-layer than the example treated here is that where a vortex of negative rotation is convected in a uniform flow; in a frame of reference which moves with the vortex, the inviscid flow is similar to that illustrated in Figure 7 except that the direction of flow is reversed. Consequently the upflow stagnation point is now the trailing relative stagnation point. The boundary-layer development that follows the sudden insertion of the plate is not identical to the present problem and will be reported elsewhere. However a dominant feature of this flow is the upflow relative stagnation point which ultimately leads to a boundary-layer eruption behind the vortex. Of course the eddy motion in the outer region of turbulent boundary layers is not two-dimensional but is highly three-dimensional; however it is conceivable that such vortex motion results in moving three-dimensional stagnation points or in moving lines of stagnation being imposed at the edge of the wall layer from time to time. A stationary axi-symmetric outflow stagnation point is known to ultimately lead to a boundary layer eruption (Howarth, 1973) giving rise to an axi-symmetric separation of the type observed behind spheres. It might be expected that if a three-dimensional stagnation point is in motion above a wall, then a boundary-layer eruption will eventually occur. This question must await further work.

Finally an important point that requires clarification is the following: In the present picture which is suggested by this study (of breakdown of the wall layer occurring due to a moving stagnation point), it is necessary to account for how the convective terms become important in the leading order wall layer equations as the quiescent period draws to a close. This requires that u >> u^*; there is evidence in Emmerling's (1973) data of intense localized disturbances having dimensions of $O(\lambda^*)$. Clearly a better theoretical understanding of the dynamics of the outer layer is required to discern how the flow field solution in the outer layer can focus to large values at very localized positions.

ACKNOWLEDGMENT

The authors are grateful for the support of this work by AFOSR under grant no. 74-2707.

REFERENCES

1. Andersen, P. S., Kays, W. M. & Moffatt, R. J. 1972 Mech. Engrg. Rept. HMT-15, Stanford University.

2. Bark, F. H. 1975 J. Fluid Mech. 70, 229-250.

3. Black, T. J. 1968 NASA CR 888.

4. Blackwelder, R. F. & Kaplan, R. E. 1972 Int. Union Theor. Appl. Mech. 12th.

5. Cebeci, T. & Smith, A. M. O. 1974 Analysis of Turbulent Boundary Layers, Academic Press.

6. Coles, D. & Barker, S. J. 1975 In Turbulent Mixing in Nonreactive and Reactive Flows, S. N. B. Murthy (ed.), Plenum Press, New York.

7. Collins, W. M. & Dennis, S. C. R. 1973 J. Fluid Mech. 60. 105-128.

8. Corino, E. R. & Brodkey, R. S. 1969 J. Fluid Mech. 37, 1-30.

9. Dennis, S. C. R. & Walker, J. D. A. 1971 J. Engrg. Math. 5, 263-278.

10. Dennis, S. C. R. & Walker, J. D. A. 1972 Phys. Fluids 15, 517-525.

11. Einstein, H. A. & Li, H. 1956 A.S.C.E. Proc. 82, paper 945.

12. Emmerling, R. 1973 Max-Planck-Institüt für Stromungsforschung, Göttingen.

13. Fendell, F. E. 1972 J. Astro. Sciences 20, 129-165.

14. Goldstein, S. & Rosenhead, L. 1936 Proc. Camb. Phil. Soc. 32, 392-401.

15. Gupta, A. K., Laufer, J. & Kaplan, R. E. 1971 J. Fluid Mech. 50, 493-512.

16. Howarth, J. A. 1973 J. Fluid Mech. 59, 769-773.

17. Kim, H. T., Kline, S. J. & Reynolds, W. C. 1971 J. Fluid Mech. 50, 133-160.

18. Kim, H. T., Kline, S. J. & Reynolds, W. C. 1968 Rep. No. MD-20, Dept. Mech. Engrg., Stanford University.

19. Kline, S. J., Reynolds, W. C., Schraub, F. A. & Rundstadler, P. W. 1967 J. Fluid Mech. 30, 741-773.

20. Kline, S. J. & Schraub, F. A. 1965 Rep. No. MD-12, Dept. Mech. Engrg., Stanford University.

21. Mellor, G. L. & Gibson, D. M. 1966 J. Fluid Mech. 24, 225-253.

22. Millikan, C. B. 1932 Proc. 5th Int. Cong. of Appl. Mech., MIT Press, Cambridge, Mass., 386-392.

23. Milne-Thomson, L. M. 1962 Theoretical Hydrodynamics, 4th edition, MacMillan, New York.

24. Nychas, S. G., Hershey, H. C. & Brodkey, R. S. 1973 J. Fluid Mech. 61, 513-540.

25. Offen, G. R. & Kline, S. J. 1975 J. Fluid Mech. 70, 209-228.

26. Offen, G. R. & Kline, S. J. 1973 Rep. No. MD-31, Dept. Mech. Engrg., Stanford University.

27. Proudman, I. & Johnson, K. 1962 J. Fluid Mech 12, 161-168.

28. Riley, N. 1975 SIAM Review, 274-297.

29. Robins, A. J. & Howarth, J. A. 1972 J. Fluid Mech. 56, 161-171.

30. Scharnhorst, R. K., Walker, J. D. A. & Abbott, D. E. 1976 CFMTR-76-6, School of Mech. Engrg., Purdue University.

31. Schubauer, G. B. 1954 J. Appl. Phys. 25, 188.

32. Van Driest, E. R. 1956 J. Aeronaut. Sci. 23, 1007-1012.

33. Walker, J. D. A., Abbott, D. E. & Scharnhorst, R. K. 1976 CFMTR-76-1, School of Mech. Engrg., Purdue University.

34. Walker, J. D. A. 1976 submitted for publication.

35. Willmarth, W. W. 1975 Advances in Appl. Mech. 15, 159-254.

DISCUSSION

COLES: (California Institute of Technology)

I would like to know if you have any idea as to the physical mechanism which causes the streaks to appear under a turbulent spot or a turbulent boundary layer.

WALKER:

The question of causation is a difficult one and at this stage all one can say is the following. In a turbulent boundary layer, during the quiescent period the integrity of the wall layer is intact and the streaks are present. During a burst there is a breakdown of the wall layer flow and the streak pattern breaks up but is quickly re-established. Since the wake layer is inviscid to leading order, it is reasonable to expect that the streaks are a response to events taking place in the outer layer. It is the question of the precise nature of these events which is difficult since the dynamics of the outer layer are poorly understood. It has been suggested by a number of authors that the large eddies which are observed to dominate the wake layer are similar to transition spots. A major analytical difficulty is that we do not completely understand the nature of the flow within a spot or even how a spot is created. Nevertheless, it is the flow relative to the spot which is relevant and if a simple U-shaped vortex moving with constant speed is assumed, then in a frame of reference moving with the spot either a point or line of stagnation should be expected underneath the spot and at the edge of the wall layer. This should act to produce a single streak in the wall layer which should be visible in the laboratory frame provided a continuous source of visualization is used (such as a hydrogen bubble wire). However this is counter to experimental evidence which indicates a number of streaks per spot. Perhaps the eddies are not simple U-shaped vortices or perhaps some instability mechanism is involved; I don't know.

COLES:

I think that maybe Steve Kline might like to be heard on this subject also. My view of the matter is that a large structure moving at some speed--there is a dispute still about what this speed is, in the neighborhood of 0.7 to 0.9--will overrun some of the flow near the surface. This overrun flow is deflected into a trajectory such that a Görtler-Taylor instability occurs. I don't know whether Kline would agree with this or not, but I think you have got to have some mechanism of this type to account for the multitude of streaks that appear in the vicinity of these active transport operators in the boundary layer.

WALKER:

You may be right. This has been conjectured before and while it certainly can't be ruled out it is as far from justification as ever. Hopefully someone will at some stage attempt flow visualization in a frame of reference moving with the spots so that we can get a clearer physical picture of what is going on. Insofar as the quiescent period within the wall layer I discussed today, it is necessary to admit that the streaks are there and further that there is a characteristic spacing which is Reynolds number dependent.

SAFFMAN: (California Institute of Technology)

I would like to ask for a little bit of caution in the free and easy use of vortices in the turbulent boundary layer. We do have a little more idea now that 50 years ago on how a vortex moves in irrotational fluid. In some ways it does not move as intuition makes you think it should. In fluid endowed with vorticity--and a boundary layer does have a lot of vorticity--it moves in ways which we just do not know and do not understand and by appealing to intuition we may be laying ourselves open to serious error.

WALKER:

I agree with what you say though I'm not clear about what you consider "free and easy." It has been assumed by many authors that the wall layer bursting is a stability problem. There is no rational theory to support this and I have difficulty in accepting that this is the only mechanism involved. The problem involving the rectilinear vortex in motion above a wall was considered as a search for another possible mechanism namely a moving outflow stagnation point. The experiments of Harvey and Perry* on the flowfield induced by aircraft

---

* Harvey, J. K. and Perry, F. J., "Flow Field Produced by Trailing Vortices in the Vicinity of the Ground," AIAA J., Vol. 9, No. 8, pp. 1659-1660, August 1971.

trailing vortices in the vicinity of the ground substantiate the theoretical findings for the rectilinear vortex. In these experiments a secondary vortex was observed to form near the ground and subsequently ejection and rapid outflow from the boundary layer occurred. Of course whether this type of mechanism is relevant to the turbulent boundary layer is conjectural and all that can be said at present is that the explosive nature of the resulting boundary layer flow bears a resemblance to the observed bursting in a turbulent boundary layer.

WALLACE: (The University of Maryland)

In reply to Dr. Coles you said there are five to six of these "things" that occur under this "thing." I'm wondering which "things" you are talking about.

WALKER:

My understanding is that under a single spot, if you take a single spot and follow it downstream, you are going to see five or six streaks.

WALLACE:

Transition spots?

WALKER:

Yes, I think it was evident in Coles' movies that there was more than one streak associated with his artificial spot. Because a continuous source of dye was not used it was difficult to get a definitive picture and it was not clear to me to what extent the streaks, observed substantially upstream of the spot, were associated with the spot.

KLINE: (Stanford University)

It depends on what you mean and where you are as Jim Wallace's question implied. If you look at Meyers' movies* or Coles movies or some of the others in the transition region then as soon as you get the Klebanoff-Tidstrom (K-T) instability you see large streaks in

---

* Meyer, K. A. and Kline, S. J., "Visual Study of the Flow Model in the Later Stages of Laminar-Turbulent Transition on a Flat Plate." Stanford ME Department, Report MD-7. Associated movie and its scenario available from ASME-ESL Film Library, Engrg. Societies Library, 345 E. 47th St., NYC, 10017; film M-3.

that region. K-T streaks are at a scale which is clearly larger than the lambda plus of 100 seen in sublayer streaks under a turbulent spot or turbulent boundary layer. When the K-T streak moves out and causes a "breakdown," which Kelbanoff and other people have now documented rather carefully with hot wires, then underneath that breakdown immediately you see a bunch of sublayer streaks. Not one but a bunch of sublayer streaks that are at the smaller scale of $\lambda^+ \sim 100$. That is what the data say. Also your conclusion that the stuff is essentially "frozen-in" seems right to me because whenever we look at a flow using a dye-injector of a bubble-wire or anything else, where the flow is fully turbulent, we see sublayer streaks in distances from the marking station which are surprisingly short in every case--many thousands of observations. In all those cases, as far as I know, this is uniform; we have never seen anything else. The question of what actually causes sublayer streaks, the causation if you want, seems to me to be the most difficult part of the whole thing. I don't even want to comment on that today. However, we do observe all those things I have just been describing; that is, data. Hence, if you consider your spot as one structure of some kind, vortical or otherwise, then there are a lot of streaks underneath it.

There are a couple of other things that the data also say. One of them is that when you get a spot, and you count the sublayer streaks under the spot after the breakdown, in what I sometimes call the fourth stage of transition (Meyer has done this and I think some other people also) then to the uncertainty of those data, you reproduce the lambda plus of 100. With the usual statistical variations, everything looks the same.

The flow in a "spot" does look like turbulence as Klebanoff and Tidstrom also concluded from the hot-wire data. The other thing is that along the sides of the "spot" where new streaks are forming, where there are new sublayer streaks, not K-T transition streaks, then the streaks are multiplying by what looks like some kind of wave action that I do not understand; this is what we call cross-contamination. New streaks then appear one by one, not in groups; this is quite clear in Meyers' movies. I think I see the same thing in Don Coles' movies. So how many sublayer streaks you have under the spot depends on how far downstream you go. However, in the movies of Jim Johnston and his student Halleen* where they put stabilizing coriolis forces on the wall layers (in a rotating system) then, in certain instances, instead of getting cross-contamination as I just described, one sees streaks dying along the edges of a spot and

---

* Halleen, R. M. and Johnston, J. P., J. Fluid Mech. 56, p. 533 ft (1972); JFE 95 p. 229 (1973); Stanford ME Department Report MD-18, 1967.

because the flow is in a "stable" or "superstable" environment, then the streaks die one by one. So that is some of the things the pictures say, but as far as the causation goes I think that this is the point about which there is still the most uncertainty and the most disagreement.

Further Investigiation of the Linear and Nonlinear Theory for Constant-Temperature Hot-Wire Anemometers

PETER FREYMUTH
University of Colorado
Boulder, Colorado (U.S.A.)

ABSTRACT

The nonlinear theory for constant-temperature hot-wire anemometers is outlined in more detail than before. It is shown that the domain of nonlinear effects is restricted to velocity fluctuations above at least 12% of the mean velocity in a frequency range larger than 10% of the cut-off frequency of the anemometer. This domain overlaps rarely, if ever, with the domain of actual turbulent internal flows.

1. INTRODUCTION

Large velocity fluctuations are encountered in internal flows. Consequently, the question whether the constant temperature hot-wire anemometer is suitable to measure large fluctuations should be examined closely by examining the nonlinear theory for constant temperature hot-wire anemometers.

Comte-Bellot (1976) in her recent survey identifies the nonlinear theory for constant-temperature anemometers as one of the pending problems in hot-wire anemometry. Although a nonlinear theory has been outlined by Freymuth (1969), it was felt that additional details were needed in order to compare results with corresponding ones for the constant current anemometer as obtained by Comte-Bellot and Schon (1969).

In particular, the nonlinear influence of large sinusoidal velocity fluctuations on power and skewness of the output signal and

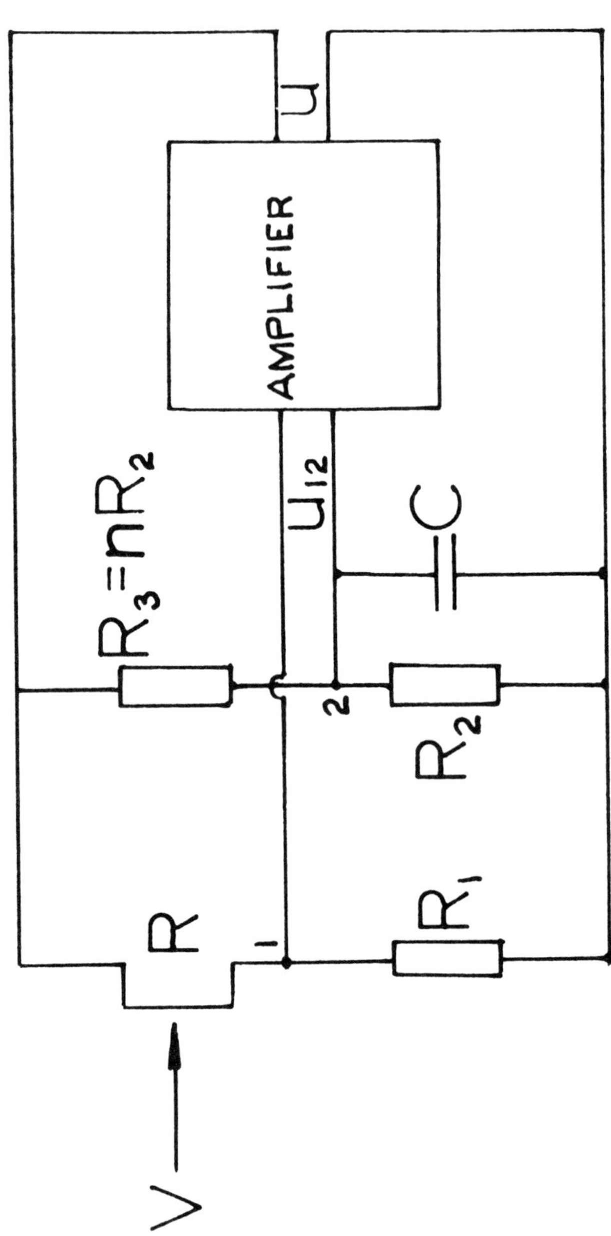

Figure 1. Diagram of the Feedback System.

an estimate of nonlinear influences on turbulent power spectra are of interest. Furthermore, Comte-Bellot and Schon (1969) pointed out that the effect of nonlinearities on the linearized output of the anemometer should be investigated. A linearizer has not been included in the theory by Freymuth (1969). Another item of concern may be added to the list. Several different frequency adjustments for a constant-temperature anemometer are of practical importance, but only the optimum aperiodic case has been discussed in the context of the linear and nonlinear theory by Freymuth (1967, 1969). In particular, the maximally flat case which represents the flattest frequency response possible at lower frequencies is of prime interest. Furthermore, an adjustment in which the response of the anemometer to a step in velocity or to an electronic step is optimized and where the real parts of the characteristic roots of the system are equal has some interest, and has been widely used by workers in the field.

It is the aim of this paper to fill in some of the remaining gaps in the linear and nonlinear theory for constant-temperature hot-wire anemometers, with emphasis on the nonlinear theory.

2. THE GOVERNING DYNAMIC EQUATIONS

Dynamic linearities and nonlinearities arise for fast velocity fluctuations from the equations for the components of the feedback system. As explained in detail by Freymuth (1967) for an equal arm bridge, the equations for the hot-wire, the bridge and the amplifier for the feedback system shown in Figure 1 read

$$U^2 - \frac{H(V)}{\alpha}\frac{R - R_0}{R}(R + R_1)^2 = \frac{c}{\alpha}\frac{(R + R_1)^2}{R}\frac{dR}{dt} \tag{1}$$

$$u_{12} = \frac{U}{n+1}\frac{nR_1 - R}{R_1 + R} + \frac{nR_2C}{(n+1)^2}\frac{dU}{dt} \tag{1a}$$

$$M''\frac{d^2U}{dt^2} + M'\frac{dU}{dt} + U = U_b + Gu_{12} \tag{1b}$$

where V is the effective velocity at the hot-wire, H(V) is a heat transfer function of the wire which in reasonable approximation has the form

$$H(V) = A\sqrt{V} + B \tag{2}$$

where A and B are empirical constants. U is the bridge output voltage, R is the resistance of the hot wire, R_0 is its cold resistance, α is the temperature coefficient of the resistivity of the wire and c is its heat capacity. R_1 is a resistance in the bridge in series with R and n is the ratio of the resistors R_3 to R_2 in the other arm of the bridge. Time is t. The amplifier of the feedback system has a gain G, an adjustable output offset voltage U_b and first and second order time constants M' and M". Combining (1a) and (1b) yields

$$\frac{GU}{n+1}\,\frac{nR_1 - R}{R_1 + R} + U_b = U + M_1\frac{dU}{dt} + M''\frac{d^2U}{dt^2} \tag{3}$$

where

$$M_1 = M' - G\frac{n}{(n+1)^2}R_2C \tag{4}$$

is a time constant which is adjustable by means of a trim condenser (or trim coil) in the bridge. Adjustable parameters have been introduced into the circuit in order to frequency optimize the feedback system.

Equations (1) and (3) are nonlinear. If these equations are combined by eliminating the resistance R, a nonlinear relation between velocity V and the output U of the anemometer results as follows. Defining a quantity ε by

$$\varepsilon = \frac{n+1}{GU}\left(M_1\frac{dU}{dt} + M''\frac{d^2U}{dt^2}\right) - \frac{n+1}{G}\frac{U_b}{U} + \frac{n+1}{G} \tag{5}$$

we obtain from equation (3)

$$R = nR_1\frac{1-\varepsilon/n}{1+\varepsilon} \tag{6}$$

and

$$\frac{dR}{dt} = -(n+1)\frac{R_1}{(1+\varepsilon)^2}\frac{d\varepsilon}{dt} \tag{7}$$

Since for a well adjusted anemometer with high gain G the quantity ε is much smaller than 1, we obtain in reasonable approximation

$$R \simeq nR_1 \tag{8}$$

$$\frac{dR}{dt} \simeq -(n+1)\,R_1\frac{d\varepsilon}{dt} \tag{9}$$

(8) and (9) into (1) yield

$$U^2 - \frac{H(V)}{\alpha}\frac{(n+1)^2}{n}R_1(nR_1 - R_0) = -\frac{c}{\alpha}\frac{(n+1)^3}{n}R_1^2\frac{d\varepsilon}{dt} \qquad (10)$$

The above equation implicitly relates the input V of the anemometer to its output U. First order approximations--for instance $R \simeq nR_1\{1 - (1 + 1/n)\varepsilon\}$ from equation (6)--could be used instead of the zeroth order approximations represented by equations (8) to (10), however, the improvement in accuracy (of order 1% or smaller) hardly justifies the increase in the amount of algebra, as far as the non-linear theory is concerned.

3. RELATION BETWEEN VELOCITY AND LINEARIZED OUTPUT

Since the quantity ε is small, equation (10) is algebraic in a frequency range far below the cutoff frequency of the anemometer and for this reason an algebraic "linearizer" is usually connected to the anemometer which gives an output voltage U_m which is proportional to V, i.e. as far as

$$\frac{d\varepsilon}{dt} \simeq 0$$

$$V = kU_m = V_m \qquad (11)$$

where V_m represents the velocity as it would be indicated by the algebraic linearizer. The connection between U and V_m is, according to (10) with

$\frac{d\varepsilon}{dt} \simeq 0$ and (11),

$$U^2 = \frac{H(V_m)}{\alpha}\frac{(n+1)^2}{n}R_1(nR_1 - R_0) \qquad (12)$$

(12) into (10) yields

$$H(V) = H(V_m) + c(n+1)\frac{R_1}{nR_1 - R_0}\frac{d\varepsilon}{dt} \qquad (13)$$

Let us introduce a time constant

$$M = \frac{(n+1)^2}{2}\frac{R_1}{nR_1 - R_0}\frac{c}{H(\overline{V})} \qquad (14)$$

where $\overline{V}$ is the average flow velocity. Substituting (14) into (13) yields

$$H(V) = H(V_m) + \frac{2}{n+1} H(\overline{V}) \, M \frac{d\varepsilon}{dt} \tag{15}$$

To proceed further we have to be specific about the functional form of H. Using the relation (3) transforms (15) after some rearrangement into

$$V = V_m + \frac{4}{n+1} \frac{A\sqrt{\overline{V}} + B}{A} M\sqrt{V_m} \frac{d\varepsilon}{dt} + \left(\frac{2M}{n+1} \frac{A\sqrt{\overline{V}} + B}{A} \frac{d\varepsilon}{dt} \right)^2 \tag{16}$$

Equation (16) relates the actual velocity V at the wire to the velocity V_m as it is measured at the output of the linearizer. For fluctuations sufficiently slow to render dynamic terms inessential, we have $V = V_m$ as desired but for fast and large fluctuations, linear and nonlinear dynamic effects appear as will be shown in detail in the following sections.

4. THE DIMENSIONLESS SECOND-ORDER APPROXIMATION

Equation (16) can be handled best if it is recast into dimensionless form. Introducing the dimensionless quantities

$$z = \frac{V}{\overline{V}} - 1 \quad (17), \quad Y = \frac{V_m}{\overline{V}} - 1 \quad (18), \quad h = \frac{A\sqrt{\overline{V}}}{A\sqrt{\overline{V}} + B} \quad (19)$$

into (16) yields

$$z = y + \frac{4M}{(n+1)h} \sqrt{1+y} \frac{d\varepsilon}{dt} + \frac{4}{(n+1)^2} \frac{M^2}{h^2} \left(\frac{d\varepsilon}{dt} \right)^2 \tag{20}$$

ε has to be expressed in terms of y which is accomplished after some elaboration as follows. According to (3) and (12)

$$U^2 = \frac{A\sqrt{V_m} + B}{\alpha} \frac{(n+1)^2}{n} R_1(nR_1 - R_0) \tag{21}$$

or

$$U^2 = U_1^2 \left[1 + h(\sqrt{1+y} - 1) \right] \tag{22}$$

where

$$U_1^2 = \frac{A\sqrt{V} + B}{\alpha} \frac{(n+1)^2}{n} R_1(nR_1 - R_0) \tag{23}$$

To keep the theory within reasonable limits we restrict ourselves to a second-order approximation such that

$$U = U_1 \left[1 + \frac{h}{4} y - \frac{h}{16}(1 + 0.5\ h)\ y^2\right] \tag{24}$$

Inserting (24) into (5) yields in second-order approximation after an additional differentiation with respect to t

$$\begin{aligned}\frac{d\varepsilon}{dt} = \frac{n+1}{G}\frac{h}{4}\Bigg[&\frac{U_b}{U_1}\frac{dy}{dt} + M_1\frac{d^2y}{dt^2} + M''\frac{d^3y}{dt^3} - \frac{U_b}{U_1}\frac{1 + 1.5h}{2}\, y\,\frac{dy}{dt} \\ &- M_1\frac{1+h}{2}\, y\,\frac{d^2y}{dt^2} - M''\frac{1+h}{2}\, y\,\frac{d^3y}{dt^3} - M_1\frac{1+h}{2}\left(\frac{dy}{dt}\right)^2 \\ &- M''(1.5 + h)\frac{dy}{dt}\frac{d^2y}{dt^2}\Bigg]\end{aligned} \tag{25}$$

and

$$\left(\frac{d\varepsilon}{dt}\right)^2 = \frac{(n+1)^2}{G^2}\frac{h^2}{16}\left[\frac{U_b}{U_1}\frac{dy}{dt} + M_1\frac{d^2y}{dt^2} + M''\frac{d^3y}{dt^3}\right]^2 \tag{26}$$

Substituting (25) and (26) into (20) yields in second-order approximation

$$\begin{aligned}z = y &+ \frac{M}{G}\frac{U_b}{U_1}\frac{dy}{dt} + \frac{MM_1}{G}\frac{d^2y}{dt^2} + \frac{MM''}{G}\frac{d^3y}{dt^3} - \frac{M}{G}\frac{U_b}{U_1}\frac{3h}{4}\, y\,\frac{dy}{dt} \\ &- \frac{MM_1}{G}\frac{h}{2}\, y\,\frac{d^2y}{dt^2} - \frac{MM''}{G}\frac{h}{2}\, y\,\frac{d^3y}{dt^2} - \frac{MM_1}{G}\frac{1+h}{2}\left(\frac{dy}{dt}\right)^2 \\ &- \frac{MM''}{G}(1.5 + h)\frac{dy}{dt}\frac{d^2y}{dt^2} + \frac{1}{4}\left[\frac{M}{G}\frac{U_b}{U_1}\frac{dy}{dt} + \frac{MM_1}{G}\frac{d^2y}{dt^2} + \frac{MM''}{G}\frac{d^3y}{dt^3}\right]^2\end{aligned} \tag{27}$$

Introducing dimensionless quantities

$$dt = \left(\frac{MM''}{G}\right)^{1/3} dx \tag{28}$$

$$a = \frac{M}{G}\frac{U_b}{U_1}\bigg/\left(\frac{MM''}{G}\right)^{1/3} \tag{29}$$

$$b = \frac{MM_1}{G}\bigg/\left(\frac{MM''}{G}\right)^{2/3} \tag{30}$$

into (27) yields as final equation

$$\begin{aligned} z = {} & y + a\frac{dy}{dx} + b\frac{d^2y}{dx^2} + \frac{d^3y}{dx^3} - \frac{3}{4}ahy\frac{dy}{dx} - \frac{bh}{2}y\frac{d^2y}{dx^2} - \frac{h}{2}y\frac{d^3y}{dx^3} \\ & + \frac{a^2 - 2b - 2bh}{4}\left(\frac{dy}{dx}\right)^2 + \frac{ab - 3 - 2h}{2}\frac{dy}{dx}\frac{d^2y}{dx^2} + \frac{a}{2}\frac{dy}{dx}\frac{d^3y}{dx^3} \\ & + \frac{b^2}{4}\left(\frac{d^2y}{dx^2}\right)^2 + \frac{b}{2}\frac{d^2y}{dx^2}\frac{d^3y}{dx^3} + \frac{1}{4}\left(\frac{d^3y}{dx^3}\right)^2 \end{aligned} \tag{31}$$

Equation (31) represents the governing equation of the anemometer in second-order approximation which relates the input z to the linearized output y. The linear part of the equation agrees with the one obtained by Freymuth (1967) for the linearized theory.

If we compare the input-output relation (31) with the corresponding relation for the anemometer without linearizer as obtained by Freymuth (1969), a considerable increase in complexity is noticed. Nevertheless, a solution for a sinusoidal input z is rather straightforward.

5. SOLUTION FOR A COSINUSOIDAL INPUT

Let us solve equation (31) for an input

$$z = \hat{z} \cos \Omega x \tag{32}$$

where $\hat{z}$ is the dimensionless amplitude of the fluctuations and Ω is a dimensionless circular frequency related to the dimensional circular frequency ω by the relation

$$\Omega x = \omega t \tag{33}$$

To solve equation (31) we adopt the same iterative procedure already used by Freymuth (1969) for the analysis of the anemometer without linearizer. The result after the second iteration is

$$y = A_0 + A_1 \cos(\Omega x + \phi_1) + A_2 \cos(2\Omega x + 2\phi_1 + \phi_2 + \phi_3) \tag{34}$$

where

$$A_0 = \frac{\overline{V}_m - \overline{V}}{\overline{V}} = -\frac{\hat{z}^2\Omega^2}{8} \frac{a^2 - 2b + (b^2 - 2a)\Omega^2 + \Omega^4}{1 + (a^2 - 2b)\Omega^2 + (b^2 - 2a)\Omega^4 + \Omega^6} \tag{35}$$

represents the relative error in the measurement of the mean value of velocity due to nonlinear effects.

$$\frac{A_1}{\hat{z}} = \frac{1}{[1 + (a^2 - 2b)\Omega^2 + (b^2 - 2a)\Omega^4 + \Omega^6]^{\frac{1}{2}}} \tag{36}$$

represents the frequency response of the anemometer in case of small fluctuations.

$$\frac{A_2}{\hat{z}} = \frac{\hat{z}\Omega}{4} \frac{\left[\left(\frac{3}{2}ah + \{ab - 3h - 3\}\Omega^2 - b\Omega^4\right)^2 + \left(2bh + b - \frac{a^2}{2} + \{a + \frac{b^2}{2}\}\Omega^2 - 0.5\Omega^4\right)^2 \Omega^2\right]^{\frac{1}{2}}}{[1 + (a^2 - 2b)\Omega^2 + (b^2 - 2a)\Omega^4 + \Omega^6][1 + 4(a^2 - 2b)\Omega^2 + 16(b^2 - 2a)\Omega^4 + 64\Omega^6]^{\frac{1}{2}}} \tag{37}$$

represents the relative amplitude of the second harmonic which vanishes for small values of $\hat{z}$.

The phase angles ϕ_1, ϕ_2, and ϕ_3 are determined by the following equations.

$$\cos\phi_1 = \frac{1 - b\Omega^2}{[1 + (a^2 - 2b)\Omega^2 + (b^2 - 2a)\Omega^4 + \Omega^6]^{\frac{1}{2}}} \tag{38}$$

$$\sin\phi_1 = \frac{-\Omega(a - \Omega^2)}{[1 + (a^2 - 2b)\Omega^2 + (b^2 - 2a)\Omega^4 + \Omega^6]^{\frac{1}{2}}} \tag{39}$$

$$\cos\phi_2 = \frac{1 - 4b\Omega^2}{[1 + 4(a^2 - 2b)\Omega^6 + 16(b^2 - 2a)\Omega^4 + 64\Omega^6]^{\frac{1}{2}}} \tag{40}$$

$$\sin\phi_2 = \frac{-2\Omega(a - 4\Omega^2)}{[1 + 4(a^2 - 2b)\Omega^2 + 16(b^2 - 2a)\Omega^4 + \Omega^6]^{\frac{1}{2}}} \tag{41}$$

$$\cos\phi_3 = \frac{-\Omega\left\{2bh+b - \frac{a^2}{2} + \{a + \frac{b^2}{2}\}\Omega^2 - 0.5\Omega^4\right\}}{\left[(1.5ah+\{ab-3h-3\}\Omega^2-b\Omega^4)^2 + \left(2bh+b - \frac{a^2}{2} + \{a + \frac{b^2}{2}\}\Omega^2 - 0.5\Omega^4\right)^2\Omega^2\right]^{\frac{1}{2}}} \tag{42}$$

$$\sin\phi_3 = \frac{1.5ah + (ab - 3h - 3)\Omega^2 - b\Omega^4}{\left[(1.5ah+\{ab-3h-3\}\Omega^2-b\Omega^4)^2 + (2bh+b - \frac{a^2}{2} + \{a + \frac{b^2}{2}\}\Omega^2 - 0.5\Omega^4\Big)^2\Omega^2\right]^{\frac{1}{2}}} \tag{43}$$

The fluctuating part y' of y reads

$$y' = A_1 \cos(\Omega x + \phi_1) + A_2 \cos(2\,\Omega x + 2\phi_1 + \phi_2 + \phi_3) \tag{44}$$

For the fluctuating intensity we therefore get

$$\overline{y'^2} = \frac{A_1^2}{2}\left\{1 + \frac{A_2^2}{A_1^2}\right\} \tag{45}$$

For the skewness S_3 we get

$$S_3 = \frac{\overline{y'^3}}{\overline{y'^2}^{3/2}} = \frac{3}{\sqrt{2}}\frac{A_2}{A_1}\left\{1 + \frac{A_2^2}{A_1^2}\right\}^{3/2} \cos(\phi_2 + \phi_3) \simeq \frac{3}{\sqrt{2}}\frac{A_2}{A_1}\cos(\phi_2 + \phi_3) \tag{46}$$

For the skewness S_5 we get

$$S_5 = \frac{\overline{y'^5}}{\overline{y'^2}^{5/2}} \simeq \frac{10}{3} S_3 \tag{47}$$

6. DISCUSSION OF RESULTS

The frequency response of interesting quantities A_0, $A_1/\hat{z}$, $A_2/\hat{z}$, $\overline{y'^2}$, and S_3 depend on parameters a, b, and h. Furthermore, A_0 and $\overline{y'^2}$ are proportional to $\hat{z}^2$ whereas A_2 and S_3 are proportional to $\hat{z}$. To limit the discussion of results manifested in equations (34)-(47) to a reasonable amount, we represent a few cases of highest interest in graphical form.

For z we choose z = 0.5 as a reference which corresponds to velocity fluctuations of 50% around the mean velocity. Rarely, if ever, are rapid velocity fluctuations of this severity encountered. Accordingly, the influence of second harmonics will be less severe in most or all practical applications.

For h we will consider the case of large mean velocity for which $h \approx 1$ and the case of small mean velocity for which $h \approx 0$. In practice, results will be in between these two cases.

As far as the anemometric parameters a and b are concerned, we select three adjustments of highest interest. Case I was previously discussed in the framework of the linear theory by Freymuth (1967), with a = b = 3. This is the optimum aperiodic case where any oscillatory behavior of the anemometer in response to a step in velocity or in response to an electronic step is avoided.

Case II: a = 2.4581 b = 2.2494. In this case the response of the anemometer to velocity or electronic steps is optimized. This case gives flattest frequency response at low frequencies under the constraint that the real parts of the characteristic roots of the system are equal.

Case III: a = b = 2. This is the case of maximally flat frequency response of the system toward lower frequencies; furthermore, the cutoff frequency is enhanced compared to the other two cases which represent an additional advantage. Case III has the highest practical potential.

Numerical computations for equations (37) to (47) can be handled rather quickly by mini computer.

Cases I, II and III can be identified in practice by means of a square wave test where an electrical step is injected into the bridge and where according to Freymuth (1967), the resulting transient pulse at the output of the anemometer is observed. Without detailing the Laplace transform techniques, we show in Figure 2 the electrical pulses obtained. Although case III seems inferior with its undershoot, it is the case of highest practical importance.

Let us discuss the relative amplitudes $A_1/\hat{z}$ and $A_2/\hat{z}$ of the fundamental component and the second harmonic of the anemometer output in case of a cosinusoidal input. The dependence of these quantities on dimensionless circular frequency Ω is represented in Figure 3 for h = 1, $\hat{z}$ = 0.5 and for all three anemometric adjustments I, II,

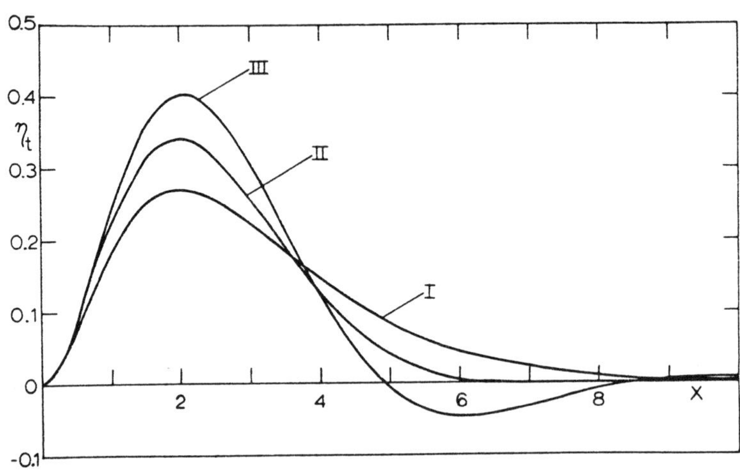

Figure 2. Response of the feedback system to an electronic step for Case I (a = b = 3), Case II (a = 2.46, b = 2.25), and Case III (a = b = 2).

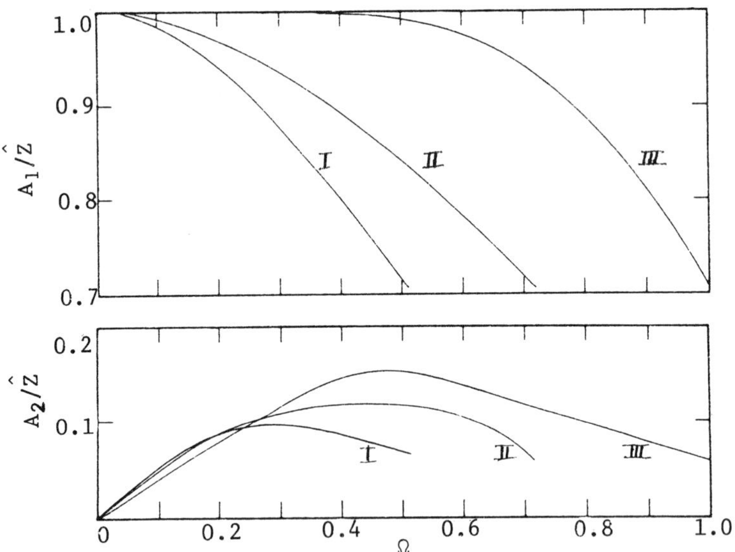

Figure 3. Normalized fundamental component $A_1/\hat{z}$ and second harmonic $A_2/\hat{z}$ of the anemometer in dependence on dimensionless frequency Ω, for Cases I, II and III, $\hat{z} = 0.5$, $h = 1$.

and III as previously explained. These quantities are represented up to the respective cutoff frequency of the anemometer (-3dB point for $A_1/\hat{z}$). The linear response is maximally flat in case III and a high cutoff frequency is incurred. For cases I and II the relation between cutoff frequency f_g and the duration time t_g of the transient pulse (down to 3% of its maximum value) reads in good approximation

$$f_g = 1/1.5\ t_g \tag{48}$$

In case III we get

$$f_g = 1/1.3\ t_g \tag{49}$$

The nonlinear response curves $A_2/\hat{z}$ reach their maxima at about half the cutoff frequency and these maxima seem quite substantial. If we tolerate a value of $A_2/\hat{z}$ of 4% as insignificant which is a most conservative estimate--as the discussion of other quantities affected by the second harmonic will show--then velocity fluctuations up to 50% of the mean velocity and up to at least 10% of the cutoff frequency do not exhibit nonlinear behavior. For instance an anemometer with a cutoff frequency of 100 kHz tolerates easily large fluctuations ($\hat{z}$ = 0.5) up to at least 10 kHz without any significant nonlinear effects. For smaller fluctuations of 25% of the mean velocity, this frequency range doubles and if fluctuations are kept below 12% the linear ceiling of 4% is never reached, over the entire frequency range. Even for velocity fluctuations up to 50%, it is possible to extend the useful frequency range to 20% of the cutoff frequency if a sharp, low pass filter with a cutoff at 20% of that of the anemometer is connected to the output of the linearizer. This filter cuts out disturbing second harmonics above 10% of the cutoff frequency of the anemometer. This leaves us with a useful frequency range of 20 kHz in the above mentioned example. The above discussion shows that nonlinearities in constant temperature anemometers have fortunately very limited practical significance since they occur at rather high frequencies only. Turbulent flows show their largest amplitudes at lowest frequencies. In contrast, nonlinearities in constant current anemometers start to become significant well below the natural cutoff frequency of the hot wire, i.e., in practice well below 1 kHz according to Comte-Bellot and Schon (1969).

From Figure 3 it may be noticed that the response in case III is not only flattest toward low frequencies for the fundamental component $A_1/\hat{z}$ but even for the second harmonic for all three cases discussed. This lends even more significance to this type of adjustment.

Figure 4 shows the relative second harmonic $A_2/\hat{z}$ in case of small mean velocity, i.e., for $h \simeq 0$. It can be seen that $A_2/\hat{z}$ never

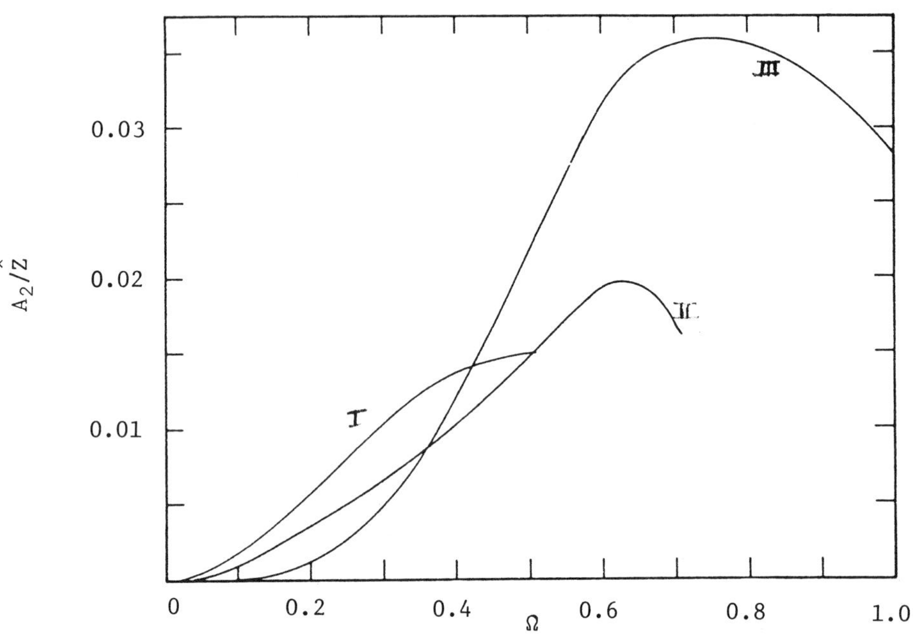

Figure 4. Normalized second harmonic $A_2/\hat{z}$ with z = 0.5, h = 0.

reaches the linear ceiling of 4%. We conclude that for low mean velocities, nonlinearities are even less important than for large mean velocities. Further discussions therefore concentrate on the most severe case, h = 1.

Figure 5 shows the relative influence A_2^2/A_1^2 of the second harmonic on the intensity of the output signal according to equation (45). This influence remains below 2.5% and for fluctuations below 10% of the cutoff frequency, it is less than 0.2%.

Figure 6 shows the relative error A_0 in the measurement of the mean velocity which remains below 2% and which is less than 0.1% at 10% of the cutoff frequency of the anemometer.

Figure 7 shows the skewness S_3 which reaches the quite significant value of 0.31 in the worst case but remains below 0.02 if the frequency of large fluctuations does not exceed 10% of the cutoff frequency. As far as maximum values of S_3 are concerned, they are quite comparable with the ones obtained by Comte-Bellot and Schon (1969) for constant current anemometers.

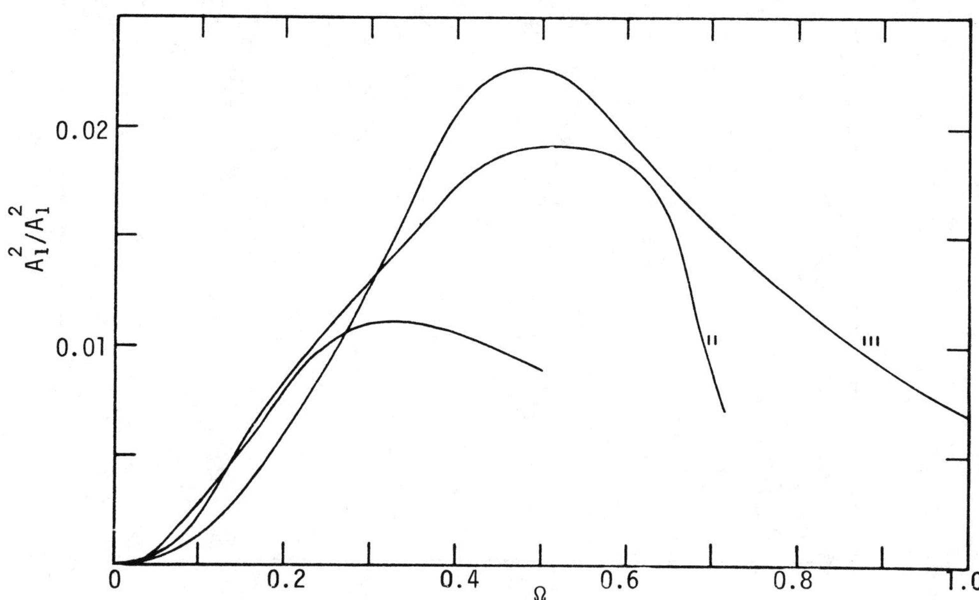

Figure 5. Ratio A_2^2/A_1^2, for $\hat{z}$ = 0.5, h = 1.

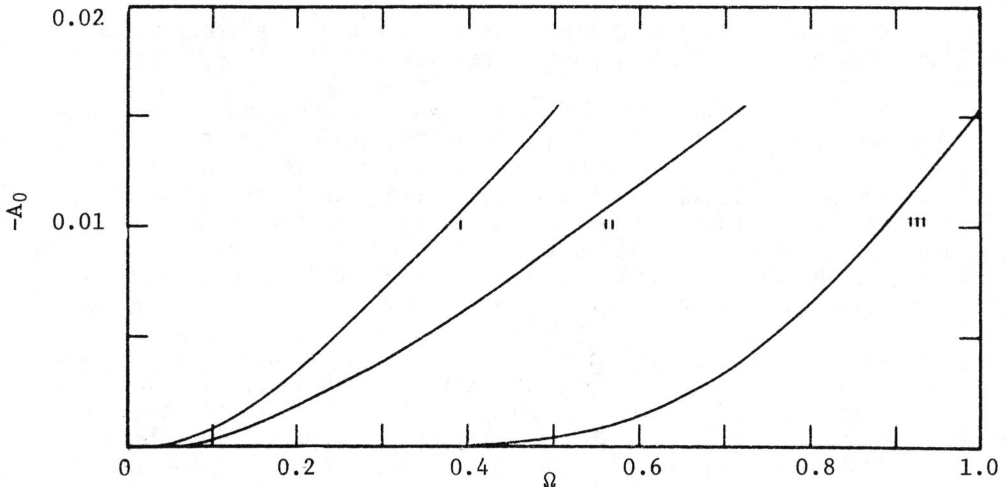

Figure 6. Relative error $A_0 = (\overline{V}_m - \overline{V})/\overline{V}$ in the measurement of the mean velocity for $\hat{z}$ = 0.5, h = 1.

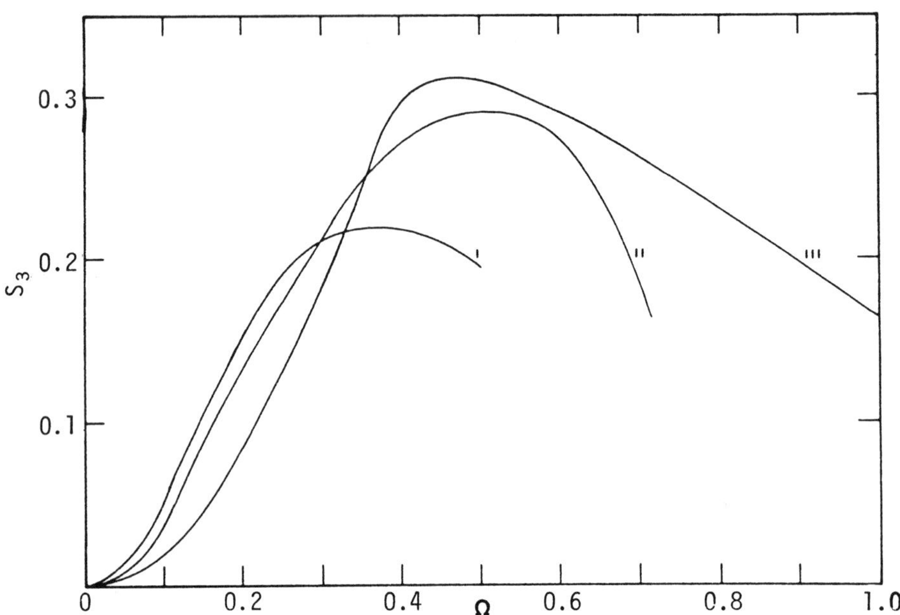

Figure 7. Skewness S_3 for $\hat{z}$ = 0.5, h = 1.

Fortunately, these maxima occur at such high frequencies that they will rarely, if ever, pose a practical problem.

To quantitatively assess the influence of nonlinearities on turbulent spectra, the use of a complicated analog computer would be unavoidable and such a computer is not available to this author. The foregoing discussion makes it seem likely that errors would reach the same order of magnitude as in constant current anemometers, around a frequency half the cutoff frequency of the anemometer. There are hardly any turbulent spectra imaginable which exhibit large fluctuations at such high frequencies.

We may conclude this discussion of results with the remark that the domain of nonlinear effects is restricted to velocity fluctuations larger than 12% of the mean velocity in a frequency range larger than 10% (or in case of the use of an additional low pass filter larger than 20%) of the cutoff frequency of the anemometer. Fortunately, this domain overlaps rarely, if ever, with the domain of actual turbulent flows.

REFERENCES

1. Comte-Bellot, G. and Schon, J. P. (1969), Int. J. Heat Mass Transfer 12, 1661.

2. Comte-Bellot, G. (1976), Annual Review Fluid Mechanics, 8, 209.

3. Freymuth, P. (1967), Rev. Sci. Instrum., 38, 677.

4. Freymuth, P. (1969), Rev. Sci. Instrum., 40, 258.

DISCUSSION

CORRSIN: (Johns Hopkins University)

I would like to remark that the last difference in error you showed is very significant for isotropic turbulence where you normally average triple correlation functions. Normally they reach the maximum skewness at the cut off frequency.

FREYMUTH:

What I mean is that when you use a constant temperature anemometer at a frequency below 10% of its cut off frequency then the induced skewness is 0.02 or smaller, which compares to a value of 0.6 for isotropic turbulence.

CORRSIN:

I am just saying that I am glad that you mentioned all of this because there is some disagreement in the measurements. For instance, if you are shown strong asymmetry from correlation measurements, they are not as antisymmetric as they are in isotropic turbulence. That could still be perhaps partly due to the measurement error and it is still controversial.

FREYMUTH:

I believe that the constant temperature anemometer has an edge over the constant current anemometer concerning this problem because the safe region is connected to the cut off frequency of the anemometer rather than to the cut off frequency of the hot wire. So I think that skewness as measured with a constant current anemometer lately are more likely to be in error than those with the constant temperature anemometer.

COLES: (California Institute of Technology)

I think there is a very similar analysis with very similar conclusions to yours by Tony Perry of Melbourne University. His manuscript* is floating around the offices of various editors. We had the same problem with very large fluctuations in making hot-wire measurements in a cylinder wake, and Perry told us how to do it. It was necessary to offset the cable inductance, to find the optimum square wave signal, and so on.

FREYMUTH:

I am very much interested in this paper.

KLINE: (Stanford University)

I want to comment on the errors in hot-film measurements because we have an important project underway, that is not yet published. The critical result is that the mean velocities are probably all right, even though these were early hot-film data taken before we had sorted out the several troubles of hot-films in water in the work of Morrow.† The mean velocity we could and did calibrate at that time. Fluctuations are a different matter. There was no accepted calibration procedure; we have now devised what we believe is a good procedure. I think this group will be interested in an idea of what we are finding. We are using a technique which is not new. John Laufer used it in his work on channels and pipes around twenty years ago. In Morrow's work we tried to do the same thing other people had done--shake the probe. However, since you need small probes, the fundamental critical frequency is say 100 or 200 Hz, and you can't get too much higher frequencies without reading probe vibrations rather than flow fluctuations. So what we have done recently was to look for a flow in which we know some honest statistics, and then use that flow as a calibration standard. Once you ask the question that way, the obvious flow is the two-dimensional, fully-established channel-flow. In this flow we know that $\partial p/\partial y$ equals $\partial \tau/\partial y$. If you then get everything checked out, which we have been able to do to one-percent statistically, then you have a calibration standard. This is, we know $\overline{uv}$, u', v', and also the main components of the spectra for those three statistics.

---

* Perry, A. E. "The Large and Small Signal Dynamics of the Constant-Temperature Hot-Wire Anemometer," Manuscript, 1976.

† Morrow, T. B. and Kline, S. J. "The Evaluation and Use of Hot-Wire and Hot-Film Anemometers in Liquids," M.E. Department, Report MD-25, Stanford, May, 1971; also ISA Procs.

When we calibrate hot-wires using this calibrating technique, we get the correct results. The X-wires check to about 1%. Using X-wires as a secondary standard, the U wires check to within about 1% or 1.25% on a 95% probability basis. When we use the same procedures on hot film probes, we find (for all the films we have measured so far) r.m.s. errors of 25% to 60% for fluctuations at normal overheat ratios. The hot-films are really off for fluctuations. These results are in air; water will not be this bad. We have not yet had time to test in water. The upshot of this is that the films are going to need calibration. The work is not finished, and we plan to recheck everything. However, I think the upshot of this will be that for reliable measurements we ought to set up such a channel in each laboratory and calibrate wires and films for fluctuations in-house so that we know that we are measuring what we intend to measure. Somebody said this morning that a lot of the scatter in the turbulence data was probably due to measurement inaccuracies. What I am saying is yes, a lot of it certainly must be, and we could eliminate much of these inaccuracies using calibration procedures that we will describe in a report shortly.

WYGNANSKI: (Tel Aviv University)

I have just a question. Those hot films you are mentioning, are they split films?

KLINE:

We have done all the TSI films but not the DISA films yet. Both companies have donated to us three sets of several kinds of probes so that we can do replication tests. We have done the split film, the cone and the wedge. We have not done the TSI solid film. We will do it in the next set of tests. We have not yet done the DISA measurements, but I don't expect to see big differences. We expect there will be some quantitative differences because we are pretty convinced that the problem is owing to the substrate, and the substrate materials are different in the DISA probes.

PART II
MODELLING PROCEDURES

Results of a Two Equation Model for Turbulent Flows and Development of a Relaxation Stress Model for Application to Straining and Rotating Flows

P. G. SAFFMAN
California Institute of Technology
Pasadena, California (U.S.A.)

ABSTRACT

Recent results of computations of turbulent flows using a two equation model are presented. They are similarity solutions for the two dimensional mixing layer, jet and wake. Profiles of passive scalars are also given.

Construction of a more general relaxation stress model is described, to be used for the study of complex flows. Application is made here to the deformation of homogeneous turbulence by uniform shear and strain, and the decay of homogeneous turbulence in a fluid in solid body rotation.

1. INTRODUCTION

The development of simple phenomenological turbulence models is of considerable importance for engineering applications. The advantages of various models can be studied by testing their ability to predict relatively simple turbulent flow configurations, for which some experimental data are available.

This paper considers first three two-dimensional turbulent flows, the mixing layer, jet and wake, using the two equation model of Saffman (1970, 1974). In each case, steady similarity solutions describing the flows are presented. Some results for the turbulent mixing layer have been given earlier (Milinazzo and Saffman, 1976). In §§ 3 and 4 results of further calculations for the inhomogeneous incompressible mixing layer are shown and effects of compressibility are described. The equations for the jet and the wake were previously solved by Govindaraju (1970). The calculations of § 5 and § 6 verify these results and provide more details.

In § 7 the results of § 3, § 5 and § 6 are used to describe the turbulent diffusion of a passive scalar in the mixing layer, jet and wake. The experimental observation that the thermal width is greater than the momentum width in each flow is explained in terms of the difference in the slopes of the temperature and velocity profiles at the turbulent-non-turbulent interfaces.

In § 8 we present a relaxation stress model to deal with flows in which the assumption of an eddy diffusivity is inadequate. The model is applied in § 9 to the deformation of homogeneous turbulence by uniform straining fields, both rotational and irrotational.

2. THE TWO-EQUATION MODEL

For the incompressible, but not necessarily homogeneous, mixing layer, jet and wake, the following equations were solved:

$$\frac{\partial}{\partial x}(\bar{\rho}U) + \frac{\partial}{\partial y}(\bar{\rho}V) = 0, \tag{2.1}$$

$$\bar{\rho}U\frac{\partial U}{\partial x} + \bar{\rho}V\frac{\partial U}{\partial y} = -\left\{\frac{\partial}{\partial x}(\overline{\rho u''u''}) + \frac{\partial}{\partial y}(\overline{\rho v''u''})\right\} \tag{2.2}$$

$$\bar{\rho}U\frac{\partial\bar{\rho}}{\partial x} + \bar{\rho}V\frac{\partial\bar{\rho}}{\partial y} = -\,\bar{\rho}^2\left\{\frac{\partial}{\partial x}\left(\frac{\overline{\rho'u'}}{\bar{\rho}}\right) + \frac{\partial}{\partial y}\left(\frac{\overline{\rho'v'}}{\bar{\rho}}\right)\right\}. \tag{2.3}$$

Here $\underset{\sim}{U} = (U,V) = \dfrac{\overline{\rho\underset{\sim}{u}}}{\bar{\rho}} = (U_i)$ is the mass averaged velocity. Single primes indicate departure of the velocity, $\underset{\sim}{u} = (u,v)$, from the mean, $\bar{\underset{\sim}{u}}$, while double primes denote fluctuation from the mass average value $\underset{\sim}{U}$.

Equation (2.1) describes mass transport and is obtained by averaging

$$\frac{\partial}{\partial x}(\rho u) + \frac{\partial}{\partial y}(\rho v) = 0.$$

Equation (2.2) is derived from the streamwise momentum equation. Equation (2.3) expresses the fact that the fluid is incompressible, and is obtained by substituting the relation $\underset{\sim}{U} = \bar{\underset{\sim}{u}} + \dfrac{\overline{\rho'(u',v')}}{\bar{\rho}}$ into $\mathrm{div}\;\bar{\underset{\sim}{u}} = 0$.

Equations (2.1)-(2.3) are closed using the Saffman model equations, in which it is postulated that the Reynolds stresses and density-velocity correlations satisfy the equations (see Saffman, 1970, 1974)

$$-\overline{\rho u_i'' u_j''} = +2A\bar{\rho}\,\frac{e}{\omega}\,S_{ij} - \frac{2}{3}\,\bar{\rho}e\,\delta_{ij}\,,$$

$$-\frac{\overline{\rho' u_j'}}{\bar{\rho}} = \frac{A}{\sigma}\,\frac{e}{\bar{\rho}\omega}\,\frac{\partial\bar{\rho}}{\partial x_j}\,,$$

where $A = .09$, $\sigma = .5$ (the turbulent Schmidt number), S_{ij} is the rate of strain tensor

$$S_{ij} = \frac{1}{2}\left\{\frac{\partial U_i}{\partial x_j} + \frac{\partial U_j}{\partial x_i}\right\},$$

and the variables e and ω satisfy the rate equations:

$$\bar{\rho}U\,\frac{\partial e}{\partial x} + \bar{\rho}V\,\frac{\partial e}{\partial y} = \bar{\rho}\alpha'' eS - \bar{\rho}e\omega + \frac{A}{2}\left\{\frac{\partial}{\partial x}\left(\frac{\bar{\rho}e}{\omega}\,\frac{\partial e}{\partial x}\right) + \frac{\partial}{\partial y}\left(\frac{\bar{\rho}e}{\omega}\,\frac{\partial e}{\partial y}\right)\right\}, \quad (2.4)$$

$$\bar{\rho}U\,\frac{\partial\omega^2}{\partial x} + \bar{\rho}V\,\frac{\partial\omega^2}{\partial y} = \bar{\rho}\alpha'\omega^2\tilde{S} - \beta'\bar{\rho}\omega^3 + \frac{A}{2}\left\{\frac{\partial}{\partial x}\left(\frac{\bar{\rho}e}{\omega}\,\frac{\partial\omega^2}{\partial x}\right) + \frac{\partial}{\partial y}\left(\frac{\bar{\rho}e}{\omega}\,\frac{\partial\omega^2}{\partial y}\right)\right\}$$

$$- A\gamma\,\frac{\omega}{\bar{\rho}}\left(\frac{\partial\bar{\rho}}{\partial x}\,\frac{\partial(\bar{\rho}e)}{\partial x} + \frac{\partial\bar{\rho}}{\partial y}\,\frac{\partial(\bar{\rho}e)}{\partial y}\right), \quad (2.5)$$

where $S = (2S_{ij})^{\frac{1}{2}}$, $\tilde{S} = [\eta(\partial U_i/\partial x_j)^2 + 2(1 - \eta)S_{ij}^2]^{\frac{1}{2}}$. As in Saffman (1974), Milinazzo and Saffman (1976), the constants α', α'', β', γ have been given the values $\alpha' = .18$, $\alpha'' = .3$, $\eta = 1$, $\beta' = 5/3$, $\gamma = 1$. This corresponds to a value of 0.4 for the Karman constant k, which is related to the other constants by $2k^2 = \beta'\alpha'' - \alpha'$.

The boundary conditions for the system of equations (2.1)-(2.5) are

$$\left.\begin{array}{llllll} U \to U_1, & e \to 0, & \omega \to 0, & \bar{\rho} \to \rho_1 & \text{as} & y \to +\infty, \\ U \to U_2, & e \to 0, & \omega \to 0, & \bar{\rho} \to \rho_2 & \text{as} & y \to -\infty, \end{array}\right\} \quad (2.6)$$

together with conditions at x = constant and $x = +\infty$. For the two-dimensional jet and wake only the case of uniform density is considered, and the constraints, valid in the boundary layer approximation,

$$\int_{-\infty}^{\infty} U^2 dy = M \,, \quad (2 - D \text{ jet}), \tag{2.7}$$

$$\int_{-\infty}^{\infty} U dy = D/U_\infty \,, \quad (2 - D \text{ wake}), \tag{2.8}$$

are imposed in addition. The equations have solutions which are not everywhere analytic and the boundary conditions (2.6) can be applied on curves $y = y_1(x)$, $y = y_2(x)$, which are the positions of the turbulent-non-turbulent interfaces.

Since e, ω and the Reynolds stresses are required to be continuous across these interfaces, e must vanish quadratically while ω vanishes linearly on $y_1(x)$, $y_2(x)$. An equivalent condition is that e vanishes together with the Reynolds stresses.

In sections 3, 4, 5, 6 the partial differential equations (2.1)-(2.5) are reduced to ordinary differential equations by considering similarity solutions of the form

$$x^p G(y/x^q). \tag{2.9}$$

For the mixing layer, equations (2.1)-(2.5) admit similarity solutions of the form (2.9) and enable a partial estimate of the error in the boundary layer approximation to be investigated since these equations contain streamwise diffusion. (The full equations also admit of a similarity solution, but were not investigated because it was thought that such a calculation should also include a more complete turbulence model allowing for anisotropy of the Reynolds stresses.) For the jet and wake, the streamwise diffusion terms were dropped.

3. EFFECT OF STREAMWISE DIFFUSION ON THE INCOMPRESSIBLE, INHOMOGENEOUS MIXING LAYER

To obtain similarity solutions it is assumed that the boundary conditions (2.6) are satisfied on the straight lines $y = 0$, $y = \lambda x$. Since $\mathrm{div}(\overline{\rho}\underset{\sim}{U}) = 0$, a streamfunction $\psi(x,y)$ can be introduced. Defining the similarity forms

$$\psi(x,y) = \rho_1 U_1 \lambda x F(\eta)$$

$$e(x,y) = U_1^2 J(\eta)$$

$$\omega(x,y) = \frac{U_1}{x} K(\eta)$$

$$\overline{\rho} = \rho_1/R(\eta)$$

where $\eta = y/\lambda x$ and substituting into equations (2.1)-(2.5) the following set of ordinary differential equations results:

$$A\left[\frac{J}{KR}\right](1 + \lambda^2\eta^2)R'' + R'\left[\sigma\lambda^2 F + A\{(1 + \lambda^2\eta^2)(\frac{J}{KR})' + \lambda^2\eta \frac{J}{KR}\}\right] = 0 \tag{3.1}$$

$$\frac{A}{2}\left[\frac{J}{KR}\right](1 + \lambda^2\eta^2)J'' + J'\left[\lambda^2 F + \frac{A}{2}\{(1 + \lambda^2\eta^2)(\frac{J}{KR})' + \lambda^2\eta \frac{J}{KR}\}\right] - \lambda \frac{J}{K}[\lambda K - \alpha'' S] = 0 \tag{3.2}$$

$$\frac{A}{2}\left[\frac{J}{KR}\right](1 + \lambda^2\eta^2)(K^2)'' + (K^2)'\left[\lambda^2 F + \frac{A}{2}\{(1 + \lambda^2\eta^2)(\frac{J}{KR})' + 5\lambda^2\eta \frac{J}{KR}\}\right] + \lambda K^2\left[\lambda A\{\eta(\frac{J}{KR})' + \frac{2J}{KR}\} - \frac{1}{R}\{\beta'\lambda K - \alpha' S\} + 2\lambda F'\right] + A\gamma \frac{K}{R}(1 + \lambda^2\eta^2)(\frac{J}{R})'R' = 0 \tag{3.3}$$

$$A\left[\frac{J}{KR}\right](1 + \lambda^2\eta^2)(RF')'' + (RF')'(\lambda^2 F + A\{(1 + \lambda^2\eta^2)(\frac{J}{KR})'\}) + \frac{2}{3}\lambda^2\eta(\frac{J}{R})' + \lambda^2 A(\frac{J}{KR}\eta FR')' = 0, \tag{3.4}$$

where

$$S = [\{(RF')'(1 + \lambda^2\eta^2) - \eta R'F\}^2 + 2\lambda^2(R'F)^2]^{\frac{1}{2}}.$$

The boundary conditions are to be satisfied at $\eta = 0$ and $\eta = 1$. The position of the dividing streamline is given by $F(\eta_s) = 0$.

The boundary conditions (2.6) become the following

$$\left.\begin{aligned} R(0) &= \frac{\rho_1}{\rho_2} = \frac{1}{s}, \qquad R(1) = 1 \\ F'(0) &= \frac{U_2}{U_1}s = rs, \quad F'(1) = 1 \\ J(0) &= J(1) = K(0) = K(1) = 0 \end{aligned}\right\} \tag{3.4a}$$

The parameters $r = U_2/U_1$ and $s = \rho_2/\rho_1$ characterize the mixing layer. The requirement that the solution be well-behaved at the interfaces gives the additional conditions

$$\left.\begin{aligned} 2\lambda^2 FRK' + AJ'' &= 0, \qquad \eta = 0 \\ 2\lambda^2 FRK' + A(1+\lambda^2)J'' &= 0, \qquad \eta = 1 \end{aligned}\right\} \tag{3.4b}$$

The conditions (3.4b) are satisfied if J vanishes quadratically while K vanishes linearly at $\eta = 0, 1$ (Saffman, 1970) or equivalently if the eddy diffusivity varies linearly with the distance from the turbulent-non-turbulent interface.

The equations were solved using the 'Box' scheme of Keller (1974), which provides a relatively simple but efficient method of solving the systems of ordinary differential equations which arise.

Equations (3.1)-(3.4) without streamwise diffusion were solved and the results have been presented elsewhere by Milinazzo and Saffman (1976). Here, we just comment on the effect of the streamwise diffusion on those results. We examined and compared two sets of data, $s = 1$, $r = 0$ to $.8$ and $s = 7$, $r = 0$ to $.8$. The computations with and without streamwise diffusion are found to be in very close agreement. The largest differences are found for $s = 7$, $0 \leq r \leq .25$, in the value of the spreading width λ. Those discrepancies are no larger than 3%. All values of the other variables are virtually unchanged. We therefore conclude that the neglect of streamwise diffusion is a good approximation.

4. THE COMPRESSIBLE HOMOGENEOUS MIXING LAYER

To model the compressible mixing layer, equation (2.3) expressing the incompressibility of the fluid is replaced by an energy equation as follows

$$\bar{\rho}U \frac{\partial}{\partial x} \left(H + \frac{1}{2}[U^2 + V^2] + \frac{1}{2}\frac{1}{\bar{\rho}} \overline{\rho[(u'')^2 + (v'')^2]} \right)$$

$$+ \bar{\rho}V \frac{\partial}{\partial y} \left(H + \frac{1}{2}[U^2 + V^2] + \frac{1}{2}\frac{1}{\bar{\rho}} \overline{\rho[(u'')^2 + (v'')^2]} \right)$$

$$= \frac{\partial}{\partial x} \left\{ -\overline{\rho H'' u''} - \frac{1}{2} \overline{\rho u''[(u'')^2 + (v'')^2]} \right\}$$

$$+ \frac{\partial}{\partial y} \left\{ -\overline{\rho H'' v''} - \frac{1}{2} \overline{\rho v''[(u'')^2 + (v'')^2]} \right\} , \qquad (4.1)$$

Here H is the mass averaged enthalpy. Neglecting the mean kinetic energy of the turbulence, relative to the kinetic energy of the mean flow, and using the boundary layer approximation, (4.1) reduces to

$$\bar{\rho}U \frac{\partial}{\partial x} \left\{ H + \frac{1}{2} U^2 \right\} + \bar{\rho}V \frac{\partial}{\partial y} \left\{ H + \frac{1}{2} U^2 \right\}$$

$$= \frac{\partial}{\partial y} \left\{ -\overline{\rho H'' v''} - \frac{1}{2} \overline{\rho v''[(u'')^2 + (v'')^2]} \right\} . \qquad (4.2)$$

Introducing the eddy viscosity approximation and using the momentum equation, equation (4.2) becomes

$$\bar{\rho}U \frac{\partial H}{\partial x} + \bar{\rho}V \frac{\partial H}{\partial y} = +A\bar{\rho} \frac{e}{\omega} \left(\frac{\partial U}{\partial y} \right)^2 + \frac{A}{\sigma} \frac{\partial}{\partial y} \left(\frac{\bar{\rho} e}{\omega} \frac{\partial H}{\partial y} \right) , \qquad (4.3)$$

where σ is now a turbulent Prandtl number.

An equation of state relating the enthalpy, h, to the density and pressure gives the required equation for $\bar{\rho}$. Since h equals $\frac{\gamma}{\gamma - 1} P/\rho$, γ = adiabatic exponent, it follows that

$$H = \frac{1}{\bar{\rho}} \frac{1}{\gamma - 1} \gamma P. \qquad (4.4)$$

If the Mach number of the turbulence is small, then the mean pressure can be taken as a constant across the shear layer, and equation (4.4) gives

$$H = \frac{a_1^2 \rho_1}{\gamma - 1} \, 1/\bar{\rho}, \qquad (4.5)$$

where a_1 is the speed of sound in the fast stream. Substituting the equation of state (4.5) into equation (4.3) gives the following.

$$\bar{\rho}U \frac{\partial}{\partial x}\left(\frac{1}{\bar{\rho}}\right) + \bar{\rho}V \frac{\partial}{\partial y}\left(\frac{1}{\bar{\rho}}\right) = + \frac{A}{\sigma}\frac{\partial}{\partial y}\left(\frac{\bar{\rho}e}{\omega}\frac{\partial}{\partial y}\left(\frac{1}{\bar{\rho}}\right)\right)$$

$$+ \frac{\gamma - 1}{\rho_1\, a_1}\frac{\bar{\rho}e}{\omega}\left(\frac{\partial U}{\partial y}\right)^2 . \tag{4.6}$$

The rate equation for e must be modified to account for the increase in turbulent energy due to isotropic compression. As in Saffman (1974), Wilcox and Alber (1972), the term

$$- \xi\bar{\rho}e\left(\frac{\partial\bar{u}}{\partial x} + \frac{\partial\bar{v}}{\partial y}\right) , \tag{4.7}$$

with $\xi = 2.5$, is added to the right-hand side of equation (2.4). Using the model equation for the transport of momentum and equation (4.6), the term (4.7) can also be expressed as

$$- \xi\frac{(\bar{\rho}e)^2}{\omega}\frac{(\gamma - 1)}{\rho_1\, a_1^2}\left(\frac{\partial U}{\partial y}\right)^2 , \tag{4.8}$$

Equations (2.1), (2.2), (2.4), (2.5), with streamwise diffusion neglected, and the term (4.8) added to the resulting form of equation (2.4), together with (4.6), are reduced to ordinary differential equations using the similarity hypothesis of § 2. This set of ordinary differential equations is then solved using the 'Box' scheme.

Calculations were done for two cases: (1) $r = 0$, $s = 1$; (2) $r = 0$, $s = s(M_1) = \left(1 + \frac{(\gamma - 1)}{2} M_1^2\right)^{-1}$. The first case corresponds to adjusting the stagnation temperatures so that at Mach number M_1 both streams have the same density. Case two corresponds to both streams having the same stagnation temperature. Equal stagnation temperatures imply

$$C_p T_0 = \frac{1}{2} U_1^2 + \frac{\gamma}{\gamma - 1}\frac{P}{\rho_1} ,$$

$$= \frac{\gamma}{\gamma - 1}\frac{P}{\rho_2} ,$$

where C_p = specific heat at constant pressure and T_0 is the stagnation temperature. Since $\gamma P = a_1^2\, \rho_1$ it follows that

$$\frac{1}{s} = \frac{1}{\rho_2/\rho_1} = 1 + \frac{\gamma - 1}{2} M_1^2 . \tag{4.9}$$

The results of the calculations are summarized in Figures 1 and 2. Figure 1 compares the thinning of the shear layer, as measured by

$$\delta'_\omega = U_1/x\left(\frac{\partial U}{\partial y}\right)_{max} ,$$

for cases 1 and 2. It can be seen that the model predicts a much sharper decrease in the width of the mixing layer when density changes are due to compressibility effects than when the density differences are due only to inhomogeneity of the fluids.

Figure 2 shows the variation of the minimum density $\bar{\rho}_{min}$ across the layer. The dependence of the maximum Reynolds shear stress, τ, with Mach number, for cases 1 and 2, is also shown, together with the dependence of τ_{max} on s for $M_1 = 0$. Although it might be expected that the thinner the layer, the lower is the value of τ_{max}, this is not the case. For corresponding values of s or M_1, the fluid density in the mixing region at the point of maximum τ is greater for case 1 than either that of case 2 or that of heterogeneous, incompressible mixing layer. (For $s \approx 7$ or $M_1 \approx 6$, the densities differ by as much as a factor of 4.) Since $\tau = \rho u''v''$ the values of τ_{max} for case 2 and the incompressible mixing

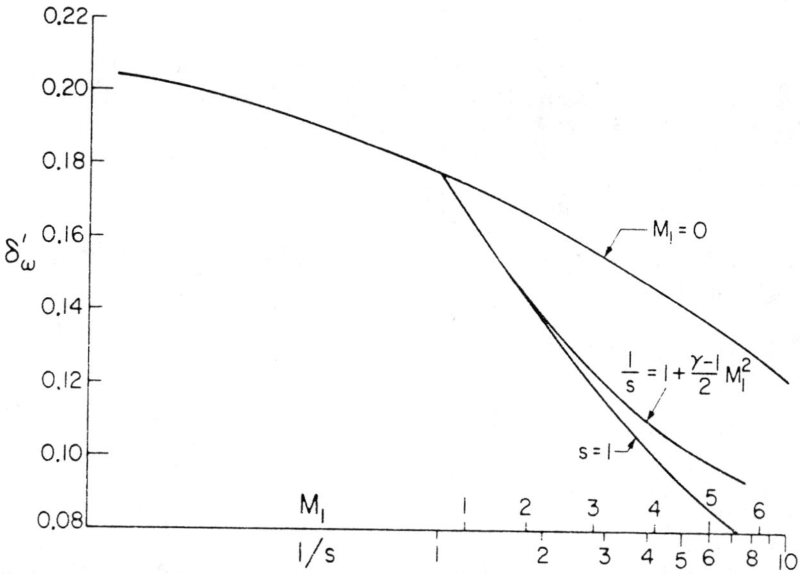

Figure 1. Variation of vorticity thickness as a function of Mach number or density ratio. Scales of s and M_1 are related on the abscissa by equation (4.9).

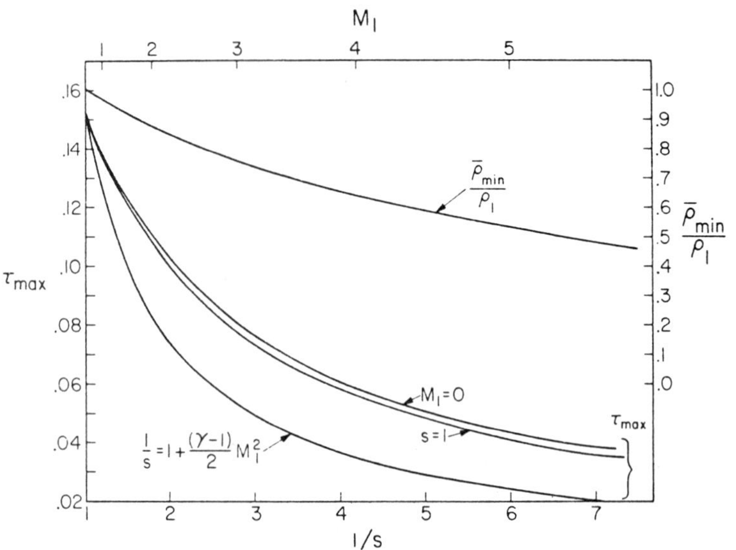

Figure 2. Variation of minimum density with Mach number for case 1, $s = 1$, and variation of $\tau_{max}/\rho_1 U_1^2$ with M_1 for cases 1 and 2. Relation between s and M_1 on abscissa again given by equation (4.9).

layer are much lower in comparison to the value of τ_{max} for case 1 than would be indicated by Figure 1. As expected, the thinner the layer, the lower is the value of $(\tau/\rho)_{max}$.

5. TWO-DIMENSIONAL JET

For a two-dimensional free jet of total kinematic momentum flux M the appropriate similarity forms are

$$\left.\begin{aligned} \psi(x,y) &= (Mx)^{\frac{1}{2}}\lambda F(\eta) \\ \rho(x,y) &= \frac{M}{x}\, J(\eta) \\ \omega(x,y) &= \left(\frac{M}{x^3}\right)^{\frac{1}{2}} K(\eta) \end{aligned}\right\} \qquad (5.1)$$

with $\eta = y/\lambda x$. Since the flow is symmetric it can be assumed that the turbulent-non-turbulent interfaces are at $\eta = \pm 1$.

Substituting the similarity forms (5.1) into (2.1), (2.3), (2.4), (2.5) gives (streamwise diffusion neglected)

$$\frac{A}{2}\left(\frac{J}{K}J'\right)' + \frac{\lambda^2}{2}J'F - \lambda J(\lambda K - \alpha''|F''| - \lambda F') = 0, \tag{5.2}$$

$$\frac{A}{2}(JK')' + \lambda^2 KK'F - \lambda^2 K^2(\beta'\lambda K - \alpha'|F''| + 3\lambda F') = 0, \tag{5.3}$$

$$A\left(\frac{J}{K}F''\right)' + \frac{\lambda^2}{2}(FF')' = 0. \tag{5.4}$$

Since the flow is symmetric about $\eta = 0$, equations (5.2)-(5.4) need only be integrated on the interval $0 \leq \eta \leq 1$. The symmetry conditions are

$$J'(0) = K'(0) = F(0) = 0, \tag{5.5}$$

while the boundary conditions at the free surface are

$$J(1) = J'(1) = K(1) = F'(1) = 0. \tag{5.6}$$

In (5.5) $F(0) = 0$ since symmetry about $\eta = 0$ requires that the transverse velocity, V, must vanish on $\eta = 0$. The additional constraint (2.7) leads to the requirement that

$$\int_0^1 [F'(\eta)]^2 d\eta = \frac{1}{2\lambda} \,. \tag{5.7}$$

Since (5.7) can be thought of as a normalizing condition it can be replaced by

$$F(1) = 1, \tag{5.8}$$

and the solution F, J, K of (5.2)-(5.7) is obtained by scaling the solution $\tilde{F}$, $\tilde{J}$, $\tilde{K}$ of (5.2)-(5.6), (5.8) according to $B^2\tilde{J} = J$, $B\tilde{K} = K$, $B\tilde{F} = F$, where $B = 1./\sqrt{2\lambda\int_0^1 [\tilde{F}'(\eta)]^2 d\eta}$.

Solutions of (5.2)-(5.7) have been found and the results are discussed in Govindaraju (1970), Saffman (1970).

The solutions obtained by Govindaraju were calculated using a parallel shooting method. It is reported (Govindaraju, 1970) that these solutions are difficult to obtain. In contrast, the 'Box' scheme gave the solutions described here relatively easily.

The results of the calculation are summarized in Table 1. Results were obtained for various values of α' and β' to study the dependence on the parameters. Experimental values of $F'(0)$ are 2.4 (Schlichting, 1960). Davies, Keffer and Baines (1974) give values of $y_{1/2}/x$ ($y_{1/2}$ is the distance from the center line where the mean velocity is one-half the center line velocity) of 0.11 when velocity profiles are measured in the usual way and 0.16 when velocities are measured with conditional sampling. Since the similarity solutions impose straight interfaces, it is more appropriate to compare the theory with measurements conditioned on turbulent fluid. The experiments of Davies et al. indicate that

$$|J_{max} - J(0)|/(J_{max} + J(0)) \approx .12$$

while Table 1 gives approximately 0.07 for the preferred value of $\beta' = 5/3$. In view of the difficulty in measuring this quantity, the agreement is reasonable.

Figure 3 shows profiles for F', λF, J, $JF''/\lambda K$ for $\alpha' = .18$, $\beta' = 1.67$. These profiles are in good qualitative agreement with those seen in experiments (e.g., Davies, Keffer and Baines, 1974).

Table 1. Results of the calculation for a 2-D jet.

β'	α'	λ	$y_{1/2}/x$	$F'(0)$	$J(0)$	J_{max}	$K(0)$	$\frac{\tau_{max}}{U^2_{max}}$
1.67	.20	.30	.14	2.21	.50	.58	4.1	.031
1.67	.18	.31	.15	2.15	.51	.59	3.9	.032
2.00	.20	.38	.20	1.87	.63	.66	2.7	.043
2.00	.18	.42	.22	1.78	.66	.68	2.5	.047

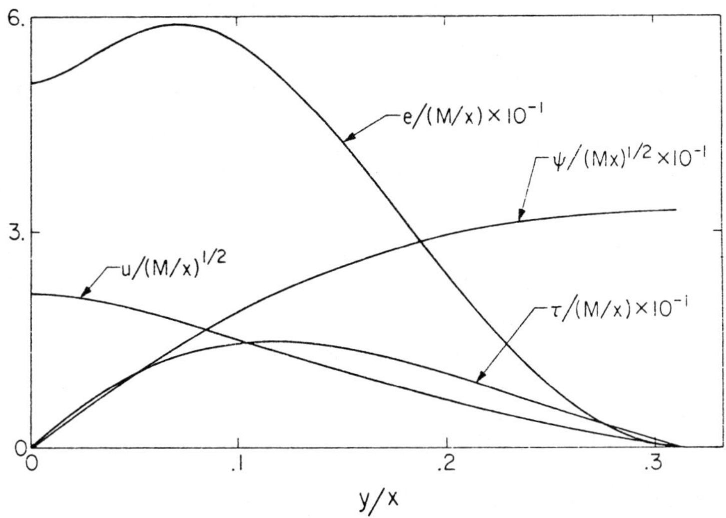

Figure 3. Profiles of $3/(M/x)$, $U/(M/x)^{\frac{1}{2}}$, $\psi/(Mx)^{\frac{1}{2}}$, $\tau/(M/x)$ as functions of y/x for the two-dimensional jet. Values of parameters are $\beta' = 5/3$, $\alpha' = .18$.

6. TWO-DIMENSIONAL WAKE

For a two-dimensional wake of total kinematic momentum defect D and free stream velocity U_∞ the appropriate similarity forms are

$$\left.\begin{aligned} u &= \left(\frac{D}{x}\right)^{\frac{1}{2}} F(\eta) \\ e &= \frac{D}{x} J(\eta) \\ \omega &= \frac{U_\infty}{x} K(\eta) \end{aligned}\right\} \qquad (6.1)$$

where $\eta = \frac{U_\infty}{D^{\frac{1}{2}}} y/\lambda x^{\frac{1}{2}}$ and u is the streamwise defect in the velocity caused by the cylinder.

Substituting the expressions (6.1) now gives the ordinary differential equations

$$\left\{A \frac{J}{K} F'\right\}' + \frac{1}{2}(\lambda^2\eta F)' = 0, \tag{6.2}$$

$$\left\{\frac{A}{2} \frac{J}{K} J'\right\}' + \frac{\lambda^2}{2}\eta J' - \lambda J(\lambda K - \alpha''|F'| - \lambda) = 0, \tag{6.3}$$

$$(AJK')' + \frac{\lambda^2}{2}\eta(K^2)' - \lambda K^2(\beta'\lambda K - \alpha'|F'| - 2\lambda) = 0. \tag{6.4}$$

As in § 5 it is only necessary to integrate equations (6.2)-(6.4) on $0 \leq \eta \leq 1$. The symmetry conditions at $\eta = 0$ are now

$$J'(0) = K'(0) = F'(0) = 0. \tag{6.5}$$

At $\eta = 0$ the boundary conditions are

$$J(1) = J'(1) = K(1) = 0. \tag{6.6}$$

The conservation of momentum condition (2.8) becomes

$$\int_0^1 F(\eta)d\eta = \frac{1}{2\lambda}. \tag{6.7}$$

Since equations (6.2)-(6.6) are invariant under the scaling $F \to B\tilde{F}$, $J \to B^2\tilde{J}$, $\lambda \to B\tilde{\lambda}$ the normalizing condition (6.7) can be replaced by

$$\tilde{F}(0) = 1. \tag{6.8}$$

The constant B can then be determined from

$$B = 1./\sqrt{2\tilde{\lambda}\int_0^1\tilde{F}(\eta)d\eta}.$$

The results of the calculation are summarized in Table 2. Values were again obtained with different values of the parameters α' and β'. Experimental values for $y_{\frac{1}{2}}/D(x)^{\frac{1}{2}}$ are about 0.35 and for F(0) about 1.4. Figure 4 shows the profiles for F, J, $JF'/\lambda K$.

7. DIFFUSION OF A PASSIVE SCALAR

We now consider the diffusion of a passive scalar, for convenience called temperature, in the incompressible mixing layer, jet and wake. By definition, a passive scalar is one which has no effect on the flow field, which we take to be those described in §§ 3, 5, 6.

Table 2. Results of the calculation for a 2-D wake.

β'	α'	λ	$\frac{y_{\frac{1}{2}}}{(Dx)^{\frac{1}{2}}}$	F(0)	J(0)	J_{max}	K(0)	$\frac{x\tau_{max}}{D}$
1.67	.21	.54	.35	1.50	.66	.68	1.30	.21
1.67	.18	.57	.36	1.43	.70	.71	1.24	.20
1.67	.15	.59	.38	1.37	.74	.75	1.19	.19
2.00	.21	.67	.45	1.17	.95	.95	.99	.16
2.00	.18	.69	.46	1.13	1.00	1.00	.96	.15
2.00	.15	.72	.48	1.09	1.05	1.05	.94	.15

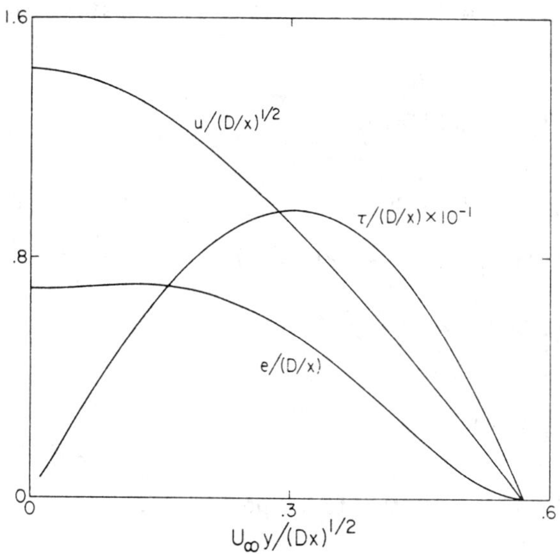

Figure 4. Profiles of $u/(D/x)^{\frac{1}{2}}$, $e/(D/x)$, $\tau/(D/x)$ as functions of y/x for the two-dimensional wake with $\beta' = 5/3$, $\alpha' = .18$.

The Reynolds equation for $\theta = \overline{\rho T}/\overline{\rho}$, in the boundary layer approximation, is

$$\overline{\rho}U \frac{\partial\theta}{\partial x} + \overline{\rho}V \frac{\partial\theta}{\partial y} = \frac{\partial}{\partial y}(-\overline{\rho v''T''}). \tag{7.1}$$

Using the gradient transport model of this work, it is postulated that θ satisfies the equation

$$\overline{\rho}U \frac{\partial\theta}{\partial x} + \overline{\rho}V \frac{\partial\theta}{\partial y} = \frac{\partial}{\partial y}\left(\frac{A}{\sigma}\frac{\overline{\rho e}}{\omega}\frac{\partial\theta}{\partial y}\right) \tag{7.2}$$

where σ is a turbulent Prandtl number.

To obtain the similarity equations for the three flows, each is considered separately.

For the mixing layer θ can be expressed as

$$\theta = (T_1 - T_2)\Theta(\eta) + T_2 \tag{7.3}$$

where $\eta = y/\lambda x$, $T_1 = T(+\infty)$, $T_2 = T(-\infty)$. Substituting (7.3) into (7.2) gives

$$\left(\frac{A}{\sigma}\frac{J}{KR}\Theta'\right)' + \lambda^2 F\Theta' = 0, \tag{7.4}$$

$$\Theta(0) = 0, \quad \Theta(1) = 1.$$

Using the results of § 3, equation (7.4) is integrated numerically for Θ.

For the 2-D jet

$$\theta = \frac{Q}{(Mx)^{\frac{1}{2}}}\Theta(\eta) \tag{7.5}$$

where $\eta = y/\lambda x$ and Q is the total heat flux,

$$\int_{-\lambda x}^{\lambda x} uT\,dy = Q, \tag{7.6}$$

which provides the constraint as follows.

$$\int_0^1 F'(\eta)\Theta(\eta)d\eta = \frac{1}{2\lambda}. \qquad (7.7)$$

The symmetry of the flow about the dividing streamline $\eta = 0$ implies

$$\Theta'(0) = 0. \qquad (7.8)$$

Substituting (7.5) into (7.2) then using (7.8) gives

$$\left(\frac{A}{\sigma}\frac{J}{K}\Theta'\right)' + \frac{\lambda^2}{2}(F\Theta) = 0. \qquad (7.9)$$

Using the momentum equation (5.4) and condition (7.7) it can be seen that

$$\Theta(\eta) = [F'(\eta)]^{\sigma}/2\lambda \int_{\lambda}^{1} [F'(\eta)]^{1+\sigma}\, d\eta. \qquad (7.10)$$

For the 2-D wake

$$\theta = \frac{Q}{(Dx)^{\frac{1}{2}}}\,\Theta(\eta) \qquad (7.11)$$

where $\eta = \frac{U_\infty}{D^{\frac{1}{2}}} y/\lambda x^{\frac{1}{2}}$. The condition

$$\int_{-\lambda x}^{\lambda x} T\, dy = Q/U_\infty \qquad (7.12)$$

provides the constraint

$$\int_0^1 \Theta(\eta)d\eta = 1/2\lambda. \qquad (7.13)$$

Symmetry about $\eta = 0$ gives

$$\Theta'(0) = 0. \qquad (7.14)$$

Substituting (7.11) into (7.3), using (7.14), then noting (6.2) and (7.13) gives

$$\Theta(\eta) = [F(\eta)]^{\sigma}/2\lambda \int_0^1 [F(\eta)]^{\sigma} d\eta \qquad (7.15)$$

For $\sigma = 1$ it is clear from the construction of the model that the temperature and velocity distributions coincide. For our calculations, we have taken in all cases the value of σ to be .5. This seems to be a representative value for the turbulent Prandtl number in a free flow away from solid walls (see Jenkins and Goldschmidt, 1974, p. 89). Some calculations done with different values of σ indicated that the dependence on σ was weak.

Figures 5, 6, 7 compare the velocity and temperature profiles for the three flows. In each case, it can be seen that the temperature profile approaches the turbulent-non-turbulent interface with a larger slope than does the velocity profile. It is easy to demonstrate analytically that the variation is like (distance)$^{\sigma}$. This difference in slope can be used to explain the observation that the temperature profile is wider than the velocity profile. If it is assumed that these steady theoretical profiles can be 'wobbled' to recover the unsteady character of the experimental profiles associated with slow variations of the interface, then this difference in slope accounts for the difference in widths. This has been noted by Davies, Keffer, Baines (1974), where it is suggested that the thermal and momentum spread are better measured by quantities conditioned on the turbulent fluid. Figures 5, 6, 7 also show the wobbled profiles, $\langle U \rangle$, $\langle \theta \rangle$, defined by

$$\langle R(y) \rangle = \int R(y')W(y - y')dy',$$

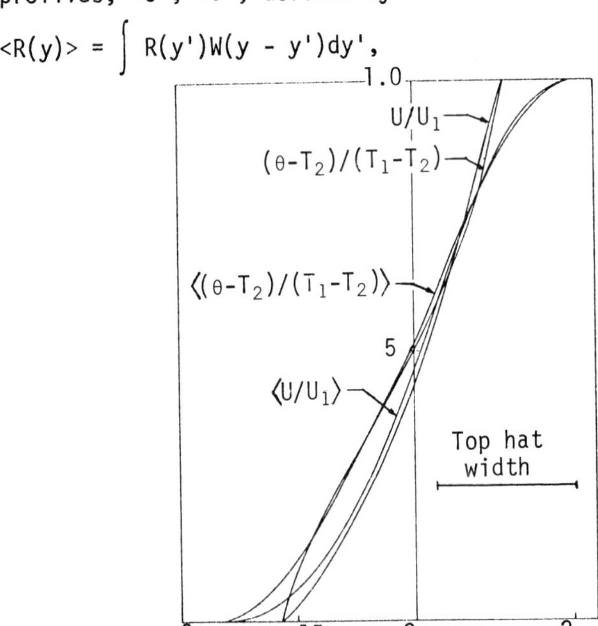

Figure 5. Effect of wobbling on profiles of mean velocity and temperature for the incompressible mixing layer $s = 1$, $r = 0$. Turbulent Prandtl number $\sigma = 0.5$.

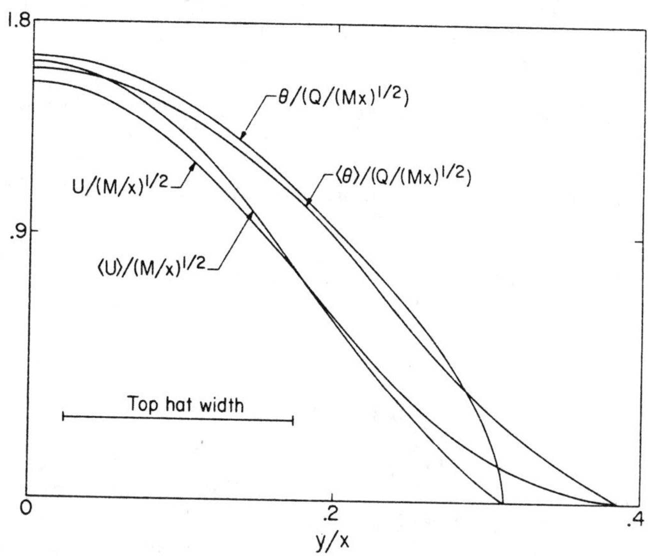

Figure 6. Effect of wobbling on profiles of mean velocity and temperature for the two-dimensional jet. $\sigma = 0.5$.

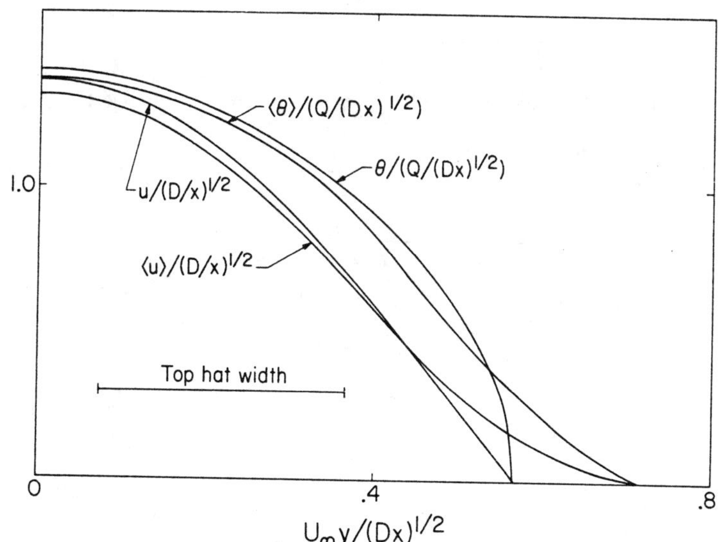

Figure 7. Effect of wobbling on profiles of mean velocity and temperature for the two-dimensional wake. $\sigma = 0.5$.

where W(y) is a top-hat function whose integral is unity. The width of the top hat was chosen to be 60%, 25% and 25% of the flow width, respectively.

The calculated profiles are in good agreement with profiles seen in experiments (see Davies, Keffer, Baines, 1974).

Table 3 compares the ratio of the calculated half-widths for temperature and velocity, $\lambda_r = \lambda_{\frac{1}{2},T}/\lambda_{\frac{1}{2},U}$ for the 2-D jet and wake.

Figures 8, 9, 10 show the Reynolds stress $\overline{u'v'}$ and the corresponding velocity-temperature correlation term $\overline{\theta'v'}$ for each of the three flows.

8. RELAXATION STRESS MODEL

The two equation model described in the previous sections has been reasonably successful in 'postdicting' the properties of turbulent mixing layers, jets and wakes. The model was also used to make an actual prediction (see Figures 1 and 2) about effects of compressibility on the mixing layer between two streams of equal density (but unequal stagnation temperatures), and it will be interesting to see how the prediction compares with experiment when these are carried out. The model also works adequately for boundary layers on flat plates, Saffman and Wilcox (1974), Knight (1975).

Table 3. Values of λ_r, theory and experiment.

Flow	λ_r			
	Calculated	Calculated with 25% wobble	Experiment	
Jet	1.5	1.2	1.3	Jenkins-Goldschmidt, 1974
			1.2	Davies-Keffer-Baines, 1974
Wake	1.3	1.2	1.2	Townsend, 1956

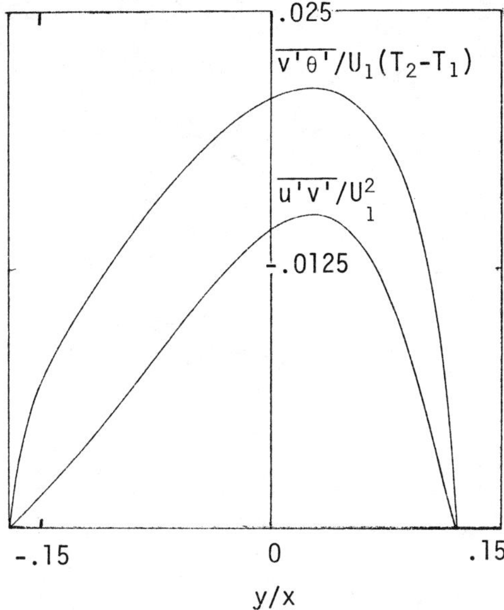

Figure 8. Comparison of Reynolds stress and transport of passive scalar for incompressible mixing layer r = 0, s = 1. Shown plotted are $\overline{u'v'}/U_1^2$ and $\overline{\theta'v'}/U_1(T_2 - T_1)$.

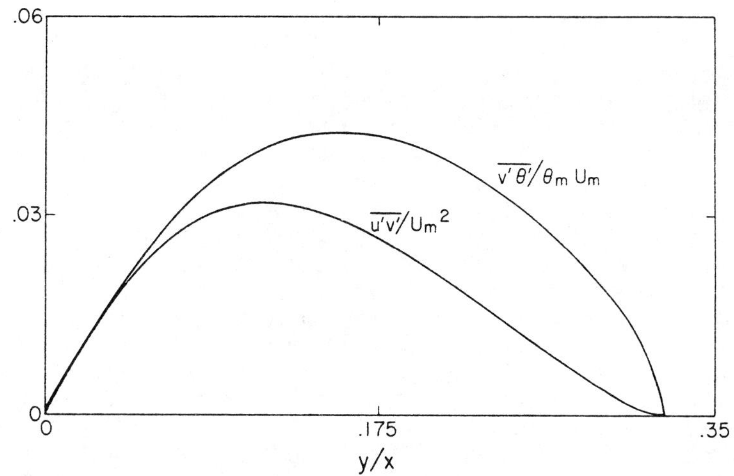

Figure 9. Comparison of Reynolds stress and scalar transport for two-dimensional jet. $\overline{u'v'}/U_m^2$, $\overline{\theta'v'}/U_m\Theta_m$. U_m = centerline mean velocity, Θ_m = centerline mean temperature.

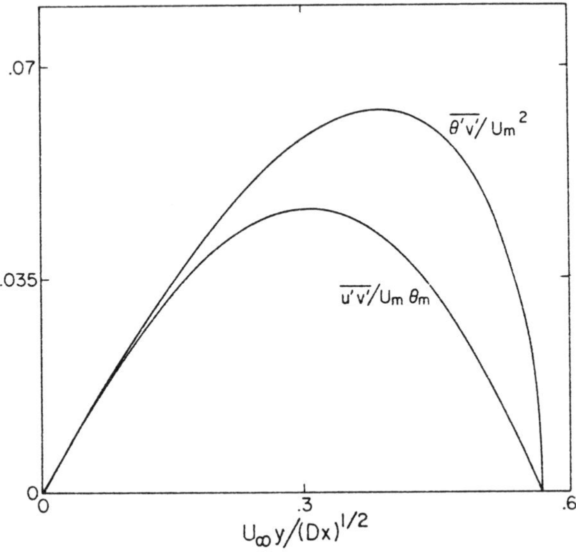

Figure 10. Comparison of Reynolds stress and scalar transport for two-dimensional wake. $\overline{u'v'}/U_m^2$, $\overline{\theta'v'}/U_m\theta_m$. U_m = centerline mean velocity defect. Θ_m = centerline mean temperature.

However, these flows have in common the property that they are essentially unidirectional and mean properties change slowly along the streamlines. There is reason to believe that the model is inadequate to describe flows in which there is significant mean streamline curvature or mean flow acceleration. The reason lies in the eddy viscosity hypothesis, according to which the principal axes of the deviatoric Reynolds stress tensor, $\overline{u_i u_j} - 1/3\, q^2\delta_{ij}$, are parallel to the principal axes of the mean rate of strain tensor, S_{ij}. The tensor S_{ij} can be changed instantaneously in incompressible fluid by the application of pressures exerted on boundaries, whereas the deviatoric Reynolds stress, being a property of the turbulent vorticity, requires a finite time to change or relax to a new value set by the new mean rate of the strain. There will also be diffusion of Reynolds stress by turbulent transport. This ensures that the shear stress doesn't automatically vanish with the mean velocity gradient. In addition, we can expect that the eddy diffusion hypothesis is but the first term in an expansion of a functional dependence of the Reynolds stress on mean flow properties. Extra terms must be incorporated because of the known anisotropy of the principal Reynolds stresses in non-axisymmetric flow, such as flow along a pipe of noncircular cross section, which is not in accord with the eddy diffusion hypothesis. Thus although the two-equation model described above has been reasonably successful in predicting the properties of a class of turbulent flows, it cannot have sufficient generality or complexity to describe flows with non negligible mean streamline curvature or acceleration (complex flows), and we shall now construct a generalization to try to remedy these deficiencies. Reynolds (1976) commented particularly on the inability of the two-equation model to describe the unsteady distortion of homogeneous turbulence by a uniform irrotational strain or uniform shear, and we shall concentrate here on the ability to predict this special class of flows with the new, 'relaxation stress' model. Application to more realistic flows is currently under active study.

For simplicity and to bring out the essential features of the approach, we shall ignore effects of compressibility and henceforth take $\overline{\rho} = 1$. Direct effects of viscosity will also continue to be neglected.

We retain equations (2.5) and (2.4) for the transport of pseudo-vorticity and pseudo-energy, but generalize the eddy viscosity hypotheses. We will also have some new values for the constants.

The first new hypothesis is that associated with a uniform mean velocity gradient tensor is an equilibrium Reynolds stress $\overline{(u_i u_j)_E}$ given by

$$\overline{(u_i u_j)_E} = \frac{2}{3} e\delta_{ij} - 2 \frac{Ae}{\omega} S_{ij} + \frac{Be}{\omega^2} S^2_{k\ell}\delta_{ij} + C \frac{e}{\omega^2} S_{ik}S_{jk}$$
$$+ \frac{De}{\omega^2}(S_{jk}\Omega_{ik} + S_{ik}\Omega_{jk}) + \frac{Fe}{\omega^2}\Omega^2_{k\ell}\delta_{ij} + \frac{Ge}{\omega^2}\Omega_{ik}\Omega_{jk}. \quad (8.1)$$

Here

$$\Omega_{ik} = U_{i,k} - U_{k,i} \quad (8.2)$$

is the rotation tensor; its components are proportional to the mean vorticity components. The numbers A, B, C, D, F, G are to be universal constants. A special case of (8.1) was proposed tentatively by Saffman (1974), and rediscovered by Pope (1975). The second hypothesis relates the actual Reynolds stress $\overline{u_i u_j}$ to the equilibrium value $\overline{(u_i u_j)_E}$ by a relaxation diffusion equation,

$$\frac{D}{Dt}\overline{u_i u_j} = -\theta\omega(\overline{u_i u_j} - \overline{(u_i u_j)_E}) + A''' \frac{\partial}{\partial x_k}\left(\frac{e}{\omega}\frac{\partial}{\partial x_k}\overline{u_i u_j}\right), \quad (8.3)$$

where θ and A''' are further universal constants describing the rate of approach to equilibrium and the turbulent diffusion of Reynolds stresses. We suppose that $\theta > 1$, so that the turbulence tends to equilibrium faster than it decays.

The new model is complete, and allows the turbulent flow to have more structure than the two-equation model, but the price is an extra number of constants which cannot all be determined by general arguments or expressed in terms of wall layer properties as was the case for the two-equation model.

To proceed, we consider special cases. First, the decay of homogeneous turbulence. When there is no mean velocity and turbulence quantities are spatially homogeneous, the solutions of the equations are

$$\omega = \frac{2}{\beta' t}, \quad e = \frac{K}{t^{2/\beta'}}, \quad \overline{u_1^2} = \frac{2}{3}\frac{K\theta}{\theta - 1} t^{-2/\beta'} + K_1 t^{-2\theta/\beta'}, \text{ etc.} \tag{8.4}$$

where K, K_1 are arbitrary constants and time is measured from a virtual origin. There are theoretical and experimental reasons (Saffman, 1967) for believing that the energy of homogeneous turbulence decays like $t^{-6/5}$, giving

$$\beta' = 5/3. \tag{8.5}$$

We now appeal to the experimental observation that in plain strain, $U_1 = \varepsilon x_1$, $U_2 = -\varepsilon x_2$, $U_3 = 0$, measured values obey $\overline{u_3^2} = \frac{1}{2}(\overline{u_1^2} + \overline{u_2^2})$. This implies that C = 0. Next we consider the decay of homogeneous turbulence in a fluid which is in rigid body rotation. Ibbetson and Tritton (1975) have recently studied such flows experimentally. The observations did not show any particular anisotropy and this suggests G = 0. These experiments also show that the value $\eta = 1$ employed previously is unacceptable, as it predicts an enhanced dissipation due to rotation orders of magnitude larger than that observed. Numerical experiments with various values and comparisons with the observations suggest that

$$\eta = \frac{1}{4} \tag{8.6}$$

is an appropriate value, but this value should be regarded as tentative until an independent verification of the experiment is carried out.

Strained homogeneous turbulence returns to isotropy when the distortion is removed. The experimental data are in conflict. For example, Grant (1958) saw no return to isotropy, the energy components all decaying at the same rate. This would imply $\theta = 1$. On the other hand, Tucker and Reynolds (1968) saw a definite return to isotropy, as do most other workers, and the rate is consistent with the value

$$\theta = 2 \tag{8.7}$$

which we shall adopt.

We now consider the statistically steady situation of a turbulent wall layer, with friction velocity u_*. In this case $\overline{u_i u_j} = \overline{(u_i u_j)}_E$, and the appropriate asymptotic solution of (2.5) and (2.4), combined with (8.1) for a pure shear and (8.3), is

$$e = \frac{\alpha''}{A} u_*^2, \quad \omega = \frac{\alpha'' u_*}{ky}, \; \tau = u_*^2, \quad \frac{dU}{dy} = \frac{u_*}{ky},$$

$$q^2 = \frac{2\alpha''}{A}\left[1 + \frac{3}{4\alpha''^2}(B + 4F)\right]u_*^2, \tag{8.8}$$

where

$$k^2 = \frac{\beta'\alpha'' - \alpha'}{4A'/A}, \tag{8.9}$$

and $\tau = -\overline{uv}$ denotes the Reynolds shear stress. The rate of production of turbulent energy by the working of the Reynolds shear stress against the mean velocity gradient is $\tau dU/dy$. Assuming that this is also equal to $q^2/2e$ times the first term on the right-hand side of equation (2.4), we have

$$u_*^2 \frac{dU}{dy} = \alpha'' \frac{q^2}{2} \frac{dU}{dy},$$

i.e.

$$\alpha'' = \Gamma, \tag{8.10}$$

where we define $\Gamma = u_*^2/\frac{1}{2} q^2$, and then from (8.8)

$$A = \Gamma^2 + \frac{3}{4} B + 3F. \tag{8.11}$$

The wall layer is a special case of simple shear, $U_1 = \kappa x_2$, for which

$$\left.\begin{aligned}\overline{(u_1^2)_E} &= \frac{2e}{3} + \left\{\frac{1}{2}B + D + 2F\right\}\frac{e\kappa^2}{\omega^2},\\ \overline{(u_2^2)_E} &= \frac{2e}{3} + \left\{\frac{1}{2}B - D + 2F\right\} e\frac{\kappa^2}{\omega^2},\\ \overline{(u_3^2)_E} &= \frac{2e}{3} + \left\{\frac{1}{2}B + 2F\right\} e\frac{\kappa^2}{\omega^2}.\end{aligned}\right\} \tag{8.12}$$

Let us now consider the flow in the vicinity of a turbulent-non-turbulent interface, which we take to lie in the plane $x_2 = Vt$, where V is the entrainment velocity or equivalently the velocity with which the interface advances into non-turbulent fluid. An analysis similar to that carried out by Saffman (1970) can be performed. From the requirements that e and ω vanish at the interface, but $e^{\frac{1}{2}}/\omega$ is finite, and that the velocity parallel to the interface and shear stress vary linearly with distance, it follows that

$$A' = A'' = \frac{1}{2}A'''. \tag{8.13}$$

In addition, it is desirable to allow $\overline{(u_2^2)_E}$ to vanish at the interface; hence

$$\frac{1}{2}B - D + 2F = 0. \tag{8.14}$$

In the wall layer, $\omega = \kappa\alpha''$, and hence the ratio of the components of turbulent energy are, using (8.11) and (8.14),

$$\overline{u^2} : \overline{v^2} : \overline{w^2} = \frac{2A}{\Gamma^2} - 1 : 1 : \frac{A}{\Gamma^2}. \tag{8.15}$$

The ratio $\overline{u^2} : \overline{v^2}$ is a measured quantity, to be denoted by r, typical values being in the range 2-3. Thus

$$A = \frac{1}{2}(r + 1)\Gamma^2 \tag{8.16}$$

is determined in terms of measurable quantities.

A further relation between the constants is obtained by considering the equilibrium stresses in a uniform irrotational distortion

$$U = \varepsilon x, \quad V = \varepsilon f y, \quad W = -\varepsilon(1 + f)z \tag{8.17}$$

where ε is the rate of extension and f is a distortion type parameter. For definiteness, we take $-\frac{1}{2} \leq f \leq 1$, so that the x and z axes are respectively the directions of greatest extension and compression. This convention will be used throughout. Then from (8.1), remembering that C = 0,

$$\left.\begin{aligned} \overline{u_E^2} &= \frac{2e}{3} - 2A\,\frac{e\varepsilon}{\omega} + 2Be\,\frac{\varepsilon^2}{\omega^2}\,(1 + f + f^2), \\ \overline{v_E^2} &= \frac{2e}{3} - 2Af\,\frac{e\varepsilon}{\omega} + 2Be\,\frac{\varepsilon^2}{\omega^2}\,(1 + f + f^2), \\ \overline{w_E^2} &= \frac{2e}{3} + 2A(1 + f)\,\frac{e\varepsilon}{\omega} + 2Be\,\frac{\varepsilon^2}{\omega^2}\,(1 + f + f^2). \end{aligned}\right\} \tag{8.18}$$

We require that these quantities be non-negative, but there is no need to exclude the possibility that $\overline{u_E^2}$ can be arbitrarily small. Hence after a little arithmetic, we see that

$$B = A^2 \tag{8.19}$$

ensures these requirements.

The anisotropy of homogeneous turbulence is measured by a distortion parameter K, defined by

$$K = \frac{\overline{w^2} - \overline{u^2}}{\overline{w^2} + \overline{u^2}}\,. \tag{8.20}$$

The dependence of K on ε and f has been the subject of experimental research, and there are many extant theories. According to the present theory, the value of K in the equilibrium state, K_E, is given by

$$K_E = \frac{A\left(1 + \frac{1}{2}\,f\right)\varepsilon/\omega}{\frac{1}{3} + \frac{1}{2}\,\frac{Af\varepsilon}{\omega} + A^2(1 + f + f^2)\left(\frac{\varepsilon}{\omega}\right)^2}\,. \tag{8.21}$$

Note that K_E attains a maximum $\left(1 + \frac{1}{2} f\right) \Big/ \left(\frac{1}{2} f + 2\sqrt{\frac{1}{3}(1 + f + f^2)}\right)$ when $\varepsilon/\omega = [3A^2(1 + f + f^2)]^{-\frac{1}{2}}$. If $f = -\frac{1}{2}$, the maximum value is 1.

It remains to relate A to A', A" and A'". We shall assume that the relation, A = 2A', still obtains, although a general argument has not been constructed.

To summarize the relaxation stress model, the Reynolds stresses, $\overline{u_i u_j}$, needed to calculate the mean velocity distribution, are assumed given by solutions of equations (2.4), (2.5), (8.1), (8.3). The constants in these equations are given the values, for reasons discussed above,

$$\left.\begin{aligned} &\eta = \frac{1}{4},\ \theta = 2,\ \beta' = \frac{5}{3},\ \alpha'' = \Gamma,\ A = \frac{1}{2}(r + 1)\Gamma^2, \\ &B = A^2,\ C = 0,\ D = \frac{1}{3}(r - 1)\Gamma^2,\ F = \frac{1}{6}(r - 1)\Gamma^2 - \frac{1}{4}A^2, \\ &G = 0,\ A' = \frac{1}{2}A,\ A'' = \frac{1}{2}A,\ A''' = A,\ \alpha' = \beta'\alpha'' - 4k^2A'/A. \end{aligned}\right\} \quad (8.22)$$

The calculations described in the following section were done with

$$k = 0.42, \quad r = 2.5, \quad \Gamma = \frac{1}{3}. \quad (8.23)$$

Boundary conditions for the equations are that e and ω and tangential Reynolds stress should vanish at the sharp interface between turbulent fluid and irrotational non-turbulent fluid; the mean velocity (but not necessarily gradients of mean velocity) is to be continuous. At a solid wall, the variables should match asymptotically to the behavior described by equation (8.8) and the velocity should match with the logarithmic law of the wall. Integration to the wall would require the introduction of viscosity into the equations.

A word concerning the philosophy of our approach is perhaps in order, as the vast majority of workers in the field of turbulence modelling prefer to follow a different method in which there is term by term modelling of the individual terms which arise when moments of the Navier Stokes equations are taken. Currently, Launder and Lumley are leading exponents of this approach. Our method follows instead the spirit of Kolmogorov (1942), and attempts to model the physical processes in the belief that if these are properly described, the predictions should be qualitatively correct and quantitatively reasonable. Term by term modelling, although conceptually easier to grasp, may be dangerous in that it may give universal local forms for terms that are perhaps determined globally, such as pressure velocity correlations. A rough analogy, drawn by Liepmann, is the difference between solving the Boltzmann equation of kinetic theory by replacing it with the Krook model or constructing a BBGKY hierarchy. But the distinction may be academic as the actual equations are often closely similar, and the practicing engineer with a real problem will probably be better off at present with a simple mixing length approximation.

9. THE DISTORTION OF HOMOGENEOUS TURBULENCE

The equations of the relaxation stress model are more complicated than those of the two-equation model, and comparison with experiment has so far been made only for a particular class of flows, namely the distortion of homogeneous turbulence by a uniform irrotational strain or uniform shear. It should be borne in mind that this class of flow is somewhat artificial. It is also uncertain how well the experiments designed to model the flow actually do so. The theoretical flow is spatially homogeneous and developing in time. The experimental flow is steady and spatially inhomogeneous; the spatial inhomogeneity of the Reynolds stresses drives secondary motions of unknown effect.

The flow fields to be considered are those in which homogeneous turbulence is subjected to a steady velocity field with uniform velocity gradient tensor

$$U_{i,j} = \begin{pmatrix} \varepsilon, & \kappa - \Omega, & 0 \\ \Omega, & \varepsilon f, & 0 \\ 0, & 0, & -\varepsilon(1+f) \end{pmatrix} \tag{9.1}$$

Table 4 lists the experimental configurations with which the theory will be compared.

The equations of the stress relaxation model reduce to

$$\frac{d\omega^2}{dt} = \mu\omega^2 - \beta'\omega^3, \quad \frac{de}{dt} = \lambda e - e\omega \tag{9.2}$$

where

$$\lambda = \alpha''[\kappa^2 + 4\varepsilon^2(1 + f + f^2)]^{\frac{1}{2}} ,$$

Table 4. Velocity gradients for experimental configurations.

Experiment	$\epsilon(\sec^{-1})$	f	$\kappa(\sec^{-1})$	$\Omega(\sec^{-1})$	T(sec)
TR	4.45	0	0	0	0.4
T	9.41	0	0	0	0.15
M	19	0	0	0	0.135
U1	7.62	$-\frac{1}{2}$	0	0	0.18
U2	18.2	$-\frac{1}{2}$	0	0	0.12
U3	32.5	$-\frac{1}{2}$	0	0	0.086
RT1	4.8	0	0	0	0.22
RT2	14	0	0	0	0.135
RT3	12	$-\frac{1}{2}$	0	0	0.145
RT4	3.25	1	0	0	0.255
C	0	0	12.9	0	0.2
ML	0	0	5.45	0	0.47
I	0	0	0	0-6	100

T is time for which strain is applied or duration of measurement.

Key:

- TR Tucker and Reynolds 1968
- T Townsend 1954
- M Marechal 1972
- U Uberoi 1956
- RT Reynolds and Tucker 1975
- C Champagne, Harris and Corrsin 1970
- ML Mulhearn and Luxton 1975
- I Ibbetson and Tritton 1975

Numbers refer to different experiments.

$$\mu = \alpha'[\kappa^2 - 2\eta\kappa\Omega + 2\eta\Omega^2 + 2(2 - \eta)\varepsilon^2(1 + f + f^2)]^{\frac{1}{2}} ,$$

and

$$\frac{d}{dt}\overline{u_iu_j} = -\theta\omega(\overline{u_iu_j} - (\overline{u_iu_j})_E) \tag{9.3}$$

where $(\overline{u_iu_j})_E$ are given by (8.1) in terms of e, ω and the $U_{i,j}$. Numerical integration of the equations is trivial. (Analytical solution is possible but more complicated.) The problem lies in selection of initial values, e_0 and ω_0, for the pseudo-energy and pseudo-vorticity. The initial values of $\overline{u_iu_j}$ are given in the experimental data. When the distorting section follows a uniform section, some information can be gained from the decay of the homogeneous turbulence, but a little trial and error was also employed. Experiments using the same wind tunnel and grid, but different distortion sections, were always compared using the same values of e_0 and ω_0.

The results are shown in Table 5. We also show calculated values of

$$K_2 = \frac{3\overline{v^2}}{q^2} - 1 \tag{9.4}$$

The agreement is not perfect but seems quite reasonable. It is no worse than that obtained by other models or by the use of subgrid modelling (see Reynolds, 1976). The qualitative features are reproduced well. The results for K and τ are sensitive to the value of A, and the constants of equation (8.23) were chosen accordingly, but otherwise the dependence on the parameters seems to be weak.

Figure 11 shows the variation of K with distance along the straining section for the experiment of Tucker and Reynolds (1968), where the straining section was followed by a uniform duct in which some return to isotropy occurred. Figure 12 shows the variation of $\overline{u^2}$, $\overline{v^2}$, $\overline{w^2}$ with distance for the same experiment.

Finally, we turn to the experiment of Ibbetson and Tritton (1975) on the decay of homogeneous turbulence in a fluid undergoing solid body rotation. The experiment presents great difficulties which were overcome in exemplary fashion, but the Reynolds number of the turbulence is of necessity relatively low and the present considerations are essentially for high Reynolds number. Figure 13 shows a calculation of $\sqrt{q^2}$ against t. Comparison with Figure 3 of Ibbetson and Tritton shows general qualitative agreement, although there is disagreement in detail. In connection with the slower

Table 5. Comparison of calculated and observed values of K, q^2 and shear stress for experiments of Table 4. q_0^2 is the initial value of q^2. Observed values are those after time T. Initial values were chosen from the data at entrance to straining section or for C and ML the values reported closest to the grid.

Experiment	$2e_0/q_0^2$	ω_0	K_{cal}	K_{obs}	$(q^2/q_0^2)_{cal}$	$(q^2/q_0^2)_{obs}$	$\tau/(\overline{u^2}\,\overline{v^2})^{1/2}_{cal}$	$\tau/(\overline{u^2}\,\overline{v^2})^{1/2}_{obs}$	$(K_2)_{cal}$
TR	.5	10	0.56	0.6	0.32	0.38	-	-	.01
RT1	.5	10	0.36	0.35	0.39	-	-	-	.04
RT2	.5	10	0.64	0.63	0.83	-	-	-	.02
RT3	.5	10	0.55	0.5	0.65	-	-	-	.31
RT4	.5	10	0.39	0.45	0.41	-	-	-	-.26
T	1	35	0.45	0.4	0.41	0.43	-	-	0
M	.8	30	0.72	0.62	0.74	0.75	-	-	.01
U1	.8	5	0.58	0.75	1.1	1.1	-	-	.32
U2	.8	5	0.78	0.77	2.2	2.9	-	-	.41
U3	.8	5	0.71	0.72	4.45	4.7	-	-	.38
C	.7	10	-.13	-.25	.74	.7	.45	.5	-.27
ML	.6	5	-.14	-.2	.56	.5	.43	.5	-.25

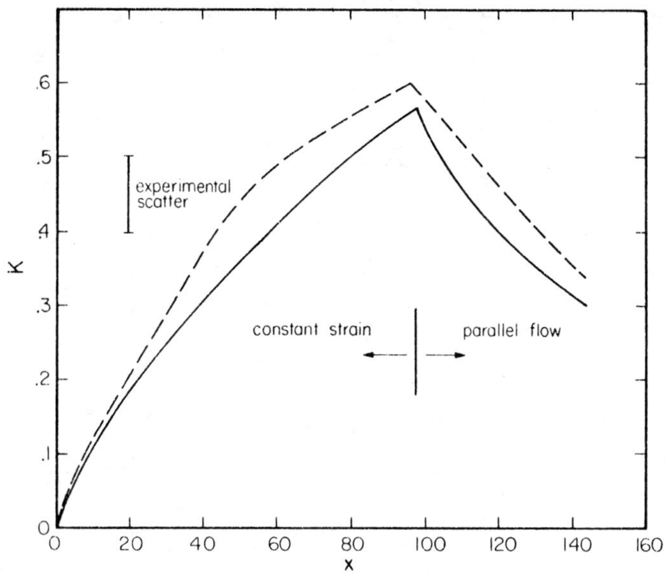

Figure 11. Variation of K with distance x from entrance to straining section for Tucker and Reynolds (1968). ———, theory. -----, experiment.

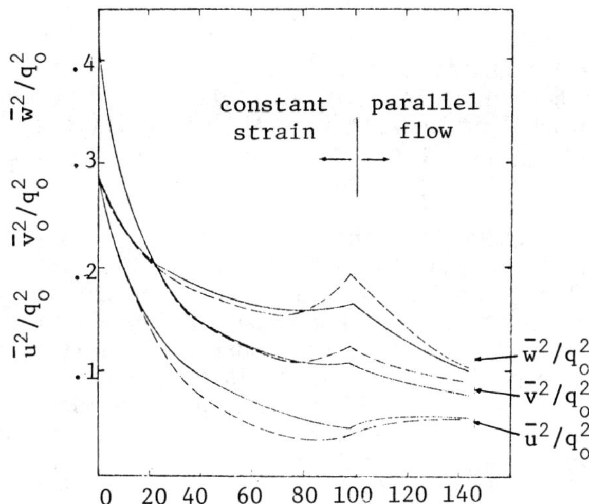

Figure 12. Variation of mean square velocity components with distance x from entrance to straining section for Tucker and Reynolds (1968). ———, theory. -----, experiment.

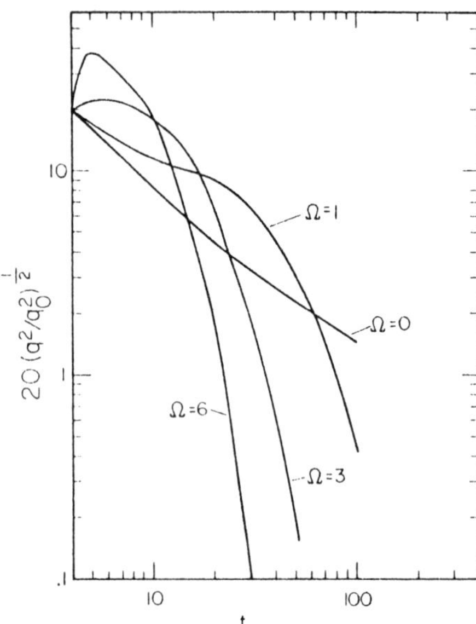

Figure 13. Decay of turbulent intensity with time, t, for homogeneous turbulence in solid body rotation with angular velocity Ω.

decay for Ω = 1 than for Ω = 0, as contrasted with faster decay for Ω = 3 and Ω = 6 after a brief initial rise, it is worth mentioning that Ibbetson and Tritton felt that this trend reversal was a possibility although they do argue against it. They attribute the extra decay to the radiation of energy to the walls by inertial waves.

The results shown here indicate that the stress-relaxation model reproduces the qualitative features of the distortion of homogeneous turbulence, and the quantitative agreement is sufficiently promising for the study to be worth continuing with the model applied to more realistic turbulent flows. Work on this is in progress and it is hoped to report on the results in the not too distant future. It should, of course, be borne in mind that the model is unlikely to be in its final form and that further development may be needed. The introduction of a tensor diffusivity in place of the scalar diffusivities proportional to e/ω may be necessary to model flows displaying substantial spatial anisotropy.

ACKNOWLEDGMENT

The calculations of §§ 3 through 7 were carried out by Dr. Fausto Milinazzo, currently at the Department of Mathematics, University of Victoria, British Columbia. Details of the numerical scheme and the techniques employed to handle indeterminate forms which arise at the boundaries are available on application to him.

The work was supported by the Energy Research Development Administration (AT 04-3-767).

REFERENCES

1. Champagne, F. H., Harris, V. G. and Corrsin, S., 1970, J. Fluid Mech., 41, 81.

2. Davies, A. E., Keffer, J. F. and Baines, W. D., 1974, Phys. Fluids, 18, 770.

3. Govindaraju, S. P., 1970. Ph.D., Thesis. Caltech.

4. Grant, H. L., 1958, J. Fluid Mech., 4, 149.

5. Ibbetson, A. and Tritton, D. J., 1975, J. Fluid Mech., 68, 639.

6. Jenkins, P. E. and Goldschmidt, V. W., 1974. Purdue University, Report HL 74-45.

7. Keller, H. B., 1974. SIAM J. Num. Anal. 11, 305.

8. Knight, D., 1975, A.I.A.A.J. 13, 945.

9. Kolmogorov, A. N., 1942. Izv. Akad. Nauk. SSR. Seria fiz. VI, p. 56.

10. Marechal, J., 1972, J. de Mec. 11, 18.

11. Milinazzo, F. and Saffman, P. G., 1976, Stud. App. Math., 55, 45.

12. Mulhearn, P. J. and Luxton, R. E., 1975, J. Fluid Mech. 68, 577.

13. Pope, S. B., 1975, J. Fluid Mech. 72, 331.

14. Reynolds, A. J. and Tucker, H. J., 1975, J. Fluid Mech., 68, 673.

15. Reynolds, W. C., 1976, Ann. Rev. Fluid Mech., 8, 183.

16. Saffman, P. G., 1967, Phys. Fluids, 10, 1349.

17. Saffman, P. G., 1970, Proc. Roy. Soc. A 317, 417.

18. Saffman, P. G., 1974, Stud. App. Math., 53, 17.

19. Saffman, P. G. and Wilcox, D. C., 1974, A.I.A.A.J., 12, 541.

20. Schlichting, H., 1960, Boundary Layer Theory. McGraw-Hill.

21. Townsend, A. A., 1954, Q. J. Mech. App. Math. 7, 104.

22. Townsend, A. A., 1956, Turbulent Shear Flow. Cambridge U.P.

23. Tucker, H. J. and Reynolds, W. C., 1968, J. Fluid Mech., 32, 657.

24. Uberoi, M. S., 1956, J. Aero. Sci., 23, 754.

25. Wilcox, D. C. and Alber, I. E., 1972. Heat Transfer and Fluid Mech. Inst. p. 231.

DISCUSSION

KOVASZNAY: (Johns Hopkins University)

In your new set of equations it appears to me that you really want to make a consistent second-order model using the correct second-order quantities. I would like you to explain to us what is the relationship of your model to a similar attempt by Lumley who somewhat earlier attempted a consistent second-order approximation with only third-order guessed terms. One of the difficulties I am facing when looking at model equations of different people is that everyone presents it separately as unique, and very little attention is paid to the relationship to other similar attempts, even though those attempts are completely in published papers.*

SAFFMAN:

I can't answer your question and the reason is that there must be at least 10 or 20 different models. If you spend time studying all of them and comparing with your own then you will do nothing but study the other models, so you have to be selective in deciding

---

* Lumley, T. L. & Knajeh-Nouri 13, 1974, "Advances in Geophysics," Vol. 18, p. 169-192, Academic Press, New York, (Int. Symp. of Turbulent Diffusion, IUTAM-IUGG, Charlottesville, VA, 1973).

which ones you will study and which ones you will not. Selections are made entirely on the grounds of prejudice and I am afraid that Lumley's model was not studied by me.

CORRSIN: (Johns Hopkins University)

Since the Reynolds stress tensor and the mean strain rate tensor are generally not oriented in the same direction, I wonder whether you couldn't just simply generalize the first model by putting in a tensor turbulent viscosity rather than going through all that elaborate additional process. Have you tried such a tensor?

SAFFMAN:

The answer is that one could do that, but there are a large number of ways that one could proceed. I went the way I've described. I could have gone the way you are suggesting, but I don't think it would be any easier. I think that my way is about the simplest that I can do and, however complicated it is, I do think it is simpler than the Reynolds stress modeling equations.

BEVILAQUA: (Rockwell International)

At first sight it seems that you may have provided a connection between the transport models and the eddy viscosity models. When the rate of turbulence production equals its rate of dissipation, the transport models reduce to a lag model for the eddy viscosity. Since your work relates the eddy viscosity to a lag model, it suggests to be a way to relate formally, or rigorously, the "constants" of the eddy viscosity models to the "universal" constants of the transport models.

SAFFMAN:

I object to the word "rigorous"!

LAKSHMINARAYANA: (Pennsylvania State University)

In your two equation model you wrote the decay rate for the wake and the spreading rate of a jet. Was this part of the solution that comes out of the equations?

SAFFMAN:

I assumed that there is a similarity solution and the solution was then calculated numerically.

LAKSHMINARAYANA:

Similarity in the solution might exist but the decay rate might be different.

SAFFMAN:

Once you assume there is a similarity solution then everything else follows automatically. There was no more arbitrariness, and no evidence of non-uniqueness, but this could not be proved rigorously.

LAKSHMINARAYANA:

The second question is related to the rotation. In the experimental data (due to Ibbetson and Tritton) you showed the rotation and velocity were in the same direction.

SAFFMAN:

There was no mean flow in their experiment. They just had the cylinder rapidly rotating and moved the grid up and down to set the fluid in turbulent motion, then let the turbulence decay.

LAKSHMINARAYANA:

How do you exactly include the rotation into the model?

SAFFMAN:

When I had quadratic terms in expressing the Reynolds stresses in terms of gradients of mean velocity, some of the quadratic terms involved the angular velocity of fluid. It also enters into the generation of pseudo-vorticity.

BRADSHAW: (Imperial College)

Just a quick comment on transport equations for eddy viscosity. I don't think I have ever seen it in the literature, but you can derive an exact transport equation for any given definition of eddy viscosity because you get a "transport equation" for a velocity gradient simply by differentiating the Navier-Stokes equations. If you do this, for instance for the thin shear layer form of eddy viscosity, you get an equation for D(eddy viscosity)/Dt, and that equation is a right can of worms. It looks very different from the eddy viscosity transport equations that one would deduce from a model like yours or even a model like Kovasznay's.* It has, for

* V. W. Nee, L. S. Q. Kovasznay, In Proc. 1968 AFOSR-1FP-Stanford Conference on Computation of Turbulent Boundary Layers (S. J. Kline et al., Eds.) Thermosciences Div., Stanford Univ., 1969.

instance, a $\partial^2 p/\partial X_1 \partial X_2$ term in it, where p is the mean pressure, and of course it also has uncomfortable quantities with zeros in the denominator.

RESHOTKO: (Case Western Reserve University)

Why is the Ekman calculation regarded as a particular test of the stress relaxation model?

SAFFMAN:

It is a simple boundary layer in a rotating fluid. I was interested in trying to see how well the effects of solid body rotation could be reproduced by this kind of turbulence model and it is something for which there is some experimental evidence. I could, I suppose, have tried a jet in a rotating fluid, but I was more interested in the Ekman layer.

RESHOTKO:

There are some scalar effective-viscosity calculations for related flows* and those turned out reasonably well. That is why I wondered if this was a test over and above that.

SAFFMAN:

On any of these flows one can construct a simpler model that will work. The object of this type of modeling is to get what you call a complete model where the form of the model is completely independent of the flow to which it is going to be applied. I think for instance, for boundary layers there is a very simple eddy viscosity or mixing length model which will work much better than this but they are limited or restricted to that kind of flow. If you go to another flow you have to use a different type of equation or a different assumption. The object here is to see if one has a predictive capacity in the form of a complete model which is independent of the type of flow under study. The object is not to predict any one particular type of flow to a high degree of accuracy.

---

* Cooper, P.: Turbulent Boundary Layer on a Rotating Disk Calculated with an Effective Viscosity, AIAA Journal Vol. 9, Feb. 1971, pp. 251-261.

Cooper, P. and Reshotko, E.: Turbulent Flow Between a Rotating Disk and a Parallel Wall, AIAA Journal, Vol. 13, May 1975, pp. 573-578.

OHRENBERGER: (TRW Systems)

With regard to the compressibility terms you added to each of the equations, are they really density-gradient terms rather than Mach number terms or do they have some special Mach number effect?

SAFFMAN:

They have a Mach number dependence because the density is a function of Mach number, but they are principally in terms of density gradients.

OHRENBERGER:

If you had an incompressible flow with density gradients, would you essentially have the same terms?

SAFFMAN:

Yes, the same terms are in there.

OHRENBERGER:

Isn't that somewhat contrary to Brown and Roshko results?

SAFFMAN:

No, I don't think so.

CORRSIN:

I just want to mention that there is an old NACA report on rigid rotation grid turbulence done by Steve Traugott on a study done about 20 years ago. You might be able to compare with that.

MELLOR: (Princeton University)

On your Ekman layer, I think it is worthwhile to point out that if you write down the Reynolds stress equation, you can readily show that the Coriolis term in the Reynolds stress equation is small compared to dissipation.

GAVIGLIO: (I.M.S.T., Marseilles)

As for inadequacy of complete eddy diffusion model, I underline that in the case of experiments performed in a separated flow at Mach Number 2.3, we found a case in which the friction actually increases although the velocity gradient decreases.

SAFFMAN:

I think that may be consistent with the solutions I have.

WYGNANSKI: (Tel Aviv University)

In the case of the mixing layer a discontinuity in the slope of the instantaneous velocity profile has been observed at the turbulent-non-turbulent interface.

SAFFMAN:

Yes, I was aware of that in constructing the model; the model was designed specifically to have these discontinuities in velocity gradient.

LUXTON: (University of Adelaide)

In one of your figures, Figure 1, where you compared your calculations with our uniform shear flow results (Mulhearn, P. J. and Luxton, R. E., Jnl. Fluid Mech. (1975), 68, 3 pp. 577-590) there was a note about the initial conditions you used. It is important to note that the conditions immediately downstream of the honeycomb in our experiments were highly anisotropic.

SAFFMAN:

I should have said that I didn't start at the grid in comparing with your experiments, I started some distance downstream, where in fact you give the data.

Application of the Turbulence-Model Transition-Prediction Method to Flight-Test Vehicles*

T. L. CHAMBERS AND D. C. WILCOX
DCW Industries, Inc.
Sherman Oaks, California (U.S.A.)

ABSTRACT

Flight conditions for six reentry vehicles have been simulated using the turbulence/transition model developed by Wilcox, et al. Material composition varies from vehicle to vehicle. The various frustum materials used are beryllium, carbon/carbon felt, carbon phenolic and teflon; five of the vehicles have graphite nosetips while one has a teflon nosetip. Twenty computations have been performed and results have been compared with measured transition points; close quantitative agreement has been obtained between computed and measures transition location. Results of the computations add to the growing list of successful applications of the method.

1. INTRODUCTION

An accurate tool for predicting boundary-layer transition onset and progression as a vehicle reenters the atmosphere is critically needed for the design of reentry vehicles. While some progress can be made in devising such tools by studying ground-test experiments,

---

* This research was jointly sponsored by the University of Dayton Research Institute and the Air Force Flight Dynamics Laboratory under Purchase Order RI-75901. The authors gratefully acknowledge the assistance of Anthony Martellucci and Dr. Charles Kyriss of the General Electric Company who supplied all pertinent input data for the flight simulations; their timely assistance and guidance were instrumental in making this a trouble-free and effective project.

validity of extrapolating ground-test data to flight-test conditions is strongly controlled by the rigor with which the extrapolation is made. Simply extending a log-log plot from the ground-test regime into the flight-test regime provides the least believable extrapolation. Nevertheless, many flight-test transition correlations are based on such extrapolations.

Increasingly, research efforts aim at establishing a reliable flight-test transition correlation by considering both ground- and flight-test data. Such an approach has the obvious advantage of including a wider data base. However, since a correlation is only as good as its data base, and since flight-test transition data are usually of questionable quality, these research efforts fail to substantially increase confidence in such a correlation's applicability beyond the established data base.

Numerical simulations of ground- and flight-test conditions would provide a major step forward in developing an engineering correlation, provided the simulations (a) faithfully reproduce carefully controlled ground-test experiments and (b) yield reasonable agreement with flight-test data. Although exact numerical simulations are ultimately needed to remove all uncertainty, such simulations are currently unfeasible. Nevertheless, the current state of affairs can be improved by use of numerical simulations based on an approximate method, provided the method has been successful in reproducing a broad data base.

Such a method, known as the turbulence-model transition-prediction technique, has been devised by Wilcox (1975, 1976, 1977). The technique is very accurate for a wide range of flows, including well-documented compressible and incompressible boundary layers and for the PANT [see Anderson (1975)] ground-test experiments on blunt bodies. The objective of this study has been to use the turbulence-model transition-prediction method to predict transition onset and progression on flight-test vehicles.

Because many aspects of the computations differ from any prior applications of the theory, Section 2 reviews the transition prediction method including the following: a description of the computer code used for solving the model equations; discussion of important assumptions incorporated into the code which are pertinent to this project; and the manner in which the transition point is determined from results of the calculations. Next, Section 3 defines the computational matrix and relevant input data. The transition predictions for the six vehicles (20 cases) and a comparison with experimental data are shown in Section 4, including effects of free-stream turbulence intensity. The concluding section summarizes results and conclusions.

2. THE METHOD

The turbulence/transition model devised by Wilcox (1975, 1976, 1977) serves as the basis of this study. For brevity the model equations are omitted; complete details are given by Wilcox (1977). In prior studies, the model equations have been used to accurately predict transition for a wide range of compressible and incompressible flows including effects of surface roughness, pressure gradient, mass injection, freestream unit Reynolds number, and surface heat transfer. Most pertinent to the present study, the theory accurately simulates ground-test data for reentry type vehicles. Thus, the turbulence/transition model consistently has been an accurate predictive method for well-documented ground-test experiments.

2.1 THE EDDYBL COMPUTER CODE

The model equations require numerical solution methods in order to predict transition. Their solution is accomplished with a boundary-layer code known as EDDYBL. The code is based on the second-order-accurate, implicit, parabolic marching method of Flugge-Lotz and Blottner (1962).

In order to achieve realistic flight-test simulations, we must account for properties of high temperature air. For the present applications, real gas effects have been assumed to be confined to the nosetip region and hence to have a negligible effect on frustum transition. Assuming this to be the case, the fluid is approximated to be a perfect gas for the entire region covered by the calculation. In addition the fluid is assumed to have a constant specific heat ratio, γ, given by

$$\gamma = 1.4 \tag{1}$$

while the gas constant, $R = p/\rho T$, is assumed to be

$$R = 1716 \text{ lb}\cdot\text{ft/slug}^\circ\text{R} \tag{2}$$

For the high temperatures prevailing for flight conditions, molecular viscosity, μ, is related to temperature, T, as follows [from Kyriss (1976)]:

$$\mu = 2.518 \cdot 10^{-10}\, T + 4.193 \cdot 10^{-7} \text{ lb}\cdot\text{sec/ft}^2 \tag{3}$$

with T given in degrees Rankine. Entropy swallowing is included in the computations under the standard approximation that the entropy in the boundary layer equals that generated by the portion of the shock through which an equivalent amount of mass flows.

Computations by Kyriss (1976) verify that ignoring real gas effects and using Equations (1), (2) and (3) to define γ, R and μ have only a slight effect on flow properties on the frustum. Hence the above assumptions, i.e., Equations (1)-(3), can be considered valid for this study.

2.2 Manner in Which Transition Point is Determined

The computer code, EDDYBL, which embodies the turbulence/transition model, begins each calculation at the stagnation point on the vehicle and marches along the vehicle surface in the s direction (see Figure 1). At the outset, the flow is laminar. However, turbulence is constantly being entrained from the freestream. Finally, at some point along the body, the entrained turbulence no longer is suppressed in the boundary layer. The turbulence then begins to be amplified and causes boundary-layer instability; the flow goes through transition and becomes turbulent. Throughout the laminar region, the heat transfer and skin friction generally decrease monotonically. However, as the flow becomes turbulent, abrupt increases in heat transfer and skin friction occur. A convenient definition of

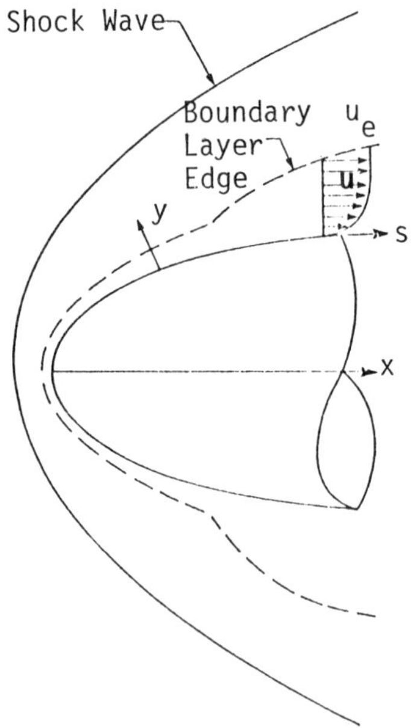

Figure 1. Schematic of Vehicle with Boundary Layer.

transition location is the point at which the heat transfer achieves a minimum (usually the same location at which minimum skin friction occurs). Figure 2 shows the variation of skin friction on a typical reentry vehicle, thus illustrating the criterion for locating transition.

3. COMPUTATION MATRIX

A total of 20 transition predictions have been made for six typical reentry vehicles. A summary of the altitudes, H, considered and the nosetip and frustum materials for each vehicle is given in Table 1; the same reference altitude, H_{ref}, has been used for all of the vehicles. EDDYBL requires input data defining the following:

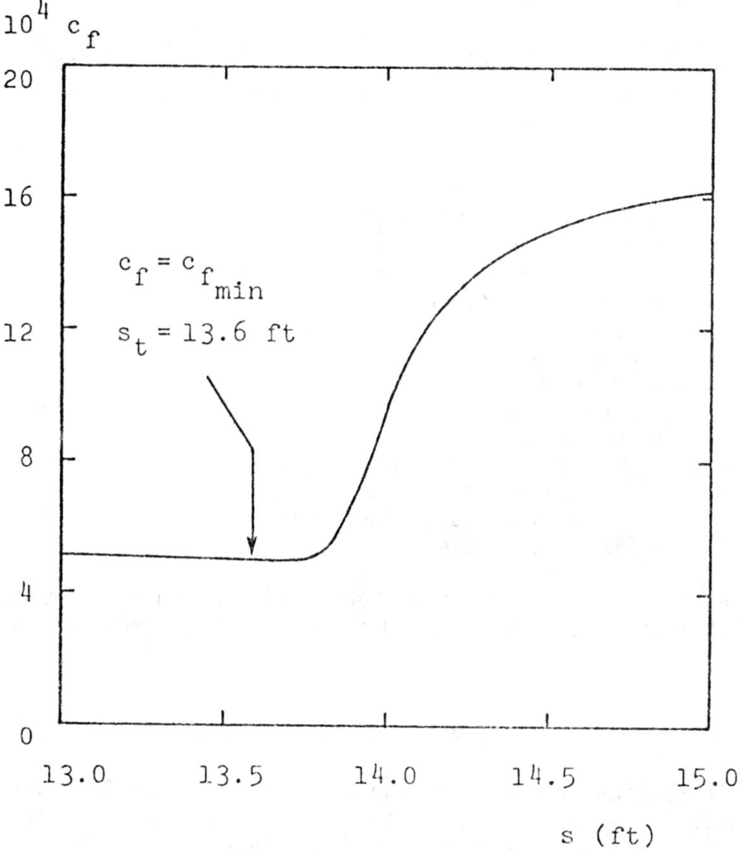

Figure 2. Computed Skin Friction for a Typical Reentry Vehicle. Transition Location Identified as Point of $c_{f_{min}}$.

Table 1. Computation Matrix

No.	Nosetip Material	Frustum Material	Altitudes, H/H_{ref}
1	ATJ	Beryllium	1.22, 1.18, .98
3	ATJ	Beryllium	1.25, 1.18, 1.07, .94
8	ATJ	Carbon Phenolic	.99, .87, .83
9	ATJ	Carbon/Carbon Felt	.36, .33, .31
10	ATJ	Teflon	1.145, 1.09, .92
12	Teflon	Teflon	1.28, 1.23, 1.17, 1.09

1. Freestream Flow Conditions
   - Mach number
   - Total pressure
   - Total temperature
2. Geometry
   - Body shape
   - Shock shape
3. Surface Conditions
   - Temperature
   - Mass-injection rate
   - Surface roughness
4. Boundary-Layer Edge Conditions†
   - Static pressure

Surface roughness effects have been neglected in the computations; all other input data for the 20 cases have been provided by Kyriss (1976).

4. TRANSITION CALCULATIONS

Transition predictions have been made for the 20 cases listed in Table 1. As mentioned in the Introduction, the calculations are sensitive to the freestream turbulence level, T'_∞. Hence, before the

† Total pressure at the boundary-layer edge is computed by EDDYBL from the specified shock shape and local mass flux in the boundary layer.

computations could be performed, an appropriate turbulence intensity level had to be established. In this section we first describe the manner in which the values of T'_∞ were established for each vehicle. Then, we present results of the computations and a correlation of the numerical data.

4.1 Sensitivity to Freestream Turbulence Intensity

In order to find the appropriate freestream turbulence intensity, T'_∞ was varied for each vehicle until close agreement between computed and measured transition points at one altitude (usually the highest altitude considered) was achieved. Vehicle No. 1 computations demonstrated that an intensity given by

$$T_\infty = 100\sqrt{\langle u'^2 \rangle}/U_\infty = .016\% \tag{4}$$

is needed to obtain close agreement with the measured transition location. In the definition of T'_∞, U_∞ is the mean streamwise velocity, u'_∞ is the fluctuating component of streamwise velocity and $\langle\ \rangle$ denotes time average. The same procedure was repeated for all six vehicles. Surprisingly, $T'_\infty = .016\%$ proved to be appropriate for all of the vehicles except for Vehicle No. 8. A freestream intensity of $T'_\infty = .005\%$ was used for Vehicle No. 8.

Figure 3 shows the effect on transition location caused by varying T'_∞. Three different turbulence intensities have been used at each altitude for Vehicle No. 12. As expected, increasing T'_∞ decreases the distance from the nose to the transition point.

4.2 Results of the Calculations

Once the proper turbulence intensity had been established for each vehicle, it was then possible to proceed with the transition calculations. Figure 4 presents results of the 20 computations; experimental data are included for comparison. As shown, the model-predicted progression of the transition point is very similar to the measured progression for all of the vehicles. Inspection of the plots indicates that if we regard altitude as a function of transition location, almost all of the transition altitudes are predicted within one mile of measured transition altitude. Furthermore in half of the cases, computed and measured transition altitudes differ by less than a half-mile. Figure 5 shows that most of the predictions fall within one foot of corresponding measurements; all but one calculation predicts transition location within two feet of the measured transition location.

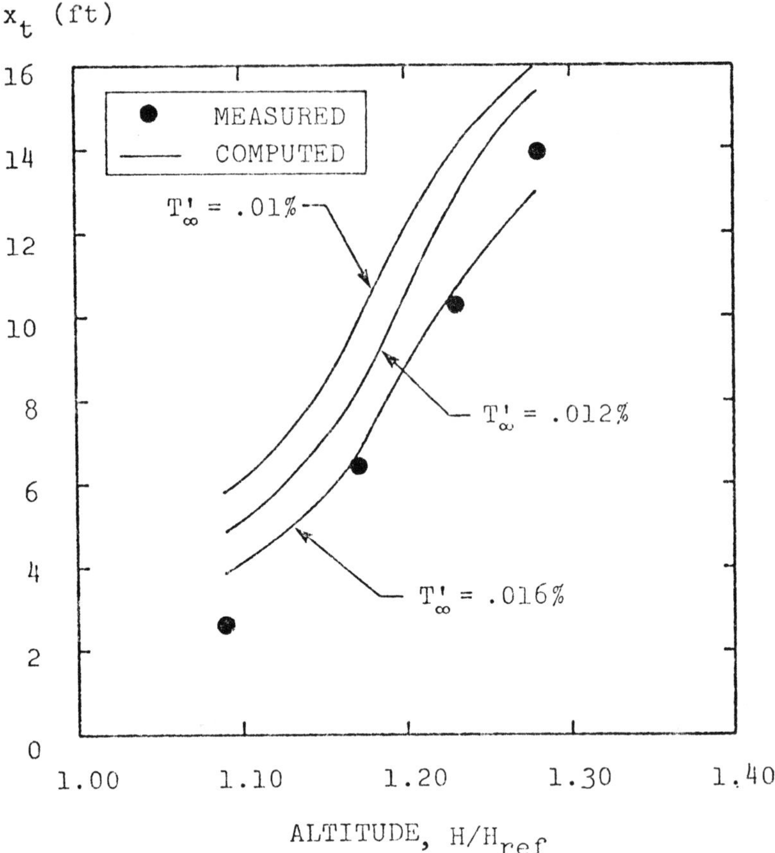

Figure 3. Effect of T'_∞ on Transition Location for Vehicle No. 12.

5. SUMMARY AND CONCLUSIONS

Results presented in Section 4 indicate that the turbulence/transition model's range of applicability includes reentry vehicles under flight-test conditions. With a single adjustable input parameter, viz, freestream turbulence intensity, the model accurately predicts transition sensitivity to surface mass injection, surface cooling, and entropy gradient.

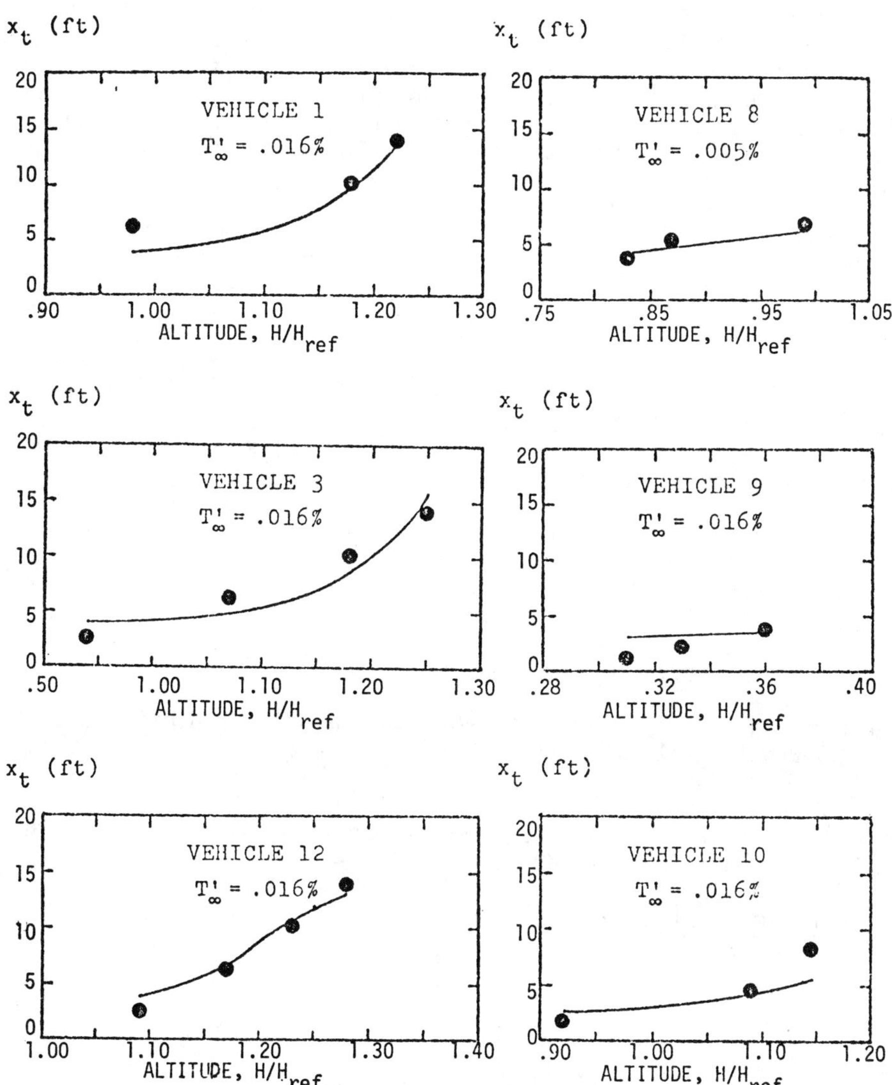

Figure 4. Comparison of Computer (——) and Measured (•) Transition Point Movement; x is Axial Distance from the Nose.

Because numerical simulations can cover a much wider range of conditions than is generally possible or feasible in experiments, numerical simulation of flight-test conditions can fill a void left by inconclusive or unreliable flight-test measurements. However, before such simulations can be done a priori, we must devise a method for specifying freestream turbulence intensity. Presumably the freestream intensity depends upon altitude, geographical locations, time of day, etc.; further research is needed to determine the dependence of freestream turbulence intensity upon such factors.

In conclusion, the results obtained add to the growing list of successful applications of the model to the problem of predicting boundary-layer transition, thus lending further confidence to the theory.

REFERENCES

1. Anderson, A. D., "Interim Passive Nosetip Technology (PANT) Program, Volume 2, Summary of Experimental Results," Appendix A, Aerotherm Report 74-100 (1975).

2. Flugge-Lotz, I. and Blottner, F. G., "Computation of the Compressible Laminar Boundary-Layer Flow Including Displacement-Thickness Interaction Using Finite-Difference Methods," AFOSR 2206 (1962).

3. Kyriss, C. L., personal communication (May 1976).

4. Wilcox, D. C., "Turbulence-Model Transition Predictions," AIAA Journal 13 (1975), 241.

5. Wilcox, D. C. and Chambers, T. L., "Numerical Simulation of Nosetip Transition: Model Refinement and Validation," AFOSR-TR-76-1112 (1976).

6. Wilcox, D. C., "A Model for Transitional Flows," AIAA Paper 77-126 (1977).

DISCUSSION

RESHOTKO: (Case Western Reserve University)

In many experiments it has been observed that there is a spectral effect, for example there are certain frequencies that are emphasized in triggering transition while others are not. How did frequency spectra enter into your calculations?

WILCOX:

That is one of the effects that we have thrown out. Spectral effects are presumably buried in the constants. Interestingly, I would have expected these flight-test calculations to exhibit strong spectral effects. Because of this intuitive notion, I am a little surprised that the same turbulence intensity does the trick for five of the six vehicles. There is some transition that doesn't care about that spectral effects. Transition that is spectrum dependent I probably won't be able to predict well because that effect is buried in the closure coefficients. We have lost spectral effects a long time ago.

WHITELAW: (Imperial College)

I have one comment and one question. The comment relates to your use, at the beginning of your talk, of the phrase "complete turbulent model." I am not sure what "complete" means.

The question stems from your reference to "messing around with the generation terms." Could you tell me what "messing around with the generation terms" means?

WILCOX:

Well, it means that I notice that for these equations as originally formulated, the constants, of course, are all determined by considering properties of fully turbulent flows. If you take the equations and you predict transition, the transition Reynolds numbers are way too low. Through numerical experimentation, I eventually found that you can get much more accurate transition if you cut down the generation terms and then I argued that you might expect this to be the case anyway because your generation mechanism ....

That's too long winded?

WHITELAW:

No, it's too short. The phrase "messing around with the generation terms" is too crude. Perhaps you mean "Let's have less generation"; but that isn't cricket. Perhaps if you included some physically based arguments....?

WILCOX:

Oh, some physics went in, yes, for sure. I have this thing that I call the neutral stability Reynolds number. I can determine the neutral stability Reynolds in closed form. The original assumption was demand that for a Blasius boundary layer the neutral stability Reynolds number match the minimum critical Reynolds number. When you do this you get most of the results that I presented. So you wind up, among other things, having a fairly sound general argument to fix one of the empirical constants that I introduce.

BIRCH: (The Boeing Company)

I would like to make a comment about the effect of freestream turbulence. The available data* for two equation turbulence models seem to suggest that the predictive effect of freestream turbulence is more sensitive to the assumed length scale than to the turbulence intensity. This means that if the length scale isn't available experimentally you leave yourself with a free constant that can be arbitrarily adjusted; you can in fact match the experimental data irrespective of freestream turbulence intensity used.

WILCOX:

There is some flexibility to do that, yes. You have two turbulence parameters and each one must have a boundary condition. Transition is sensitive to the length scale but it is not as sensitive to the length scale as it is to the intensity. It is much much more sensitive to the intensity. All of the computations I have done have been with the same value of the length scale. I haven't messed around with it. I have used the value that is appropriate for fully turbulent flows.

BIRCH:

Do the experiments you compare this with have the same length scale?

WILCOX:

That is a little tough, because now you require me to define exactly what the length scale is. Maybe the thing is the integral length but maybe it is not, so I don't know. I don't know exactly what that quantity is at this point.

KOVASZNAY: (Johns Hopkins University)

Just a little question. If you go to a simpler case, flat plate, incompressible, very clean, let's suppose you get the turbulence so low that transition is not longer functional. Such experiments have been done at the Bureau of Standards and elsewhere. What do you predict then?

---

* Oh, Youn H. and Bushnell, Dennis M., "Influence of External Disturbances and Compressibility on Free Turbulent Mixing," NASA SP-347 (1975).

WILCOX:

I would predict that you would keep getting transition all the way until it is zero. It just gets higher and higher and higher.

KOVASZNAY:

This means that you have a floating parameter that adjusts according to taste? Let's suppose that the turbulence is so low that you no longer affect the transition, that is the case I am speaking about.

WILCOX:

I don't know that I believe that there is a finite transition Reynolds number as the intensity goes to zero. If that is the case then the theory is stuck. I don't believe that is true, however.

KLEBANOFF: (National Bureau of Standards)

I would like to take the other view in terms of the influence of freestream turbulence, that is the effect with increasing level. One then finds a very large disturbance in the laminar boundary layer that is higher in intensity than the freestream level, and which, for the most part, consists of frequencies that are not unstable according to stability theory. Their intensity is scale dependent, that is scale dependent on the freestream scale, and on the intensity of the freestream and yet the transition occurs, as indicated by Reshotko, from a mechanism involving frequencies that are unstable according to stability theory, and which have been excited by the freestream turbulence. You will not predict that at the critical position the overall disturbance energy in the boundary layer which increases with increasing freestream turbulence will be higher than that in the freestream and it does go up considerably.

WILCOX:

All of these things of course are nice esoteric properties that a theory this simple hasn't got a prayer of predicting. But I don't think that we should lose sight of the fact that I have been able to go over a wide range of applications all the way from incompressible to Mach 20, with pressure gradients, and surface roughness. In all these applications I have shown that I can predict many properties of transition. There must be a message in that. While some of the detailed nitty-gritty mechanisms of transition are complicated, there are some properties of transition which you can predict very simply, because I don't adjust parameters or anything.

KLEBANOFF:

I just want to expand on the comment Kovasznay made about the low turbulence experiments. The matching between freestream disturbances and boundary layer disturbances is not a one to one match, and the laminar boundary layer has a certain impedance to the freestream, even as the freestream turbulence is going to lower and lower levels. As a result your model, I believe, would not take into account other types of disturbances which may become dominant, for example, sound at certain frequencies to which the boundary layer still reacts and although the overall intensity of the freestream disturbances is decreasing, transition does not behave as you expect. It would seem that any model that doesn't have the impedance behavior characterized loses much of the physics of the problem.

MELLOR: (Princeton University)

To reiterate Whitelaw's comment, any model has its limitations and I think it is up to the modeler to try and frame out what those limitations are.

But at one point, you said you messed around with the generation terms and I think that a lot of us feel that that is the only term that you don't need to worry about; that is, they are exact and need not be altered.

WILCOX:

The generation terms in these equations are not $\tau(\partial u/\partial y)$. They are very similar to the generation terms in one of Professor Bradshaw's formulations.

KLINE: (Stanford University)

I just want to say that I don't agree with your statement that in the limit of infinite Reynolds number the flow can remain laminar. We have to distinguish, I think, between essential instability and metastability. Essential instability means mathematically that any infinitely small disturbance will kick you over the cliff to turbulent flow. In metastable situations, that is at lower Reynolds number, the flow requires a larger disturbance to get turbulence started, but if you have essential instability then you have to visualize absolutely no perturbation of any kind if the flow is to remain laminar. It is very hard to visualize a situation that has no perturbations whatsoever. At infinite Reynolds number, the flows we know become <u>essentially</u> unstable. Hence if you are getting laminar flow in that limit, then it seems to me that you have the wrong limit.

MOREL: (General Motors Research Laboratories)

I believe that as you go to very low limits of disturbances then you get into the linear regime where the disturbance amplifies exponentially. Then you would expect that as you reduce the excitation level the position of the transition point will become less and less sensitive to what the excitation level is. In addition a reduction in the excitation level means that the linear regime extends farther downstream where the disturbance growth rate is usually higher. This increase in the exponential growth rate will further reduce the influence of the excitation level on the transition point location. That might explain the difference and if your model can calculate the exponential growth of very small disturbances, then it might actually have this in it.

WILCOX:

My guess is, however, that it is a nice continuous function that goes to infinity as the intensity goes to zero.

A Second Moment Turbulence Model Applied to Fully Separated Flows*

M. BRIGGS, G. MELLOR, AND T. YAMADA
Princeton University
Princeton, New Jersey (U.S.A.)

ABSTRACT

A turbulence model which has previously been applied primarily to boundary layer flows including geophysical flows is here applied to a fully separated flow. The full elliptic problem is solved and differential equations for all components of the Reynolds stress tensor are involved. All constants in the model have been determined from other more simple flows so that we need not appeal to "computer optimization."

Calculations are compared to data obtained by Abbott and Kline. It is our understanding that the turbulence data are not accurate. Nevertheless, these data and the predictions compare reasonably well. The predicted mean velocity profiles and the point of reattachment agree very well with the data.

1. INTRODUCTION

Our experience at Princeton University with second moment turbulence modeling began at the time of the 1968 Stanford Conference on Turbulent Boundary Layer Computation (Kline et al., 1968) with a Prandtl type model based on the solution of the turbulent kinetic energy. At that time, this model, Bradshaw's model and our older eddy viscosity model did very well in predicting all of

---

* This work was supported by the Air Force Office of Scientific Research Grant No. 75-2756. Some computer time was supplied by the NOAA/Geophysical Fluid Dynamics Laboratory.

the data compiled for the conference. Thus it was not clear that more complicated models based on hypotheses by Rotta (1951) and Kolmogorov (1941) and requiring consideration of all components of the Reynolds stress tensor could be justified on the basis of improving predictions.

The first results from the second moment calculations of Donaldson and Rosenbaum (1968) (using Rotta's energy redistribution hypothesis but not the Kolmogorov isotropic dissipation hypothesis) when applied to wakes and jets and subsequent studies by Hanjalic and Launder (1972) were encouraging but the newer model involved more empirical constants than did the older models which, on this basis alone, would facilitate agreement with data.

Our interest in the Rotta-Kolmogorov model was greatly enhanced by data obtained by So and Mellor (1973, 1975) which demonstrated the effect of wall curvature on turbulent flow. On stable, concave walls the Reynolds stress was virtually extinguished in the outer half portion of a turbulent layer and reduced significantly in the inner portion. With no empirical adjustment involving curvature, the model quantitatively predicted this rather dramatic observation (So, 1975; Mellor, 1975). The same model, when extended in--it now seems--a rather straightforward manner, predicted the observed stabilizing and destabilizing effect of density gradients in a gravity field (Mellor, 1973). Again, no adjustment of the model to specifically accommodate stratification was required. Yamada and Mellor (1975) and Lewellen et al. (1976) have now made persuasive predictions of stratified laboratory flows and atmospheric boundary layers which are strongly influenced by diurnally varying density stratification.

The empirical content of the models described above resides in various turbulent length scales; a necessary assumption is that all of the scales are proportional to a single master length scale. The experience generated thus far indicates that this is a viable assumption.

Surprisingly, although the length scale proportionality constants are important, many results are either independent or not very sensitive to willful variations in the master length scale. Thus, we have tried to separate our study of the turbulence model into considerations of the single point moment equations, their various length scales and attendant constants of proportionality and consideration of the master length scale itself.

We have been uncomfortable in dealing with a master length scale equation and for some time were content with algebraic prescriptions. However, following Rotta once again, a turbulence scale equation can be approached via the equations for the turbulence spectra or the equations for the two-point correlation functions. When the separation distance, r, approaches zero we obtain the single point, moment equations discussed previously and lose all length scale information. Thus the approach suggested by Rotta was to consider the integral of the correlation functions over r-space. Conceptually complicated assumptions are required to close the resultant equations. Nevertheless, in the present paper we have applied the full model with a differential, master length scale equation to the classic fluid dynamic problem of flow in a channel with a sudden expansion. The results compare well with data by Abbott and Kline (1962) which includes mean velocity profiles and turbulent energy profiles. The point of reattachment is predicted well. Again, there were no empirical adjustments in the turbulence model to accommodate this problem. The stabilizing or destabilizing effects of curvature are automatically taken into account with no additional empiricism.

The numerical algorithm applied to the sudden expansion flow is an explicit, time marching scheme* and all components of the mean Reynolds stress tensor are involved. By matching the calculated velocity near a surface to the law of the wall it is apparent that the problem can be handled with fairly crude resolution. Associated with grid resolution is an uncertainty in the amount of vorticity advected into the flow at the sudden expansion corner. What appears to be a simple solution to this problem will be discussed.

Thus, it appears that the very wide range of fluid dynamic problems involving turbulent separating flows can be attacked with the help of second moment turbulence models.

2. THE DIFFERENTIAL EQUATIONS OF MOTION

We denote mean dependent variables by capital letters and turbulent variables by lower case letters. The mean equations of motion for incompressible flow are then,

$$\frac{\partial U_k}{\partial x_k} = 0 \tag{1}$$

---

* Similar or alternate schemes applied to laminar flow or calculations using a simple eddy viscosity model have been presented by a number of others, some of whom are Thoman and Szewczyk (1966), Gosman et al. (1969), Kirkpatrick and Walker (1972), Fromn and Harlow (1963), Roach, (1972).

$$\frac{\partial U_j}{\partial t} + \frac{\partial}{\partial x_k}(U_k U_j) = -\frac{\partial P}{\partial x_j} + \frac{\partial}{\partial x_k}(-\overline{u_k u_j}) \tag{2}$$

where an overbar represents the mean of products of turbulent variables. P, the kinematic pressure, is the pressure divided by density.

The turbulent model equations are based on hypotheses by Rotta (1951) and Kolmogorov (1941); other details are discussed by Mellor and Herring (1973) and Mellor (1973). The equations are

$$\frac{\partial}{\partial t}(\overline{u_i u_j}) + \frac{\partial}{\partial x_k}(U_k \overline{u_i u_j}) = \frac{\partial}{\partial x_k}\left[\frac{3}{5}\ell q \tilde{S}_q\left(\frac{\partial \overline{u_i u_j}}{\partial x_k} + \frac{\partial \overline{u_i u_k}}{\partial x_j} + \frac{\partial \overline{u_j u_k}}{\partial x_i}\right)\right]$$

$$- \frac{q}{3A_1 \ell}\left(\overline{u_i u_j} - \frac{\delta_{ij}}{3} q^2\right) + C_1 q^2 \left(\frac{\partial U_i}{\partial x_j} + \frac{\partial U_j}{\partial x_i}\right)$$

$$- \overline{u_k u_i}\frac{\partial U_j}{\partial x_k} - \overline{u_k u_j}\frac{\partial U_i}{\partial x_k} - \frac{2}{3}\frac{q^3}{B_1 \ell}\delta_{ij} \tag{3}$$

It should be noted that the stabilizing or destabilizing effect of streamline curvature is automatically included in (3). Of course, the equations would have to be written in streamline coordinates (So, 1975; Mellor, 1975) for the "effect" to be easily identified.

It is useful to note that the trace of (3) is

$$\frac{\partial}{\partial t} q^2 + \frac{\partial}{\partial x_k}(U_k q^2) = \frac{\partial}{\partial x_k}\left[\frac{3}{5}\ell q \tilde{S}_q\left[\frac{\partial q^2}{\partial x_k} + 2\frac{\partial \overline{u_i u_k}}{\partial x_k}\right]\right]$$

$$= -2\,\overline{u_k u_i}\frac{\partial U_i}{\partial x_k} - 2\frac{q^3}{B_1 \ell} \tag{4}$$

such that $q^2 \equiv \overline{u_i^2}$ is twice the turbulent kinetic energy. Variants of the above model have been applied with good success in the aforementioned papers. The terms in which the constants, $\tilde{S}_q$, A_1, C_1, B_1, appear are, of course, model terms for turbulent diffusion, Rotta's energy redistribution model and a Kolmogorov isotropic dissipation model. In the past, we have supplemented the above equations with rather simple algebraic master length equations which seem to work rather well. In the present work it is actually simpler to solve a differential length scale equation which has a property that it is invariant to coordinate orientation and should accommodate arbitrary boundary conditions. The equation we use here is a variant of one proposed by Rotta (1951) and is

$$\frac{\partial}{\partial t}(q^2\ell) + \frac{\partial}{\partial x_k}(U_k q^2 \ell) = \frac{\partial}{\partial x_k}\left[q\ell\tilde{S}_\ell \frac{\partial(q^2\ell)}{x_k}\right]$$

$$- E\ell\overline{u_i u_j}\frac{\partial U_i}{\partial x_j}\left\{1 - \left(k^{-1}\frac{\partial \ell}{\partial x_k}\right)^2\right\} - \frac{q^3}{B_1} \qquad (5)$$

All terms on the right side of (5) are model terms.

3. DISCUSSION OF THE MODEL EQUATIONS AND THEIR ATTENDANT EMPIRICAL CONSTANTS

Readers who are not interested in detailed modeling considerations may wish to skip this section.

Temporarily consider a surface normal to the y-axis. Near the surface let $W = 0$ and $V = 0$. In this region $\ell = ky$, where k is the von Kármán constant, and it may be shown that the tendency, advection and diffusion terms in (3) may be neglected. As a consequence, it has also been shown (Mellor and Herring, 1973; Mellor, 1973) that if we let

$$\gamma \equiv \frac{1}{3} - 2\frac{A_1}{B_1} \qquad (6a)$$

then

$$\overline{u^2} = (1 - 2\gamma)q^2 \qquad (6b)$$

$$\overline{v^2} = \gamma q^2 \qquad (6c)$$

$$\overline{w^2} = \gamma q^2 \qquad (6d)$$

$$u_\tau^2 \equiv |-\overline{uv}| = \frac{q^2}{B_1^{2/3}} \qquad (6e)$$

$$C_1 = \gamma - \frac{1}{3A_1 B_1^{1/3}} \tag{6f}$$

and

$$\frac{\partial U}{\partial y} = \frac{u_\tau}{ky} \tag{7}$$

All turbulence quantities are evaluated near the wall just outside of the viscous sublayer (e.g. $u_\tau y/\nu \simeq 30$). Based on law of the wall data we decided that $B_1 = 15.0$ from (6e) and $\gamma = 0.23$ from (6b). The sum of (6c) and (6d), $\overline{v^2} + \overline{w^2} = 2\gamma q^2$, is correctly obtained but the separate components are in error more or less dependent on which data set is examined. To correct this error one can complicate the modeling in ways suggested by Launder, Reece and Rodi (1975) or Monin (1965), but thus far we have been inclined to keep the model simple and accept the error.

With γ and B_1 determined, $A_1 = 0.78$ is obtained from (6a) and $C_1 = 0.056$ from (6f).

By comparing constant pressure, boundary layer q^2 profiles with model computations we have obtained $\tilde{S}_q = 0.45$, a relatively less important constant compared to the others as discussed below.

The constants in equation (5) have also been determined from simple considerations. First, a constant should appear as a coefficient of the last term in (5). However, it has been set equal to unity such that (4) and (5) when specialized to homogeneous decaying turbulent yields the high Reynolds number, initial period decay law, $q^2 \propto x^{-1}$.* By considering the law of the wall again when $\ell = ky$ and $q^2 \simeq$ constant, we obtain

$$\tilde{S}_\ell = \frac{1}{k^2 B_1} \quad . \tag{8}$$

---

* Decaying homogeneous turbulent highlights a limitation of any model characterized by a single length scale. Turbulence theory and data (Batchelor and Stewart, 1950) and some new unpublished but corroborative Princeton data indicate that decaying turbulence can become more anisotropic during the decay process since the smaller more isotropic eddies decay faster than the larger, more anisotropic eddies. The present model--since it describes the turbulence through only one length scale--would predict that the turbulence tends only to isotropy during decay.

For $k = 0.40$ and $B_1 = 15.0$, we have $\tilde{S}_\ell = 0.42$. Note that it appears that $\tilde{S}_\ell \simeq \tilde{S}_q$. A consideration of length scale growth in the homogeneous shear flow data of Champagne, Harris and Corssin (1970) indicates very approximately that $E \simeq 1.3 - 1.5$.

At this point we note that, whereas in many flows (e.g., fully developed channel flow) the diffusion terms in (3) or (4) are quite small, the diffusion term in (5) is a dominant term and near a wall is balanced by the last term in (5). Rotta (1972) has expressed some misgivings about length scale models wherein diffusion terms are dominant. However, Thompson and Turner (1975) have performed an experiment whereby a vibrating grid creates a plane source of turbulence in a stationary field and where the diffusion and dissipation terms in (4) and (5) are exactly in balance. They obtain the result that $q^2 \propto y^{-1.5}$ and $\ell \propto y$ where y is the distance from the grid. It is noteworthy that the present model also predicts this behavior. Thus, when the model reduces to $\partial[q\ell\tilde{S}_q\partial q^2/\partial y]/\partial y = 2q^3/B_1\ell$ and $\partial[q\ell\tilde{S}_\ell\partial(q^2\ell)/\partial y]/\partial y = q^3/B_1$ one obtains as a solution $q^2 \propto y^{-n}$ and $\ell = by$ where $2(n-1)(3n-2)/3n^2 = \tilde{S}_q/\tilde{S}_\ell$ and $S_q = 4/(3n^2b^2B_1)$.* We can only roughly obtain an estimate of $b \approx 0.07$ from the data; this and the value, $n = 1.5$, yield $\tilde{S}_q \approx 8$ and $\tilde{S}_\ell \approx 22$.

Thus, considering flow fields where production is important and the self diffusing flow field where production is zero, we find large differences in $\tilde{S}_q$ and $\tilde{S}_\ell$. However, this is not unexpected. As discussed by Mellor and Yamada (1974) our model may be simplified such that the Reynolds stress is related to the mean velocity shear via a conventional, flux-gradient, eddy viscosity coefficient which, however, takes the form, $q\ell\tilde{S}_M$. $\tilde{S}_M$ emerges primarily as a known function of $G \equiv \ell^2(\partial U_i/\partial x_j)^2/q^2$. Therefore, one should also expect $\tilde{S}_q = \tilde{S}_q(G)$ and $\tilde{S}_\ell = \tilde{S}_\ell(G)$. Although the experimental base is thus far fragile, one might tentatively propose a simple linear relation between the values given above for $G = 0$ and the law of wall value, $G = B_1^{2/3}$. However, in the calculations discussed below we have maintained $\tilde{S}_q = \tilde{S}_\ell = 0.42$.

Further discussion of equation (5) is included in §9.

---

* Since two boundary conditions may be imposed on ℓ and q^2, the solution is not unique. However, away from boundaries, numerical experiments indicate that this algebraic decay behavior does prevail for a wide variety of boundary conditions.

4. THE EQUATIONS FOR TWO-DIMENSIONAL, MEAN, PLANAR FLOW

We now restrict attention to planar flows where the mean velocity, $U_i = (U, V, 0)$, and all properties are invariant in the z-direction. If we further define a mean vorticity and stream function,

$$\xi \equiv \frac{\partial V}{\partial x} - \frac{\partial U}{\partial y} \tag{9}$$

$$U \equiv \frac{\partial \Psi}{\partial y}, \quad V \equiv -\frac{\partial \Psi}{\partial x} \tag{10a,b}$$

then (1) is satisfied and (9) may be written

$$\frac{\partial^2 \Psi}{\partial x^2} + \frac{\partial^2 \Psi}{\partial y^2} = -\xi \tag{11}$$

The curl of (2) for two-dimensional flow is

$$\frac{D\xi}{Dt} = \frac{\partial^2}{\partial x^2}(-\overline{uv}) - \frac{\partial^2}{\partial x \partial y}(\overline{v^2} - \overline{u^2}) - \frac{\partial}{\partial y^2}(-\overline{uv}) \tag{12}$$

The equations for the Reynolds stresses may be written

$$\frac{D}{Dt}\begin{bmatrix} \overline{u^2} \\ \overline{v^2} \\ \overline{w^2} \\ \overline{uv} \end{bmatrix} = \frac{\partial}{\partial x} K \begin{bmatrix} 3\partial\overline{u^2}/\partial x \\ \partial\overline{v^2}/\partial x + 2\partial\overline{uv}/\partial y \\ \partial\overline{w^2}/\partial x \\ \partial\overline{u^2}/\partial y + 2\partial\overline{uv}/\partial x \end{bmatrix} + \frac{\partial}{\partial y} K \begin{bmatrix} \partial\overline{u^2}/\partial y + 2\partial\overline{uv}/\partial x \\ 3\partial\overline{v^2}/\partial y \\ \partial\overline{w^2}/\partial y \\ \partial\overline{v^2}/\partial x + 2\partial\overline{uv}/\partial y \end{bmatrix}$$

$$- \frac{q}{3A_1 \ell}\begin{bmatrix} \overline{u^2} - q^2/3 \\ \overline{v^2} - q^2/3 \\ \overline{w^2} - q^2/3 \\ \overline{uv} \end{bmatrix} + C_1 q^2 \begin{bmatrix} 2\partial U/\partial x \\ 2\partial V/\partial y \\ 0 \\ \partial U/\partial y + \partial V/\partial x \end{bmatrix}$$

(continued)

$$+\begin{bmatrix} -2\overline{u^2}\,\partial U/\partial x - 2\overline{uv}\,\partial U/\partial y \\ -2\overline{uv}\,\partial V/\partial x - 2\overline{v^2}\,\partial V/\partial y \\ 0 \\ -\overline{u^2}\,\partial V/\partial x - \overline{v^2}\,\partial U/\partial y \end{bmatrix} - \frac{2}{3}\frac{q^3}{B_1\ell}\begin{bmatrix} 1 \\ 1 \\ 1 \\ 1 \end{bmatrix} \qquad \begin{matrix} (13a) \\ (13b) \\ (13c) \\ (13d) \end{matrix}$$

where $Df/Dt \equiv \partial f/\partial t + \partial(Uf)/\partial x + \partial(Vf)/\partial y$ and $K \equiv 3\ell q\tilde{S}_q/5$.

Finally, the master length scale equation is

$$\frac{D}{Dt}(q^2\ell) = \frac{\partial}{\partial x}\left[q\ell\tilde{S}_\ell \frac{\partial(q^2\ell)}{\partial x}\right] + \frac{\partial}{\partial y}\left[q\ell\tilde{S}_\ell \frac{\partial(q^2\ell)}{\partial y}\right]$$

$$+ E\ell\left[-\overline{uv}\left(\frac{\partial U}{\partial y} + \frac{\partial V}{\partial x}\right) - \overline{u^2}\frac{\partial U}{\partial x} - \overline{v^2}\frac{\partial V}{\partial y}\right]$$

$$\times\left[1 - \left(\frac{1}{k}\frac{\partial \ell}{\partial x}\right)^2 - \left(\frac{1}{k}\frac{\partial \ell}{\partial y}\right)^2\right] - \frac{q^3}{B_1} \qquad (14)$$

5. FINITE DIFFERENCE EQUATIONS

The differential equations are differenced in a conventional way. The time differencing is forward explicit. Advective terms are all "upwind" differenced* whereas the Laplacian operator in (11) and the Reynolds stress terms or the mean vorticity diffusion terms in (12) are centrally differenced as are the diffusion terms in (13) and (14). Poisson's equation, equation (11), is solved by an alternating-direction implicit (ADI) method. Further details are provided in the Appendix.

6. BOUNDARY CONDITIONS

At some intermediate time in the course of a calculation, the Reynolds stress from the previous time is known and is used to

---

* Which is a stable scheme but which introduces some numerical viscosity which we believe not to be significant. Nevertheless, we should but have not yet carried out a calculation where advective terms are centrally differenced.

evaluate new values of the vorticity according to (12). Except at singular corners, which we shall later discuss in some detail, surface values of vorticity are not required since, in the advection terms, the surface values are multiplied by the surface normal component of velocity which is zero. Inflow values of vorticity are prescribed by data.

Once the vorticity is everywhere known, (11) is solved for the stream function. On solid walls we impose constant boundary values of Ψ that differ on opposite walls by an amount determined by the volume flow rate. On upstream flow surfaces, Ψ is determined by inflow velocity data. On downstream surfaces we set $\partial^2\Psi/\partial x^2$ equal to zero.

For the type of time marching, fully elliptic algorithm adopted here, one cannot afford very fine grid resolution and one cannot resolve the viscous sublayer or a large portion of the logarithmic layer. We will, therefore, match our numerical solution to the law of the wall which is the integral of (7)

$$U = u_\tau\left[\frac{1}{k}\ln y^+ + 4.9\right] \tag{15}$$

where $y^+ = u_\tau y/\nu$. (In §9, we discuss possible error incurred by using (15) near the reattachment point.)

Here we restrict discussion to the case where the wall is normal to the y-coordinate and the flow is in the positive x-direction as illustrated in Figure 1a. Now during the course of a calculation the stream-function will be calculated and the role of (15) will be to provide Reynolds stress wall boundary conditions for equations (13a, b, c, d). The integral of (15) for a point at y_1, where y_0 is on the wall, is

$$|\Psi_1 - \Psi_0| = u_\tau\,\Delta y\left[\frac{1}{k}\ln\frac{u_\tau\Delta y}{\nu} + 4.9 - \frac{1}{k}\right] \tag{16}$$

where $\Delta y = |y_1 - y_0|$. Thus, as the calculation proceeds, values of $\psi_1 - \psi_0$ and (16) provide a value of u_τ (inside the log term, u_τ may be reckoned at the previous time step since ln x is a weak function of x). Values of $\overline{u^2}$, $\overline{v^2}$, $\overline{w^2}$ and $\overline{uv}$ are then obtained from (6a, b, c, d, e).

For the more general boundary illustrated in Figure 1b we can write

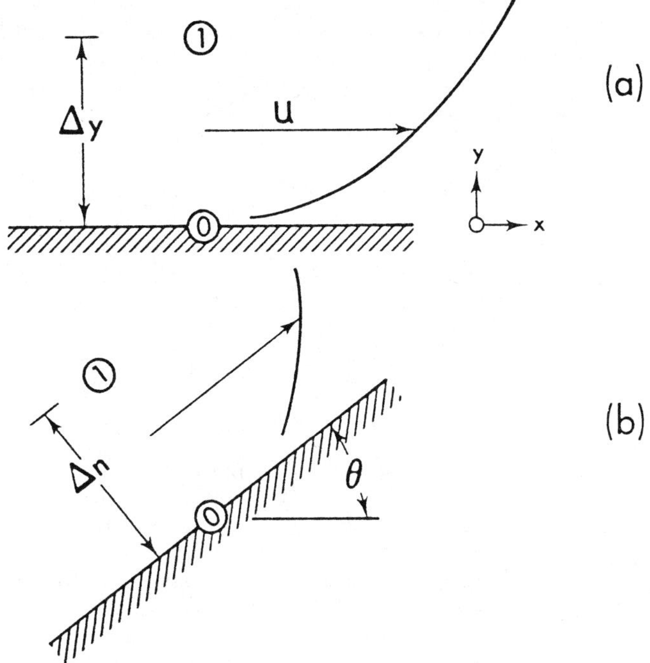

Figure 1. Illustration of the Geometry in the Discussion of Wall Boundary Conditions. Numerically, the mean velocity parallel to a wall is never defined at the wall whereas the Reynolds stresses are evaluated as finite values at the wall; the viscous sublayer is therefore considered to be indefinitely thin.

$$\begin{bmatrix} \overline{u_s^2} \\ \overline{v_s^2} \\ \overline{u_s v_s} \end{bmatrix} = \begin{bmatrix} n_y^2 , & n_x^2 , & 2n_x n_y \\ n_x^2 , & n_y^2 , & -2n_x n_y \\ -n_x n_y , & n_x n_y , & n_x^2 - n_y^2 \end{bmatrix} \begin{bmatrix} (1 - 2\gamma) q_s^2 \\ \gamma q_s^2 \\ u_\tau^2 \operatorname{sgn}(\psi_1 - \psi_0) \end{bmatrix} \qquad (17)$$

where $(n_x, n_y) \equiv (-\sin\theta, \cos\theta)$ are the components of a unit vector normal to the wall. Δn, the distance from the wall, replaces Δy in (13) and as before $q_s^2 = u_\tau^2 B_1^{2/3}$. For the simple rectangular geometry discussed below, θ only takes on the values 0°, 90°, 180°, and 270°. It should be noted that the velocity is never defined numerically at the wall. On the other hand, the surface values of the Reynolds stress are evaluated at the wall itself, the thickness of the viscous sublayer thus considered to be negligibly small. The subscripts in (17) denote the surface values.

7. FULLY DEVELOPED STEADY CHANNEL FLOW

A fully developed channel flow calculation was carried out and compared with data by Laufer (1951) and Hanjalic (1970). Calculations for two values of E = 1.3 and 1.5 are shown in Figure 2. Henceforth, we have used the value E = 1.5. Laufer's data when plotted in law of the wall coordinates yield the constant, 5.5, instead of 4.9 in (16) and this probably accounts for the discrepancy in Figure 2a.

8. SEPARATED FLOW BEHIND A STEP

The major effort in this paper has been the numerical solution of flow from one channel into a larger channel caused by a step on one wall. The particular geometry is identical to an experiment by Abbott and Kline (1961). If h is the step height and u_0 is the small channel centerline velocity, then the Reynolds number, $u_0h/\nu = 3 \times 10^4$ for both the experiments and the numerical simulations.

The initial calculations shown in Figures 3 and 4 are trials with crude resolution and are not intended to be an accurate simulation of the experiment. Figure 3 illustrates the temporal evolution of the flow after which a steady flow is established. At t = 0, the flow field is irrotational whereas at $tu_0/h = 200$ the flow is very nearly steady. The calculation time step, $\Delta tu_0/h = 0.10$, is governed by stability considerations.

At x = -2h, the upstream velocity profile is matched to the data whereas the turbulence quantities are obtained from the fully developed channel flow solution discussed previously. A mistake was made in transcribing the upstream length scale profile such that at x=0 the initial profile is about 40% smaller than the true profile. The mistake was corrected for results shown after Figure 4.

At $x \simeq 10h$, the length scale on the separating wall was quite large and there was no region where $\ell \simeq ky$. The resolution was crude but still we wanted to be certain that the model could produce a far downstream, fully developed flow. The calculation was accomplished by piecewise matching the steady profiles at x = 14h to the inflow condition of a second channel and so on. The Δx spacing was allowed to increase with each channel length. A composite result is shown in Figure 4 so that at $x/h \simeq 200$ it can be reasonably concluded that a fully developed channel flow has been reestablished.

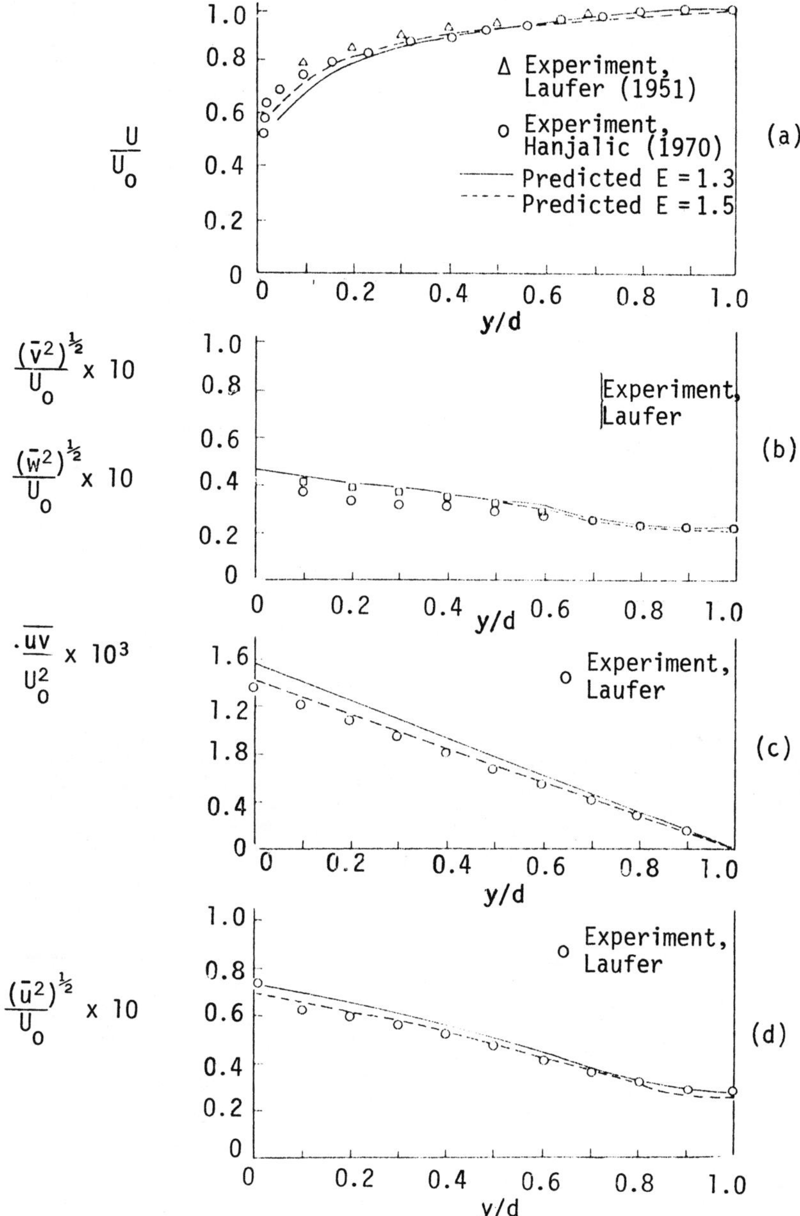

Figure 2. Calculations of Fully Developed Channel Flow Compared with Data.

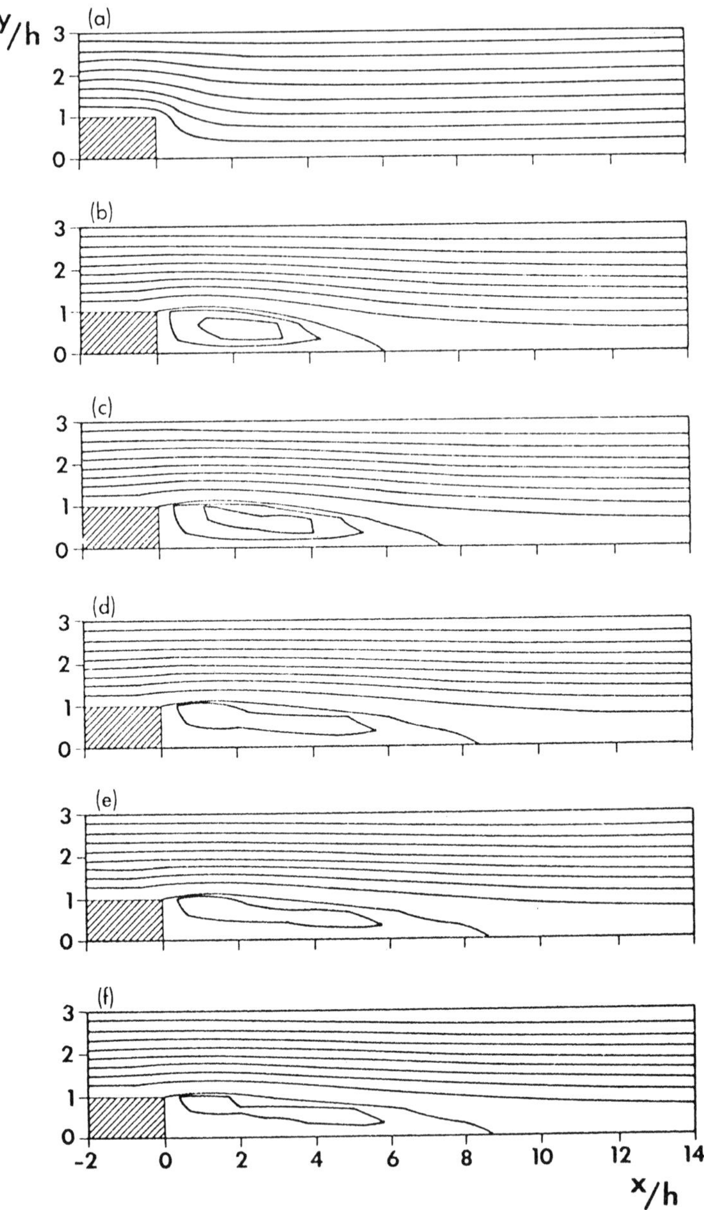

Figure 3. The Time Evolution of the Separated Flow. tu_0/h = 0 (the initial irrotational flow) (a); 40 (b); 80 (c); 120 (d); 160 (e); 200 (f). The resolution was crude.

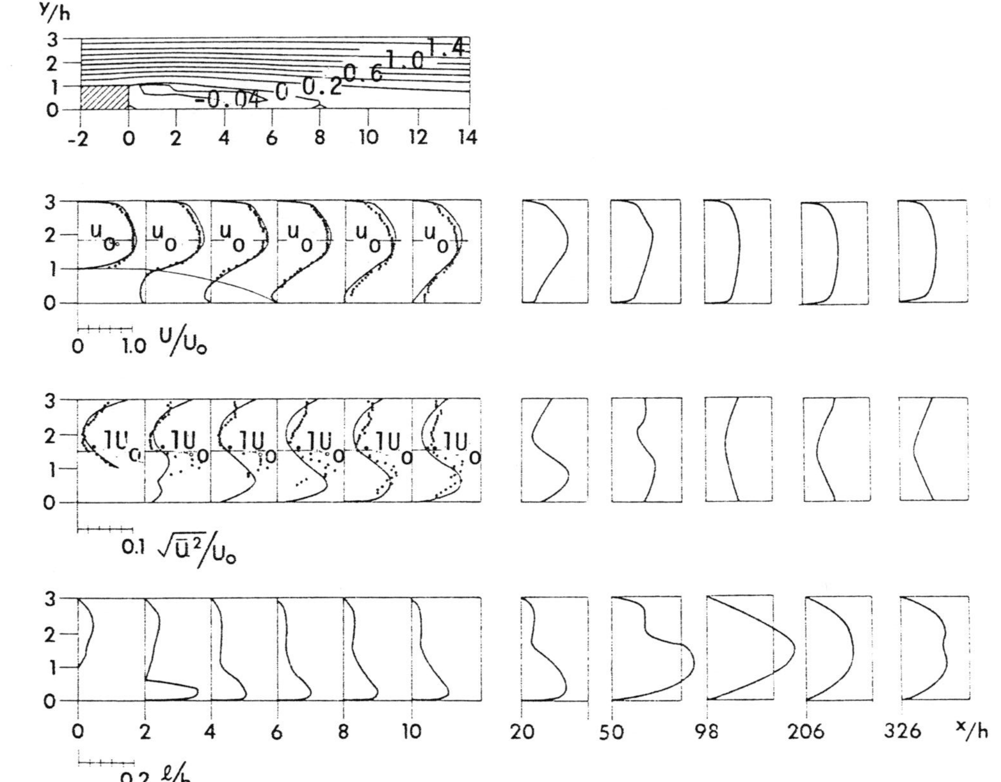

Figure 4. The Downstream Reestablishment of Fully Developed Channel Flow. The resolution was crude. The initial values of ℓ were low by 40%. This was corrected for all subsequent calculations.

A problem was encountered which was resolved for the present geometry but which may require other strategies for other geometries. As previously mentioned the vorticity advection terms in (12) do not require solid surface boundary conditions since the velocity normal to the surface is zero. However, at a corner, as illustrated in Figure 5, there is no unique normal velocity and, in terms of our finite difference scheme the upwind contribution to vorticity advection at this point is

$$(\psi_1 - \psi_5)\xi_0/(2\Delta y)$$

where

$$\xi_0 = -U_{\frac{1}{2}}/\Delta y = -(\psi_1 - \psi_0)/\Delta y^2$$

The above is, of course, a crude approximation to the vorticity <u>at</u> the corner and in the viscous sublayer. With no further alteration, calculated flows turned the corner in an unrealistic fashion. We have therefore set

$$\xi_0 = -\beta \frac{U_{\frac{1}{2}}}{\Delta y} \tag{18}$$

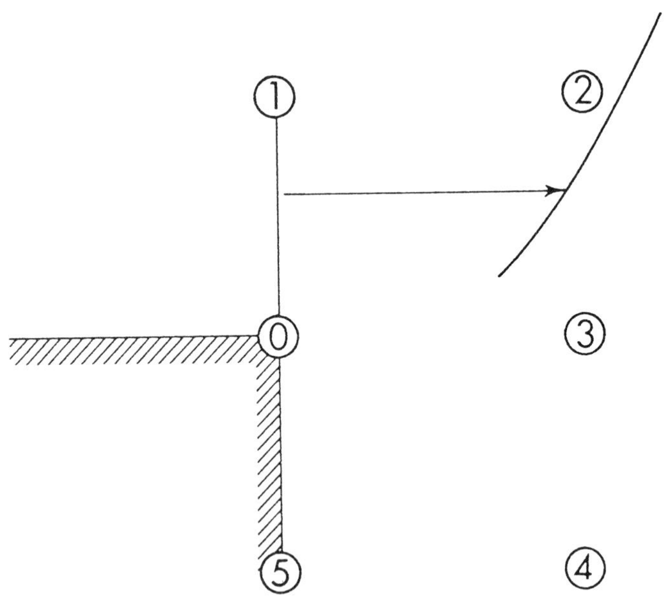

Figure 5. Illustration of the Flow at a Corner.

and then have experimented with variations in β. For crude resolution the calculations for β = 1.30, 1.59, 2.67 and 5.63 were performed, the first three of which are shown in Figure 6 whereas for the finer resolution calculations for β = 1.25, 1.56, and 2.45 produced the results shown in Figure 7. All of these calculations produced different values of the reattachment length, x_R, which are plotted in Figure 8 as a function of β. Noteworthy features of Figure 8 are as follows:

1) The sensitivity of x_R (and the entire solutions) to β decreases as the resolution is increased, a comforting result.

2) Solutions yielding the observed reattachment length correspond closely to those cases where the separation streamline is parallel to the entrance wall at the corner.

Following up on the second point, we have therefore made a calculation where β is adjusted during the calculation so that the parallel streamline condition is exactly satisfied. These are the solid symbols shown in Figure 8 corresponding to β = 1.30 for the crude resolution and to β = 1.25 for the fine resolution.

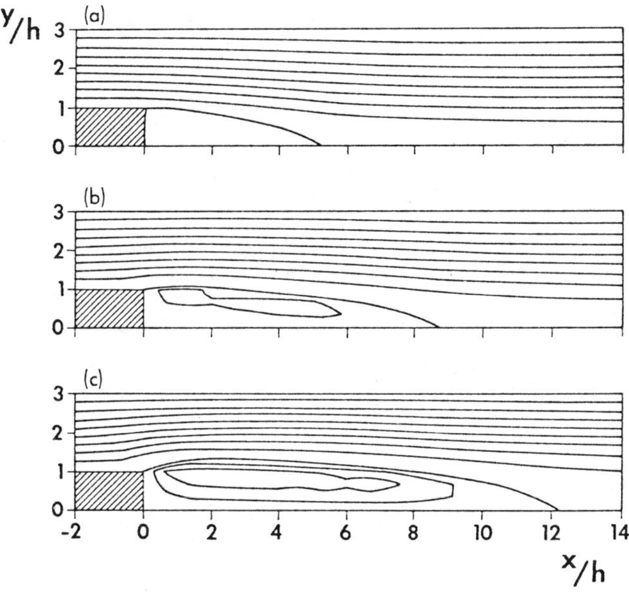

Figure 6. Stationary Flow Streamlines for Crude Resolution, $\Delta x/h$ = 2/3, $\Delta y/h$ = 1/3 and β = 1.30 (a); 1.59 (b); 2.67 (c).

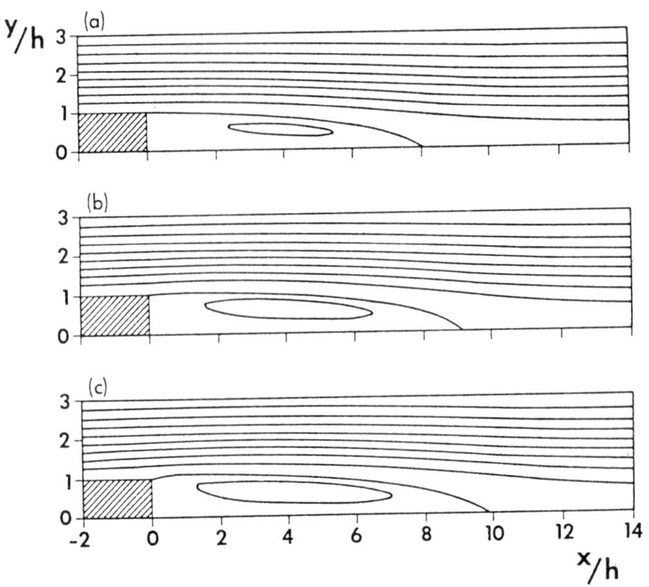

Figure 7. Stationary Flow Streamlines for the Finer Resolution $\Delta x/h = 2/3$, $\Delta y/h = 1/12$ and $\beta = 1.25$ (a); 1.56 (b); 2.45 (c).

In Figure 9 some detailed profiles of mean velocity and turbulence intensity are illustrated for the crude resolution whereas in Figures 10 and 11 the fine resolution results are shown. Figure 11 is the result that should most nearly match the data. It does appear from Figures 9 and 11 that some improvement in mean profile prediction might be realized with still finer resolution.

The turbulent intensity comparison of prediction with data would probably not improve. On the other hand, it is our understanding (verbal communication, S. J. Kline) that hot film turbulence measurement in 1960 was not accurate. In a separated flow region this would be true for hot film or hot wire measurements even in 1976.

9. DISCUSSION

A turbulence closure model has been applied to a fully separated flow. All of the constants had been established by previous

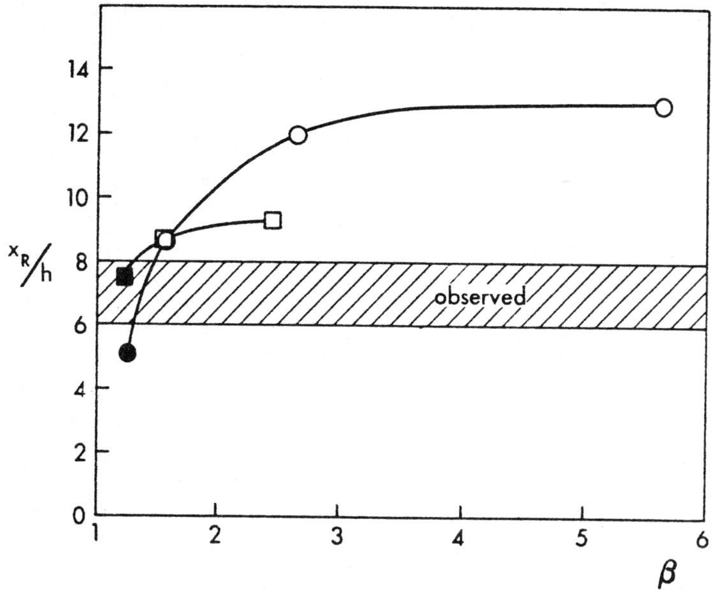

Figure 8. Summary Plot of the Reattachment Length, x_R, versus β. The circles correspond to $\Delta x/h = 2/3$, $\Delta y/h = 1/3$ and the squares to $\Delta x/h = 2/3$, $\Delta y/h = 1/12$. The solid symbols are cases where the value of β was adjusted during the calculation so that the separating streamline was parallel to the entrance channel wall.

reference to more simple experiments. A possible exception is the constant E in equation (5), where some a priori uncertainty existed. We have, therefore, tried two values yielding slightly different solutions to fully developed channel flow. The value, E = 1.5, was chosen for all of the separated flow cases.

The final comparison of predicted flow properties in Figure 11 is quite favorable although the accuracy of the turbulence intensity data is questionable.

The problem of how much vorticity is shed from a sharp corner appears to disappear as resolution is increased and, for finite resolution, may be resolved by insisting on the parallel streamline condition (a kind of Kutta condition?). The problem should, in principle, not emerge in the case of flow separation from smooth boundaries. However, difficulties may reappear in other flow cases in which the experience generated here may be useful.

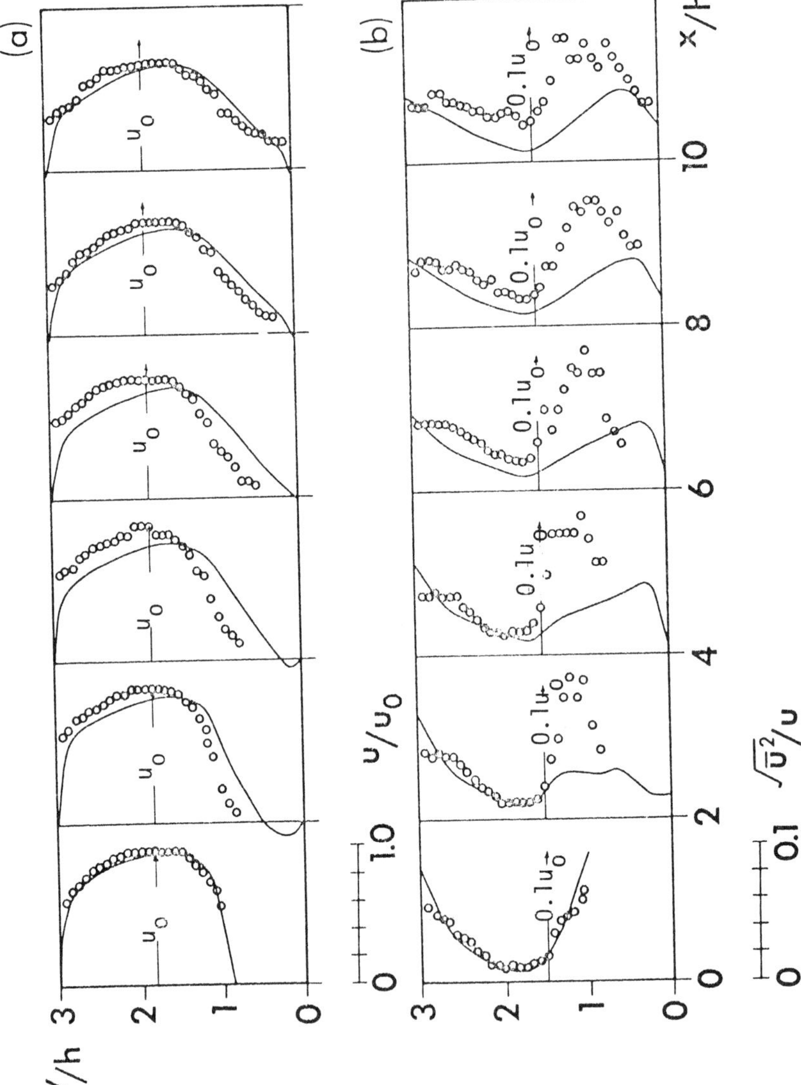

Figure 9. Calculations for $\Delta x/h = 2/3$, $\Delta y/h = 1/3$ and $\beta = 1.30$, the Parallel Separating Streamline Case.

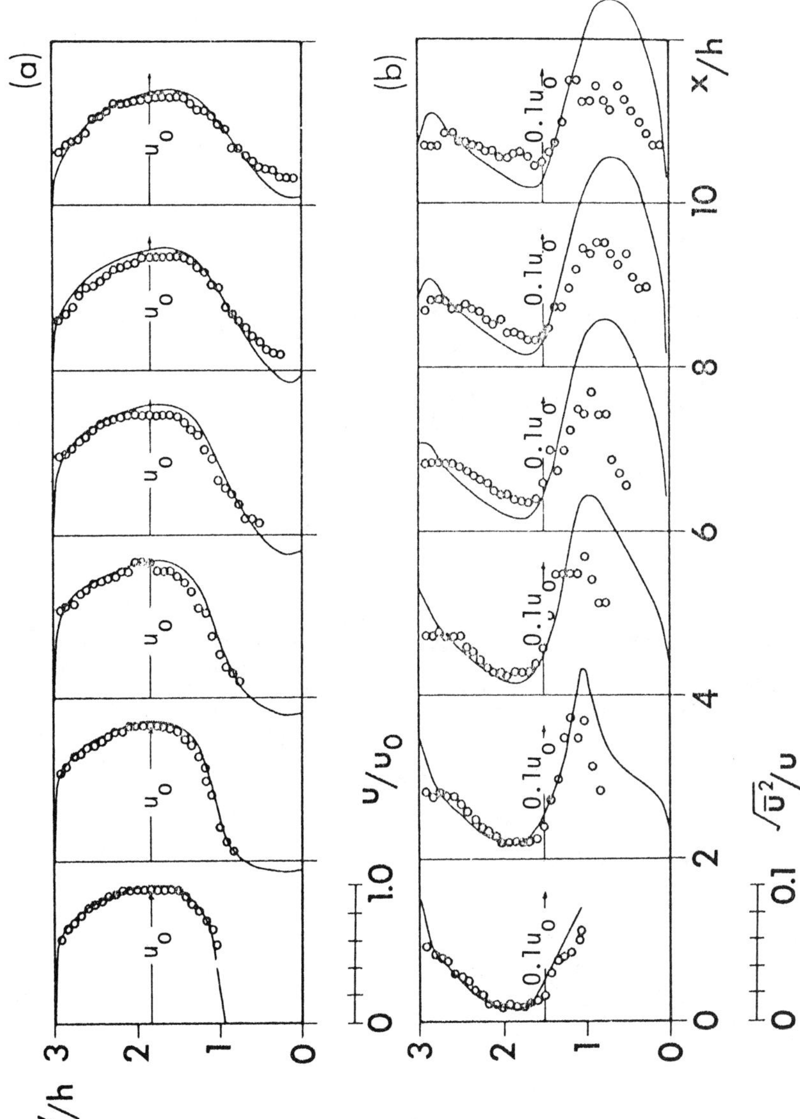

Figure 10. Calculations for $\Delta x/h = 2/3$, $\Delta y/h = 1/12$ and for $\beta = 1.56$.

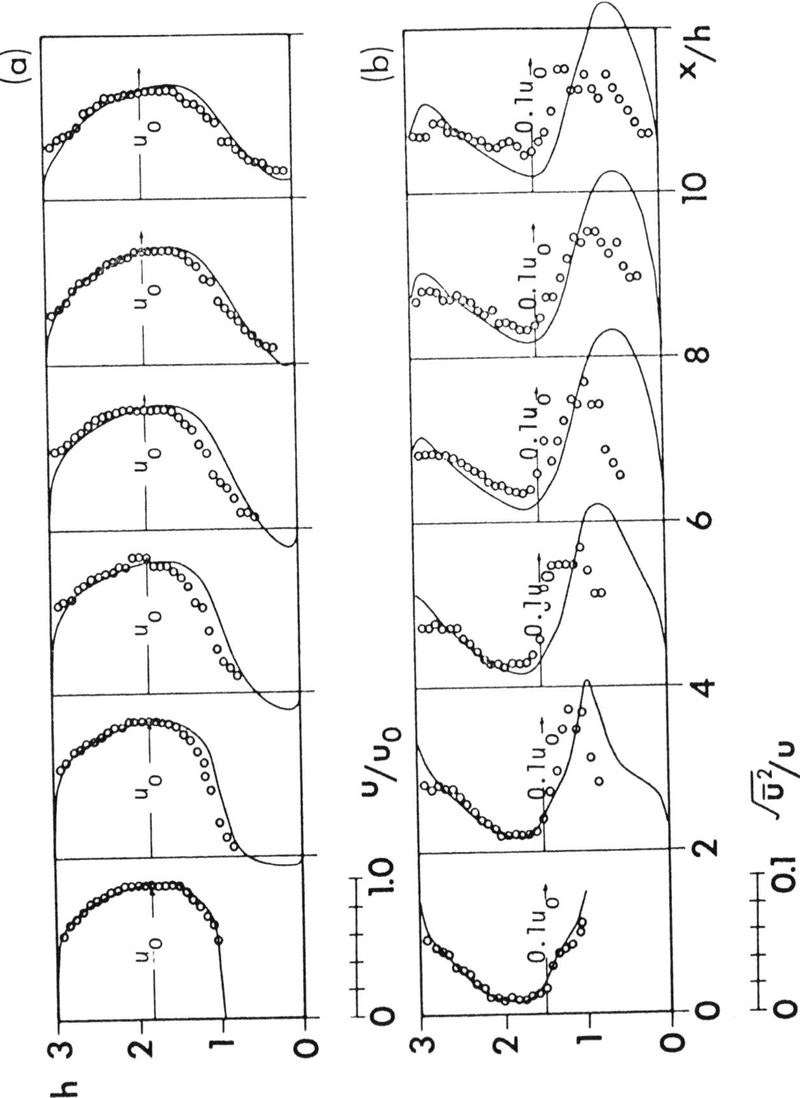

Figure 11. Calculations for $\Delta x/h = 2/3$, $\Delta y/h = 1/12$ and $\beta = 1.25$, the Parallel Separating Streamline Case.

It is known that close to a separation point or, in our case, close to a reattachment point, equation (15) should not apply and it is possible to apply a more general boundary condition along lines suggested by Townsend (1961) and Mellor (1966). Our guess is that the overall result will not be significantly altered. However, we do plan to repeat the calculations with the improved boundary condition in the near future.

Finally, a comment on the master length scale equation (5). As previously mentioned, Rotta (1951) interpreted the equation as an equation for the integral of the two point, correlation function over the separation distance. This is certainly a proper interpretation as the equation liable to produce a macro-length scale. Nevertheless, the "wall correction" term, the term in curly brackets in (5), is completely ad hoc. Some such term is absolutely necessary,* but the one chosen here is only one of several alternatives (Ng and Spalding, 1972; Wolfstein, 1970; Mellor and Herring, 1973; Lewellen et al., 1976; Rotta, 1972). All one can say is that it seems to work well in the present case and in other cases we have tried.

It should be noted that terms like the term in question is automatically included in the somewhat popular equation for the dissipation (Daly and Harlow, 1970; Hanjalic and Launder, 1972) which we here model as $q^3/(B_1\ell)$. However, the dissipation is the curvature of the two point velocity correlation function as the separation distance approaches zero. Conceptionally such an equation is inappropriate to the purpose of providing a macro-length; in the large Reynolds limit where the correlation function asymptotes to the two-thirds power law, its curvature limits to infinity as the separation distance approaches zero. This singularity is, of course, corrected by considering the small scale, viscosity dependent, turbulence; the latter is not relevant to the modeling of high Reynolds number turbulence, however.

APPENDIX: FINITE DIFFERENCE EQUATION

We here list the finite differences version of the differential equations to be solved numerically.

---

* Consider the case of fully developed channel flow where dissipation very nearly balances production. Then, without the "wall correction" term equation (5) may be written

$$\partial[q\ell\tilde{S}_\ell \partial(q^2\ell)/\partial x_k]/\partial x_k = (E - 1)q^3/B_1$$

If we integrate this equation from one wall at $y = 0$ to the other at $y = H$, we obtain $(E - 1)\int_0^H q^3 dy/B_1 = 0$, an impossible result.

Equation (11) is represented as

$$\frac{\psi^n_{i+1,j} - 2\psi^n_{i,j} + \psi^n_{i-1,j}}{\Delta x^2} + \frac{\psi^n_{i,j+1} - 2\psi^n_{i,j} + \psi^n_{i,j+1}}{\Delta y^2} = -\xi^n_{i,j} \quad \text{(A1)}$$

which is solved iteratively by an alterating direction implicit (ADI) method.

The vorticity is then advanced in time according to (12) which is represented as

$$\xi^{n+1}_{i,j} = \xi^n_{i,j} - \left\{ \left[\frac{\partial(U\xi)^n}{\partial x} + \frac{\partial(V\xi)^n}{\partial y}\right]_{i,j} + \left(\frac{\partial^2\overline{uv}}{\partial x^2}\right)^n_{i,j} + \left(\frac{\partial^2\overline{v^2}}{\partial x\partial y}\right)^n_{i,j} \right.$$

$$\left. - \left(\frac{\partial^2\overline{u^2}}{\partial x\partial y}\right)^n_{i,j} - \left(\frac{\partial^2\overline{uv}}{\partial y^2}\right)^n_{i,j} \right\}\Delta t \quad \text{(A2)}$$

Upwind advective differencing is used such that

$$\left[\frac{\partial(U\xi)}{\partial x}\right]_{i,j} = \left[\frac{(U\xi)_{i+1,j} - (U\xi)_{i,j}}{2\Delta x}\right]\left[1 - \frac{U_{i,j}}{|U_{i,j}|}\right]$$

$$+ \left[\frac{(U\xi)_{i,j} - (U\xi)_{i-1,j}}{2\Delta x}\right]\left[1 + \frac{U_{i,j}}{|U_{i,j}|}\right] \quad \text{(A3a)}$$

$$\left[\frac{\partial(V\xi)}{\partial x}\right]_{i,j} = \left[\frac{(V\xi)_{i,j+1} - (V\xi)_{i,j}}{2\Delta y}\right]\left[1 - \frac{V_{i,j}}{|V_{i,j}|}\right]$$

$$+ \left[\frac{(V\xi)_{i,j} - (V\xi)_{i,j-1}}{2\Delta y}\right]\left[1 + \frac{V_{i,j}}{|V_{i,j}|}\right] \quad \text{(A3b)}$$

and, here

$$U_{i,j} \equiv \frac{\Psi_{i,j+1} - \Psi_{i,j-1}}{2\Delta x} \tag{A4a}$$

$$V_{i,j} \equiv - \frac{\Psi_{i+1,j} - \Psi_{i-1,j}}{2\Delta y} \tag{A4b}$$

It should be noted that except for the above definitions for advection, velocities are otherwise evaluated more accurately at staggered grid points such that $U_{i,j+\frac{1}{2}} = (\Psi_{i,j+1} - \Psi_{i,j})/\Delta y$, $V_{i+\frac{1}{2},j} = -(\Psi_{i+1,j} - \Psi_{i,j})/\Delta x$.

The Reynolds stress gradient terms are evaluated like

$$\left(\frac{\partial^2 \overline{uv}}{\partial x^2}\right)_{i,j} = \frac{\overline{uv}_{i+1,j} - 2\overline{uv}_{i,j} + \overline{uv}_{i-1,j}}{\Delta x^2} \tag{A5a}$$

$$\left(\frac{\partial^2 \overline{v^2}}{\partial x \partial y}\right)_{i,j} = \frac{\overline{v^2}_{i+1,j+1} - \overline{v^2}_{i+1,j-1} - \overline{v^2}_{i-1,j+1} + \overline{v^2}_{i-1,j-1}}{\Delta x \Delta y} \tag{A6b}$$

The other terms are evaluated in a similar manner.

All of the terms in equation (13a, b, c, d) and (14) are differenced in a manner analogous to those cited above.

REFERENCES

1. Abbott, D. E. and S. J. Kline, (1962): Experimental investigation of subsonic turbulent flow over single and double backward facing steps, J. Basic Eng., Trans. ASME, Series D, 317-325.

2. Batchelor, G. K. and R. W. Stewart, (1950): Anisotropy of the spectrum of turbulence at small wave-numbers, Quart. J. Mech. and Applied Math., III, 1-8.

3. Champagne, F. H., V. G. Harris, and S. Corrsin, (1970): Experiments on nearly homogeneous turbulent shear flow, Journal of Fluid Mech. 41, 31-141.

4. Donaldson, C. duP and H. Rosenbaum, (1968): Calculation of the turbulent shear flows through closure of the Reynolds equations by invariant modeling, ARAP Rept. 127, Aeronautical Research Associates of Princeton, Princeton, N.J.

5. Daly, B. J. and F. H. Harlow, (1970): Transport equations of turbulence, Physics of Fluids, 13, 2634.

6. Fromn, J. E. and F. H. Harlow, (1963): Numerical solution of the problem of vortex street development, Physics of Fluids, 6, 972-982.

7. Gosman, A. D., W. M. Pun, A. K. Runchal, D. B. Spalding and M. Wolfshtein, (1969): Heat and mass transfer in recirculating flows, Academic Press.

8. Hanjalic, K., (1970): Two-dimensional flow in an axisymmetric channel, Ph.D. Thesis, University of London.

9. Hanjalic, K., and B. E. Launder, (1972): Fully developed asymmetric flow in a plane channel, Journal of Fluid Mechanics, 52, 609-638.

10. Kirkpatrick, J. R. and W. F. Walker, (1972): Numerical method for describing turbulent, compressible, subsonic separated jet flows, J. Computational Physics, 10, 185-201.

11. Kline, S. J., M. V. Morkovin, G. Sovran, D. J. Cockrell, editors (1968): Proceedings, computation of turbulent boundary layers, 1968 AFOSR-IFP-Stanford Conference. Stanford University.

12. Kolmogorov, A. N., (1941): The local structure of turbulence in incompressible viscous fluid for very large Reynolds number, C. R. Academy of Sciences, USSR, 30, 301. (Translation Friedlander, S. K. and L. Topper, ed., Turbulence: Classic Papers on Statistical Theory, Intersciences, New York, 1961).

13. Laufer, J., (1951): Investigation of turbulent flow in a two-dimensional channel, NACA Report 1053.

14. Launder, B. E., C. J. Reece and W. Rodi, (1975): Progress in the development of a Reynolds-stress turbulence closure, J. of Fluid Mech., 68, 537-566.

15. Lewellen, W. S., M. E. Teske and C. duP. Donaldson, (1970): Variable density flows computed by a second-order closure description of turbulence, AIAA Journal, 14, 382-387.

16. Mellor, G. L., (1966): The effects of pressure gradients in turbulent flow near a smooth wall, J. Fluid Mech., 24, 255-274.

17. Mellor, G. L., (1973): Analytic prediction of the properties of stratified planetary surface layers, J. of Atmos. Sci., 30, 1061-1069.

18. Mellor, G. L., (1975): A comparative study of curved flow and density stratified flow, J. Atmos. Sci., 32, 1278-1282.

19. Mellor, G. L. and H. J. Herring, (1973): A survey of mean turbulent field closure models, AIAA Journal, 11, 590-599.

20. Mellor, G. L. and T. Yamada, (1974): A hierarchy of turbulence closure models for planetary boundary layers, J. of Atmos. Sci., 31, 1791-1806.

21. Monin, A. S., (1965): On the symmetry properties of turbulence in the surface layer of air, Izvestiya Atmospheric and Oceanic Physics, 1, 45-54.

22. Ng, K. H. and D. B. Spalding, (1972): Turbulence model for boundary layers near walls, Physics of Fluids, 15, 20-30.

23. Roache, P. J., (1972): Computational Fluid Dynamics, Hermose Publishers, Albuquerque, N.M. 434 pages.

24. Rodi, W. and B. Spalding, (1970): A two parameter model of turbulence and its application to free jets, Warme and Staffibertragung 3, 85.

25. Rotta, J. C., (1951): Statistische theorie nicthomogener turbulenz, Zeitsch Fur Physik, 129, 547 and 131, 51.

26. Rotta, J. C., (1972): Turbulent shear layer prediction on the basis of the transport equations for the Reynolds stresses, Proceedings of the 13th International Congress of Theoretical and Applied Mechanics, Moscow University, 295-308.

27. So, R. M. C., (1975): A turbulent velocity scale for curves shear flows, J. Fluid Mech., 70, 37-57.

28. So, R. M. C. and G. L. Mellor, (1973): Experiments on convex curvature effects in turbulent boundary layers, J. Fluid Mech., 60, 43-62.

29. So, R. M. C. and G. L. Mellor, (1975): Experiments on turbulent boundary layers on a concave wall, Aero. Quart. 26, 25-40.

30. Thoman, D. C. and A. A. Szewczyk, (1969): Time dependent viscous flow over a circular cylinder. Physics of Fluids Supplement, 12, 76-86.

31. Thompson, S. M. and J. S. Turner, (1975): Mixing across an interface due to turbulence generated by an oscillating grid. J. Fluid Mech., 67, 349-368.

32. Townsend, A. A., (1961): Equilibrium layers and wall turbulence, J. Fluid Mech., 11, 97-120.

33. Wolfshtein, M., (1970): On the length-scale-of-turbulence equation, Israel Journal of Technology, 8, No. 1-2, 87-99.

34. Yamada, T. and G. Mellor, (1975): A simulation of the Wangara atmospheric boundary layer data, J. Atmos. Sci., 32, 2309-2329.

DISCUSSION

KLINE: (Stanford University)

I want to note that δy has to do with the grid size, not the physics of the problem.

OHRENBERGER: (TRW Systems)

Just out of curiosity, have you considered doing a round-the-shoulder to eliminate this spherical problem?

MELLOR:

Yes, we really ought to do that and that is the next step. Of course, we are exercising the turbulence model and we wanted the simplest geometry we could find and this was it. But, sure, that is right. In principle that takes away the corner problem.

OHRENBERGER:

Also, is it clear that separation occurs at the corner and not down from the corner on the back face of the step?

MELLOR:

It is pretty clear from the data. Do you mean does it ever go around the corner? I should say that in some of the calculations with the wrong beta, the flow does go around the corner, but that again it is a numerical thing; that is, if the beta is too big then it goes around the corner. I have heard that debate with respect to laminar flow but I can't imagine it being significant.

COMTE-BELLOT: (University of Lyon)

It seems that you have used two length scales at the beginning of your turbulence modeling, an integral length scale and a Taylor length scale. Later on, however, you have only one equation. I presume an assumption has been made. Would you comment on that point? In particular, how does this assumption limit the types of flow, which can be studied with your modeling?

MELLOR:

The dissipation term is labeled q^3/λ, q^2 being twice the turbulent energy and λ is one of the length scales. The term is considered to be isotropic. Now I think, that is the one modeling term that everybody sort of agrees on. So, that gives you one length scale. Incidentally, q^3/λ can be defined by the way, by equating it to the derivative of the triple correlation with respect to r for small r. The other place where the significant modeling term comes in is in the Rotta assumption* that says the pressure-velocity correlations are proportional to some function of the Reynolds stress tensor. Another inverse length scale is the proportionality coefficient. There are therefore two prominent length scales in this whole problem and they are both assumed to be proportional to each other. Obviously when people are thinking of spectra then it is a simplification to say that one single length scale to which these two are proportional governs the flow. You just simply have to see how many flows this predicts and it predicts a lot of flows. But, there is a limit somewhere!

CORRSIN: (Johns Hopkins University)

You say that the dissipation expression q^3 over λ is generally agreed upon. I have not followed the model business very closely, of course. But, that tends to be proportional to the rate at which the large structure feeds energy into the shear turbulence. The turbulence gets rid of it and adjusts the fine scale to get rid of it. But, that is true only if the turbulence is dissipated where it is produced. Any situation where there is appreciable transport laterally, so that the energy is not dissipated where it is produced, doesn't allow you to use that equality and I am wondering why that is commonly accepted.

MELLOR:

I don't understand. We have transport terms in the rest of the equation, of course. So this is supposed to be the local dissipation at a point.

---

* Rotta, J. C., 1951, J. Phys., 129, 547-572; 131, 51-77.

CORRSIN:

The rigorous expression is the one you get from the Navier-Stokes equation. That can be represented by defining a "micro-scale" in the manner of G. I. Taylor, if you have local isotrophy.* If you don't, you can do some kind of directional averaging. The expression you equate to dissipation is really an energy production.

MELLOR:

That is a dissipation term.

CORRSIN:

That is rationalized by the fact that it is the amount of energy of the mean flow dumps into the turbulence, and then the turbulence dissipates. But, it doesn't have to dissipate at the same place. You wouldn't expect that to be equal to dissipation at the same point, when there is appreciable transport.

MELLOR:

I would expect it. That is the production, is the correct transfer from the mean flow into the turbulence?

CORRSIN:

That is approximately q^3 over λ.

MELLOR:

Sure, that is what the model says, if you neglect the transport terms.

CORRSIN:

I am telling you where q over λ came from. It came from that expression you just wrote. It is really rationalized by turbulence production. Now, as a local function it would be true only if it is dissipated where it is produced.

MELLOR:

You are saying that, but I don't agree with it.

---

* See, for example, Batchelor's book on Homogeneous Turbulence.

BRADSHAW: (Imperial College)

I think possibly the best way to keep one's brains unscrambled is to regard the equation as a definition of λ.

MELLOR:

Yes, and I say it can be defined according to the slope of the triple correlation in r space.

BRADSHAW:

What this will mean is that if you have a transport equation for this scale, λ, that transport equation will have a turbulent transport term in it. So in that sense, λ is not determined by local conditions although you then insert it into an expression that is locally bound.

MELLOR:

For the moment that is all right. It is a definition of λ. Stan, if you were going to model dissipation how would you do it?

CORRSIN:

I use the viscous expression for the local dissipation: $\approx \nu(\overline{u_i u_i})/\lambda^2$, where ν is Kinerrahi viscosity and λ is a directionally averaged "Taylor microscale"; then use the Karman-Howarth (isotropic) connection between λ and integral scale L (Proc. Roy. Soc., 1938): $\lambda/L \sim 1/R_\lambda = \nu/(\sqrt{\overline{u_i u_i}} \cdot \lambda)$. Especially, this has been found to be roughly valid in shear flows too.

MELLOR:

So in the large Reynolds member limit you come to something that is independent of Reynolds numbers, obviously.

CORRSIN: (Modified after further discussion during coffee break)

I see your point. I had forgotten that the substitution actually gives a local estimate for dissipation roughly equal to production.

MELLOR:

I agree with that but in a large Reynolds number limit you have to come to this kind of asymptotic result.

CORRSIN:

I think your expression is plausible. But I'm puzzled because, although the integrated dissipation equals integrated production, in principle they aren't locally equal unless transport is absent.

MELLOR:

Of course, we have the other terms in the equation that say just what you are saying.

LAUFER: (University of Southern California)

I think actually this is more than a definition and while Corrsin is perfectly correct there is no theoretical basis for that kind of formulation. But this is again one of those little understood features in turbulence. There are large numbers of turbulent shear flows for which this formulation--Townsend is really the one who pointed out this--seems to work very well and I don't think really there is any theoretical basis for it.*

MELLOR:

Only dimensional analysis. If you say you have a dissipation equation which is $\overline{\nu(\partial u_i/\partial x_j)^2}$ and then you say that cannot be dependent on viscosity and it has to be isotropic, there is no other choice.

LAUFER:

One can use hand-waving, of course.

MELLOR:

I don't think there is any other choice.

LUXTON: (University of Adelaide)

I feel that a very good piece of work on curvature effects on boundary layers, which has recently been completed in Bangalore, may be relevant to Professor Mellor's work.† It is for very mild

---

* Townsend, A. A., The Structure of Turbulent Shear Flow, Cambridge University Press, 1956.

† B. G. Shiva Prasad, "An experimental study of the effect of 'mild' longitudinal curvature on the turbulent boundary layer," Indian Institute of Science, Bangalore, Department of Aeronautical Engineering, Ph.D. thesis, Feb., 1976.

curvatures where the ratio of the boundary layer thickness to the radius of curvature of the surface is of the order of 1/75. There are quite marked effects on the structure due to that mild curvature, as Peter Bradshaw had predicted there probably would be.* Bradshaw's model is quite well justified for regions of the boundary layer right up to $y \simeq 0.6\delta$, but it is less realistic near the outer edge of the layer. The departures are more marked for concave curvature than for convex curvature. Unfortunately I missed the first part of your talk and I am not quite sure which beta you are referring to. If it was the Bradshavian beta, which relates the Curvature Richardson number to the length scale, then, where Bradshaw suggested a value of about 7, it would seem from the Bangalore work that a value more like 10 provides the best fit to the experimental data.

MELLOR:

I had a couple of slides (which I guess you missed) which had the effect of a stratification Richardson number and also a curvature which is derived from this model. You might enjoy seeing whether or not those results fit your idea. From my point of view there is certainly an overall analogy between curvature and stratification but when you get into the details, it is quite a bit different. I mean, the terms that are involved in stratification involve heat flux and all kinds of interconnecting terms.

---

* Bradshaw, P. (1973), AGARDograph No. 169.

The Modelling of a Turbulent Near Wake Using the Interactive Hypothesis

B. S. NG AND G. DAVID HUFFMAN
University of School of Science
Indianapolis Center for Advanced Research & Purdue
Indianapolis, Indiana (U.S.A.)

ABSTRACT

A semi-empirical mathematical model for a turbulent near wake has been developed within the framework of an "interaction hypothesis" suggested by Bradshaw. The near wake behind an airfoil has been treated as a "complex" shear flow consisting of two neighboring simple shear layers with distinct but overlapping shear stress profiles of opposite signs. The present model utilizes the mean momentum and continuity equations together with two shear stress transport equations derived from the turbulent kinetic energy equation. By relating the shear stresses to the local turbulence quantities, closure for the governing systems is achieved without the use of the eddy viscosity concept. The shear stress is therefore no longer required to vanish at the velocity extremum. The model has been compared to the experimental data of Chevray and Kovasznay and it appears to be superior to other calculation schemes in its ability to match the measured results. Further comparison with the asymmetric cascade wake data of Raj and Lakshminarayana is presently underway. Our preliminary calculations have validated the basic philosophy of an interactive approach in the study of near wakes. However, it also clearly demonstrates that the accuracy of certain empirical functions used to define the turbulence structure has a direct impact on the success of any calculation method.

1. BASIC FORMULATION: INTERACTIVE APPROACH

Despite recent advances in numerical methods dealing with turbulent boundary layers,[1] little progress has been made in the modelling of an asymmetric turbulent near wake. This is in part due to the fact that the state of the art in turbulent modelling is such that a large amount of empiricism is often required to insure the success of any prediction method. In the case of a near wake formed by the coalescence of two turbulent boundary layers with opposite shear stresses, very little is known about the mechanism of their interaction. Moreover, for an asymmetric wake, the point of zero Reynolds stress does not necessarily coincide with the point of zero velocity gradient. Any attempt to obtain closure for the governing equations by using a simple eddy viscosity model for the shear stress will therefore encounter difficulties. It is the purpose of this work then to outline a method by which these problems can be resolved.

An important element in our present analysis is the "interaction hypothesis" first proposed by Bradshaw, et al. in conjunction with the study of duct flow.[2] Within the context of this hypothesis, a near wake can be regarded as an interaction between two neighboring simple shear layers with distinct but possibly overlapping shear stress profiles of opposite signs (cf. Figure 1). If the interaction is sufficiently weak so that the *turbulence structure* in each layer is essentially unaffected by the presence of the adjacent layer, a superposition of the two shear stress fields for the purpose of calculating the net Reynolds stress is then possible. Moreover, by relating the shear stresses to the local turbulent quantities, as it has been

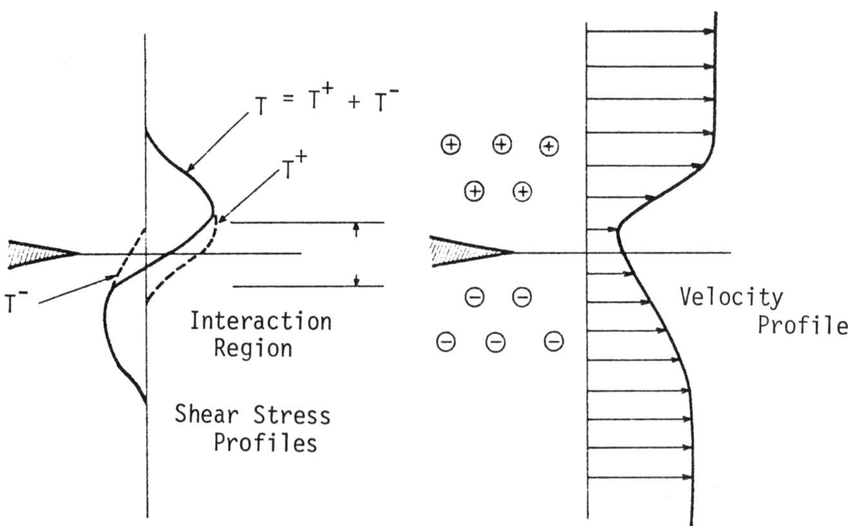

Figure 1. The "Interaction Hypothesis."

suggested by Bradshaw, Ferriss and Atwell,[3] two shear stress transport equations can be derived from the turbulent kinetic energy equation. Closure for the governing system is therefore achieved without making use of the eddy viscosity concept.

In order to explore the validity of the interaction approach and to gain numerical experience, our preliminary investigation has been confined to a prototype model which includes only the essential physical details. The fluid is therefore assumed to be incompressible. Furthermore, due to the magnitude of the Reynolds number usually encountered in turbomachine internal flow applications, the flow is also expected to satisfy the boundary layer approximation and the viscous stresses have been ignored when compared to the Reynolds stresses. The momentum and continuity equations are then given by

$$\left(U \frac{\partial}{\partial x} + V \frac{\partial}{\partial y}\right) U = U_\infty \frac{dU_\infty}{dx} + \frac{\partial \tau}{\partial y} , \tag{1}$$

and

$$\frac{\partial U}{\partial x} + \frac{\partial V}{\partial y} = 0 . \tag{2}$$

The Cartesian coordinate system for these equations is chosen such that the x-axis is tangent to the mean camber line of an upstream airfoil at the trailing edge; x is equal to 0 at the trailing edge and positive in the downstream direction. The time-averaged velocity components along the x and y axes are then given by U and V, respectively. The freestream velocity U_∞ in the pressure gradient term of equation (1) is to be replaced by $U_{+\infty}$ or $U_{-\infty}$, depending on whether $y > y_c$ or $y < y_c$, where $y_c(x)$ is the locus of the velocity minimum. Physically, our prototype problem can be envisioned as that of a two-dimensional asymmetric wake behind an airfoil or a flat plate, formed by the coalescence of the upper and lower surface boundary layers having possibly different characteristics. In our present notation, the usual shear stress is replaced by τ = (shear stress)/(density) = $-\overline{uv}$, where u and v are velocity fluctuations and the overbar denotes a time average.

According to the interaction hypothesis, we can further write

$$\tau = \tau^+ + \tau^-$$

such that τ^+ is the dominant shear stress in the region of positive shear--nominally where $y > y_c$--and vice versa for τ^-. In order to obtain a tractable system, certain assumptions must be made to secure closure for equations (1) and (2). Using the Bradshaw-Ferriss-Atwell turbulence model[3] we define two sets of empirical functions, (a_1^+, G^+, L^+) and a_1^-, G^-, L^-), such that

$$a_1^{\pm} = \frac{\tau^{\pm}}{\overline{q^2}} , \tag{3}$$

$$G = \frac{(\overline{p'v} + \frac{1}{2}\,\overline{q^2 v})}{|\tau^{\pm}|\,|\tau_m^{\pm}|^{\frac{1}{2}}} , \tag{4}$$

and

$$L^{\pm} = \frac{\tau^{\pm}|\tau^{\pm}|^{\frac{1}{2}}}{\varepsilon} . \tag{5}$$

The relations (3)-(5) are equivalent to assuming that (i) the local shear stress is proportional to the turbulent intensity $\overline{q^2}$; (ii) the energy diffusion is directly proportional to the local shear stress with a factor depending on the maximum of the shear stress, τ_m, and (iii) the dissipation rate ε is determined by local shear stress and a length scale L. With equations (3), (4) and (5), the structure of turbulence is defined in terms of relations between turbulence quantities and it is thus independent of the mean flow. It has been observed that the above assumptions are quite valid over a wide range of pressure gradients, Mach number, Reynolds number and appear to be rather insensitive to rapid changes in the mean flow.[4,5] This will, of course, enable the present method to be extended to a wide variety of mean flow conditions, often without the need for extra data. On substituting (3), (4) and (5) into the exact turbulent energy equation, we can derive two transport equations for τ^+ and τ^- of the form

$$\left(U \frac{\partial}{\partial x} + V \frac{\partial}{\partial y}\right)\left(\frac{\tau^{\pm}}{2a_1^{\pm}}\right) = \tau^{\pm}\frac{\partial U}{\partial y} \mp \frac{\partial}{\partial y}\left(G^{\pm}\tau^{\pm}|\tau_m^{\pm}|^{\frac{1}{2}}\right) \mp \frac{\tau^{\pm}|\tau^{\pm}|^{\frac{1}{2}}}{L^{\pm}} \tag{6}$$

The system consisting of equations (1), (2) and (6) is now closed provided that the empirical functions can be determined experi mentally. A schematic representation of these functions is given in Figure 2. Conceptually, the two sets of empirical functions are to be regarded as distinct, but in fact they are qualitatively the same except for changes in signs and their dependence on y as illustrated in Figures 2 and 3. It must be emphasized that the success of the present approach, regardless of the method used in solving the governing system, is then dependent on the accuracy of ($a_1^{\pm}$, $G^{\pm}$, $L^{\pm}$).

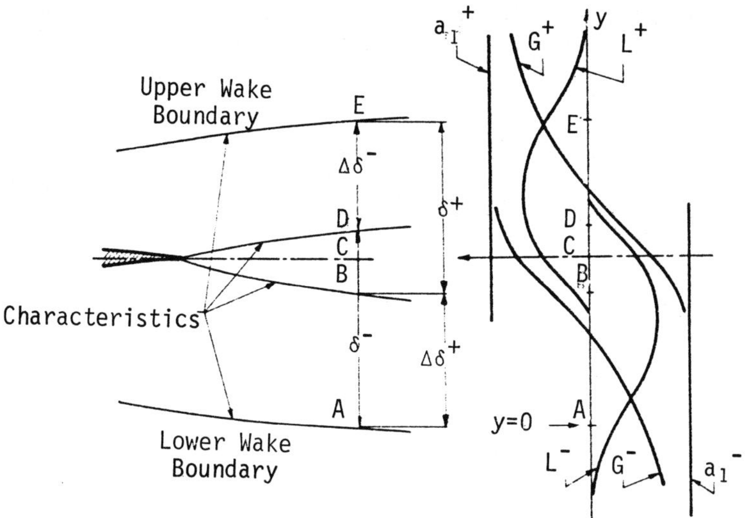

Figure 2. A Schematic Representation of the Empirical Functions.

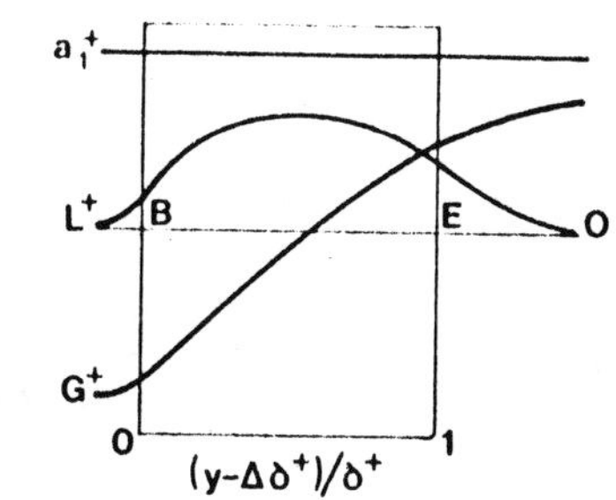

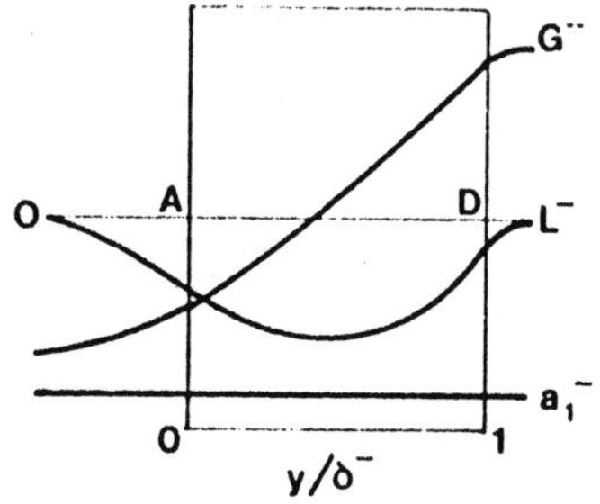

Figure 3. The Relation Between (a_1^+, G^+, L^+) and (a_1^-, G^-, L^-).

The details of the empirical functions used in our calculations are discussed in the Results and Discussion section. In addition to equations (1), (2) and (6), certain initial and boundary conditions must be specified.

On noting that the genesis of a near wake is effected by the merging of two surface boundary layers at the trailing edge where $x = 0$, the natural initial conditions for an airfoil with zero trailing edge thickness are therefore given by

$$U = U(0,y) \ , \tag{7}$$

$$V = V(0,y) \ , \tag{8}$$

$$\tau^+ = \begin{Bmatrix} \tau^+(0,y) \ , & y \geq 0 \\ 0 \ , & y < 0 \end{Bmatrix} \tag{9}$$

and

$$\tau^- = \begin{Bmatrix} 0 \ , & y > 0 \\ \tau^-(0,y) \ , & y \leq 0 \end{Bmatrix} \tag{10}$$

where the velocity profiles $U(0,y)$ and $V(0,y)$, and the shear stress profiles $\tau^+(0,y)$ and $\tau^-(0,y)$ are ideally supplied by an upstream boundary layer calculation or measured directly from experiments. For $x > 0$, we shall also require that $V(x,y_c) = 0$, when $y_c = 0$ for the symmetric case.

The boundary conditions for the present problem are

$$\begin{Bmatrix} U \rightarrow U_{+\infty} \\ \tau^+ \rightarrow 0 \end{Bmatrix} \quad \text{as } y \rightarrow +\infty \ , \tag{11}$$

and

$$\begin{Bmatrix} U \rightarrow U_{-\infty} \\ \tau^- \rightarrow 0 \end{Bmatrix} \quad \text{as } y \rightarrow -\infty \ . \tag{12}$$

A careful examination of the turbulent shear stress equations also indicates that τ^+ should vanish almost immediately after crossing into the region dominated by negative shear. In fact, on noting that the governing system is hyperbolic, it is possible to determine two characteristics, Γ^+ and Γ^-, emanating from the trailing edge such that $\tau^+ \equiv 0$ in a region bounded by Γ^- and the lower edge of the wake, and vice versa for τ^-. In differential form, these characteristics are given by

$$\Gamma^{\pm} : \frac{dy}{dx} = \tan \gamma_{\pm} \ , \quad \text{and} \quad y(0) = 0 \ , \tag{13}$$

where

$$\tan \gamma_{\pm} = \frac{V \pm a_1^{\pm} G^{\pm} |\tau_m^{\pm}|^{\frac{1}{2}} \pm [(a_1^{\pm} G^{\pm})^2 |\tau_m^{\pm}| + 2a_1^{\pm} \tau^{\pm}]^{\frac{1}{2}}}{U} \ .$$

Hyperbolicity has indeed been fully exploited in the work of Bradshaw, et al. (see, e.g., references 2, 3 and 4). In the present analysis, however, the characteristics associated with the governing system play a relatively minor role. They are useful insofar as providing a conceptual framework within which the precise boundaries of the interacting shear layers can be defined for the purpose of scaling the empirical functions (cf. Figures 2 and 3). On the other hand, the conditions that τ^+ and τ^- be respectively zero in most of the regions of negative and positive shear will be automatically satisfied if we require that

$$\tau^+ \to 0 \quad \text{as} \quad y \to -\infty \ , \tag{14}$$

and

$$\tau^- \to 0 \quad \text{as} \quad y \to +\infty \ . \tag{15}$$

The explicit use of characteristics in specifying boundary conditions is therefore unnecessary. It is useful to note here a few key points of our present interactive analysis:

(i) Due to the inclusion of the boundary layer profiles at the trailing edge as initial conditions, the observed flow asymmetry is being dealt with directly.

(ii) Since the near wake is highly dependent upon the structure of the lower and upper surface boundary layers, the interaction approach is expected to produce better results than any similarity methods in which the loss of initial conditions is implicit in the zero initial wake-width assumption.

(iii) The (indirect) use of the turbulent energy equation leads to the incorporation of advective and diffusive effects in the energy budget, in addition to the usual balance between production and dissipation of turbulence. The "past history" of the turbulence is therefore explicitly taken into account. In a highly turbulent and rapidly varying flow region, such as a near wake, this consideration will have a direct bearing on the accuracy of the present model.

(iv) The decomposition of the net shear stress profile into τ^+ and τ^-, each of which satisfies a separate transport equation, enables us to circumvent the usual difficulties associated with any complex shear flow having an extremum in its velocity profile.

(v) It is also of interest to point out that past experience with boundary layers and duct flow indicates that the empirical functions exhibit a remarkable measure of universality.[5] It is expected that the corresponding functions for near wakes, once determined, will be canonical, to some extent, to the present class of flow. In any internal flow applications, a particular cascade geometry and the angle of attack or incidence of the airfoil will only influence the near wake solution through the initial and boundary conditions.

2. SOLUTION ALGORITHM

A number of very effective algorithms have been developed in recent years for solving quasi-linear hyperbolic systems of the form given by equations (1), (2) and (6). In the work of Bradshaw, et al. on turbulent boundary layers,[3] the method of characteristics was adopted to solve a somewhat simpler system of three equations involving the dependent variables U, V and τ. The advantages of such an approach are that the resulting numerical scheme is essentially explicit, and that the well-known Courant-Friedrichs-Lewy criterion provides a clearcut limit to the x-step that can be taken to insure numerical stability. The uncoupling of the continuity equation from the remaining system in that case also effectively reduces the problem to one of only two unknowns--namely, U and τ--so that the characteristic angles at each point in the flow can be conveniently determined by a second-order algebraic equation.

For our present purpose, a direct generalization of the method of characteristics to deal with equations (1), (2) and (6) is, of course, possible but has been found to be somewhat cumbersome. Despite the fact that V can again be uncoupled, the determination of the characteristic angles now requires the solution of essentially a third-order algebraic equation. An attempt to circumvent this difficulty has since been discussed by Bradshaw, Dean and McEligot in conjunction with their work on duct flow.[2] Due to the rather weak coupling of the governing system, it has been suggested that the two turbulent shear stress equations can be solved separately using the velocity profile at the previous x-step as a first approximation. The resulting τ^+ and τ^- are then summed and a new velocity profile is then calculated. Accuracy can presumably be improved through iteration. For our purpose, however, we find it convenient to formulate a more direct numerical scheme based on the general finite difference approach similar to the one devised by Ferriss.[6]

In anticipation of the rapidly varying velocity profiles in the inner wake region similar to those encountered in boundary layers, a mesh of variable grid size in the normal directions has been adopted for the present analysis. The y-steps are gradated outward from the x-axis in the form of a geometric sequence such that

$$(\Delta y)_j = \alpha(\Delta y)_{j-1} = \alpha^{|j|}(\Delta y)_0 ,$$

where

$$(\Delta y)_j = y_{j+1} - y_j , \tag{16}$$

and

$$y_1 = -y_0 = \frac{(\Delta y)_0}{2} .$$

The geometric ratio α is usually taken to be of the order of 1.05. We shall also denote the boundary grid points for the lower and upper wake edges respectively by y_{-L} and y_M. The total number of interior grid points at any given x-station is then given by

$$J = L + M - 1 .$$

The required numerical scheme for solving equations (1), (2) and (6) can now be formulated in terms of difference quotients within a typical computation cell consisting of two adjacent mesh rectangles as shown in Figure 4. For any function f(x,y) which is at least twice differentiable in y, the second-order-correct finite difference approximation to the partial derivative of f with respect to y at (x_i, y_i) is then given by

$$\left.\frac{\partial f}{\partial y}\right|_{i,j} = \frac{f_{i,j+1} - (1 - \lambda^2)f_{i,j} - \lambda^2 f_{i,j-1}}{\alpha^{|j|}(1 + \lambda)(\Delta y)_0} . \tag{17}$$

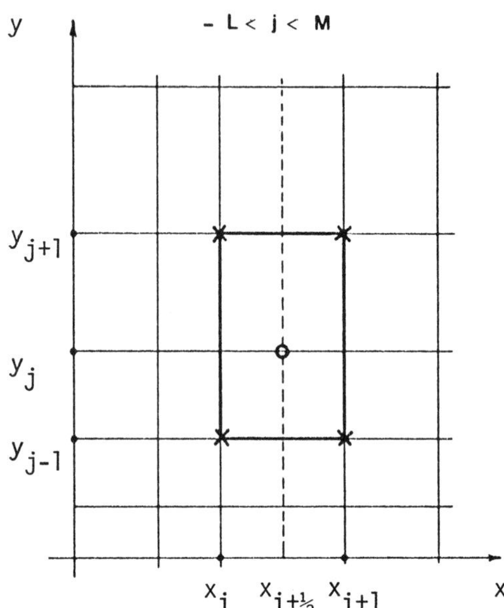

Figure 4. A Typical Computation Cell for Difference Approximations (17) and (18).

where

$$\lambda = \begin{Bmatrix} \alpha & & j \geq 1 \\ \alpha^{-1} & \text{for} & j < 1 \end{Bmatrix} .$$

In the case of a logarithmic function of y, it can be shown that the error associated with an approximation of the form given by (17) remains constant throughout the entire grid despite the increase in y-step size in the outer regions (e.g., reference 7). A "centered" difference equation is used to approximate the partial derivative of f with respect to x. Thus we have

$$\left(\frac{\partial f}{\partial x}\right)_{i+\frac{1}{2},j} \approx \frac{f_{i+1,j} - f_{i,j}}{(\Delta x)_i} , \tag{18}$$

where

$$x_{i+\frac{1}{2}} = \frac{x_{i+1} + x_i}{2} ,$$

and

$$(\Delta x)_i = x_{i+1} - x_i .$$

Moreover, it is useful to note that

$$\left(\frac{\partial f}{\partial y}\right)_{i+\frac{1}{2},j} \approx \left\{\left(\frac{\partial f}{\partial y}\right)_{i+1,j} + \left(\frac{\partial f}{\partial y}\right)_{i,j}\right\} / 2 . \qquad (19)$$

The finite difference analog to the governing system can now be derived by replacing the partial derivatives in equations (1), (2) and (6) with their corresponding approximations given by (18) and (19). For this purpose, we now define

$$E_{i,j}^{(r)} = \begin{bmatrix} e_{i,j,r}^{+} & -\tau_{i,j}^{+} & 0 \\ 1 & -V_{i,j} & 1 \\ 0 & -\tau_{i,j}^{-} & e_{i,j,r}^{-} \end{bmatrix} , \qquad (20)$$

$$H_{i,j} = \begin{bmatrix} h_{i,j}/a_1^{+} & 0 & 0 \\ 0 & 2h_{i,j} & 0 \\ 0 & 0 & h_{i,j}/a_1^{-} \end{bmatrix} , \qquad (21)$$

where

$$e_{i,j,r}^{\pm} = \frac{V_{i,j}^{\pm}}{2a^{\pm}} \pm G_{i+1,j+r}^{\pm} |\tau_m^{\pm}|_i^{\frac{1}{2}} \qquad r = -1, 0, 1, \qquad (22)$$

and

$$h_{i,j} = \frac{(\Delta y)_0}{(\Delta x)_i} \alpha^{|j|} (\lambda + 1) U_{i,j} . \qquad (23)$$

It can then be shown that around the point $(x_{i+\frac{1}{2}}, y_j)$, equations (1), (2) and (6) are reducible to

$$A_{i,j}\overleftarrow{s}_{i+1,j-1} + B_{i,j}\overleftarrow{s}_{i+1,j} + C_{i,j}\overleftarrow{s}_{i+1,j+1} = \overleftarrow{d}_{i,j} \tag{24}$$

where

$$A_{i,j} = -\lambda^2 E^{(-1)}_{i,j} , \tag{25}$$

$$B_{i,j} = -(1 - \lambda^2)E^{(0)}_{i,j} - H_{i,j} , \tag{26}$$

$$C_{i,j} = E^{(1)}_{i,j} \tag{27}$$

and

$$\overleftarrow{s}_{i,j} = \begin{bmatrix} \tau^+_{i,j} \\ U_{i,j} \\ \tau^-_{i,j} \end{bmatrix} \tag{28}$$

The inhomogeneous term $d_{i,j}$ is given by

$$\begin{aligned} \overleftarrow{d}_{i,j} = -A_{i,j}\overleftarrow{s}_{i,j-1} &+ [(1 - \lambda^2)E^{(0)}_{i,j} - H_{i,j}]\overleftarrow{s}_{i,j} \\ &- C_{i,j}\overleftarrow{s}_{i,j+1} + \overleftarrow{g}_{i,j} \end{aligned} \tag{29}$$

where

$$\overleftarrow{g}_{i,j} = (\Delta y)_0 \alpha^{|j|}(\lambda + 1) \begin{bmatrix} \tau^+_{i,j}|\tau^+_{i,j}|^{\frac{1}{2}}/L^+_{i,j} \\ U_{i+\frac{1}{2},\infty}(U_{i+1,\infty} - U_{i,\infty})/\Delta x \\ \tau^-_{i,j}|\tau^-_{i,j}|^{\frac{1}{2}}/L^-_{i,j} \end{bmatrix} . \tag{30}$$

In equation (30), the edge velocity $U_{,\infty}$ must be replaced by $U_{,+\infty}$ or $U_{,-\infty}$, depending on whether $y_j > y_c$ or $y_j < y_c$. In the course of deriving equation (24), we have also linearized the corresponding equations by approximating the coefficients involving U, V, τ^+ or τ^- by their values at x_i, which are taken to be known. The non-linear version may be recovered, if needed, by replacing i + ½ for the i-indices of those dependent variables appearing in equations (20)-(23). At any given x-station, the J interior grid points then generate 3J equations with the same number of unknowns. The resulting system can be written in the following compact form as indicated in equation (31), where the i-indices have been omitted for clarity. The boundary conditions at y_{-L} and y_M are incorporated through the inhomogeneous vector $\overleftarrow{d}$. Before the above linear system is actually solved, however, a modification is desirable. On noting that the downstream shear stress distributions of both the top and bottom shear layers should vanish in most of the regions of opposite shear due to zero upstream conditions, it is possible to calculate τ^+ and τ^- only at those grid points where they are non-zero. The advantage of this approach is that it results in a reduction of the number of equations and therefore the size of the matrix M. A more important consequence is that we are dispensed with the need to define and justify the empirical functions too far beyond the boundaries within which the top and lower shear layers are supposed to be confined. On the other hand, care must be exercised to allow for the growth of the shear layers. For this purpose, whenever a shear stress profile is computed, the magnitude of the shear stress at an edge grid point is checked to determine whether certain limits (usually 0.01 $\tau_m^{\pm}$) have been exceeded. The boundary for the downstream profile will then be moved outward by one grid point if the tolerance has not been met.

The reduced form of the coefficient matrix M is of band type with nine non-zero diagonals. The linear system can therefore be solved efficiently by one of the many standard routines available through various scientific subroutine packages. The values of U, τ^+ and τ^- at the interior grid points are then contained in the solution vector $\overleftarrow{s}$.

Our linearization of the numerical scheme also has the effect of uncoupling the continuity equation, and therefore V, from the remaining system. Using equation (2) to eliminate ($\partial U/\partial x$) from the momentum equation, we obtain

$$-U^2 \frac{\partial}{\partial y}\left(\frac{V}{U}\right) = U_\infty \frac{dU_\infty}{dx} + \frac{\partial \tau}{\partial y} , \tag{32}$$

where all quantities, except V, are now known. Using the initial condition $V(y_c) = 0$, equation (32) can then be integrated outwards from y_c to determine the vertical velocity profile V.

$$\underbrace{\begin{pmatrix} B_{-L+1} & C_{-L+1} & & & & & & & & \\ A_{-L+2} & B_{-L+2} & C_{-L+2} & & & & & \mathbf{0} & & \\ & A_{-L+3} & B_{-L+3} & C_{-L+3} & & & & & & \\ & & \ddots & \ddots & \ddots & & & & & \\ & & & A_{-1} & B_{-1} & C_{-1} & & & & \\ & & & & A_0 & B_0 & C_0 & & & \\ & & & & & A_1 & B_1 & C_1 & & \\ & & \mathbf{0} & & & & \ddots & \ddots & \ddots & \\ & & & & & & & A_{M-3} & B_{M-3} & C_{M-3} \\ & & & & & & & & A_{M-2} & B_{M-2} \; C_{M-2} \\ & & & & & & & & & A_{M-1} \; B_{M-1} \end{pmatrix}}_{\mathcal{M}} \underbrace{\begin{pmatrix} s_{-L+1} \\ s_{-L+2} \\ s_{-L+3} \\ \vdots \\ s_{-1} \\ s_0 \\ s_1 \\ \vdots \\ s_{M-3} \\ s_{M-2} \\ s_{M-1} \end{pmatrix}}_{s} = \underbrace{\begin{pmatrix} d_{-L+1} - A_{-L+1} s_{-L} \\ d_{-L+2} \\ d_{-L+3} \\ \vdots \\ d_{-1} \\ d_0 \\ d_1 \\ \vdots \\ d_{M-3} \\ d_{M-2} \\ d_{M-1} - C_{M-1} s_M \end{pmatrix}}_{d} \qquad (31)$$

The values of U, τ^+, τ^- and V now being known at x_i, say, the implicit finite difference scheme can be used to advance U, τ^+ and τ^- to x_{i+1}; V is then calculated by equation (32). This process is repeated until a solution over the entire domain of interest is obtained. Alternately, before an increment in the x-direction is made, the current velocity and shear stress profiles can be used to update the linearized coefficients and to improve the solution by iteration.

The proposed numerical scheme is believed to be quite stable, but caution must be exercised to insure the accuracy of the solution. Although the effects of a Goldstein-type singularity at the trailing edge is expected to vanish quickly,[8] a recent study by Burggraf[9] suggests that an extremely fine mesh must be used to insure accuracy in the very near wake region. Typically an x-step size of the order of $R^{-3/5}$, R being the Reynolds number based on chord length, must be used for a mixing length model. Our numerical experiments indicate a comparable x-step size must also be used (for the first 2 or 3% chord length downstream) in the present calculation. Moreover, it is useful to note that the rates of growth for the lower and upper wake edges can be determined by the maximum characteristic angles $\gamma_{-\infty}$ and $\gamma_{+\infty}$ at y_{-L} and y_M. From equation (13) we derive

$$\tan \gamma_{\pm\infty} = \frac{V_{\pm\infty} \pm 2a_1^{\pm} G^{\pm} |\tau_m^{\pm}|^{\frac{1}{2}}}{U_{\pm\infty}} . \qquad (33)$$

Thus, the x-step size must also in general be restricted such that the wake width is allowed to expand by no more than one grid point in either normal direction for each x-increment.

3. RESULTS AND DISCUSSIONS

As the first test case for our model, we have chosen the measurements by Chevray and Kovasznay of a symmetric wake behind a thin flat plate.[10] The data available in this case are perhaps the most extensive and best documented from both the experimental and the theoretical points of view. The study by Burggraf[8] adapted the Cebeci-Smith and Glushko models to compute the velocity and shear stress profiles and it thus provides a basis of comparison between the present model and other prediction methods now available. It should be noted, however, that due to the use of the eddy viscosity concept, the Cebeci-Smith and Glushko models used by Burggraf are restricted to symmetric flows. The present approach does not assume symmetry in its formulation and it is currently being tested against the asymmetric cascade wake data of Raj and Lakshminarayana.[11] The results of this second test case are not yet complete but they will be presented in the near future.

As repeatedly noted in our previous discussions, one of the most crucial steps in the application of the present scheme--or, in fact, any other calculation method--is the choice of empirical functions which define the turbulence structure. In a recent interactive study of symmetric jets and wakes by Morel,[12] a set of empirical functions (see Figure 5) were suggested for wake calculations starting at a point well downstream of the trailing edge. The present solution algorithm has been used to reproduce Morel's results as a check and good agreement was obtained; but the functions used therein are not valid in the near wake region.

It can be argued that the evolution of two coalescing boundary layers into a near wake is initially confined to a region bounded by the two characteristics emanating from the trailing edge.[12] As a first approximation, the boundary layer length scale L is expected to be valid except near the wake centerline where it has been assumed to be constant and proportional to the width of the "inner region" (see Figure 6). We have tentatively used Morel's function G and the point where G = 0 is correlated with the shear stress maximum. With this normalization, the function G closely approximates its boundary layer counterpart in the outer wake edges but assumes negative values near the wake center. This allows the shear stresses to be diffused away from their maxima. The value for a_1 is again taken to be 0.15. The empirical functions used in our preliminary calculations are therefore necessarily crude pending more thorough experimental investigations. In particular, the effects of the assumed "interaction" between the two shear layers must be clarified and incorporated as adjustments in the empirical functions. Nevertheless, the results obtained so far are encouraging and they are presented in Figures 7-12 along with results obtained by Burggraff based on the Cebeci-Smith and Glushko models. The present method produces good agreements with the experimental velocity profiles at x/θ_0 = 8.6, 34.4 and 86.2--corresponding to x/c = 0.0208, 0.083 and 0.208 respectively. The predicted shear stress maxima also show improvements over those of Cebeci-Smith and Glushko. Thus, our investigation so far has validated the basic philosophy of the interactive approach for near wake calculations, but it also clearly suggests that further refinements in the empirical functions based on reliable experimental data are needed before the present method can be used with confidence as a design tool for asymmetric flows.

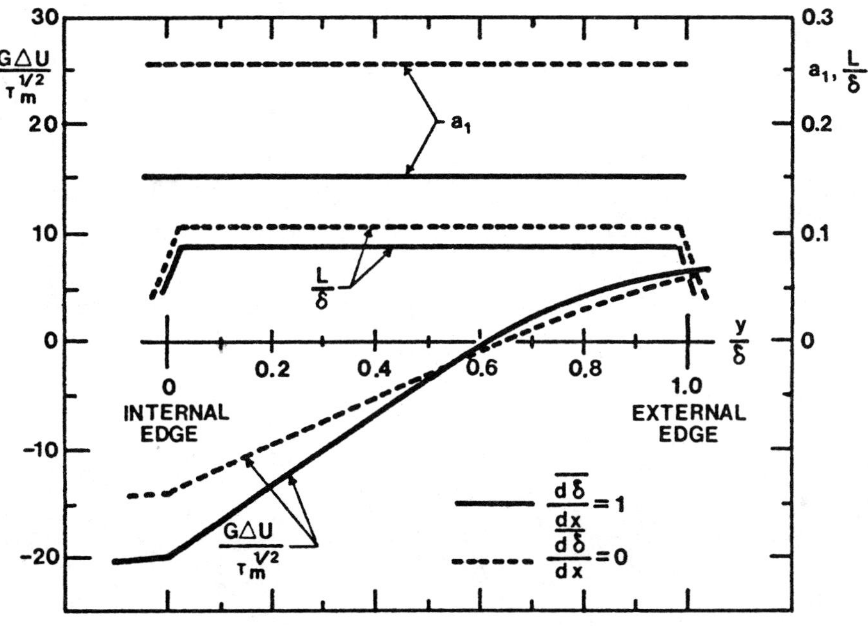

Figure 5. Empirical Functions--Jets and Wakes (from Morel, 1972).

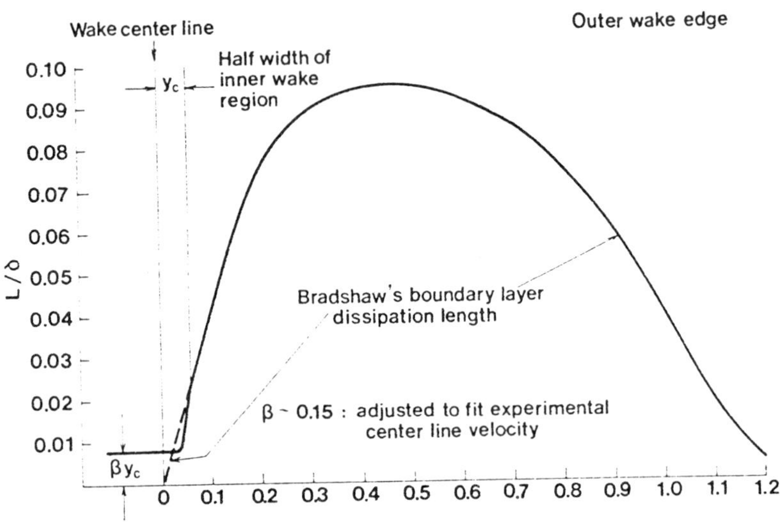

Figure 6. The Empirical Function L.

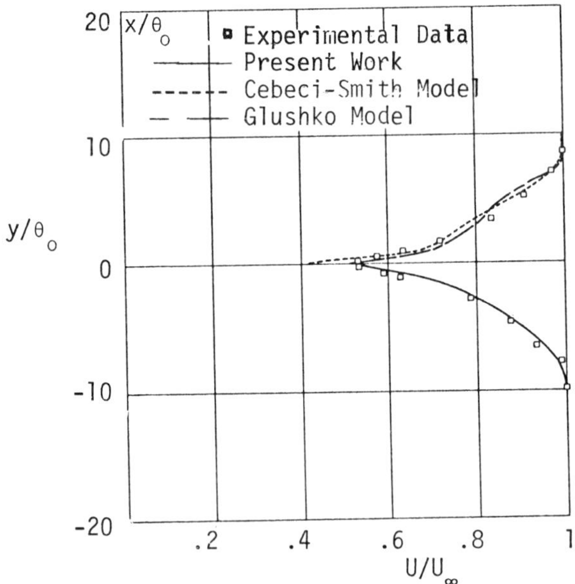

Figure 7. Mean Velocity Distribution in Wake.

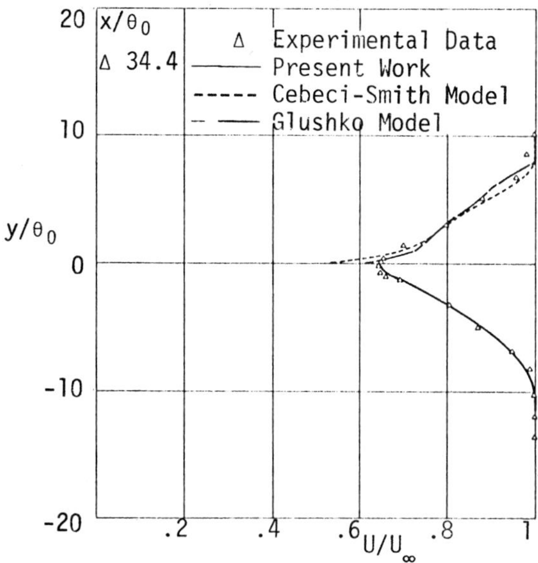

Figure 8. Mean Velocity Distribution in Wake.

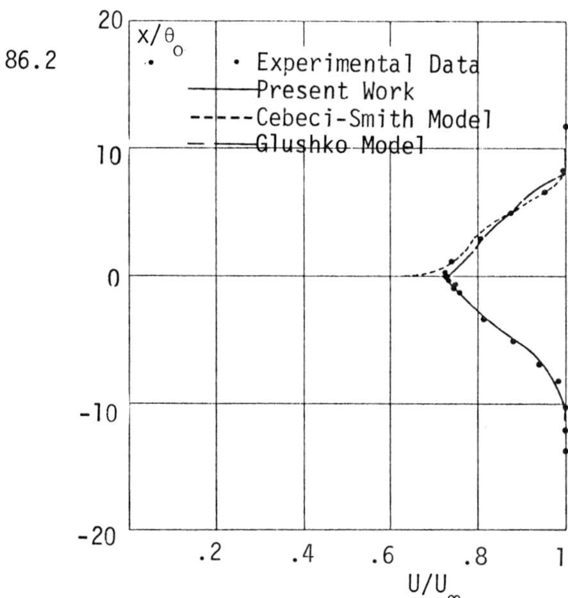

Figure 9. Mean Velocity Distribution in Wake.

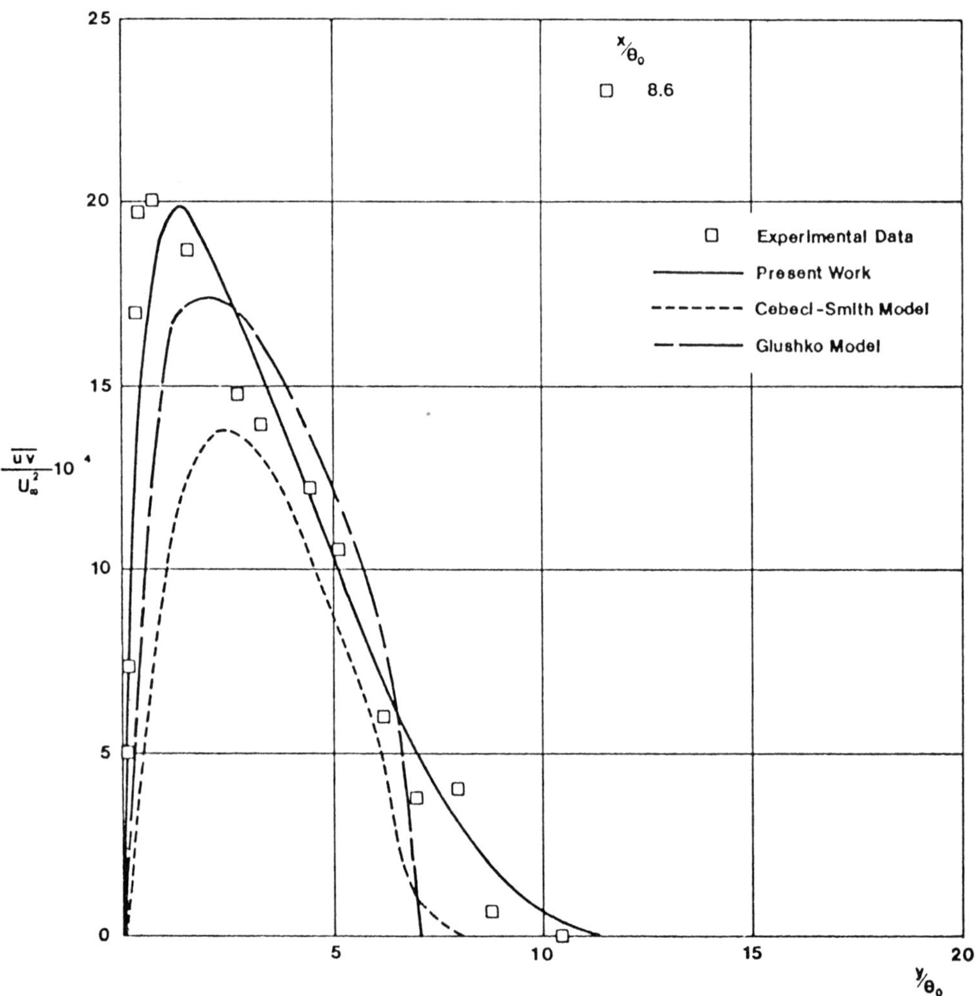

Figure 10. Distribution of Reynolds Stress.

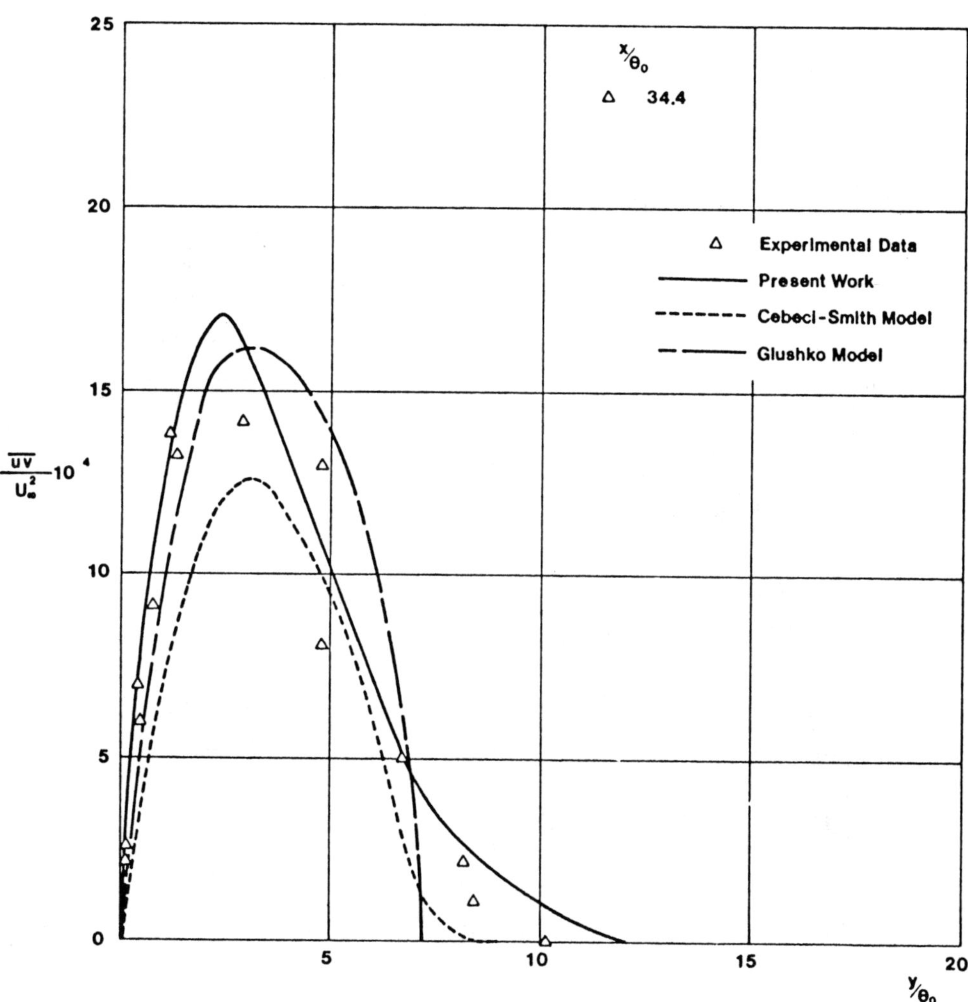

Figure 11. Distribution of Reynolds Stress.

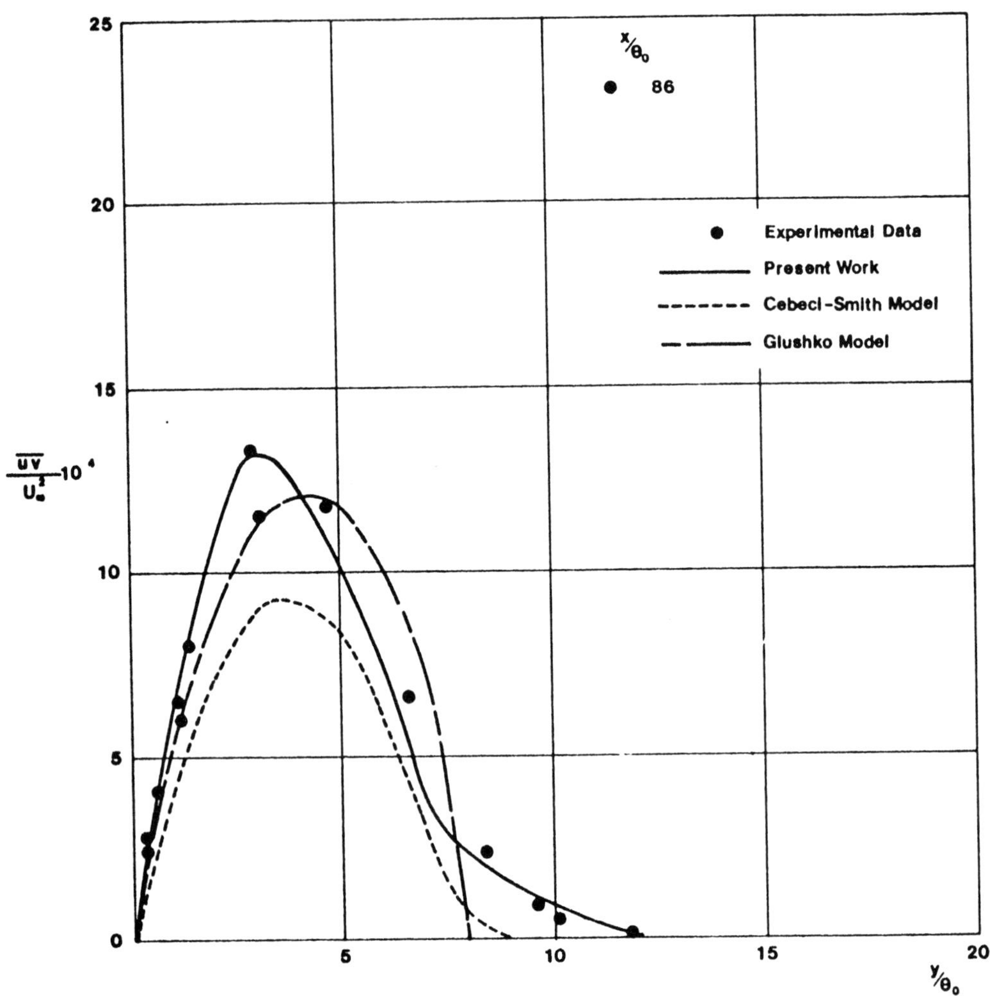

Figure 12. Distribution of Reynolds Stress.

4. REFERENCES

1. Kline, S. J., et al., (ed.) Proceedings, Computation of Turbulent Boundary Layers - 1968 AFOSR - IFP - Stanford Conference, Vol. 1, Stanford University, Thermosciences Division, 1969.

2. Bradshaw, P., et al. "Calculation of Interacting Turbulent Shear Layers - Duct Flow," Journal of Fluids Engineering, Trans. ASME, Series I, Vol. 95, No. 2, 1972, pp. 214-220.

3. Bradshaw, P., et al. "Calculation of Boundary Layer Development Using the Turbulent Energy Equation," Journal of Fluid Mechanics, Vol. 28, Part 3, 1967, pp. 593-616.

4. Bradshaw, P. and Ferriss, D. H. "Applications of a General Method of Calculating Turbulent Shear Layers," Journal of Basic Engineering, Trans. ASME, Series D, Vol. 94, No. 2, 1972, pp. 345-352.

5. Bradshaw, P. "Calculation of Boundary-Layer Development Using the Turbulent Energy Equation IX Summary," NPL Report 1287, Jan., 1969.

6. Ferriss, D. H. "An Implicit Numerical Method for the Calculation of Boundary Layer Development Using the Turbulent Energy Equation," NPL Aero Report 1295, 1969.

7. Cebeci, T. and Smith, A. M. O. Analysis of Turbulent Boundary Layers, Academic Press, New York, 1974.

8. Bradshaw, P. "Prediction of the Turbulent Near Wake of a Symmetrical Airfoil," AIAA Journal, Vol. 8, 1970, pp. 1507-1508.

9. Burggraf, O. R. "Comparative Study of Turbulence Models for Boundary Layers and Wakes," Aerospace Research Laboratories, TR-74-0031, March 1974.

10. Chevray, R. and Kovasznay, L. S. G. "Turbulent Measurements in the Wake of a Thin Flat Plate," AIAA Journal, Vol. 7, No. 8, 1969, pp. 1641-1643.

11. Raj, R. and Lakshminarayana, B. "Characteristics of the Wake Behind a Cascade of Airfoils," J. Fluid Mechanics, Vol. 61, 1973, pp. 707-730.

12. Morel, T. "Calculation of the Free Turbulent Mixing: Interaction Approach," Ph.D. Thesis, Ill. Inst. of Tech., Chicago, 1972.

DISCUSSION

WHITELAW: (Imperial College)

We have recently solved differential equations in elliptic form for real wakes with and without separation using a couple different stress models and also a two equation model.* My recollection is that, for the near wakes without recirculation, the initial conditions were rather important. Also, once separation occurs, it is a new ball game and boundary-layer techniques won't work.

HUFFMAN:

Dr. Whitelaw's question is: How do you deal with boundary layer separation in the vicinity of the trailing edge? I might point out that this situation does occur in many turbine engines as a result of the mechanical constraints which tend to dictate a relatively thick trailing edge. The initial conditions for the wake calculations are being generated by using the boundary layer calculations along the pressure and suction surfaces of the airfoil in conjunction with an inviscid program. In the cases analyzes to date we have been able to reproduce the measured boundary layer thickness on the airfoil surface. Depending upon the airfoil geometry and/or diffusion factor, the boundary layer can obviously separate. We have not dealt with this situation in any detail and I am not prepared to comment further aside from the fact that I don't totally eliminate the interaction model in this case provided the extent of the separated region is small.

LAKSHMINARAYANA: (Pennsylvania State University)

I would like to congratulate you on a fine piece of work. One thing we notice in cascades is that the wake which is asymmetrical, initially, shows a tendency to become symmetrical far downstream. Does your calculation show that?

HUFFMAN:

We have never really carried the calculations that far downstream. We are currently in the development phase with the program and have been trying to conserve computer costs. As a result, we have limited calculations to about 20% of the chord downstream of the trailing edge. We have been carrying out calculations on your data and in general have reproduced the observable experimental traces of streamline displacement. Since the coordinate system is

---

* S. B. Pope and J. H. Whitelaw, J. Fluid Mech. 73, 9, 1976.

not attached to the minimum velocity streamline when the interaction hypothesis is used, this streamline is located by interpolation once the velocity field is computed. The location of the minimum velocity streamline and the magnitude of the velocities do seem to agree with the data at this point.

LAKSHMINARAYANA:

The second question is about initial conditions. Do you go through a boundary layer program to predict the growth of the boundary layer and then go on to the wake?

HUFFMAN:

Conceptually we would analyze the cascade and/or blade flow using a boundary layer program in conjunction with an inviscid core calculation. In the series of comparisons shown in the presentation we did not employ this technique to generate initial conditions but rather used the experimental data which were available. In the case of the asymmetric cascade flow we are using the Bradshaw boundary layer calculation as presented in reference 1, in conjunction with the inviscid calculation of Katsanis, reference 2. We are in essence carrying out a first-order interaction between the boundary layer and the inviscid core, i.e., calculating the boundary layer displacement thickness with a known core velocity distribution and then rechecking our calculation with a new airfoil shape corrected via the displacement thickness. We have matched the trailing edge conditions reasonably well for the airfoils analyzed.

Reference 1. Bradshaw, P. and K. Unsworth. "An Improved FORTRAN Program for the Bradshaw-Ferriss-Atwell Method of Calculating Turbulent Shear Layers," Imperial College Aero Report 74-02, February 1974.

Reference 2. Katsanis, T. "FORTRAN Program for Calculating Transonic Velocities on a Blade-to-Blade Stream Surface of a Turbomachine," Lewis Research Center, Cleveland, Ohio, September 1969.

LAKSHMINARAYANA:

One last question: How do you check on the details of the turbulence, the three functions you have in the equation?

HUFFMAN:

We have generated the empirical functions for the symmetric case using the boundary layer model as a guide. We are currently using the same functions for the asymmetric case. We think it is very important to generate a data set for an asymmetric turbulent wake that has sufficient spatial resolution to determine the empirical functions with accuracy. The suitability of the symmetric function for asymmetric calculations remains to be shown and until a sufficient data base is generated this uncertainty cannot be resolved.

WILCOX: (DCW Industries)

In view of the fact that the equations of motion are hyperbolic, would you actually be able to get reverse flow solutions if you did them as part of a characteristic solution?

HUFFMAN:

I would rather have Peter Bradshaw answer that.

BRADSHAW: (Imperial College)

The simplest way of looking at the inclined characteristics, which are the ones that Dave Huffman mentioned, is that they would--roughly speaking--outline a smoke plume put into the flow. So if you have a reversed flow then the characteristics would, I guess, be going upstream. I think it is possible in principle to do solutions of this sort; it might turn out to be very messy in the computer logic.

SAFFMAN: (California Institute of Technology)

I think it might be a bit worse than that. Since the equations are nonlinear your characteristics are going to start running together and you will get mathematical shocks in the system which are physically meaningless.

BRADSHAW:

Yes, I agree entirely. The hyperbolic model comes from a simple assumption for turbulent diffusion which seems to work remarkably well in boundary layers. There are occasions where it would be nice to have a bit of parabolic gradient type diffusion in order to smear out the "shock waves." This is very possibly one of those situations in which the hyperbolic model is just a useful first approximation.

HUFFMAN:

The solution algorithm that we use now does not make specific use of the method of characteristics. It is an implicit finite difference scheme.

BRADSHAW:

This is absolutely true but you will still run into the same mathematical difficulties.

SIMPSON: (Southern Methodist University)

We tried to do exactly what has been talked about here and the first problem that you run into is that in the equation for the direction of the characteristic you have U in the denominator, and as U goes to zero things blow up and numerical instabilities occur. What we ended up doing (Collins and Simpson (1976))* was using the characteristics in the outer part of separated flow and then putting a velocity profile model, a "law of the wall" so to speak, near the wall.

---

* Collins, M. A. and Simpson, R. L. (1976), "Flowfield Prediction for Separating Turbulent Boundary Layers," Report WT-4, Dept. of Civil and Mechanical Engrg., Southern Methodist University; to appear in NTIS series; see the paper by Simpson in this volume.

Some Important Physical Phenomena in Flows with Separated Turbulent Boundary Layers

ROGER L. SIMPSON
Southern Methodist University
Dallas, Texas (U.S.A.)

ABSTRACT

Using experimental results and the results from prediction and simulation efforts, several important physical phenomena associated with two-dimensional incompressible turbulent boundary layer separation from a body such as an airfoil are examined. The separated zone, the attached boundary layers, and the near wake downstream of a body interact with the inviscid freestream flow in determining the final pressure distribution that produces the lift and drag forces. Turbulent boundary layer separation begins at intermittent separation, or where backflow occurs near the wall on an intermittent basis. Pressure gradient relief follows downstream of intermittent separation. It appears that for separation that occurs well upstream of the trailing edge of the body, the separated flow diverges from the body to minimize the streamwise pressure gradient; when separation occurs near the trailing edge, the velocity and length scales of both pressure and suction side turbulent boundary layers are important in determining the pressure gradient relief. Downstream these shear layers interact with the freestream to determine the near wake pressure relaxation.

In predicting the boundary layer in the vicinity of separation it is observed that the turbulent normal stresses, which are usually neglected, are important. The traditional law of the wall velocity profile appears to closely hold up to intermittent separation. The low velocity backflow downstream appears only to serve the purpose of satisfying the continuity equation after pressure gradient relief determines the freestream flow. The prediction procedures of Collins and Simpson for the regions near and downstream of separation are outlined and compare fairly well with available experimental data.

1. INTRODUCTION

The general problem of flow separation from a body is an old, but still important, topic in fluid mechanics. It is particularly important for the aerodynamic design of aircraft and flow passages of machines and devices. For an airfoil the maximum lift always occurs with the onset of separation of the boundary layer. In a positive pressure gradient flow passage, the maximum pressure recovery also occurs near the boundary layer separation condition.

A great deal of current interest and effort on this type of problem centers around the development of high lift wings for short-take-off-and-landing (STOL) aircraft at low speeds. Several ways of controlling the flow to postpone strong separation (breakdown of lift) to higher angles of attack and to higher lift coefficients have been suggested: multi-element airfoils with slots and different kinds of flaps and boundary layer control by blowing and suction (Lachmann, 1961; Schlichting and Truckenbrodt, 1969; AGARD 1972; AGARD, 1974). In addition, because of dynamic or unsteady stall (Crimi, 1975) on helicopter rotors and the blades of turbomachines, there is interest in understanding "steady" turbulent boundary layer separation phenomena to possibly provide some physical insight on the behavior of the unsteady turbulent motions.

In this paper the discussion is largely limited to the case of a steady freestream, incompressible, two-dimensional mean flow over a streamlined or gently curved body or surface with a developed turbulent boundary layer upstream of the separation zone. Thus, in essence, the separation of the boundary layer is due to an adverse pressure gradient. The main purpose of this paper is to point out important phenomena that influence the flowfield behavior. We will present a physical picture of this class of flows which is supported by experimental observations and by flow prediction and simulation efforts. A summary of the results of some very recent prediction efforts is given and recommendations of major areas for future research are made.

2. OBSERVATIONS OF THE INVISCID FLOW BEHAVIOR

Figure 1 shows experimental chordwise distributions of the suction side velocity just outside the boundary layer U_e for an airfoil at several angles of attack α. This figure was taken from Cebeci, Mosinskis, and Smith (1972) and shows the predicted separation locations of the four prediction methods considered by those authors. The Head (1960), Stratford (1959) and Cebeci-Smith methods are in closest agreement with one another for this airfoil and predict separation in the region where the streamwise pressure gradient is rapidly decreasing. Downstream of this zone the velocity and

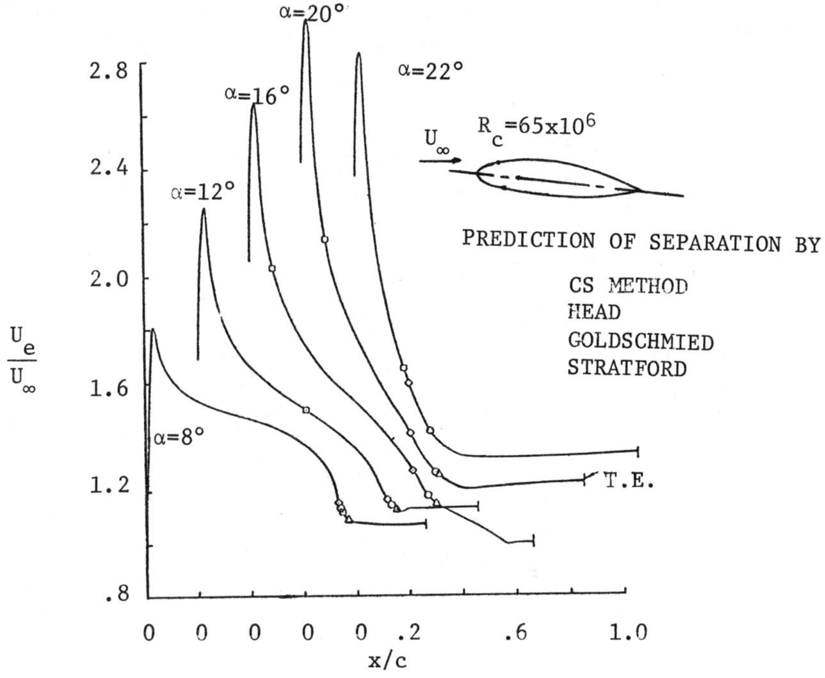

Figure 1. U_e/U_∞ experimental distribution for the NACA 66, 2-420 airfoil with predicted separation points; after Cebeci et al. (1972).

pressure appear to remain nearly constant until the trailing edge of the airfoil. In general, one will observe this same behavior for a variety of bodies, including a circular cylinder and many different airfoil designs, several examples of which are presented by Cebeci et al. For these cases, one must conclude from these observations that in the separated flow zone the velocity and pressure just outside the shear layer approach the free-streamline condition--constant pressure and velocity.

Downstream of the trailing edge, U_e must eventually return to U_∞ in both magnitude and direction, since this irrotational flow outside the shear layer obeys Bernoulli's equation. In cases where separation occurs close to the trailing edge, no constant pressure region is observed and the freestream velocity continues to decrease, sometimes to below the U_∞ value. In these cases the velocity downstream of the trailing edge must increase to U_∞. Figures 2 and 3 show schematically these two cases, which we will label free-streamline separation and trailing edge separation. The data of McDevitt et al. (1976) for transonic flow over an airfoil also show

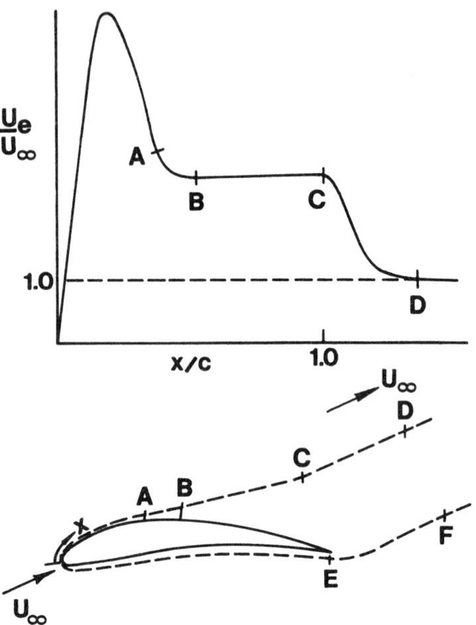

Figure 2. Conceptual schematic of free-streamline separation: top figure - characteristic velocity distribution just outside the shear layer on the suction side: pressure gradient relief region between A and B, free-streamline region between B and C, wake relaxation region between C and D; lower figure - body (solid line) and the effective body (dashed line) that consists of the real body plus the displacement thickness, with comparable A, B, C, and D suction side locations and pressure side wake relaxation region between E and F. Suction side wake velocity and length scales much larger than those for the pressure side wake.

these two classes of separated flow behavior for shock-induced separation at high Reynolds numbers.

In the case of trailing edge separation, there is apparent strong interaction between the wakes of the suction and pressure sides, since the thickness and velocity scales are not extremely different. Thus the freestream velocity distribution in the region between A and C on Figure 3 is controlled by both shear layers. We might say that free-streamline separation occurs when the velocity and length scales of the suction side shear layer are much larger

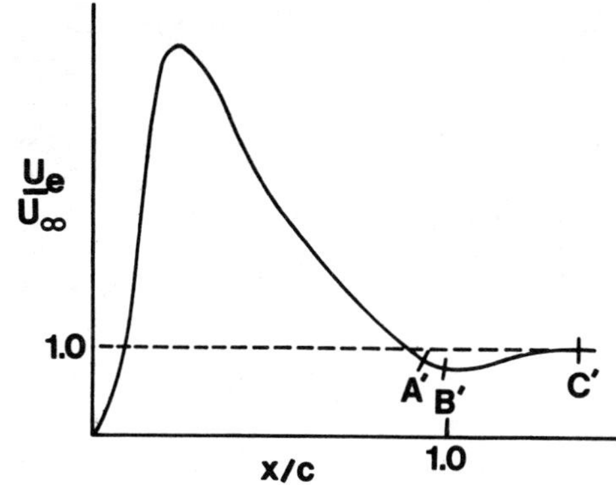

Figure 3. Conceptual schematic of trailing edge separation: top figure - characteristic velocity distribution just outside the shear layer on the suction side: small separated zone between A' and B', wake relaxation region between B' and C'; lower figure - body (solid line) and effective body (dashed line) that consists of the real body plus the displacement thickness, with comparable A', B' and C' locations and pressure side wake relaxation region between D' and E'. Velocity and length scales comparable for the wakes from the two sides.

than those found on the pressure side, as shown in Figure 2. In other words, when the suction side separation occurs sufficiently far upstream of the trailing edge so that the pressure side shear flow only very weakly interacts with the suction side shear flow, then a free-streamline region between B and C is possible.

3. SOME EXPERIMENTAL OBSERVATIONS ON THE SEPARATED SHEAR LAYER BEHAVIOR

Aside from all of this experimental evidence that free-streamline separation occurs, one would at first glance only take the free-streamline condition between B and C as a tentative assumption, assuming that the separated shear layer beneath the potential flow at least partly controls the potential flow. This caution would probably be guided by some laminar flow calculations, in which the shear layer strongly interacts with the potential flow, and by a few backward-facing pitot tube measurements that showed a rather strong backflow. In general one cannot transfer conclusions for laminar flows to turbulent flows since the transport mechanisms are grossly different.

One cannot believe any results from a backward-facing pitot tube or hot-wire anemometer when the flow is changing direction. For example, Simpson, Strickland, and Barr (1973) presented backward-facing pitot tube results that show backflow velocities up to 0.2 U_e but are much too large when compared to the directionally sensitive laser anemometer results of Simpson, Strickland, and Barr (1974, 1976) for the same flow. It is well known that pitot tubes are sensitive to flow direction, Reynolds number, and turbulence intensity so it is only wishful thinking that backward-facing pitot tubes provide accurate information when the flow direction is changing. Hot wires are directionally insensitive to the cooling velocity, as discussed by Simpson (1976a), so one cannot distinguish backflow from flow moving downstream. The directionally sensitive laser anemometer results are believed to be the most reliable. Simpson (1976b) presented a review of experimental techniques for separated flows.

The directionally-sensitive laser anemometer results of Simpson et al. (1974, 1976) for the mean velocity are shown in Figure 4 for the freestream velocity and pressure gradient distributions shown in Figure 5. Figure 6 is a side view schematic of the 16 foot long, three foot wide test section of the blown wind tunnel used for this experiment. The test boundary layer on the wind tunnel floor was an airfoil type with first flow acceleration and then deceleration as shown in Figure 5a. To eliminate preferential separation of the curved top wall boundary layer, this layer was removed through a scoop prior to the last eight feet of test section. To provide the necessary backpressure to blow out this flow, a perforated metal plate was located at the exit.

In analysis of these data, Simpson et al. (1974, 1976) observed that the time-averaged mean pressure gradient dropped off rapidly after the beginning of the backflow on an intermittent basis ($\gamma_p < 1$) <u>or intermittent separation</u>. Here γ_p is the fraction of time that the flow moves downstream. According to Sandborn and Kline (1961) this first location where $\gamma_p < 1$ is where turbulent boundary layer separation begins. Downstream where the average wall shearing stress is zero is the so-called <u>fully-developed separation</u> point or time-averaged separation point. It is clear (Simpson, 1976a) that with an observed Gaussian probability distribution for the streamwise

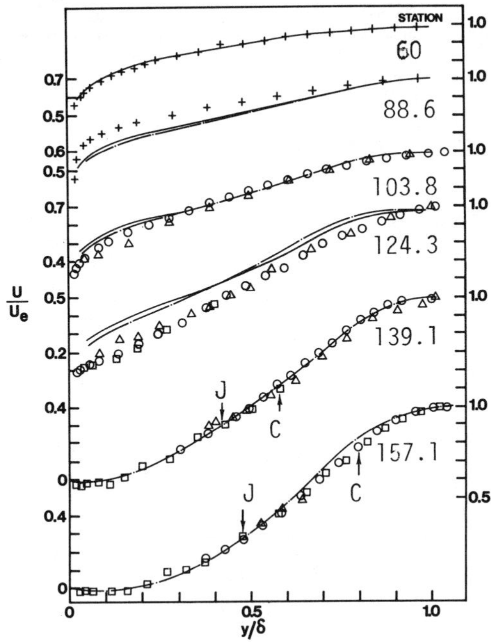

Figure 4. Mean velocity profiles: + impact probe, o normal hot film, Δ laser anemometer (Strickland and Simpson, 1973), □ laser anemometer (Simpson et al., 1974). Solid lines - upstream predictions by unmodified Bradshaw et al. method. Broken lines - model 1 predictions upstream of separation; smoothed shear profile predictions for two downstream separated flow profiles. J and C denote locations of match point and separation characteristic II', respectively. Note displaced ordinates.

velocity at a point, intermittent separation occurs at the first streamwise location where $\sqrt{\overline{u^2}}/U > 1/3$. This would appear to be a good criterion for the beginning of intermittent separation. The location of fully-developed separation is where $\gamma_p = 1/2$ at the wall. These data are in good agreement with the separation criteria of Sandborn as shown in Figure 7 and discussed by Simpson et al. (1976). The data of Sandborn and Liu (1968), which agree with these criteria, indicate that $\gamma_p \approx 0.7$ near the wall at intermittent separation, somewhat agreeing with the observation of Simpson et al. that $\gamma_p \approx 0.8$ on the wall at intermittent separation. The point here is that the freestream pressure gradient relief appears to begin close to where intermittent separation begins. This is

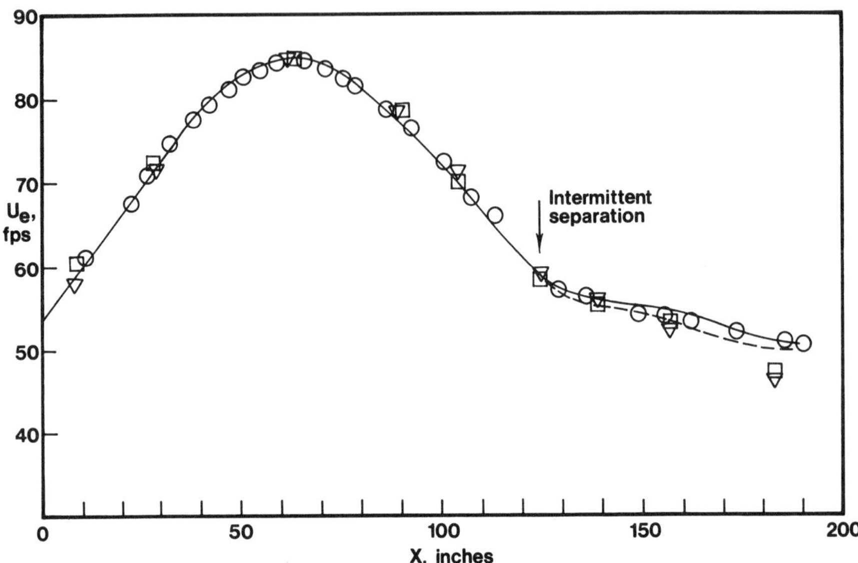

Figure 5a. Freestream velocity distribution at boundary layer outer edge: o from bottom wall static taps; □ from bottom boundary layer pitot probe; ∇ from top boundary layer pitot probe. Solid line denotes D = $-0.005\ ft^{-2}$ results from minimum pressure gradient model of the separated flow freestream predictions of chapter 4; dashed line for D = $-0.004\ ft^{-2}$. Arrow denotes intermittent separation at 124.3 inches.

significant because we must know where the pressure gradient relief begins in order to properly calculate the entire flowfield. It also appears that the rapid pressure gradient relief occurs between the intermittent separation point and the fully-developed separation point. Further insight on the flow in this region is given below.

Figure 4 shows rather flat mean velocity profiles for $y/\delta < 0.15$ downstream of the beginning of intermittent separation. As shown by Simpson et al. (1976) the local turbulence intensity $\sqrt{\overline{u^2}}/U$ is very large but still the momentum and mean kinetic energy of this backflow is still very small compared to the outer shear flow and the free-stream. This low velocity region evidently just serves the function of providing just the small amount of net backflow required to satisfy the continuity requirement after the energetic flow near the

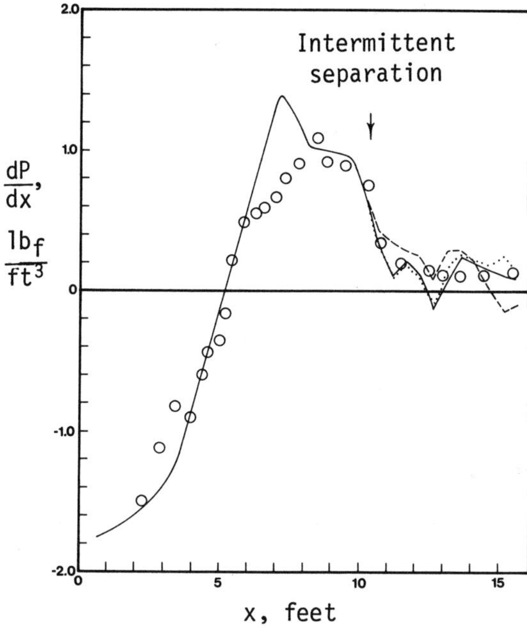

Figure 5b. o pressure gradient along bottom wall; solid line denotes results at boundary layer outer edge for D = $-0.005\ ft^{-2}$ in the minimum pressure gradient model of the separated flow freestream predictions of Collins and Simpson. Dotted line for D = $-0.006\ ft^{-2}$; dashed line for D = $-0.004\ ft^{-2}$.

freestream has deflected away from the wall upon separation. After separation the freestream flow seeks to reduce pressure gradients, so the freestream flow deflects from the wall to reduce further divergence of the streamlines. In the case of free-streamline separation this pressure gradient is reduced to zero. The consistency of this picture is presented in the calculations of Collins and Simpson (1976) that are outlined below. In essence, the freestream condition determines the behavior of the shear flow near the wall.

4. INSIGHTS FROM SOME PREVIOUS PREDICTION AND SIMULATION RESULTS USING FREE-STREAMLINE SEPARATION

To further emphasize the applicability of free-streamline separation and to examine the important details, the airfoil flow

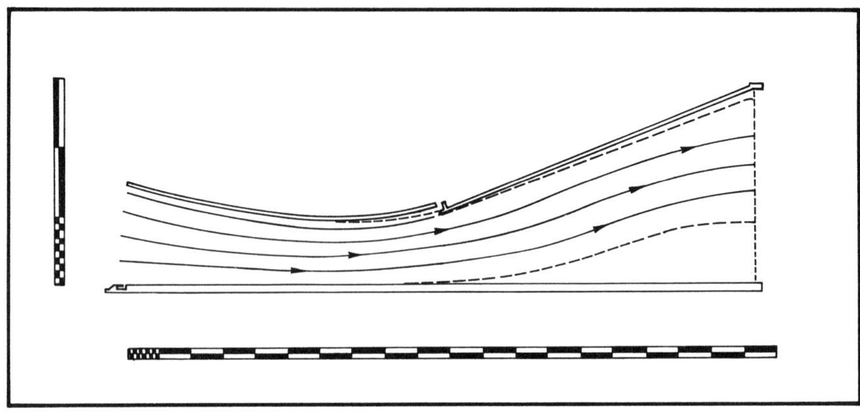

Figure 6. Distribution of streamlines for the minimum pressure gradient computation with D = $-0.005\ \text{ft}^{-2}$. Sideview schematic of the test section for the Simpson et al. flow. Major divisions on scales: 10 inches. Note baffle plate upstream of blunt leading edge on bottom test wall, upper wall boundary layer scoop, and perforated exit plate.

prediction efforts of Bhateley and Bradley (1972) and Jacob (1969, 1974, 1975) will be briefly reviewed.

Bhateley and Bradley used an equivalent airfoil system consisting of a linearly varying vorticity distribution over the surface of each airfoil element to simulate the separated wake. The computed boundary-layer displacement thickness was superimposed on the airfoil contour to form an equivalent airfoil surface for each element. This procedure was iterated until convergence occurred. The flow downstream of a separation point was allowed to develop as a free-streamline flow with no surface boundary conditions. There was tangential flow on only that part of the equivalent airfoil having attached flow. The pressure distribution downstream of the separation was assumed constant and equal to that value of pressure obtained by linear extrapolation of the equivalent body boundary

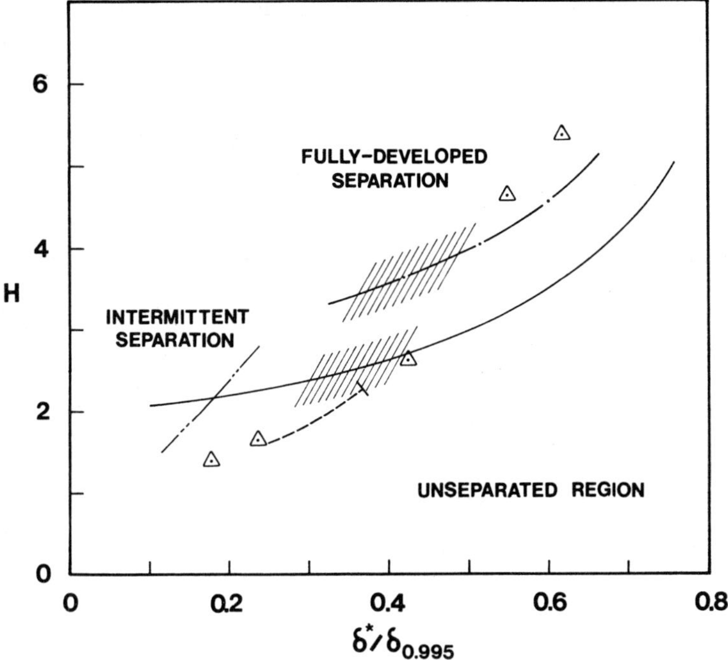

Figure 7. H vs. $\delta^*/\delta_{0.995}$. Δ Simpson et al. (1974): stations 88.2, 103.8, 124.3, 139.1 and 157.1, respectively, for increasing H. Shaded areas - data for intermittent and fully-developed separation (Sandborn and Kline, 1961). Path predicted from Perry and Schofield velocity profiles ---. Solid line —— intermittent separation; —·— fully developed separation (Sandborn); —··— data of Sandborn and Liu (1968).

point pressures to the separation point. They used the experimentally obtained separation point.

Very good predictions of lift and pressure coefficient were made with this method as long as the free-streamline model satisfied the data. For low angles of attack, trailing edge separation was present for their test cases and pressure coefficient predictions in this region were not good. When there was a long relaxation zone (AB on Figure 2) their estimate of the free-streamline pressure was also in error. They point out the deficiency of not having a wake model. In summary, their method did not include any pressure gradient relief model (for AB on Figure 2), no wake model, and used experimental data to locate separation. It still did a good job in many cases of predicting the pressure coefficient, which basically supports the free-streamline idea.

Jacob presented a similar type method for single airfoils (1969) and for multiple-element airfoils with the capability of inclusion of ground effects (Jacob and Steinbach, 1974). Vortex and source distributions on the contour were used and a boundary layer calculation was made for the attached flow. The separation point was predicted to be where H = 4, which, as observed from Figure 7, is in good agreement with Sandborn's criterion for fully-developed separation for low curvature bodies such as in the flow of Simpson et al. The displacement thickness effect was described as an outflow from the airfoil. The "dead air" or separated zone was simulated as shown in Figure 8a with a separation streamline SU. This separating streamline was required to be tangent to the surface at S, rather than tangential to the superimposed displacement thickness distribution, as it should be. The pressure was required to be equal at three special points of the separating streamlines, at the beginning locations S and T and at point U above the trailing edge. In addition, the pressure was allowed to vary "very little" between points S and U. Thus, the

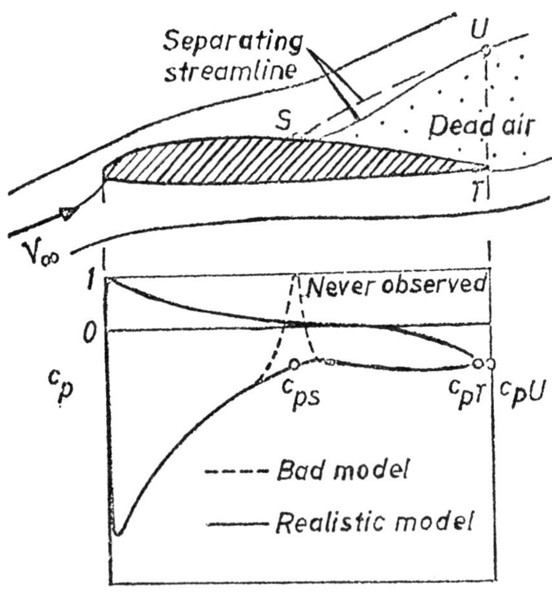

Figure 8a. Pressure distribution for flow with a separated wake (Jacob and Steinbach, 1974).

separation streamline SU is not exactly a free-streamline, but in practice is close to being one. A source distribution along the body in the dead air region provided the outflow in this region. The circulation-contributing part of the potential flow and the outflow were adjusted to obtain the equal pressures at the three points. Geller's (1976) method for cascade flow is basically very similar to Jacob's (1969) procedure. The boundary layer displacement effect was assumed small and the simulated wake was assumed to have an infinite length.

This model of Jacob gives good predictions for the lift coefficient for free-streamline separation. Like Bhateley and Bradley, this method did not include any pressure gradient relief model at separation and any wake model. These authors pointed out that the pressure drag calculation is very sensitive to the dead air pressure value, much more so than the lift. They concluded that their dead air pressure prediction needed improvement to improve pressure drag calculations.

Jacob (1975) modified his method to simulate the effect of the wake on the drag and lift, as shown in Figure 8b. From the symmetric flow over various shaped ellipses, he determined that $a_S \approx S^*$ for body thickness to length ratios less than 1/2, where a_S is the x-coordinate location of a sink S downstream of the body and S* is the length along the body surface between the two separation points. The strength of the sink S equals the strength of the source along the separated flow surface AU. Since Y_S was zero for these symmetric cases, he could only empirically determine the sink location that produced the best drag result. The Y_S for cases with non-zero angles of attack was assumed to be given by

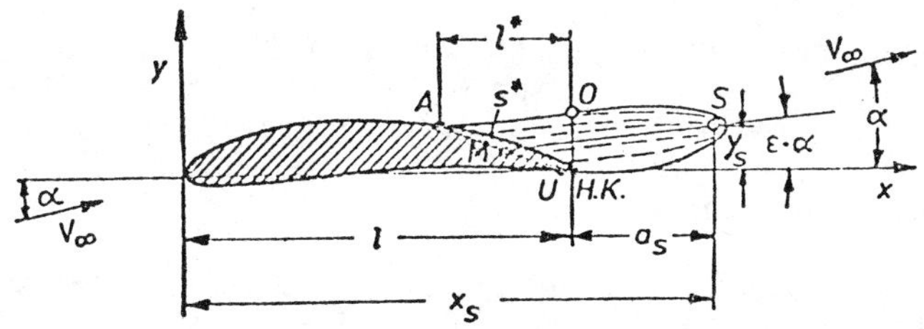

s* = Conture length between A and U

Figure 8b. Sketch of the new Jacob dead air flow region model with unsymmetric separation (Jacob, 1975).

$$Y_S = Y_M + (1 - X_M + a_S) \tan (\varepsilon\alpha)$$

where $\varepsilon = 0.5$ and X_M and Y_M are the coordinates of the midpoint of a straight line connecting the separation points. This reduces to zero for symmetric flow over a symmetric body.

There was considerable improvement over the low drag predictions of the 1974 version. For the cases presented the lift-drag polar plots were in very good agreement with measurements even though in some cases the predicted lift coefficients were considerably different from the measurements. The effect of varying ε from 0.3 to 0.7 was said to be small on the drag prediction while the selection of laminar-turbulent transition "point" was found to be important. A very important conclusion from this work is that the wake flow behavior strongly influences the drag.

5. INSIGHTS FROM SOME RECENT SEPARATING TURBULENT BOUNDARY LAYER PREDICTION EFFORTS

Figure 9 shows the several regions of a separating turbulent boundary layer: the upstream attached turbulent flow, the freestream flow, and the separated turbulent flow. Collins and Simpson (1976) modified the Bradshaw, Ferris and Atwell (1974) turbulent boundary layer prediction method to predict the turbulent flow regions. The outer freestream condition on the separated shear flow was predicted using the condition of a minimum pressure gradient after the beginning of intermittent separation. The Bradshaw et al. method was selected and used for two reasons: (1) this method is regarded as one of the more general and better two-dimensional methods (Kline et al., 1968) and (2) a hyperbolic set of equations with real characteristics with some important physical significance for separation is used.

The observations of Simpson et al. (1974, 1976) were used in the program modifications. Simpson et al. observed that the <u>normal stress</u> terms of the momentum and turbulence energy equations were significant near separation, as Rotta (1962) had already pointed out. However, no prediction method has previously accounted for these important terms.

Consider the time-averaged momentum equations for the streamwise or x direction and the y direction or direction normal to the wall (Rotta, 1962):

$$\text{(x momentum)} \qquad U\frac{\partial U}{\partial x} + V\frac{\partial U}{\partial y} + \frac{\partial \overline{u^2}}{\partial x} + \frac{\partial \overline{uv}}{\partial y} = \frac{-1}{\rho}\frac{\partial P}{\partial x} + \nu\frac{\partial^2 U}{\partial y^2} \qquad (1)$$

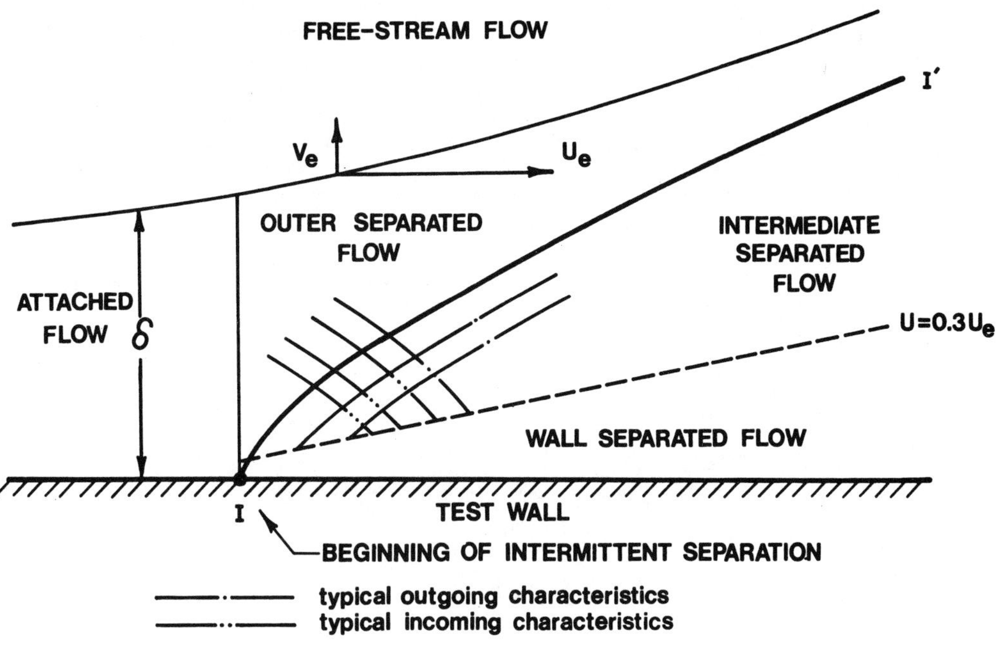

Figure 9. Schematic of the several mean velocity flow regions of the separated flow model.

(y momentum) $$\frac{\partial \overline{uv}}{\partial x} + \frac{\partial \overline{v^2}}{\partial y} = \frac{-1}{\rho} \frac{\partial P}{\partial y} \tag{2}$$

Since $\partial \overline{uv}/\partial x$ is at least one order of magnitude smaller than $\partial \overline{v^2}/\partial y$, equation (2) can be integrated to produce

$$\frac{P}{\rho} = \frac{P_e}{\rho} - \overline{v^2} \tag{3}$$

since the pressure outside the boundary layer is P_e. Equation (3) can then be used in equation (1) to produce an equation with only the pressure gradient just outside the boundary layer

$$U \frac{\partial U}{\partial x} + V \frac{\partial U}{\partial y} = \frac{-1}{\rho} \frac{\partial P_e}{dx} + \underbrace{\frac{\partial(-\overline{uv})}{\partial y}}_{\text{I}} + \underbrace{\nu \frac{\partial^2 U}{\partial y^2}}_{\text{II}} - \underbrace{\frac{\partial}{\partial x} (\overline{u^2} - \overline{v^2})}_{\text{III}} \tag{4}$$

The last three terms are the Reynolds (I) and viscous (II) shearing stress gradients and the normal stresses gradient (III). In most boundary layer cases term III is negligible, but near separation Simpson et al. found it to be about 1/4 of term I in magnitude and therefore not negligible. From the data of Sandborn and Slogar (1955), Rotta (1962) also noted the need to retain that term near separation. The unmodified Bradshaw et al. method does not contain that term in its momentum equation.

Rotta (1962) also presented the turbulence energy equation with all terms retained that were of the same order as those retained in equation (4)

$$\underbrace{\left(\frac{U}{2}\frac{\partial \overline{q^2}}{\partial x} + \frac{V\partial \overline{q^2}}{2\partial y}\right)}_{A} + \underbrace{\frac{\partial}{\partial y}\left(\frac{\overline{pv}}{\rho} + \frac{1}{2}\overline{q^2 v}\right)}_{B} - \underbrace{\nu\frac{\partial^2}{\partial y^2}\left(\frac{1}{2}\overline{q^2} + \overline{v^2}\right)}_{C} + \underbrace{\varepsilon}_{D}$$

$$= \underbrace{-\overline{uv}\frac{\partial U}{\partial y}}_{E} - \underbrace{(\overline{u^2} - \overline{v^2})\frac{\partial U}{\partial x}}_{F'} \qquad (5)$$

$$\overline{q^2} \equiv \overline{u^2} + \overline{v^2} + \overline{w^2}, \qquad \tau = -\rho\overline{uv}, \qquad \varepsilon \approx \nu\overline{\left(\frac{\partial u_i}{\partial x_j}\right)^2}$$

where the terms are advection (A), turbulent diffusion (B), viscous diffusion (C), dissipation (D), shear stress production (E), and normal stresses production (F'). Figure 10 shows estimates of the ratio of normal stress to shear stress turbulence energy production for the several stations of the Simpson et al. flow on a flat wall, along with the results of Schubauer and Klebanoff (1951) near separation. It is clear that near separation, normal stresses production can contribute up to 1/3 of the turbulence energy production and therefore is not negligible. The unmodified Bradshaw et al. method does not contain this term either. The curvature turbulence production term $-\overline{uv}\,\partial V/\partial x$ (Bradshaw, 1973) has been neglected from the right side of the equation (5), since in the Simpson et al. flow the largest value of $(\partial V/\partial x)/(\partial U/\partial y)$ was crudely estimated to be less than 0.03. Thus in that case, the effect of curvature attributable production was negligibly small as compared to shear production.

In addition to the normal stresses effect explicitly contained in equations (4) and (5), the turbulence structure represented by

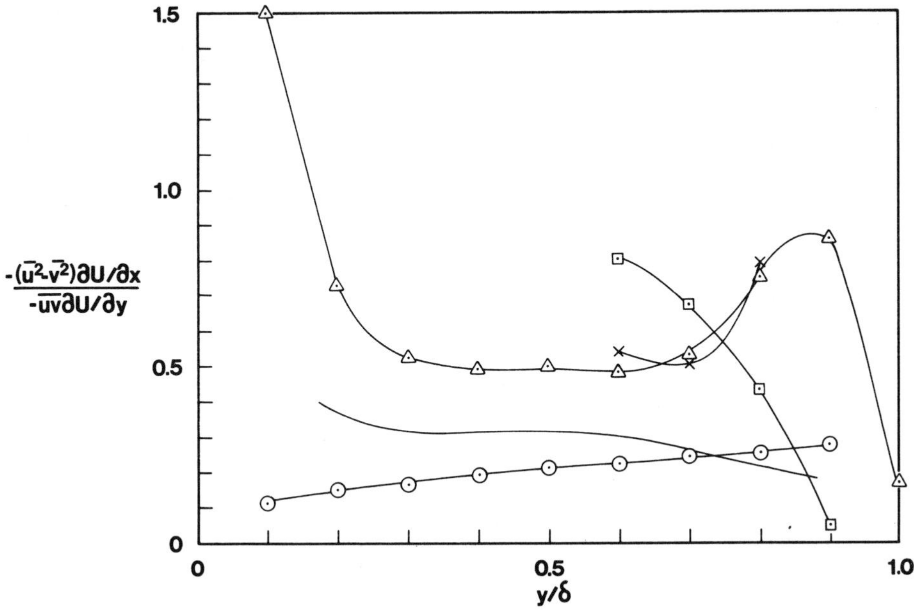

Figure 10. Ratio of normal stress production to shear stress production for the Simpson et al. flow at the several streamwise locations: o 103.8, Δ 124.3, □ 139.1, × 157.1. Solid line shows results of Schubauer and Klebanoff flow near separation, 24.5 feet.

Bradshaw's empirical functions is also influenced by this effect near separation. Simpson (1975a) and Collins and Simpson (1976) accounted for this effect on the structure by the factor

$$F = 1 - \frac{(\overline{u^2} - \overline{v^2})\ \partial U/\partial x}{-\overline{uv}\ \partial U/\partial y} \tag{6}$$

which is the ratio of total turbulence energy production to the shear production.

With the recent publication by Bradshaw (1973) on the effects of streamline curvature, it has become popular to seek to explain complex turbulent flows by any possible extra rates of strain. Bradshaw suggested that the Reynolds stress changed by a factor $1 \pm 10\ e/(\partial U/\partial y)$, where e is the small extra rate of strain, which would be $\partial U/\partial x$ in our case near separation. As long as $\partial U/\partial y >> e$, the mean flow equations were suggested to be adequate without the normal stress terms. It has been informally suggested by several people that that approach is the more correct one since the mean streamlines deflect from the surface upon separation. However, the normal stresses effect arises naturally in the governing equations and the turbulence structure and thus has a more physically sound basis.

Simpson (1975b) illustrated the current approach in discussion of the three-dimensional separating turbulent boundary flow of Elsenaar et al. Elsenaar et al. had used the extra rate of strain approach ($e = \partial U/\partial x$) to account for reduced turbulence structure function values near separation. Simpson obtained almost the same reduction in these quantities using the above turbulence energy approach. Since application of both the present turbulence kinetic energy approach and the extra rate of strain approach could be redundant, the present physically based method is preferred. To avoid confusing normal stress effects with wall curvature effects, only separating turbulent boundary layers on zero or small curvature surfaces were considered by Collins and Simpson in their test cases.

For the unseparated flow upstream of intermittent separation several different alternative models were tested for the wall region since there are several different formulations on the influence of strong adverse pressure gradients on the law-of-the-wall velocity profile. One model is that contained in the unmodified Bradshaw et al. (1974) version with the McDonald (1969) type wall law

$$U_1 = \frac{U_\tau}{K}\left[\ln\left(\frac{U_\tau y_1}{\nu}\right) + C\right] + \frac{U_\tau}{K}\,\phi\left(\frac{\tau_1}{\tau_W}\right) \quad (7)$$

where

$$\phi(z) = \ln\left[\frac{4((1+z)^{\frac{1}{2}} - 1)}{z((1+z)^{\frac{1}{2}} + 1)}\right] + 2((z+1)^{\frac{1}{2}} - 1) \quad (8)$$

and $z = \tau_1/\tau_W$. y_1, τ_1, and U_1 are the distance from the wall, the shearing stress, and velocity at the first mesh location from the wall, respectively. $U_\tau = \sqrt{\tau_W/\rho}$; K is the von Karman constant. These equations, the equation on the incoming characteristic, and the shear stress equation

$$\tau_1 = \tau_W - y_1 U_e \frac{dU_e}{dx} + 0.3\, y_1 \frac{dU_1^2}{dx} \quad (9)$$

are used to compute τ_W, τ_1, and U_1 at the first mesh point at the next downstream mesh location. This latter equation comes from using a one-fifth power law velocity profile in the momentum equation without the normal stresses term and integrating between the wall and the first mesh point.

A second model is simpler and is based on equation (9) and the traditional law of the wall, i.e., $\phi(\tau_1/\tau_W) = 0$ in equation (7)

$$U_1 = \frac{U_\tau}{K}\left[\ln\left(\frac{U_\tau y_1}{\nu}\right) + C\right] \qquad (10)$$

While equations (7) and (8) supposedly account for pressure gradient effects, there is a great deal of experimental evidence for equation (10). The reanalysis of a large number of adverse pressure gradient velocity profiles by Perry and Schofield (1973), the experimental results of Simpson et al. (1974, 1976), and the results of Samuel and Joubert (1974) all support equation (10), at least up to intermittent separation.

While $\partial\tau/\partial y = dP_e/dx$ on the wall, the experimental results of Newman (1951) and Simpson et al. support the evidence of a very low shear stress gradient near the wall for near intermittent separation conditions. As pointed out by Simpson et al. (1974, 1976) the normal stresses term in equation (4) is still significant near the wall near intermittent separation and can cause the near wall shearing stress gradient to be lower than that implied by equation (9). Spangenberg et al. (1967) and Stratford, in discussion of their work, also support this view. Thus with a lower stress gradient, τ_1/τ_W would approach unity and equations (7) and (8) would degenerate to equation (10).

These latter observations lead to third and fourth wall region models used by Collins and Simpson: equation (9) modified to account for the normal stresses term downstream of intermittent separation and either equations (7) and (8) or the traditional law, equation (10). In this modification the pressure gradient term of the shear stress equation (9) was heuristically reduced to model the effect of the normal stresses by the following reasoning.

The experimental results of Spangenberg et al. show significant normal stresses effects near intermittent separation. In those experiments the normal stresses term did not immediately rise to totally balance the pressure gradient but increased somewhat gradually along the flow. Eventually $\partial(\overline{u^2} - \overline{v^2})/\partial x$ almost balanced $-dP_e/dx$, in effect almost eliminating the last two terms of equation (9). These observations suggest a heuristic damping factor on the pressure gradient term in equation (9) when $\sqrt{\overline{u^2}}/U_1 > 1/3$, namely

$$\exp\left(1 - 9\,\overline{u^2}/U_1^2\right) \tag{11}$$

Simpson et al. showed that $\overline{u^2}/\overline{u_M^2}$ vs. y/y_M profiles for their flow were similar downstream of intermittent separation, where the subscript M denotes the y location where $\overline{u^2}$ is a maximum for a given profile. The Spangenberg et al. data in the intermittently separating region indicate that the near wall $\overline{u^2}$ is about the same as τ_M/ρ, so that these data indicate that the large eddies which produce τ_M must also contribute to the near wall $\overline{u^2}$. Thus the factor (11) can be written as

$$\exp\left(1 - \frac{B\,\tau_M}{\rho U_1^2}\right) \tag{12}$$

Equation (9) remains unchanged when $\tau_M/\rho U_1^2 \leq 1/B$ but becomes

$$\tau_1 = \tau_w - y_1\,U_e\,\frac{dU_e}{dx}\exp\left(1 - \frac{B\,\tau_M}{\rho U_1^2}\right) + 0.3\,y_1\,\frac{dU_1^2}{dx} \tag{13}$$

when $\tau_M/\rho U_1^2 > 1/B$. Several values of $B \geq 9$ were tested with several test cases. In all of these models, as in the unmodified Bradshaw et al. program, fully-developed separation is predicted when the incoming characteristic cannot intersect the wall. In other words, the differential equation along the incoming characteristic cannot be solved with the wall model to produce a non-zero skin friction value.

Predictions were made for five attached adverse pressure gradient boundary layer flows on flat walls with available experimental data: Simpson et al., Spangenberg et al., Schubauer and Spangenberg flow E (1960), Perry (1966), and Newman (1951). Each test flow was computed with the unmodified Bradshaw program as well as with each of the four modified prediction models. The degree of prediction improvement, if any, was primarily judged by comparisons with the experimental skin friction coefficient, $C_f/2$, shape factor, H, displacement thickness, δ^*, and F values, and the location of separation, quantities that are of primary significance in predicting flow over a body with separation. The location of intermittent separation was predicted when $\tau_M/\rho U_1^2 = 1/B$. Mean velocity and shearing stress profiles were compared with experimental results near separation.

For brevity, the detailed results from only one test flow are presented here. Figure 11 shows the results from the several models

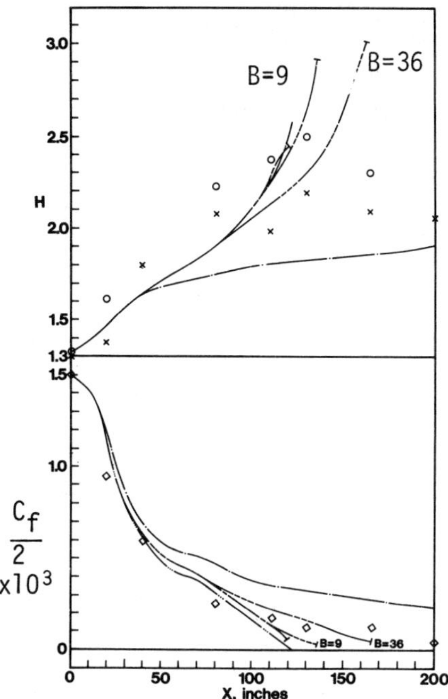

Figure 11a. Spangenberg et al. (1967) flow: $C_f/2$, H, F, and δ^* experimental (flow B) and predicted distributions. Results from hot-wire data: o; from pitot data: ×. Experimental $C_f/2$ by originators from Clauser plot of average between pitot and hot-wire velocity profile results: ◇. Estimates of F from experimental data: × . Predicted curves legend: —·— Unmodified Bradshaw et al.,; —···— Model 1; ——— Model 2; —–-— Model 3; —---— Model 4.

for the Spangenberg et al. flow. The approximate curve fit free-stream velocity given by those authors was used for x > 35 inches while the experimental velocities were used upstream. Figure 12 shows the experimental pressure gradients obtained for their A and B flows and the distribution implied by their curve fit distribution. Since in general the flow B experimental pressure gradient distribution was close to the curve fit result, the predictions were compared to the experimental quantities deduced for that flow. Both flows A and B indicated that $\sqrt{\overline{u^2}}/U_1$ was first greater than one-third at about x = 80 inches, indicating $\gamma_p < 1$ and intermittent separation. The flow A boundary layer was highly perturbed downstream but apparently recovered somewhere downstream of 130 inches. Flow B had

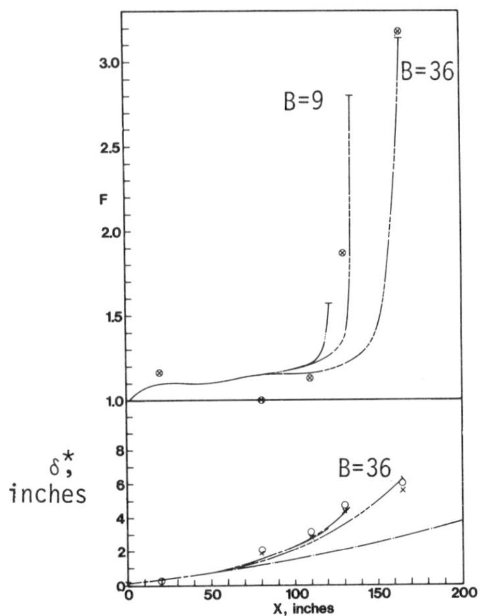

Figure 11b.

small-scaled intermittent transitory stall between 110 and 165 inches, as revealed by a chemical smoke. It was estimated that this stall was present about half of the time at 130 inches, decreasing to about 15% at 165 inches as the boundary layer recovered somewhat.

The pressure gradient is overwhelmingly the most important parameter that influences the predictions. For 35 inches < x < 60 inches, the tabulated data dP/dx is less than the curve fit pressure. Hence, $d(C_f/2)/dx$ is more negative for the curve fitted dP/dx case. For 60 inches < x < 80 inches, the tabulated dP/dx result is greater than the curve fitted result so $d(C_f/2)/dx$ becomes more negative and separation is predicted at about 70 inches for both A and B flows. H is better predicted by the experimental dP/dx distribution for x < 60 inches, while there is no appreciable difference in the predictions of F.

Using the pressure gradient implied by the curve-fitted free-stream velocity distribution, Model 1 shows a significant improvement over the unmodified Bradshaw program as observed in comparisons of $C_f/2$, H, and δ^*. Model 2 is an improvement, but not as good as Model 1, especially for x < 80 inches. Both models always predict lower $C_f/2$ and higher H as compared to the unmodified Bradshaw case. Since

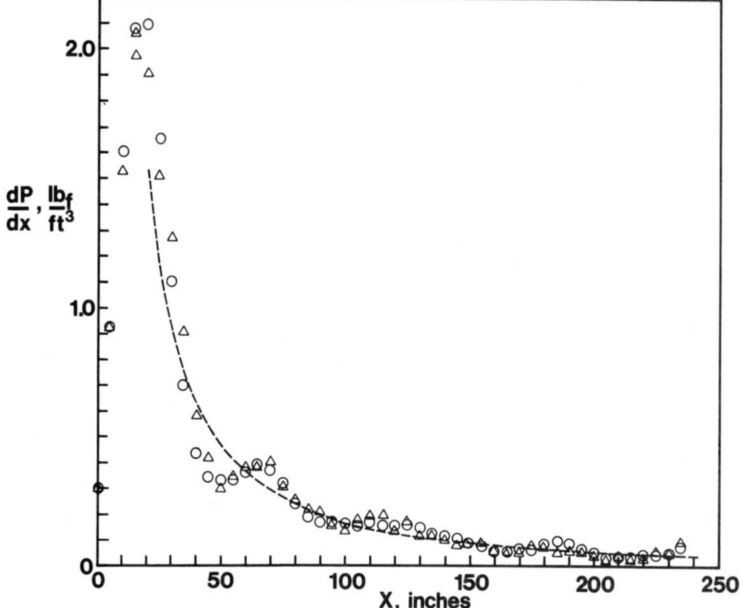

Figure 12. Experimental pressure gradient distributions for the Spangenberg et al. (1967) flows: o flow A; Δ flow B. Pressure gradient implied by their curve-fitted free-stream velocity distribution

$$\frac{U_e}{U_{e_{initial}}} = 1.13 \left(\frac{x}{12} + 0.83\right)^{-0.33}$$

shown by the dashed line; $U_{e_{initial}}$ = 84 fps, x in inches.

all of the modified methods predict premature fully-developed separation as compared to the flow B experimental results, the unmodified Bradshaw method predicts the general shape of $C_f/2$ and H distributions better.

Model 4 with B = 36 properly predicts intermittent separation at about 80 inches, where experimentally observed, and does not predict fully-developed separation until x = 165 inches, where the velocity profile near the wall quickly developed much scatter and the prediction became unstable. Predictions of δ^* and F are not extremely far from the flow B experimental results. Smoothing of the near wall velocity profile was tried, but did not greatly improve the results. The source of this difficulty was traced to the fact that as U_1, the velocity value at the first point off the wall, drops below 0.1 U_e, the equation for the direction of the incoming characteristic greatly amplifies numerical errors. Similar behavior was observed for Model 3 and B = 36. In Models 1 and 2 fully-developed separation soon followed after $U_1 < 0.1\ U_e$, so when dP/dx is a little large, as shown on Figure 12 for the experimental dP/dx at 60 inches < x < 80 inches, separation was predicted. The major effect of the damping model is believed to be prevention of U_1 falling below 0.1 U_e. The experimental velocity profiles agreed well within 3% of the predictions for x < 40 inches, with the predictions near the wall becoming progressively poorer downstream.

On the whole, the Collins and Simpson model of the normal stresses effect in the outer region shows improved predictions of $C_f/2$, H, δ^*, and F at least up to the beginning of intermittent separation for the test cases. Upstream of the predicted location of intermittent separation for these test cases, the Model 1 and Model 2 predictions were within about 10% of one another. This indicates that $\phi(z)$ contributes less than about 8% to equation (7). In other words, $\alpha_0 = (\nu/\rho U_\tau^2)dP/dx < 0.02$, according to McDonald (1969), and there is only a so-called weak pressure gradient effect on the law of the wall. Since the normal stresses appear to be important downstream of intermittent separation, there is some question as to whether McDonald's law is a good model downstream of intermittent separation. Hot-wire and pitot tube results obtained downstream of intermittent separation are suspect, so equation (7) cannot really be verified by these types of data. More data in the intermittent separation zone using a directional laser anemometer would be useful in eliminating this uncertainty.

The wall damping model for the normal stress effect on the net pressure gradient near the wall was only partially successful. With B = 36, the location of the beginning of intermittent separation for each test flow was in good agreement with available experimental results. Improved predictions using Model 3 or 4 were achieved in several flows but produced poorer results for the Simpson et al. flow. More work is needed on modeling the normal stress relief of the pressure gradient after the beginning of intermittent separation. Clearly more experimental data in this region are needed.

For the separated shear flow, Collins and Simpson began computations at the intermittent separation (I in Figure 9) since this is where the pressure gradient begins to be relieved. The outgoing characteristic II' of the Bradshaw et al. model which originates at the wall at intermittent separation has the physical significance of dividing the downstream outer separated flow from the intermediate flow. As pointed out by Bradshaw et al. (1967) and by Lister (1960), only information upstream of a hyperbolic characteristic can influence the computed quantities along that characteristic. Since there is some intermittent backflow that influences the time-averaged flow structure downstream of intermittent separation, the outgoing characteristic of intermittent separation is the last one that is directly independent of the downstream turbulence structure. Thus given U_e and V_e at the outer edge of the shear flow and the flow structure at intermittent separation, the downstream outer separated flow can be calculated.

The incoming characteristics from the downstream outer separated flow cross the outgoing characteristic II' to bring outer region information into the intermediate region. Because the directions of the characteristics become increasingly uncertain as $U \to 0$ in the finite difference form, the Bradshaw et al. method is matched at $U = 0.3\ U_e (y \equiv y_J)$ with the separated flow wall model described below. The same modified Bradshaw et al. model as used in the outer region unseparated flow is used when $U > 0.3\ U_e$. However, V is calculated from the outer and intermediate regions inward to the match location y_J since V_e is known. V_J at the match location is then used to determine how much backflow is required in the separated wall region.

The separated flow wall model basically consists of a parabolic mean velocity profile with the option that it can degenerate to a straight line just after intermittent separation. Since V is calculated inward to the wall, the backflow can be calculated by the velocity at y_J, the no slip condition at the wall, and the continuity condition. The continuity condition satisfies the only strong requirement of any backflow since a negligibly small amount of momentum and kinetic energy were experimentally observed by Simpson et al. (1974) in this region.

For free-streamline separation, Collins and Simpson demonstrated that U_e and V_e could be calculated using the minimum freestream pressure gradient principle for the Simpson et al. flow. While the idea of minimizing the pressure gradient downstream of separation is not new, the new observation is that the minimum pressure gradient requirement must begin at the beginning of intermittent separation. Upstream the attached boundary layer controls the growth of the displacement thickness. Downstream the minimum pressure gradient condition determines the required displacement thickness distribution of the separated shear flow when conditions on unseparated boundaries of the potential flow are known.

In determining the location of the separated flow displacement thickness distribution for the Simpson et al. flow, Collins and Simpson used a rather crude cubic polynomial model

$$\delta^* = \delta^*_I\left(1 - \frac{\xi^2}{L^2}\right) + \delta^*_E\left(\frac{\xi}{L}\right)^2 + \left(\frac{d\delta^*}{dx}\right)_I\left(\xi - \frac{\xi^2}{L}\right) + DL^3\left(\frac{\xi^3}{L^3} - \frac{\xi^2}{L^2}\right) \qquad (14)$$

where $\xi = x - x_I$ and $L = x_E - x_I$. At x_I (intermittent separation point I), δ_I^* and $(d\delta^*/dx)_I$ are known from the upstream attached boundary layer flow while δ_E^* is determined from exit flow continuity requirements at x_E and D is selected so that the pressure gradient for $x_I < x < x_E$ is minimized.

Figures 5 show that the results for $D = -0.005\ ft^{-2}$ appear to minimize the pressure gradient slightly better than either $D = -0.004\ ft^{-2}$ or $-0.006\ ft^{-2}$. Figure 6 shows the streamlines for the potential flow and the separated flow displacement thickness. Figure 13 shows that the predicted V_e distribution upstream of 140 inches is in agreement with the experimental results within 10%. Downstream the disagreement is considerable although the shape of the curve is proper. Naturally the shape of this curve is dependent on the assumed form of the δ^* distribution, equation (14). The shape of the predicted V_e distribution is very sensitive to the value of D. For $D = -0.004\ ft^{-2}$, V_e is up to a 1/2 fps lower upstream of 150 inches and up to a 1/2 fps higher downstream as compared with the $-0.005\ ft^{-2}$ results. The $D = -0.006\ ft^{-2}$ results show about the opposite effect, V_e being up to 1/2 fps higher till 150 inches and 1/2 fps lower downstream.

Figures 4 and 14 show the separated shear flow prediction results using the experimental U_e and V_e data. Several of these predicted quantities are in very good agreement with the available experimental data of Simpson et al. (1974): (1) the mean velocity profiles, although the high prediction of $\delta_{0.99}$ at 157.1 inches makes the non-dimensional profile appear to be in slightly poorer agreement; (2) $\delta_{0.99}$; (3) the location of the point where the velocity is half the local freestream velocity U_e, $y_{1/2}$; (4) the shape factor H; (5) the location y_b where the mean backflow is divided from the forward flow; and (6) the magnitude level of the backflow. The shearing stress profiles are not in good agreement, although it may be possible that the experimental hot-film values shown nearest that wall may be uncertain due to the presence of a very small amount of the intermittent backflow (Simpson et al., 1974). Without some smoothing of the shear profile, oscillations in τ values occurred in the intermediate separated flow zone. Since the effect of the intermittent backflow in the shearing stress deduced from hot-film signals has not been investigated, this uncertainty cannot be resolved here. However, the locus of the maximum $\overline{u^2}$ fluctuation locations as determined by the reliable directionally-sensitive laser anemometer is in fair agreement with the predicted maximum shear locations in Figure 14a. Since $\overline{u^2}$ and $-\overline{uv}$ maxima are almost at the same location in a profile for a mixing layer, the predicted maximum shear locations may be good for $x < 12$ feet. Without some smoothing of the shear profile, oscillations in τ values occurred in the intermediate separated flow zone between y_J and the outgoing characteristic II'. Predictions without smoothing are in

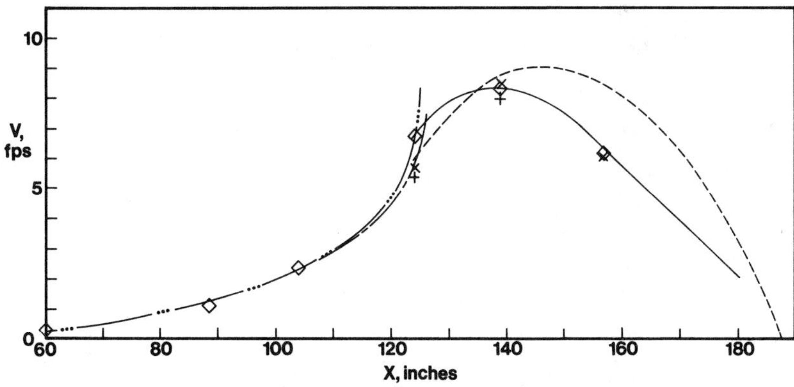

Figure 13. V at outer boundary layer edge for the Simpson et al. flow. Values from experimental results: ×, from U similarity; +, from integral continuity equation with linear $U_e\delta^*$ vs. x curve fit; ◊ from integral continuity equation with $\ln|U_e\delta|$ vs. x curve fit. Solid line is faired curve used in downstream separated flow predictions. Attached flow prediction: —···— Model 1; ——— Model 2. Dashed line is distribution predicted from the minimum pressure gradient separated freestream calculations with D = -0.005 ft^{-2}.

slightly better magnitude agreement with the experimental results although streamwise oscillations in y_M and the maximum shear are produced.

This flow was also calculated using the U_e and V_e distribution computed using the minimum freestream pressure gradient principle with the parameter D = -0.005 ft^{-2} and shown in Figures 5 and 13. Figures 14 show these calculation results using shear stress profile smoothing. Significantly, δ and $y_{1/2}$ are predicted to be about the same as for the other shown predictions up to 13.5 feet and with a smoother distribution for δ downstream. In particular this V_e distribution has the correct magnitude at intermittent separation and approximately the correct distribution shape. Thus if one were

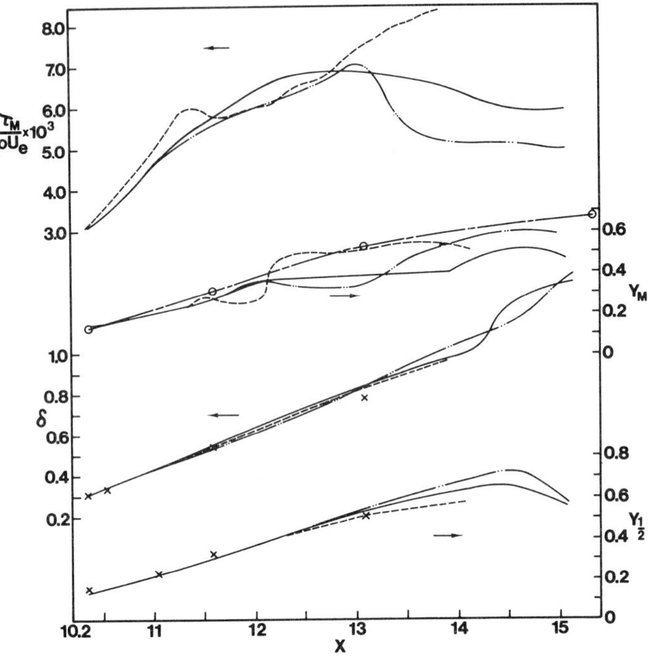

Figure 14a. Comparison of predicted and experimental results for the Simpson et al. flow downstream of intermittent separation: × - experimental data; ⊙ - experimental values for location of maximum streamwise fluctuation as determined by the directional laser anemometer, points connected by —-—. Predictions using faired solid curve in Figure 13 for V_e: solid lines - smoothed shearing stress profile predictions; dashed lines - unsmoothed shearing stress profile predictions. Predictions using dashed curve in Figure 13 for V_e - dotted broken line.

computing δ from this separated shear flow analysis with U_e and V_e as input information, the initial estimate of $\delta(x)$ used to pick U_e and V_e from the potential flow for the first trial calculation would not be critical. The locations of y_b and y_c are predicted to be closely the same as for the other shown predictions. H is also crudely predicted, but not quite as well as by other shown predictions when compared to the available experimental results.

The maximum shearing stress is predicted to be about the same up to 13 feet, but undergoes a dip downstream due to smoothing of the shear stress profile. For the same reason the location of the

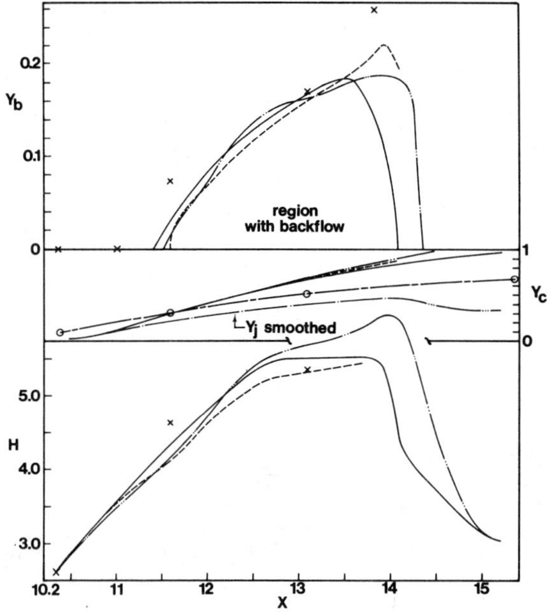

Figure 14b.

maximum shearing stress y_M increases after 13 feet to be in closer agreement with the locus of the maximum streamwise fluctuation. In fairness one cannot say that these latter predictions using the U_e and V_e distributions predicted with a minimum freestream pressure gradient are any worse than the predictions using experimental U_e and V_e.

6. SUMMARY OF OBSERVATIONS AND RECOMMENDATIONS FOR FURTHER WORK

At this point, we wish to summarize the important effects that must be accounted for in order to closely predict the two-dimensional flow over an airfoil or body with turbulent boundary layer separation. Since our limitations on understanding this type flow are largely due to the lack of sufficient reliable experimental data, several areas are recommended for future research.

A. Attached Boundary Layer

To accurately predict the important quantities of laminar-turbulent transition location, the displacement thickness, and the intermittent and fully-developed separation locations, one must have a flexible model that incorporates all of the important physical phenomena. It is well known (Schlichting, 1968) that the transition location is influenced by pressure gradient, surface roughness, and freestream turbulence, although exact prediction of this location for a real case is not easy as pointed out by Jacob (1975). We did not address this problem here, but considerable research has and is being done on this important subject.

Predictions for favorable pressure gradient turbulent boundary layers, such as found on the pressure side and on the front part of the suction side of an airfoil, are already very good (Kline et al., 1968), and are not treated here. For adverse pressure gradient boundary layers, the work of Collins and Simpson (1976) shows the importance of accounting for the normal stresses effect just upstream of intermittent separation. Predictions of the important parameters $C_f/2$, δ^*, and the beginning of intermittent separation were improved by accounting for this effect for the several experimental cases tested.

It is clear that experimental velocity profile data available downstream of intermittent separation are suspect if obtained by hot wires or pitot tubes. Thus, any law of the wall modified to account for streamwise pressure gradients using these data is also suspect. One other thing is clear: experimentalists should stop attempting to use directionally insensitive pitot-static tubes or hot-wire anemometers in the region with intermittent backflow that is downstream of intermittent separation. Directionally sensitive laser anemometers should be used to provide further data on this intermittent region. Extreme care should be taken to measure the pressure gradients as accurately as possible since these parameters influence predictions the most.

B. Separated Shear Flow

It is clear that the pressure gradient relief region (AB on Figure 2) must be predicted in order to predict the freestream velocity for free-streamline separation (BC on Figure 2). The Collins and Simpson model for the prediction of the intermittent separation location was in good agreement with available experimental data. Intermittent separation is <u>the</u> proper separation location for mean steady turbulent boundary layers since this is where the boundary layer is first detached on an intermittent basis. The pressure gradient relief between intermittent separation and fully-developed separation needs further experimental research to relate the behavior of the freestream and shear flows in this vicinity. Wooley and Kline (1973) in their diffuser prediction procedure also support these views, although in their procedure they use the Sandborn and Kline intermittent separation model shown in Figure 7, which also is in agreement with experimental data.

In fact the lack of a clear understanding of the interaction between the shear and freestream flows after intermittent separation prevents accurate modeling of flows that maintain long regions between intermittent and fully-developed separation. For example in the Spangenberg et al. flow, intermittent separation occurs rather quickly in the presence of a strong adverse pressure gradient, but fully-developed separation is prevented by steadily reducing the freestream pressure gradient downstream. As noted by Spangenberg et al., it is rather difficult to maintain this type of flow experimentally since if the pressure gradient is slightly too large fully-developed separation will occur, resulting in a fully-stalled flow downstream. The normal stress modified program with wall region normal stress damping of the effective pressure gradient (Model 4) predicted this flow for some distance after intermittent separation, but that same model produced poorer results for the Simpson et al. flow. It appears that a more fundamental model of the turbulence structure in this region is in order. It will most likely follow after more detailed laser anemometer measurements.

For fully-stalled flows, such as the Simpson et al. flow, which have relatively short distances between intermittent and fully-developed separation, one can effectively use the flow conditions at intermittent separation to predict the downstream flow. Given $d\delta^*/dx$ and δ^* at intermittent separation and an assumed form for $\delta^*(x)$ downstream, the downstream location of $\delta^*(x)$ can be computed subject to the minimum freestream pressure gradient condition, if all other boundary conditions on the elliptic freestream flow domain are known. Naturally the displacement effects on all surfaces with attached boundary layers must be accounted for. This procedure could be improved without a preselected assumed form for $\delta^*(x)$ if the pressure gradient relief after intermittent separation were properly modeled. The procedure used by Collins and Simpson in effect accounts for pressure gradient decay.

The internal consistency of allowing the freestream and upstream conditions to solely determine the downstream shear layer and backflow was demonstrated by Collins and Simpson. Having computed the freestream flow, U_e and V_e at the outer edge of the separated shear would be known, so the procedure can be used to predict the separated flow. While there are difficulties in predicting the shearing stress profile that need to be eliminated, the important parameters of δ and the mean velocity profile are well predicted. This means that an initial trial distribution of $\delta(x)$ could be used in an iterative procedure to evaluate U_e and V_e from the freestream potential flow. Then $\delta(x)$ would be predicted from the separated shear flow and new estimates of U_e and V_e would be obtained from the freestream flow and the procedure repeated until sufficient computational convergence is achieved.

These computational procedures appear useful for prediction of airfoil flows with free-streamline separation. The procedure of Bhateley and Bradley (1972) could be used for the potential flow in an iterative procedure that accounts for displacement effects. The procedures of Collins and Simpson are directly applicable in their current form for low curvature surfaces. Trailing edge separation with its associated strong interaction of the shear flows from both sides of the body was not treated here because of the lack of experimental data on the turbulence phenomena. Clearly, considerable experimental work is needed on trailing edge separation to determine its structure.

C. Wake Flow

The near wake region (CD on Figure 2, BC on Figure 3) is the critical part since it is characterized by strong interaction of both separated shear layers with the potential flow and controls the downstream distance to where the pressure is uniform. It is clear from the work of Jacob (1975) that the prediction of this region is important to the overall drag prediction. Unfortunately, we do not know of any detailed experimental flow structure data for the near wake shear flow of a lifting airfoil with turbulent separation. These data are needed in order to examine the turbulent transport processes, if any fundamental improvements are to be made over Jacob's (1975) procedure. For this reason, this problem was not treated here. The far wake does not play any important role in the lift or drag prediction since the pressure is uniform.

ACKNOWLEDGMENT

The author would like to thank the U.S. Army Research Office for support of his experimental and prediction research on separating turbulent boundary layers.

REFERENCES

1. AGARD (1972) Fluid Dynamics of Aircraft Stalling, AGARD-CP-102, Proceedings of Specialists Meeting, April 25-28, 1972, Lisbon, Portugal.

2. AGARD (1972) V/STOL Aerodynamics, AGARD-CP-143, Proceedings of Specialists Meeting, April 1974, Delft, The Netherlands.

3. Bhateley, I. C. and Bradley, R. G. (1972), "A Simplified Mathematical Model for the Analysis of Multi-element Airfoils near Stall," paper 12, AGARD-CP-102.

4. Bradshaw, P. (1973), "Effects on Streamline Curvature on Turbulent Flow," AGARD-AG-169.

5. Bradshaw, P., Ferris, D. H. and Atwell, N. P. (1967), "Calculation of Boundary-Layer Development Using the Turbulent Energy Equation," J. Fluid Mech., 28, Part 3, pp. 593-616; (1974) revised version, Imperial College Aero. Rept. 74-02.

6. Cebeci, T., Mosinskis, G. J., and Smith, A. M. O. (1972), "Calculation of Separation Points in Incompressible Turbulent Flows," J. Aircraft, 9, pp. 618-624.

7. Collins, M. A., and Simpson, R. L. (1976), "Flowfield Prediction for Separating Turbulent Boundary Layers," Report WT-4, Dept. Civil and Mechanical Engineering, Southern Methodist University, Feb., 1976, to appear in NTIS AD Series.

8. Crimi, P. (1975), Dynamic Stall, AGARD-AG-172.

9. Geller, W. (1976), "Calculation of the Turning Angle of Two-Dimensional Incompressible Cascade Flow," AIAA J., 14, pp. 297-298.

10. Head, M. R. (1960), "Entrainment in the Turbulent Boundary Layer," ARC R&M 3152.

11. Jacob, Klaus (1969), "Berechnung der abgelösten incompressiblen Strömung um Tragflügelprofile und Bestimmung des maximalen Auftriebs, Z. für Flugwiss, 17, pp. 221-230.

12. Jacob, Klaus (1975), "Weiterentwicklung eines Verfahrens zur Berechung der abgelösten Profilströmung mit besonderer Bërucksichtigung des Profilwiderstandes," DFVLR-AVA-Internal Report 251-75A16 June 1975.

13. Jacob, K. and Steinbach, D. (1974), "A Method for Prediction of Lift for Multi-Element Airfoil Systems with Separation," paper 12, AGARD-CP-143.

14. Kline, S. J., Morkovian, M. V., Sovran, G., and Cockrell, D. S. (1968), Computation of Turbulent Boundary Layers - 1968 AFOSR-IFP-Stanford Conference, Vol. I, Stanford Univ. Dept. Mechanical Engineering.

15. Lachmann, G. V. (1961), Boundary Layer and Flow Control, Vol. 1, Pergamon Press, Oxford.

16. Lister, M. (1960), "The Numerical Solution of Hyperbolic Partial Differential Equations by the Method of Characteristics," Mathematical Methods for Digital Computers, A. Ralston and H. S. Wilf, ed., Wiley.

17. McDevitt, J. B., Levy, L. L., and Deiwert, G. S. (1976), "Transonic Flow about a Thick Circular-Arc Airfoil," AIAA J., 14, pp. 606-613.

18. McDonald, H. (1969), "The Effect of Pressure Gradient on the Law of the Wall in Turbulent Flow," J. Fluid Mech., 35, pp. 311-336.

19. Newman, B. G. (1951), "Some Contributions to the Study of the Turbulent Boundary-Layer Near Separation," Report ACA-53, Dept. of Supply Aeronautical Research Consultative Committee, Univ. of Sydney, Australia.

20. Perry, A. E. and Schofield, W. H. (1973), "Mean Velocity and Shear Stress Distributions in Turbulent Boundary Layers," Phys. Fluids, 16, pp. 2068-2074.

21. Rotta, J. C. (1962), "Turbulent Boundary Layers in Incompressible Flow," Progress in Aeronautical Sciences, Vol. 2, Pergamon Press.

22. Samuel, A. E. and Joubert, P. N. (1974), "A Boundary Layer Developing in an Increasingly Adverse Pressure Gradient," J. Fluid Mech., 66, pp. 481-505.

23. Sandborn, V. A. and Kline, S. J. (1961), "Flow Models in Boundary-Layer Stall Inception," Journal of Basic Engineering, Trans. ASME, 83, pp. 317-327.

24. Sandborn, V. A. and Liu, C. Y. (1968), "On Turbulent Boundary-Layer Separation," J. Fluid Mech., 32, pp. 293-304.

25. Sandborn, V. A. and Slogar, R. J. (1955), "Study of the Momentum Distribution of Turbulent Boundary Layers in Adverse Pressure Gradients," NACA TN 3264.

26. Schlichting, H. (1968), Boundary Layer Theory, 6th Edition, McGraw-Hill.

27. Schlichting, H. and Truckenbrodt, E. (1969), Aerodynamik des Flugzeuges, Vol. 2, Springer Verlag, Berlin.

28. Schubauer, G. B. and Klebanoff, P. S. (1951), "Investigation of Separation of the Turbulent Boundary Layer," NACA Report 1030.

29. Schubauer, G. B. and Spangenberg, W. G., (1960), "Forced Mixing in Boundary Layers," J. Fluid Mech., 8, pp. 10-32.

30. Simpson, R. L. (1975a), "Characteristics of a Separating Incompressible Turbulent Boundary Layer," paper 14, AGARD-CP-168.

31. Simpson, R. L. (1975b), Discussion of "Three-Dimensional Separation of an Incompressible Turbulent Boundary Layer on an Infinite Swept Wing," by A. Elsenaar, B. van der Berg, and J. P. F. Lindhout, paper 34, AGARD-CP-168.

32. Simpson, R. L. (1976a), "Interpreting Laser and Hot-Film Anemometer Signals in a Separating Boundary Layer," AIAA J., 14, pp. 124-126.

33. Simpson, R. L. (1976b), "Experimental Techniques for Separated Flow," review paper presented at Lockheed-Georgia Viscous Flow Symposium, June 22-23, Marietta, Ga.

34. Simpson, R. L., Strickland, J. H., and Barr, P. W. (1973), "Features of a Separating Turbulent Boundary Layer as Revealed by Laser and Hot-Film Anemometry," Third Biennial Symposium in Turbulence in Liquids, pp. 151-170, University of Missouri-Rolla.

35. Simpson, R. L., Strickland, J. H., and Barr, P. W. (1974), "Laser and Hot-Film Anemometer Measurements in a Separating Turbulent Boundary Layer," Thermal and Fluid Sciences Center, Southern Methodist University, Report WT-3; NTIS, AD-A001115.

36. Simpson, R. L., Strickland, J. H., and Barr, P. W. (1976), "Features of a Separating Turbulent Boundary Layer in the Vicinity of Separation," accepted for publication, J. Fluid Mech.

37. Spangenberg, W. G., Rowland, W. R. and Mease, N. E. (1976), "Measurements in a Turbulent Boundary Layer in a Nearly Separating Condition," pp. 110-151, Fluid Mechanics of Internal Flow, Edited by G. Sovran, Elsevier Publishing Co., New York.

38. Stratford, B. S. (1959), "An Experimental Flow with Zero Skin Friction Throughout its Region of Pressure Rise," J. Fluid Mech., 5, pp. 17-35.

39. Strickland, J. H. and Simpson, R. L. (1973), "The Separating Turbulent Boundary Layer: An Experimental Study of an Airfoil Type Flow," Thermal and Fluid Sciences Center, Southern Methodist University, Report WT-2; also AD-771170/8GA.

40. Woolley, R. L. and Kline, S. J. (1973), "A Method for Calculation of a Fully Stalled Flow," Report MD-33, Thermosciences Div., Dept. Mech. Engr., Stanford University.

Turbulence Velocity Scales for Swirling Flows

RONALD M. C. SO
General Electric Company
Schenectady, New York (U.S.A.)

ABSTRACT

The energy redistribution terms in the Reynolds-stress equations for swirling flows are modeled by the simple return to isotropy terms proposed by Rotta. Simplifications of the equations are then obtained by assuming local equilibrium to exist, neglecting the diffusion and advection terms and by invoking the boundary-layer approximations. The resulting equations are algebraic in the shear stresses and can be solved in terms of a gradient Richardson number and the mean flow quantities. Results show that for moderate Richardson number, the eddy viscosity is not isotropic and the flow is characterized by two velocity scales. However, for small Richardson number, the eddy viscosity is isotropic and a Monin-Oboukhov formula for the velocity scale is again obtained. The free parameter in the formulae can be determined from non-swirling plane flow data. Consequently, the only empiricism that enters into the present analysis is that required in two-dimensional plane flows.

1. INTRODUCTION

The prediction of three-dimensional turbulent flows has been the subject of continued interest for many fluid mechanics researchers. Among the various different three-dimensional flows, the rotating turbulent flows with or without axisymmetry are of particular interest because of their inherent importance in many areas of engineering applications. Early attempts to predict rotation turbulent flows usually followed the integral approach, that is, velocity profiles were assumed for the flow and the integral equations of motion were solved. Such an approach was restricted to flows with simple boundary conditions, simple boundary geometries and constant fluid properties (Dorfman, 1963; Krieth, 1968). In order to develop a more general prediction method, later attempts have been concentrated on the solution of the differential equations of motion rather than the integral equations. As such, they involve modelling the Reynolds stresses which appear in the mean momentum equations. Among the many different closure models for the Reynolds stresses, the least sophisticated is the mixing-length or eddy viscosity model. This simple model for three-dimensional, rotating turbulent flows forms the subject of the present investigation, while the interested reader can refer to a recent book by Launder and Spalding (1972) for a discussion of other turbulence models.

In order to extend the mixing-length concept of Prandtl to three-dimensional rotating flows, some assumptions concerning the length scales in the flow have to be made. The general approach is to assume an isotropic length scale and that rotation exerts its influence on the length scale through a characteristic dimensionless parameter of the flow in question. Bradshaw (1968) drew an analogy between the effects of buoyancy and streamline curvature and argued heuristically that the characteristic dimensionless parameter is the gradient Richardson number and that the effects of rotation can be best described by the use of a rotating flow analogue of the Monin-Oboukhov (1954) formula for the mixing-length. Since then, the Monin-Oboukhov formula has been used by Cham and Head (1970), Koosinlin and Lockwood (1971) and Koosinlin et al. (1974) in their study of rotating turbulent flows. Although they found good agreement between calculations and measurements, they also discovered that the empirical constant in the Monin-Oboukhov formula has to be varied to accommodate changing experimental conditions. Other modifications to the mixing-length have been investigated by Koosinlin and Lockwood (1971), but here again, they found that the empirical constants in the mixing-length formulation have to be varied for different experimental conditions considered. They attributed this difficulty to the assumption of an isotropic length scale which is not consistent with the fact that swirling flows usually exhibit a significant anisotropy of the eddy viscosity (Lilly & Chigier, 1971; Lilly, 1973). Because of this, Koosinlin and Lockwood (1971) concluded that the mixing-length approach is inadequate and closure of the mean momentum equations should be effected by properly modelling the various terms in the Reynolds-stress equations and then solving the equations simultaneously with the mean momentum equations.

The present study investigates the assumption of an isotropic length scale for three-dimensional, rotating turbulent flows and hopes to establish the validity and extent of such an assumption.

In addition, it also addresses the question concerning the proper choice of dimensionless parameter for the Monin-Oboukhov formula and the value of the constant in the formula. The approach taken is similar to that used by So (1975) in his analysis of two-dimensional curved shear flows. Closure of the Reynolds-stress equations is again effected by suitably modelling the energy redistribution and dissipation terms, by assuming the advection and diffusion terms to be small and by limiting the consideration to thin shear layers. The resultant Reynolds-stress equations are algebraic in the shear stresses and can be solved in terms of the mean flow quantities.

Rotating turbulent flows can be divided into two classes; those whose axes of rotation are normal to and those whose axes lie in the same plane as the plane that contains the direction of the streamwise flow. Since flows belonging to the first category have been investigated by So (1975), therefore, the following analysis will concentrate on the effects of rotation on the turbulence field in flows belonging to the second class. Such flows are generally known as swirling flows.

2. THE REYNOLDS-STRESS EQUATIONS

For turbulent swirling flows, the Reynolds-stress equations in generalized tensor form are

$$\underbrace{\frac{\partial}{\partial t}\overline{u_i u_j} + (U^k\,\overline{u_i u_j})_{,k}}_{\text{Advection}} + \underbrace{\left[\overline{u^k u_i}\,U_{j,k} + \overline{u^k u_j}\,U_{i,k}\right]}_{\text{Production}}$$

$$= \underbrace{\left[\overline{pu_{j,i}} + \overline{pu_{i,j}}\right]}_{\text{Energy Redistribution}} - \underbrace{\left[\overline{\tau_i^k\,u_{j,k}} + \overline{\tau_j^k\,u_{i,k}}\right]}_{\text{Dissipation}}$$

$$- \underbrace{\left[\overline{u^k u_i u_j} + \delta_i^k\,\overline{pu_j} + \delta_j^k\,\overline{pu_i} - \overline{u_j\tau_i^k} - \overline{u_i\tau_j^k}\right]}_{\text{Diffusion}} \qquad (1)$$

where x_i is the ith component of the generalized coordinates, u_i and U_i are the ith components of the fluctuating and mean velocities respectively, $p = p'/\rho$ is the fluctuating pressure, ρ is the fluid density, $\tau_i^k = \nu g^{k\ell}(u_{\ell,i} + u_{i,\ell})$ is the viscous stress tensor, ν is the fluid kinematic viscosity and $g^{k\ell}$ is the metric tensor. The set of equations (1) has more unknowns than equations; however, (1) can be closed by writing it in terms of single-point double velocity correlations $\overline{u^k u_i}$. This requires modelling the terms on the right-hand side of (1).

If local equilibrium in the turbulence field is assumed, that is, production of turbulent energy balances viscous dissipation except in a very thin layer next to the surface and the advection and diffusion terms are assumed to be small, then it is only necessary to model the energy redistribution and dissipation terms (So, 1975). Recently, Peskin and So (1976) have examined the importance of including additional terms for rapid distortion and return to isotropy in the modelling of the energy redistribution terms in turbulent shear flows under the influence of external body forces. They found that for small external body forces, a Monin-Oboukhov (1954) formula for the mixing-length is again obtained; however, the effects of the additional rapid distortion and return to isotropy terms on the formula are negligible when compared with the use of the simple return to isotropy terms proposed by Rotta (1951). In view of this, the model terms proposed by Rotta (1951) for energy redistribution are adopted here and they can be written as

$$\overline{pu_{i,j}} + \overline{pu_{j,i}} = \frac{q}{3\ell_1}\left(\overline{u_i u_j} - \frac{1}{3} g_{ij} q^2\right) \tag{2}$$

where $q = (\overline{u^i u_i})^{\frac{1}{2}}$ and ℓ is a turbulent length scale which is as yet undefined. The dissipation terms can be written in terms of a dissipation length scale Λ if Kolmogorov's (1941) hypothesis of local small-scale isotropy is assumed. Therefore,

$$\overline{\tau_i^k u_{j,k}} + \overline{\tau_j^k u_{i,k}} = \frac{2}{3} g_{ij} \frac{q^3}{\Lambda} \tag{3}$$

When (2) and (3) are substituted into (1), the Reynolds-stress equation in generalized tensor form becomes

$$\frac{\partial}{\partial t}\overline{u_i u_j} + U^k(\overline{u_i u_j})_{,k} = -\left[\overline{u^k u_i}\, U_{j,k} + \overline{u^k u_j}\, U_{i,k}\right] - \frac{q}{3\ell_1}\left[\overline{u_i u_j} - \frac{1}{3} g_{ij} q^2\right] - \frac{2}{3} g_{ij}\frac{q^3}{\Lambda} + [\text{Diffusion Terms}] \tag{4}$$

If an orthogonal coordinate system that is symmetric with respect to the axis of rotation (which is parallel to the mean flow direction) is chosen (Figure 1), then $\{x^i\} = (x, r, \theta)$, $\{u_i\} = (u, v, w)$ and $\{U_i\} = (U, V, W)$. The rotational velocity, W, creates a "centrifugal" force which is balanced by the normal pressure gradient and gives rise to two significant shear stress components, $-\overline{uv}$ and $-\overline{vw}$, in the turbulence field. This makes the flow akin to three-dimensional shear flows, hence complicates the analysis. However, the set of equations (4) can be further simplified by assuming the flow to be statistically steady, the advection and diffusion terms to be small and by invoking the boundary-layer approximations and local equilibrium assumption. The resultant component equations written with respect to the coordinate system shown in Figure 1 are

$$-\frac{q}{3\ell_1}\left(\overline{u^2} - \frac{q^2}{3}\right) - \frac{2}{3}\frac{q^3}{\Lambda} - 2\overline{uv}\frac{\partial U}{\partial r} = 0 \tag{5}$$

$$-\frac{q}{3\ell_1}\left(\overline{v^2} - \frac{q^2}{3}\right) - \frac{2}{3}\frac{q^3}{\Lambda} + 4\overline{vw}\frac{W}{r} = 0 \tag{6}$$

$$-\frac{q}{3\ell_1}\left(\overline{w^2} - \frac{q^2}{3}\right) - \frac{2}{3}\frac{q^3}{\Lambda} - 4\overline{vw}\frac{W}{r} - 2\overline{vw}\left(\frac{\partial W}{\partial r} - \frac{W}{r}\right) = 0 \tag{7}$$

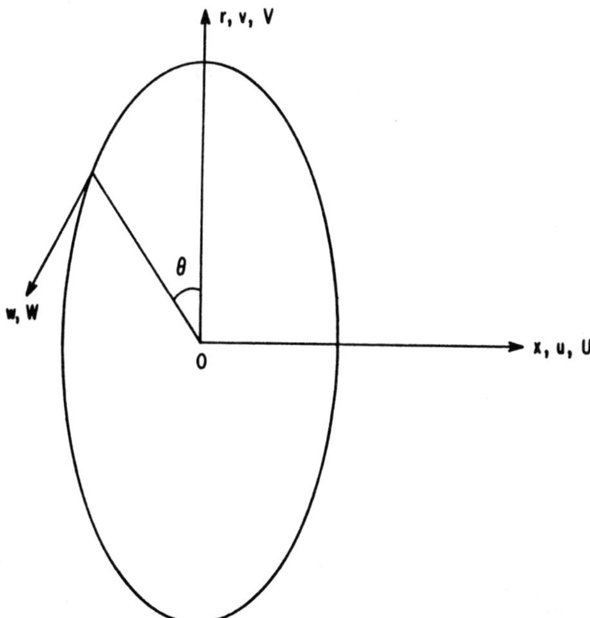

Figure 1. Definition sketch for coordinate system.

$$-\frac{q}{3\ell_1}\overline{uv} - \overline{v^2}\frac{\partial U}{\partial r} + 2\overline{uw}\frac{W}{r} = 0 \tag{8}$$

$$-\frac{q}{3\ell_1}\overline{vw} - \overline{v^2}\left(\frac{\partial W}{\partial r} - \frac{W}{r}\right) + 2(\overline{w^2} - \overline{v^2})\frac{W}{r} = 0 \tag{9}$$

$$-\frac{q}{3\ell_1}\overline{uw} - \overline{vw}\frac{\partial U}{\partial r} - \overline{uv}\left(\frac{\partial W}{\partial r} - \frac{W}{r}\right) - 2\overline{uv}\frac{W}{r} = 0 \tag{10}$$

3. THE TURBULENCE VELOCITY SCALES

The simplified Reynolds-stress equations (5)-(10) are algebraic in the shear stresses and can be solved in terms of the mean flow quantities. This set of equations was first analyzed by Bissonnette and Mellor (1974) but they concluded that for $W/r \neq 0$ and if $W/r \ll \partial W/\partial r$ cannot be assumed, then the algebra becomes extraordinarily complicated. Because of this, they expressed the resultant shear $\tau = [(-\overline{uv})^2 + (-\overline{vw})^2]^{\frac{1}{2}}$ in terms of $\varepsilon = (W/r)(\partial W/\partial r)^{-1}$ and $\phi = (3\ell_1/q)(\partial W/\partial r)$ which is not independent of the turbulence quantities. In the following, it will be shown that the algebraic complexity can be overcome and the shear stresses can indeed be solved in terms of the mean flow quantities alone.

To solve for $-\overline{uv}$ and $-\overline{vw}$, an equation between q, $-\overline{uv}$ and $-\overline{vw}$ is first obtained by summing (5)-(7) and the result is

$$q^3 = a\tau_x + b\tau_\theta \tag{11}$$

where

$$\tau_x = -\overline{uv} \tag{12a}$$

$$\tau_\theta = -\overline{vw} \tag{12b}$$

$$a = \Lambda\frac{\partial U}{\partial r} \tag{12c}$$

$$b = \Lambda\left(\frac{\partial W}{\partial r} - \frac{W}{r}\right) \tag{12d}$$

Two other expressions relating q, τ_x and τ_θ are then obtained from (6)-(10) and the result is

$$dq^3 = \tau_\theta(q^2 + c) \tag{13}$$

$$gq^3 = \tau_x\left(q^2 + \frac{c}{4}\right) + \tau_\theta f \tag{14}$$

where

$$c = 72\ \ell_1^2\ \frac{W}{r}\left(\frac{\partial W}{\partial r} + \frac{W}{r}\right) \tag{15a}$$

$$d = \ell_1\left(1 - \frac{6\ell_1}{\Lambda}\right)\left(\frac{\partial W}{\partial r} - \frac{W}{r}\right) \tag{15b}$$

$$f = 54\ \ell_1^2\ \frac{W}{r}\frac{\partial U}{\partial r} \tag{15c}$$

$$g = \ell_1\left(1 - \frac{6\ell_1}{\Lambda}\right)\frac{\partial U}{\partial r} \tag{15d}$$

By eliminating τ_x and τ_θ from (11), (13) and (14), a quadratic equation for q^2 is obtained. This can then be solved to give

$$q^2 = \frac{1}{2}\left\{ag + bd - \frac{5c}{4}\right\} \pm \frac{1}{2}\left[\left(ag + bd - \frac{3c}{4}\right)^2 + a(3cg - 4df)\right]^{\frac{1}{2}} \tag{16}$$

Equation (16) can be written in a more familiar form if the following definitions are adopted, namely

$$Ri = \frac{2\ \frac{W}{r}\left(\frac{\partial W}{\partial r} + \frac{W}{r}\right)}{\left(\frac{\partial U}{\partial r}\right)^2 + \left(\frac{\partial W}{\partial r} - \frac{W}{r}\right)^2} \tag{17}$$

$$\sigma = \frac{\frac{\partial U}{\partial r}}{\frac{\partial W}{\partial r} + \frac{W}{r}} \tag{18}$$

$$\beta = \frac{72\ \ell_1/\Lambda}{1 - \frac{6\ell_1}{\Lambda}} \tag{19}$$

The meaning of Ri and the significance of β will be discussed later. For the present, (16) can be rewritten with the help of (12), (15) and (17)-(19). Omitting all the algebra, the result is

$$q^2 = \ell_1\Lambda\left(1 - \frac{6\ell_1}{\Lambda}\right) Z(Ri) \left[\left(\frac{\partial U}{\partial r}\right)^2 + \left(\frac{\partial W}{\partial r}\right) - \frac{W}{r}\right)^2\right] \tag{20}$$

where

$$2Z = 1 - \frac{5}{8}\beta Ri \pm \left[\left(1 - \frac{3}{8}\beta Ri\right)^2 + \frac{3}{2}\beta Ri^2\sigma^2\right]^{\frac{1}{2}} \tag{21}$$

The correct sign to take in (21) can be determined by examining (20) for the case of a two-dimensional non-rotating plane shear flow. For this flow, the expression for q^2 deduced by So (1975) is

$$q^2 = \ell_1\Lambda\left(1 - \frac{6\ell_1}{\Lambda}\right)\left(\frac{\partial U}{\partial r}\right)^2 \tag{22}$$

Therefore, it can be seen that only the positive sign in (21) will give a q^2 that approaches (22) when $W = 0$.

Finally, expressions for τ_x and τ_θ can be obtained by solving (11), (13) and (14) simultaneously with the help of (12), (15) and (20). The result is

$$\tau_x = \ell_0^2 \frac{Z^{3/2}}{Z + 1/2\ \beta Ri}\left[\left(\frac{\partial U}{\partial r}\right)^2 + \left(\frac{\partial W}{\partial r} - \frac{W}{r}\right)^2\right]^{\frac{1}{2}}$$

$$\times\left[Z + \frac{1}{2}\beta Ri + \gamma^2\left(Z + \frac{1}{2}\beta Ri - 1\right)\right]\left(\frac{\partial U}{\partial r}\right) \tag{23}$$

$$\tau_\theta = \ell_0^2 \frac{Z^{3/2}}{Z + 1/2\ \beta Ri}\left[\left(\frac{\partial U}{\partial r}\right)^2 + \left(\frac{\partial W}{\partial r} - \frac{W}{r}\right)^2\right]^{\frac{1}{2}}\left(\frac{\partial W}{\partial r} - \frac{W}{r}\right) \tag{24}$$

where

$$\ell_0^2 = \left[\ell_1\Lambda^{1/3}\left(1 - \frac{6\ell_1}{\Lambda}\right)\right]^{3/2} \tag{25a}$$

$$\gamma = \frac{\dfrac{\partial W}{\partial r} - \dfrac{W}{r}}{\dfrac{\partial U}{\partial r}} \tag{25b}$$

The total shear stress τ can be written as

$$\tau = \ell_0^2 \frac{Z^{3/2}}{Z + 1/2\ \beta Ri}\left[\left(\frac{\partial U}{\partial r}\right)^2 + \left(\frac{\partial W}{\partial r} - \frac{W}{r}\right)^2\right]^{\frac{1}{2}}\left[\left(Z + \frac{1}{2}\beta Ri\right)^2\right.$$

$$\left. + \gamma^2\left(Z + \frac{1}{2}\beta Ri - 1\right)^2\right]^{\frac{1}{2}}\left[\left(\frac{\partial U}{\partial r}\right)^2 + \left(\frac{\partial W}{\partial r} - \frac{W}{r}\right)^2\right]^{\frac{1}{2}} \tag{26}$$

In the absence of swirl, $\tau_\theta = 0$ and $\tau = \tau_x$ becomes

$$(\tau)_0 = \ell_0^2 \left(\frac{\partial U}{\partial r}\right)_0^2 \tag{27}$$

where the subscripts zero denote properties of a corresponding two-dimensional non-rotating plane shear flow. So (1975) interpreted ℓ_0 as the familiar Prandtl mixing-length and defined the eddy viscosity ν_0 in terms of the turbulence velocity scale v_0' and ℓ_0 so that

$$\nu_0 = v_0' \ell_0 = \ell_0^2 \left(\frac{\partial U}{\partial r}\right)_0 \tag{28}$$

Since (23) and (24) are derived from the Reynolds-stress equations, it is reasonable to assume that rotation only affects the turbulence velocity scales and not the length scales of the flow. Therefore, analogous to (28) the eddy viscosities ν_{e_x} and ν_{e_θ} can be defined as $\nu_{e_x} = v_x' \ell_0$ and $\nu_{e_\theta} = v_\theta' \ell_0$ where v_x' and v_θ' are the turbulence velocity scales.

Equations (23) and (24) suggest that τ_x and τ_θ be related to the mean velocity gradients as

$$\tau_x = \nu_{e_x} \left(\frac{\partial U}{\partial r}\right) \tag{29}$$

$$\tau_\theta = \nu_{e_\theta} \left(\frac{\partial W}{\partial r} - \frac{W}{r}\right) \tag{30}$$

The relation (30) for τ_θ is consistent with the definition proposed by Keyes (1960), Ragsdale (1960) and Reynolds (1961) in their investigation of vortex flows. Also, with τ_x and τ_θ defined by (29) and (30), the denominator in (17) equals the square of the resultant mean shear. Since the numerator of (17) represents the square of the Brunt-Väisälä frequency evaluated from the rotational motion only, (17) can be interpreted as the definition for the gradient Richardson number, Ri for turbulent swirling flows [given in equation (39)]. Thus, the meaning of (17) is clear and the dependence of the shear stresses on the gradient Richardson number is established.

The expressions for v_x' and v_θ' can now be obtained from (23) and (24) with the help of (28)-(30). They are

$$\frac{v'_x}{v'_0} = \frac{Z^{3/2}}{Z + 1/2\ \beta Ri} \frac{\left[\left(\frac{\partial U}{\partial r}\right)^2 + \left(\frac{\partial W}{\partial r} - \frac{W}{r}\right)^2\right]^{1/2}}{\left(\frac{\partial U}{\partial r}\right)_0}$$

$$\times \left[Z + \frac{1}{2}\beta Ri + \gamma^2\left\{Z + \frac{1}{2}\beta Ri - 1\right\}\right] \qquad (31)$$

$$\frac{v'_\theta}{v'_0} = \frac{Z^{3/2}}{Z + 1/2\ \beta Ri} \frac{\left[\left(\frac{\partial U}{\partial r}\right)^2 + \left(\frac{\partial W}{\partial r} - \frac{W}{r}\right)^2\right]^{1/2}}{\left(\frac{\partial U}{\partial r}\right)_0} \qquad (32)$$

Once again, the turbulence velocity scales for swirling flows are shown to be functions of Ri alone. However, contrary to the beliefs of other investigators (Cham & Head, 1970; Bradshaw, 1973; Koosinlin et al., 1974), the turbulence velocity scales are not isotropic even when the assumption of local equilibrium is made. The eddy viscosity ratio $\nu_{e_x}/\nu_{e_\theta} = v'_x/v'_\theta$ can be written as

$$\frac{\nu_{e_x}}{\nu_{e_\theta}} = Z + \frac{1}{2}\beta Ri + \gamma^2\left\{Z + \frac{1}{2}\beta Ri - 1\right\} \qquad (33)$$

which is also a function of Ri alone. By expanding Z as an ascending series in Ri, it can be easily shown that to the first order of Ri

$$\frac{\nu_{e_x}}{\nu_{e_\theta}} = 1 + \frac{3}{4}\beta\left(\frac{\frac{W}{r}}{\frac{\partial W}{\partial r} + \frac{W}{r}}\right) Ri \qquad (34)$$

which is similar to the expression obtained by Koosinlin and Lockwood (1974). On the basis of this expression for the eddy viscosity ratio, they have concluded that if the external body force is due to swirl, the modification is more properly applied to the ratio ν_{e_x}/ν_{e_θ}. However, the present analysis shows that the modification should be applied to the eddy viscosities themselves and (34) follows naturally.

Evidence in support of (33) can be obtained from the rotating cylinder measurements of Bissonnette and Mellor (1974). From their measured mean velocities and shear stresses data obtained for two different cylinder rotational speeds, they evaluated ν_{e_θ}/ν_{e_x} for $y/\delta_\theta = .1$ to $\simeq .9$, where δ_θ is the circumferential displacement thickness. They found that although the scatter in the data is large, no trend of ν_{e_θ}/ν_{e_x} with respect to x or r can be detected.

Consequently, the mean ratio of ν_{e_θ}/ν_{e_x} is calculated to be .7 with a standard deviation of ~20%. Using the measured mean velocity profiles obtained by Bissonnette and Mellor (1974), the ratio ν_{e_θ}/ν_{e_x} is calculated from (33). In the range, y/δ_θ = .1 to .6, where local equilibrium assumption is approximately valid, it is found that the mean ratio of ν_{e_θ}/ν_{e_x} is .8 with a standard deviation of ~2%. Again, no definite trend of ν_{e_θ}/ν_{e_x} with respect to x or r can be evidenced. The slight difference between the mean value reported by Bissonnette and Mellor (1924) and that calculated from (33) is probably due to the error involved in their measurement of the shear stresses. In view of this, the general agreement can be considered very good.

4. THE VALUE OF β

Although two length scales, ℓ_1 and Λ, are introduced in the closure model for (1), they only appear as a ratio in the final expressions for v_x' and v_θ'. Therefore, (31) and (32) will be completely defined if the value of β is known. Empirical information is extracted from non-rotating, constant-pressure plane flows, and on this basis alone, a value for β will be determined. As a result, no empiricism saved that required in non-rotating, two-dimensional plane flows is introduced into the present formulation for the shear stresses.

The Reynolds-stress equations for a non-rotating, two-dimensional plane flow can be obtained from (5) to (10) by setting W = 0 and $-\overline{uw} = -\overline{vw} = 0$. After rearrangement, they can be written as

$$\begin{vmatrix} \frac{\overline{u^2}}{q^2} - \frac{1}{3} & -\frac{\overline{uv}}{q^2} & 0 \\ -\frac{\overline{uv}}{q^2} & \frac{\overline{v^2}}{q^2} - \frac{1}{3} & 0 \\ 0 & 0 & \frac{\overline{w^2}}{q^2} - \frac{1}{3} \end{vmatrix} = \begin{vmatrix} 4\alpha & (\alpha - 6\alpha^2)^{\frac{1}{2}} & 0 \\ (\alpha - 6\alpha^2)^{\frac{1}{2}} & -2\alpha & 0 \\ 0 & 0 & -2\alpha \end{vmatrix} \quad (35)$$

where $\alpha = \ell_1/\Lambda$. The flat plate boundary layer measurements of Klebanoff (1955) and So and Mellor (1973) give

$$\begin{vmatrix} .24 & .16 & 0 \\ .16 & -.18 & 0 \\ 0 & 0 & -.06 \end{vmatrix} \quad \text{and} \quad \begin{vmatrix} .20 & .17 & 0 \\ .17 & -.13 & 0 \\ 0 & 0 & -.06 \end{vmatrix}$$

respectively for the stress tensor, while the homogeneous turbulence data of Rose (1966) and Champagne et al. (1970) give

$$\begin{vmatrix} .16 & .17 & 0 \\ .17 & -.09 & 0 \\ 0 & 0 & -.05 \end{vmatrix}$$

If α is chosen to be .055, then (35) will take on a value of

$$\begin{vmatrix} .22 & .19 & 0 \\ .19 & -.11 & 0 \\ 0 & 0 & -.11 \end{vmatrix}$$

which closely represents the average of the above data. Although other values of α are also possible, the resultant stress tensors calculated from them differ considerably from the reported data. Also α = .055 compares favorably with the values of .053 and .052 obtained by So (1975) and Mellor (1973) respectively. Since these values have been demonstrated by So (1975) and Mellor (1973) to give good agreement with a wide range of curved shear flow and atmospheric boundary layer data, it will be adopted for swirling flows too.

From (19), the value of β is then determined to be 6.

5. THE MONIN-OBOUKHOV FORMULA

The mixing-length correction formula for a stratified atmosphere was first deduced by Monin and Oboukhov (1954) assuming a constant stress layer to exist in the atmospheric surface layer. This is the equivalent of the local equilibrium assumption, therefore, it is little wonder that a curved flow analogue of the Monin-Oboukhov formula can be derived if the production of turbulent energy is assumed to balance dissipation and the external body force effects are small (So, 1975). In the case of swirling flows, Bradshaw (1973) postulated that $\overline{vw}/\overline{uv} = (\partial W/\partial r)/(\partial U/\partial r)$ when local equilibrium is assumed and he again argued that the Monin-Oboukhov formula can be applied to account for the effects of swirl if $W/r \ll (\partial W/\partial r)$. The purpose of this Section is to show that Bradshaw's (1973) hypothesis is derivable from the present analysis and that for small swirl, the assumption of an isotropic eddy viscosity is valid.

If $W/r \ll (\partial W/\partial r)$, Ri will be much smaller than one. Therefore, to first order of Ri, Z can be written as

$$Z = 1 - \frac{1}{2}\beta Ri$$

When this result is substituted into (31) and (32) they become

$$\frac{v'_x}{v'_0} = \frac{v'_\theta}{v'_0} = \left(1 - \frac{3}{4}\beta Ri\right) \frac{\left[\left(\frac{\partial U}{\partial r}\right)^2 + \left(\frac{\partial W}{\partial r} - \frac{W}{r}\right)^2\right]^{\frac{1}{2}}}{\left(\frac{\partial U}{\partial r}\right)_0} \tag{36}$$

which gives the isotropic behavior of the turbulence velocity scales. Defining the mixing-length, ℓ, for swirling flows as

$$v'_x = v'_\theta = \ell\left[\left(\frac{\partial U}{\partial r}\right)^2 + \left(\frac{\partial W}{\partial r} - \frac{W}{r}\right)^2\right]^{\frac{1}{2}} \tag{37}$$

it follows from (28) and (36) that

$$\frac{\ell}{\ell_0} = 1 - \frac{3}{4}\beta Ri \tag{38}$$

which is the swirling flow analogue of the Monin-Oboukhov formula. With $\beta = 6$, the coefficient for Ri takes on a value of 4.5 and this is in excellent agreement with a value of 5 used by Koosinlin et al. (1974) in their prediction of swirling flows. This result, together with that obtained earlier by So (1974) for curved shear flows, demonstrates that if the external body force effects are small, then the Monin-Oboukhov formula for the mixing-length is valid for different kinds of flows.

6. DISCUSSION

6.1 Description of the Effects of Swirl

In stratified flows, two dimensionless parameters can be defined to describe the effects of buoyancy. If the oscillation of a fluid element displaced in the direction of the buoyancy force is considered, a dimensionless parameter known as the gradient Richardson number is obtained. It can be defined as follows.

$$Ri = \left[\frac{\text{Brunt Väisälä frequency}}{\text{Typical turbulence frequency}}\right]^2$$

$$= \frac{\text{Typical body force}}{\text{Typical inertia force}}$$

$$= \frac{-(g/\bar{\rho})(\partial\bar{\rho}/\partial z)}{(\partial U/\partial z)^2} \tag{39}$$

where $\bar{\rho}$ is the mean density, g the gravitational force and z is the vertical coordinate. Ri is positive for stable stratification, zero for neutral flow and negative for unstable stratification. On the other hand, if consideration is directed to the production terms in the turbulent energy equation, a flux Richardson number Ri_f, can be defined. This is given by

$$Ri_f = -\frac{\text{v-component energy production due to buoyancy}}{\text{u-component energy production}}$$

$$= -\frac{g\,\overline{\rho' v}/\bar{\rho}}{(\overline{uv})(\partial U/\partial z)} \tag{40}$$

where ρ' is the fluctuating density. In general, Ri and Ri_f are different. However, under the assumption of local equilibrium and small buoyancy effect, the same mixing-length can be used to describe $\overline{\rho' v}$ and $\overline{uv}$, then Ri = Ri_f.

The similarity in effects between buoyancy and streamline-curvature in two-dimensional flows was recognized by Prandtl (1929) and Bradshaw (1969) and the curved flow analogue of Ri and Ri_f was defined by Bradshaw (1969). Although uncertainty exists in the choice of velocity gradient for the denominator in the definition of Ri, there is no ambiguity in the meaning of Ri. This is because in two-dimensional curved shear flows, the plane of surface curvature, the plane of rotation and the plane of the mean shear all coincide and a typical inertial force can be defined without ambiguity.

This is not the case in three-dimensional and axisymmetric swirling flows. For such flows, the plane of rotation, the plane of curvature of the mean streamline and the plane of the mean shear do not all coincide. Therefore, it is not clear whether Ri should be defined with respect to the velocity and curvature in the plane of curvature of the mean streamline or whether it should be defined with respect to the resultant mean shear. This difficulty led

Bradshaw (1973) to abandon Ri in favor of Ri_f. However, in his analysis of an axisymmetric swirling flow, Bradshaw (1973) has to assume that $\overline{vw}/\overline{uv} = (\partial W/\partial r)/(\partial U/\partial r)$ and $W/r \ll \partial W/\partial r$, then he obtained

$$Ri = Ri_f = \frac{2\,\frac{W}{r}\,\frac{\partial W}{\partial r}}{\left(\frac{\partial U}{\partial r}\right)^2 + \left(\frac{\partial W}{\partial r}\right)^2} \tag{41}$$

and concluded that the effects of swirl can again be equally described by either Ri or Ri_f. For small swirl, (41) is indeed consistent with the present result because it is the first approximation to (17) and the eddy viscosity is indeed isotropic.

For larger Ri, the eddy viscosity is no longer isotropic and the present analysis shows that the gradient Richardson number as defined in (17) is the proper dimensionless parameter to use to describe the effects of swirl, and Ri_f is an ill-defined parameter. From (5)-(7) and (40), the swirling flow analogue of Ri_f can be defined. This is given by

$$Ri_f = \frac{2\overline{vw}\,\frac{W}{r}}{\overline{uv}\,\frac{\partial U}{\partial r} + \overline{vw}\left(\frac{\partial W}{\partial r} + \frac{W}{r}\right)} \tag{42}$$

With the results given in (23) and (24), (42) can be written as

$$Ri_f = \frac{2\,\frac{W}{r}\left(\frac{\partial W}{\partial r} - \frac{W}{r}\right)}{\left(\frac{\partial U}{\partial r}\right)^2 \left[Z + \frac{1}{2}\beta Ri + \gamma^2\left(Z + \frac{1}{2}\beta Ri - 1\right)\right] + \left(\frac{\partial W}{\partial r}\right)^2 - \frac{W^2}{r^2}} \tag{43}$$

which gives the relation between Ri and Ri_f in swirling flows rather than a definition of Ri_f in terms of the mean flow quantities. This is because there is no simple way to partition the effects of swirl between $\overline{uv}$ and $\overline{vw}$. Also, there is no a priori reason to suggest that the shear stress vector should be skewed in the direction of the velocity gradient. Instead, the swirling flow data of Bissonnette and Mellor (1974) and the three-dimensional data of Johnston (1970) showed that the shear stress vector is actually skewed in the direction of the resultant velocity. Therefore, it can be concluded that Ri as defined in (17) is the proper dimensionless parameter to use to describe the effects of swirl.

6.2 Stability of Swirling Flows

Just as in the case of stratified flows and curved shear flows, some upper limit on the critical Richardson number beyond which turbulence cannot exist can be obtained by considering the simplified turbulent energy equation. For swirling flows, this equation is given by

$$(\overline{u^2} + \overline{w^2} \text{ production})\ (1 - Ri_f) = \text{Dissipation} \tag{44}$$

where Ri_f is defined by (43). Since the critical flux Richardson number Ri_{fcr} can be interpreted as corresponding to the total production of turbulent energy in the flow becoming very small while viscous dissipation continues with the result that the whole level of turbulence decreases, it follows from (44) that Ri_{fcr} has an absolute upper limit of one. This condition implies the vanishing of the resultant mean velocity gradient and this leads to a corresponding limit of infinity for the critical gradient Richardson number, Ri_{cr}. Since production of turbulent energy also ceases when τ_x and τ_θ vanish, therefore, it can be seen from (23) and (24) that the shear stresses also vanish when $Z \rightarrow 0$. This gives

$$Ri_{cr} \left[1 - \frac{\left(\frac{\partial U}{\partial r}\right)^2_{cr}}{\left(\frac{\partial W}{\partial r} + \frac{W}{r}\right)^2_{cr}} \right] = \frac{1}{3} \tag{45}$$

However, lack of data prevents (45) from being verified and the determination of a critical flux or gradient Richardson number for swirling flows.

6.3 Solution of the Mean Momentum Equations

The results given in (23) and (24) for τ_x and τ_θ respectively can be used to close the equations of mean motion for an axisymmetric turbulent swirling thin shear layer. If the fluid properties are assumed uniform, then the mean momentum equations become

$$\frac{\partial U}{\partial x} + \frac{1}{r}\frac{\partial}{\partial r}(rV) = 0 \tag{46}$$

$$U\frac{\partial U}{\partial x} + V\frac{\partial U}{\partial r} = -\frac{\partial P}{\partial x} + \frac{1}{r}\frac{\partial}{\partial r}(r\tau_x) \tag{47}$$

$$\frac{W^2}{r} = \frac{\partial P}{\partial r} \tag{48}$$

$$U \frac{\partial W}{\partial x} + \frac{V}{r} \frac{\partial}{\partial r} (rW) = \frac{1}{r^2} \frac{\partial}{\partial r} (r^2 \tau_\theta) \tag{49}$$

where $P = P'/\rho$ is the mean static pressure. The only unknown in (23) and (24) is ℓ_0 which is the mixing-length of a corresponding non-rotating plane flow. Although ℓ_0 is defined by (25a) and α is determined to be .055, the individual variation of ℓ_1 and Λ across the shear layer is not known. Hence, ℓ_0 cannot be evaluated.

There are many ways to evaluate ℓ_0, but none employs as little empiricism as the equilibrium layer technique of Mellor and Gibson (1966). In their analysis, an eddy viscosity is hypothesized and the mean momentum equations are solved for non-rotating plane flows where the pressure gradient parameter $\delta^*(dP/dx)/\tau_W$, with δ^* denoting the displacement thickness and τ_W the wall shear stress, is held constant. They found that their calculations agree very well with other measured equilibrium profiles (Clauser, 1954, 1956; Stratford, 1959). Later, Mellor (1967) applied the same eddy viscosity hypothesis to calculate flows with arbitrary streamwise pressure gradient and again he found good agreement with measured data. Therefore, once the pressure gradient of the flow in question is known, the method of Mellor and Gibson (1966) can be used to calculate the variation of ℓ_0 across the layer. Unlike other empirical formulae for ℓ_0, this method employs the least empiricism saved that of constant pressure flat plate flows and it accounts for the pressure gradient effects automatically. Hence, for the present calculation, this technique is used to obtain ℓ_0 for any given dP/dx.

With $\ell_0(r)$ known, (46)-(49) together with (23) and (24) can be solved by a number of finite difference techniques. For convenience, the integration technique employed by Mellor (1967) is adopted here. The calculated velocity profiles and momentum thicknesses are compared with the rotating cylinder data of Furuya et al. (1966). Good agreement is obtained (Figures 2-4). Further calculations were made assuming the eddy viscosity to be isotropic and hence the use of (36) to evaluate τ_x and τ_θ. The results are essentially the same as those shown in Figures 2-4, and are no different from those presented by Koosinlin et al. (1974). Therefore, the present results provide direct evidence in support of the use of a Monin-Oboukhov formula for swirling flows when Ri is small.

Although (36) is derived assuming the axis of rotation to be parallel to the direction of the streamwise flow, the investigation of Koosinlin et al. (1974) shows that it is just as applicable to

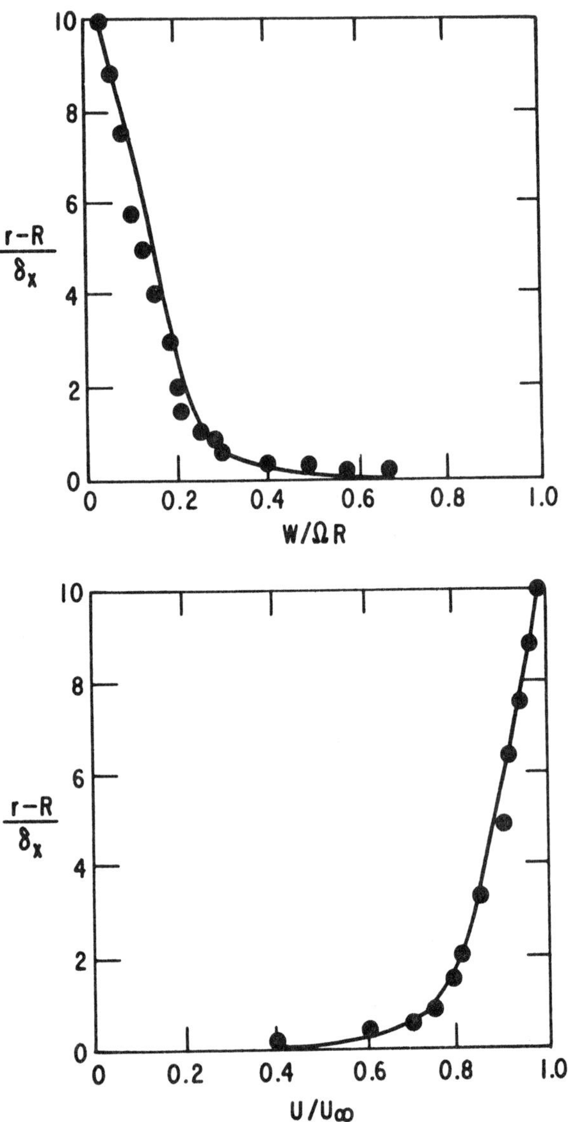

Figure 2. A comparison of calculated and measured velocity profiles in flow past rotating cylinder. ●, data of Furuya et al. (1966), $Re_\infty = U_\infty R/\nu = 6 \times 10^4$ and $S_\Omega = \Omega R/U_\infty = 2$; ———, present calculation. δ_x is the boundary layer thickness in the axial direction.

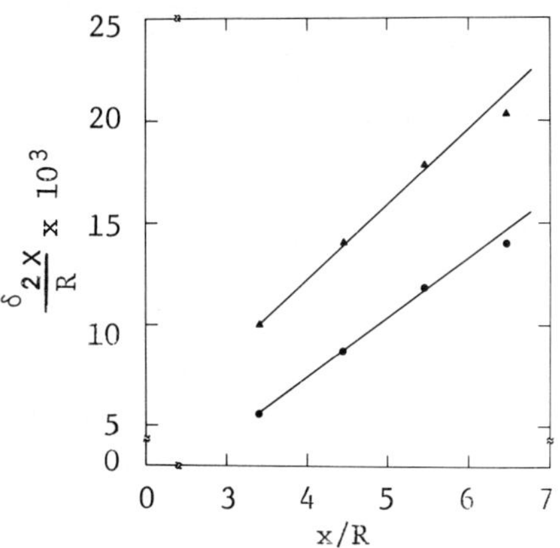

Figure 3. A comparison of calculated and measured axial momentum thickness in flow past rotating cylinder. ●, $S_\Omega = 1$ and ▲, $S_\Omega = 2$ ($Re_\infty = 6 \times 10^4$) are data of Furuya et al. (1966); ——, present calculation

$$\delta_{2X} = \int_R^{R+\delta_X} (U/U_\infty)\ (1 - U/U_\infty)dr$$ is the axial momentum thickness. Definitions for Re_∞ and S_Ω are the same as those given in Figure 2.

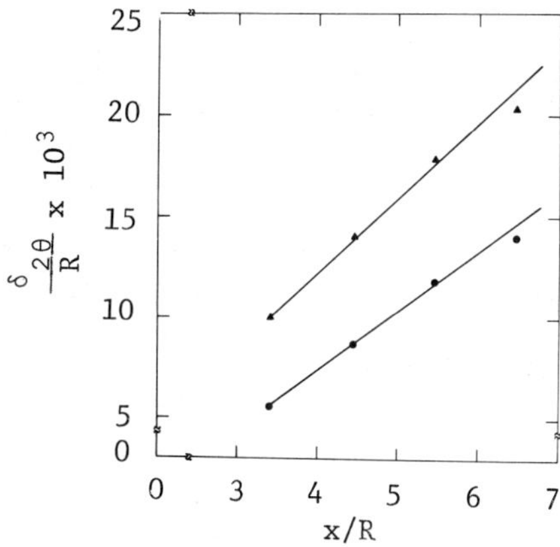

Figure 4. A comparison of calculated and measured circumferential momentum thickness in flow past rotating cylinder. Symbols as in Figure 3.

$$\delta_{2\theta} = \int_r^{R+\delta_X} (U/U_\infty)\ (W/\Omega R)dr$$ is the circumferential momentum thickness.

flows whose axis of rotation makes an angle with the direction of the mean flow. The present calculations for rotating cylinders do not differ from those of Koosinlin et al. (1974), therefore, it is expected that the calculations for spinning cones would also be the same. This is not surprising because the coefficient of Ri in the Monin-Oboukhov formula used by Koosinlin et al. and the present calculation are essentially the same. In view of this, a comparison with spinning cone data was not made.

Finally, it should be pointed out that the mixing-length formula used by Cham and Head (1970) is different from (38) and that used by Koosinlin et al. (1974). The difference is in the definition of Ri. Koosinlin et al. (1974) defined Ri just as it is given in (17), however, Cham and Head (1970) defined Ri with respect to U_m, ΩR and $\partial U/\partial r$, where U_m is the free stream velocity and ΩR is the circumferential velocity of the cylinder. On this basis, Cham and Head (1970) concluded that β is not a constant and its value changes from .25 to 60 across the thin shear layer. From the calculations of Koosinlin et al. (1974) and the present analysis, it is obvious why the β used by Cham and Head (1970) varies across the thin shear layer. This is because the variation of the circumferential velocity W across the shear layer was not properly taken into account in the definition of Ri.

7. CONCLUSIONS

Formulae for the variation of the turbulent velocity scales with Richardson number in a swirling flow have been derived from the Reynolds-stress equations by assuming production of turbulent energy balances viscous dissipation. The resultant formulae show that the eddy viscosity is not isotropic. Also, the formulae only contain one free parameter, but it can be determined from data for non-rotating plane flows. Consequently, there is no free constant in the formulae. For small Richardson number, the eddy viscosity is isotropic and a Monin-Oboukhov formula for the velocity scale is again obtained. Therefore, this provides direct evidence supporting the application of the Monin-Oboukhov formula to swirling flows as suggested by Bradshaw (1973).

The result also indicates that the proper dimensionless parameter to use for the description of swirl effects on the turbulence field is the gradient Richardson number as defined in (17).

A major part of this work was carried out while the author was with the Geophysical Fluid Dynamics Program, Rutgers University, New Brunswick, New Jersey. Research support by the Atmospheric Sciences Section, National Science Foundation, NSF Grant GA-33421 is gratefully acknowledged. The author also wishes to thank Professor R. L. Peskin, Director, Geophysical Fluid Dynamics Program, for his encouragement and support. Finally, the expert typing of Ms. Beverly Holcomb of General Electric Corporate Research and Development is acknowledged.

REFERENCES

1. Bissonnette, L. & Mellor, G. L. 1974 J. Fluid Mech., 63, 369-413.

2. Bradshaw, P. 1969 J. Fluid Mech., 36, 177-191.

3. Bradshaw, P. 1973 AGARDograph, no. 169.

4. Cham, T. S. & Head, M. R. 1970 J. Fluid Mech., 42, 1-15.

5. Champagne, F. H., Harris, V. G. & Corrsin, S. 1970 J. Fluid Mech., 41, 31-141.

6. Clauser, F. 1954 J. Aero. Aci., 21, 91-108.

7. Clauser, F. 1956 Adv. Appl. Mech., 4, 1-51.

8. Dorfuman, L. A. 1963 Hydrodynamic Resistance and Heat Loss of Rotating Solids, Oliver and Boyd, London, England.

9. Furuya, Y., Nakamura, I. & Kawacki, H. 1966 Bulletin JSME, 9, 702-710.

10. Johnston, J. P. 1970 J. Fluid Mech., 42, 823-844.

11. Keyes, J. J. 1960 Proc. Heat Transfer & Fluid Mech. Inst., Palo Alto, Calif. Stanford University Press, 31-46.

12. Klebanoff, P. S. 1955 N.A.C.A. Rept. no. 1247.

13. Kolmogorov, A. N. 1941 C. R. Akad. Nauk. S.S.S.R., 30, 301-305. (Trans. Turbulence, Classic Papers on Statistical Theory (ed. S. K. Friedlander & L. Topper). Interscience, 1961.)

14. Koosinlin, M. L. & Lockwood, F. C. 1971 Rept. BL/TN/B/42, Mech. Eng. Dept., Imperial College, London, England.

15. Koosinlin, M. L. & Lockwood, F. C. 1974 A.I.A.A.J., 12, 547-554.

16. Koosinlin, M. L., Launder, B. E. & Sharma, B. I. 1974 J. Heat Transfer, Trans. ASME, 96, 204-209.

17. Kreith, F. 1968 Adv. Heat Transfer, 5, 129-251.

18. Launder, B. E. & Spalding, D. B. 1972 Mathematical Methods of Turbulence, Academic Press, London/New York.

19. Lilly, D. G. & Chigier, N. A. 1971 International J. Heat and Mass Transfer, 14, 573-585.

20. Lilly, D. G. 1973 A.I.A.A.J., 11, 955-960.

21. Mellor, G. L. & Gibson, D. M. 1966 J. Fluid Mech., 24, 225-253.

22. Mellor, G. L. 1967 A.I.A.A.J., 5, 1570-1579.

23. Mellor, G. L. 1973 J. Atmos. Sci., 30, 1061-1069.

24. Monin, A. S. & Oboukhov, A. M. 1954 Trudy. Geofiz. Inst. AN S.S.S.R., 24, 163-187.

25. Peskin, R. L. & So, R. M. C. 1976 Submitted to J. Fluid Mech.

26. Prandtl, L. 1929 Sonderdruck aus Vortrage aus dem Gebiete der Aerodynamik and Verwandter Gebiete. Aachen. (Trans. N.A.C.A. Tech. Memo no. 625).

27. Ragsdale, R. G. 1960 N.A.S.A. TN D-288.

28. Reynolds, A. J. 1961 ZAMP, 12, 149-158.

29. Rose, W. G. 1966 J. Fluid Mech., 25, 97-120.

30. Rotta, J. C. 1951 Z. Phys. 120, 547-572; 131, 51-77.

31. So, R. M. C. & Mellor, G. L. 1973 J. Fluid Mech., 60, 43-62.

32. So, R. M. C. 1975 J. Fluid Mech., 70, 37-57.

33. Stratford, B. S. 1959 J. Fluid Mech., 5, 17-35.

DISCUSSION

CORRSIN: (Johns Hopkins University)

Since you take into account the nonisotropy of the eddy viscosity have you considered also including it even with the Richardson number? It doesn't matter that in general in shear flows it would have to be a fourth rank tensor. You could probably get a lot of the components from existing data.

SO:

Yes, I have been working on that and I don't have the answer yet. It turns out to be a lot more complicated than I expected.

Effects of Freestream Turbulence and Initial Boundary Layers on the Development of Turbulent Mixing Layers

OTTO LEUCHTER
Office National d'Etudes et de Recherches Aerospatiales
Chatillon, France

ABSTRACT

Confluence configurations of practical interest in turbomachinery and external aerodynamics are discussed in this paper, the main interest being focused on the effects of initial boundary layers and freestream turbulence on the development of the turbulent mixing layer. The prediction of those effects is provided by a numerical method using a two-parameter turbulence model. This is based on the eddy viscosity concept and the transport of the kinetic energy and Rotta's integral length scale. The empirical constants of the model are determined from the behavior of the asymptotic free shear layer. Satisfactory agreement between numerical predictions and experimental results are achieved for the mean flow and for some typical turbulence parameters, especially in the configurations of plane wakes and coaxial jets.

1. INTRODUCTION

The behavior of turbulent mixing layers initiated by the confluence of two parallel streams is discussed in this paper in the general configuration of unequal velocities on each side and with initial boundary layers and/or freestream turbulence present in both flows. Typical features of this basic flow configuration (Figure 1) are the presence of a region of rapidly varying flow properties immediately downstream of the confluence and the possibility of a strong interaction effect resulting in an enhancement of the turbulent activity in the shear layer and in a consecutive increase of the overall spreading rate.

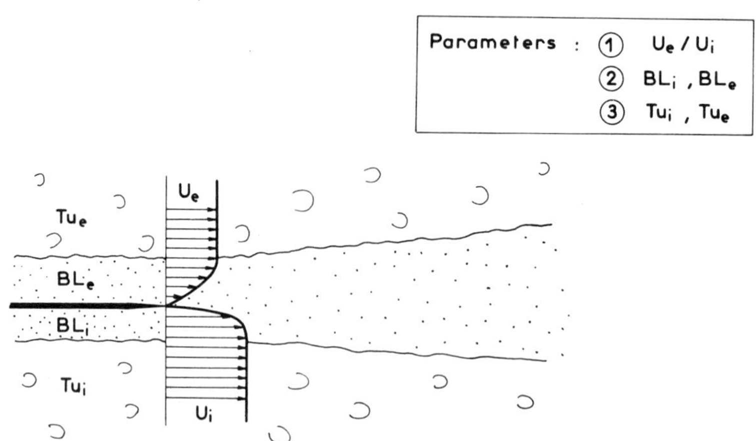

Figure 1. The Basic Flow Configuration: Confluence of Two Dissimilar Flows.

The configuration of Figure 1 may be considered as representative for various flow situations encountered frequently in external and internal aerodynamics, especially in turbomachinery. There it applies to wakes downstream of turbine or compressor blades, to internal mixing layers and to the outer jet mixing layer between the exhaust jet and the ambient air, where the effects of initial boundary layers and freestream turbulence on the aeroacoustic characteristics of the mixing layer are of primary importance.

Little information is available on these effects and, considering their importance for a wide range of practical applications, it became necessary to develop at ONERA a fundamental research program comprising detailed experimental investigations of subsonic two-stream mixing layers under various initial conditions as well as the development of numerical prediction methods. Theses include appropriate turbulence modelling able to take into account initial boundary layer and freestream turbulence effects.

The purpose of the present paper is to discuss some typical results obtained recently for this type of flow, especially for the configurations of plane wakes and coaxial jets. Emphasis is placed on the presentation of a workable calculation method capable to predict correctly the dominant features of these flows. The turbulence model retained for this purpose is based on the eddy viscosity concept and includes the transport of two characteristic turbulence parameters, namely a turbulent velocity scale and a turbulent length scale. This model was found to yield a reasonable compromise between the complexity of the present flow situations to be described and the simplicity of the prediction method required for practical engineering applications. Simpler models have also been used in the context of the present research, but they are restricted to the particular case of simple free shear layers with one of the flows at rest, for which the velocity profiles are monotonic throughout the flow. The mixing length concept, which have been extended to supersonic speeds and the concept of eddy viscosity transport have been successfully applied to this case (Leuchter, 1973).

The first part of this paper is devoted to the description of the predictive method and to the determination of the empirical constants involved in the turbulence model. Some experimental data are presented in the second part and compared there with the numerical results.

2. NUMERICAL DESCRIPTION OF THE FLOW FIELD

2.1 Basic Equations

The mean flow field is governed by the continuity and the streamwise momentum equations, which are used here within the approximations of two-dimensional incompressible boundary layer flows:

$$\frac{\partial \bar{u}}{\partial x} + \frac{1}{y^n}\frac{\partial}{\partial y}(y^n\,\bar{v}) = 0 \tag{1}$$

$$\bar{u}\frac{\partial \bar{u}}{\partial x} + \bar{v}\frac{\partial \bar{u}}{\partial y} = -\frac{1}{e}\frac{\partial \bar{p}}{\partial x} - \frac{1}{y^n}\frac{\partial}{\partial y}(y^n\,\overline{u'v'}) \tag{2}$$

with $n = 0$ for plane flows
and $n = 1$ for axisymmetrical flows.

High Reynolds numbers are assumed allowing the viscous term in the momentum equation to be neglected.

The eddy viscosity concept is used throughout the present work, relating the Reynolds stress to the transverse gradient of the mean velocity

$$\overline{u'v'} = -\nu_T\frac{\partial \bar{u}}{\partial y} \tag{3}$$

so that the eddy viscosity ν_T remains the only turbulent quantity to be specified in order to achieve closure for the mean field. The mixing length model

$$\nu_T = \ell^2 \left|\frac{\partial \bar{u}}{\partial y}\right| \tag{4}$$

relating ν_T to only mean flow quantities had been successfully used for the description of free shear layers, even in the presence of initial boundary layers (Leuchter, 1973), provided that the mean velocity profiles are monotonic. Only the mixing length ℓ needs then to be specified as a function of space. If the velocity profiles are not monotonic throughout the flow, as is the case in the present configurations, and if in addition interactions between turbulent flows of different structures occur, a more versatile viscosity model has to be adopted. Therefore the Prandtl-Kolmogorov model

$$\nu_T = \sqrt{E}\ L \tag{5}$$

relating the eddy viscosity to only turbulent parameters has been retained for this study. These parameters are the turbulent kinetic energy E and a typical turbulent length scale L representative for the large scale motions of the turbulence. Closure is achieved then by means of empirical balance equations for both quantities. In practice composite quantities containing both, the energy E and the length scale L are considered as the dependent variables for the second equation instead of the variable L itself, so that L results from the simultaneous solutions of both balance equations. Typical examples are the dissipation rate of the kinetic energy of turbulence (Davidov, 1961; Harlow & Nakayama, 1968) which stands for $E^{3/2}/L$, the Kolmogorov frequency $\sqrt{E}/L$ (Kolmogorov, 1942) or its square E^3/L^2 (Spalding, 1969; Saffman, 1970) and the EL product of Rotta (1951) which may be defined from the trace of the two point correlation tensor

$$EL(\vec{x}) \sim \iiint \sum_{i=1}^{3} R_{ii}(\vec{x},\vec{r})\ \frac{d\vec{r}}{r^2} \tag{6}$$

with $R_{ii}(\vec{x},\vec{r}) = \overline{u_i'(\vec{x})\ u_i'(\vec{x}+\vec{r})}$

or alternatively in the wave number space from the spectral density F(k) (Rotta, 1951)

$$EL(\vec{x}) \sim \int_0^\infty \frac{F(\vec{x},k)}{k}\ dk \tag{7}$$

with $E = \int_0^\infty F(k)dk$.

More recently Rotta (1971, 1975) had proposed an alternative version of Equ. (6), more appropriate for the treatment of two-dimensional shear flows, in which EL is defined from the lateral (y-direction) two-point correlations

$$EL \sim \int_{-\infty}^{\infty} \sum_{i=1}^{3} R_{ii}(\vec{x},y)dy \tag{6'}$$

but the physical meaning is basically the same as that expressed in Equ.'s (6) and (7). Rotta's concept of the transport of the EL-product which has been successfully applied to wall-boundary layers (Ng & Spalding, 1972), to free shear layers (Rodi & Spalding, 1970) and to the description of initial boundary layer effects on mixing layers and jets (Leuchter, 1973, 1975) has also been retained in the present investigation.

The semi-empirical balance equations for the two dependent variables E and EL are:

$$\bar{u}\frac{\partial E}{\partial x} + \bar{v}\frac{\partial E}{\partial y} = \frac{C_1}{y^n}\frac{\partial}{\partial y}\left(\nu_T\, y^n \frac{\partial E}{\partial y}\right) - \overline{u'v'}\frac{\partial \bar{u}}{\partial y} - \varepsilon \tag{8}$$

$$\bar{u}\frac{\partial EL}{\partial x} + \bar{v}\frac{\partial EL}{\partial y} = \frac{C_2}{y^n}\frac{\partial}{\partial y}\left(\nu_T\, y^n \frac{\partial EL}{\partial y}\right) - C_p L\,\overline{u'v'}\frac{\partial \bar{u}}{\partial y} - C_D L\varepsilon \tag{9}$$

Both equations, which are similar in structure, state that advection is balanced by the sum of (diffusion + production - dissipation). The dissipation rate ε can be expressed in terms of the energy and the length scale alone, as is the case for the eddy viscosity, Equ. (4):

$$\varepsilon = C\,\frac{E^{3/2}}{L} \tag{10}$$

where C is an empirical constant for high turbulent Reynolds numbers (Rotta, 1951). The diffusion terms (first terms on the right-hand sides) represent only rough approximations for the actual diffusion of E and EL arising from the triple velocity and the pressure-velocity correlations. The number of the empirical constants contained in the model is thus reduced to five.

2.2 Determination of the Constants

The set of empirical constants to be specified comprises the dissipation constant c of Equ. (10), the two diffusion constants C_1 and C_2, the production constant C_P and the dissipation constant C_D appearing in the transport equation for EL. The latter constant results immediately from the empirical decay data of homogeneous turbulence if the model is applied to this particular case. Then the production and diffusion terms cancel out and C_D is directly related to the decay rates of the turbulent energy and the length scale:

$$C_D = 1 + \frac{d \ln L}{d \ln E} \tag{11}$$

which is of the order of 0.6 as can be deduced from experimental decay data.

The four remaining constants of the model can be determined entirely from the behavior of the asymptotic state of the plane free mixing layer (Leuchter, 1973). In this particular case the assumption of Prandtl (1945) applies, namely that the turbulence length scale L is proportional to the mixing length l:

$$L = \beta \ell \tag{12}$$

where β is an empirical constant which may depend on the configuration of the flow. Relation (12) can be shown to be equivalent to the assumption that the module of the shear stress is proportional to the turbulence energy:

$$|\overline{u'v'}| = \beta^2 E \tag{13}$$

which is the experimentally well supported Bradshaw (1967) hypothesis. If Equ. (13) is introduced into the balance equations (8) and (9), the two source terms (production and dissipation) collapse and the auxiliary constants A and A' appear:

$$\underbrace{\bar{u}\,\frac{\partial E}{\partial x} + \bar{v}\,\frac{\partial E}{\partial y}} = \underbrace{C_1\,\frac{1}{y^n}\,\frac{\partial}{\partial y}\left(\nu_T\, y^n\,\frac{\partial E}{\partial y}\right)} + \underbrace{A\,\nu_T\left(\frac{\partial \bar{u}}{\partial y}\right)^2} \tag{14}$$

$$\underbrace{\bar{u}\,\frac{\partial EL}{\partial x} + \bar{v}\,\frac{\partial EL}{\partial y}}_{\text{I}} = \underbrace{C_2\,\frac{1}{y^n}\,\frac{\partial}{\partial y}\left(\nu_T\, y^n\,\frac{\partial EL}{\partial y}\right)}_{\text{II}} + \underbrace{A' L \nu_T\left(\frac{\partial \bar{u}}{\partial y}\right)^2}_{\text{III}} \tag{15}$$

with $A = 1 - \frac{c}{\beta^4}$ (16)

with $A' = C_P - C_D \frac{c}{\beta^4}$ (17)

A and A' are determined from Equ's. (14) and (15) by integrating both equations over the shear layer and by taking into account the similarity properties of the mean velocity field expressed by

$$\frac{\bar{u}(x,y)}{U_0} = \phi(\eta) \quad (18)$$

with

$$\eta = \frac{y - y_{0.5}}{b} \quad (19)$$

$y_{0.5}$ is the half-velocity ordinate and b is the thickness of the mixing layer. Both quantities are linear functions of x. The integration of Equ. (14) yields then:

$$A = \frac{b'}{\beta^2} \frac{\int \bar{u}\,\overline{u'v'}\,d\eta}{\int \frac{d\bar{u}}{d\eta}\,\overline{u'v'}\,d\eta} \quad (20)$$

where b' is the growth rate of the mixing layer and where the integrals are taken over the width of the shear layer. The shear stress distribution is determined prior to the integration from Equ's. (1) and (2) by taking into account for the similarity properties of the mean velocity profiles. If the relative variation of the turbulent length scale in the transverse direction is supposed to be small for the asymptotic state of the mixing layer, then the integration of Equ. (9) yields:

$$A' = 2A \quad (21)$$

The unknown constants c and C_P can thus be evaluated from Equ's. (16), (17), (20) and (21), with the appropriate empirical inputs for β and b' (or for the mixing parameter, $\sigma \approx 1.79/b'$).* Values for σ and β^2 of respectively 10.5 and 0.38 as suggested by various experimental results on plane or axisymmetrical jet mixing layers result in values for c and C_P of approximately 0.1 and 1.0

---

* The analytical profile of Tollmien (1926) has been retained here as the reference profile.

respectively, as can be seen from Table I, where different similarity profiles have been considered for the evaluation of A from Equ. (20). They comprise two experimental profiles, namely that of Wygnanski and Fiedler (1970) and that of Patel (1973) together with the analytical profile of Tollmien (1926). The dissipation constant C_D has been fixed for these evaluations at 0.58.

The second step in the determination of the constants concerns the two diffusion constants D_1 and D_2. These can be evaluated either locally or globally from Equ's. (14) and (15) together with Equ. (18) and the previously determined constants A and A'. The local determination is made on the dividing streamline since the turbulent shear stress reaches there its maximum and the contributions of the diffusion terms in Equ's. (14) and (15) are there most significant. C_1 and C_2 result then directly from the ratios

$$\frac{I - III}{II}$$

relative to Equ's. (14) and (15) respectively and evaluated at the position of the dividing streamline. The comparison of the values of C_1 and C_2 thus determined (summarized in Table I) shows that they are almost independent on the form of the profiles and that the diffusion constant of the energy-length product equation is slightly higher than that of the energy equation. The same trend was observed from the evaluation of C_1 and C_2 made globally over the whole mixing layer yielding slightly higher levels for C_1 and C_2 that those determined locally of approximately 1.2 and 1.4 respectively (Leuchter, 1973).

The values of the constants finally retained for the calculations are also given in Table I.

Table I. Empirical Constants of the Model.

	experimental		analytical	retained value
	WYGNANSKI - FIEDLER	PATEL	TOLLMIEN	
$\sigma \beta^2 A$	1.17	1.11	1.18	
A	0.29	0.28	0.30	
c	0.10	0.10	0.10	0.10
C_D	1.00	0.98	1.00	0.98
C_1	1.06	1.10	1.06	1.10
C_2	1.12	1.20	1.12	1.20

2.3 Application to Simple Shear Layers

In Figure 2 it is shown how the model predicts the growth rate of equilibrium two-stream mixing layers as a function of the velocity ratio. The growth rate is represented here by the (non-normalized) mixing parameter σ. The calculated evolution of $1/\sigma$ shows reasonable agreement with the experimental data collected in the literature (Brown & Roshko, 1971; Liepmann & Laufer, 1947; Miles & Shih, 1968; Patel, 1973; Spencer & Jones, 1971; Wygnanski & Fiedler, 1970; Yule, 1971). Some results obtained at ONERA on coaxial jet mixing layers are also included. It may be noted that the relation proposed by Sabin (1963):

$$\frac{\sigma}{\sigma_0} = \frac{U_i + U_e}{U_i - U_e} \tag{22}$$

fits almost exactly the calculated curve.

The transverse variation of the length scale L in the one-stream ($U_e = 0$) mixing layer as resulting from the numerical solution of Equ's. (1), (2), (8) and (9) is shown in Figure 3. There it is compared to the length scale deduced from Equ. (5) by using the experimental profiles of the mean velocity (from which the eddy viscosity was deduced) and of the turbulent energy measured by Wygnanski and Fiedler (1970). The experimental length scales deduced from the streamwise two-point correlation functions measured by the same authors are also reported on this figure. For this purpose a scaling

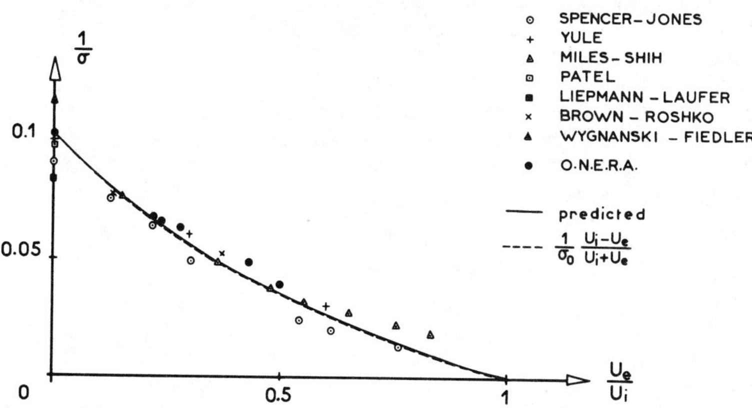

Figure 2. Evolution of the Mixing Parameter σ for the Two-Stream Mixing Layer.

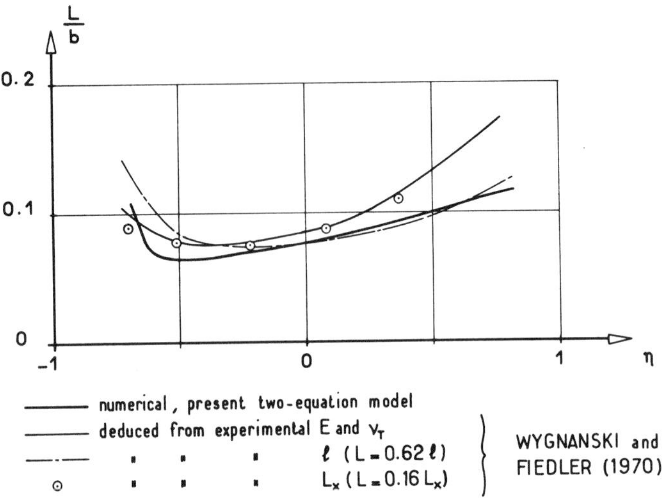

Figure 3. The Length Scale in the Asymptotic Free Shear Layer.

factor of 0.16 was applied to the measured integral scales L_X, which results from the comparison between L_X and $L = \nu_T/\sqrt{E}$ in the central part of the mixing layer. An alternative evaluation of the length scale can also be obtained from the mixing length profile by using Equ. (12). The mixing length is deduced from the mean velocity profiles by applying Equ's. (1) to (4). The corresponding L-profile is also shown on Figure 3, where the value of β was the same as that used above for the determination of the constants. The dominant features of these comparisons illustrated on Figure 3 are the satisfactory degree of consistency between the calculated length scale level and that of the various empirically determined scales over the major part of the mixing layer and the common trend of only moderate transverse variations in its central part.

2.4 Extension to the Reynolds Stresses

Experimental surveys of the fluctuating velocity field are usually performed with the aid of hot wire or laser doppler anemometry and the most commonly measured turbulent quantity is then the r.m.s. value of the fluctuating streamwise component of the velocity (or the corresponding normal stress $\overline{u'^2}$). On account of the highly anisotropic nature of the shear flow turbulence we are dealing with, this quantity cannot be related directly to the turbulent energy E implied in the turbulence model. In order to achieve more significant comparisons between experimental and numerical results concerning the turbulent field, the present two-equation model has therefore been extended to include the non-vanishing components of the Reynolds stress tensor. This extension was primarily intended for getting complementary informations about the anisotropy of the stress tensor rather than for achieving an improved closure for the shear stress.

The modeled terms of the corresponding balance equations, summarized in Table II, are similar in structure to those of Equ's. (8) and (9), except for two additional terms arising from the interaction between the fluctuating pressure and the fluctuating strain rate: the first term is due to the interaction of only fluctuating velocities, whereas term II simulates the interaction between mean velocity gradients and the fluctuating velocity field. Following the early proposal of Rotta (1951) the first contribution is taken to be proportional to the anisotropy level of the Reynolds stresses, whereas the second part is supposed to be proportional to the anisotropie of the corresponding production tensor, following current practice of modelling (Launder, 1975; Launder et al., 1975). For this simplified version of the simulation of the pressure-strain correlation only two new constants appear, namely ω and γ. Typical values proposed in the literature range from 1.5 to 2.8 for ω (Launder et al., 1975; Rodi, 1972, Hanjalic & Launder, 1972), with subsequent variations of γ from approximately 0.6 to 0.3. It can be observed that an approximate relation between γ and ω may be derived again from the asymptotic behavior of the free shear layer yielding:

$$\gamma = 0.86 - 0.2\,\omega \qquad (23)$$

Table II. Modeled Transport of the Reynolds Stress Tensor.

Convection =	Diffusion	+ Production	+ Dissipation	+ Pressure-strain I	+ Pressure-strain II
$\frac{D\overline{u'^2}}{Dt}$	$C_4 \frac{\partial}{\partial y}\left(\nu_T \frac{\partial \overline{u'^2}}{\partial y}\right)$	$-2\,\overline{u'v'}\,\frac{\partial \bar{u}}{\partial y}$	$-\frac{2}{3}\varepsilon$	$-\omega \frac{\varepsilon}{E}\left(\overline{u'^2} - \frac{2}{3}E\right)$	$\frac{4}{3}\gamma\,\overline{u'v'}\,\frac{\partial \bar{u}}{\partial y}$
$\frac{D\overline{v'^2}}{Dt}$	$C_5 \frac{\partial}{\partial y}\left(\nu_T \frac{\partial \overline{v'^2}}{\partial y}\right)$	0	$-\frac{2}{3}\varepsilon$	$-\omega \frac{\varepsilon}{E}\left(\overline{v'^2} - \frac{2}{3}E\right)$	$-\frac{2}{3}\gamma\,\overline{u'v'}\,\frac{\partial \bar{u}}{\partial y}$
$\frac{D\overline{w'^2}}{Dt}$	$C_6 \frac{\partial}{\partial y}\left(\nu_T \frac{\partial \overline{w'^2}}{\partial y}\right)$	0	$-\frac{2}{3}\varepsilon$	$-\omega \frac{\varepsilon}{E}\left(\overline{w'^2} - \frac{2}{3}E\right)$	$-\frac{2}{3}\gamma\,\overline{u'v'}\,\frac{\partial \bar{u}}{\partial y}$
$\frac{D\overline{u'v'}}{Dt}$	$C_7 \frac{\partial}{\partial y}\left(\nu_T \frac{\partial \overline{u'v'}}{\partial y}\right)$	$-\overline{v'^2}\,\frac{\partial \bar{u}}{\partial y}$	0	$-\omega \frac{\varepsilon}{E}\,\overline{u'v'}$	$\gamma\,\overline{v'^2}\,\frac{\partial \bar{u}}{\partial y}$

so that only one of the two constants, say ω, must be specified. Computer experiments run with this model have shown however that in the case of simple shear layers the calculated stress levels are only weakly affected by the choice of ω, if ω varies between the before mentioned limits and if Equ. (23) is applied to adjust the values of γ. More elaborate expressions have been proposed by Launder et al. (1975) for the simulation of the second part of the pressure-strain correlation; in the context of the present work however further sophistication of the stress equations has not appeared to be necessary.

The turbulent diffusion of the Reynolds stresses is approximated here again, as in Equ's. (8) and (9), by a gradient type form. The corresponding constants are determined from a more elaborate expression for the triple correlation tensor (Hanjalic & Launder, 1972):

$$\overline{u_i'u_j'u_k'} \sim \frac{E}{\varepsilon}\left[\overline{u_i'u_e'}\,\frac{\partial \overline{u_j'u_k'}}{\partial x_e} + \overline{u_j'u_e'}\,\frac{\partial \overline{u_k'u_i'}}{\partial x_e} + \overline{u_k'u_e'}\,\frac{\partial \overline{u_i'u_j'}}{\partial x_e}\right]$$

by comparison with the diffusion constant of the energy equation and by taking into account the anisotropy properties of the Reynolds stress tensor in the asymptotic free mixing layer. The corresponding diffusion constants C_3 through C_6 are then:

$C_3 = 0.98$

$C_4 = 1.86$

$C_5 = 0.62$

$C_6 = 1.86$

It can be observed from Table II that with the presently adopted closure assumptions the equations for the normal stresses are completely uncoupled allowing each of them to be integrated independently. Thus in most of the following examples only the equation for the normal stress in the longitudinal direction ($\overline{u'^2}$) has been integrated, since in most of the experimental investigations only this turbulent field quantity has been measured.

Figure 4 illustrates the degree of agreement which may be achieved with this simple version of the Reynolds stress model in the case of an equilibrium two-stream mixing layer of velocity ratio 0.3. The experimental data are from Spencer and Jones (1971). As can be seen from this figure, the level of the nonzero components of the Reynolds stress tensor as well as the shapes of their profiles are reasonably well predicted here, with the pressure-strain constants of $\omega = 2.8$ and $\gamma = 0.3$.

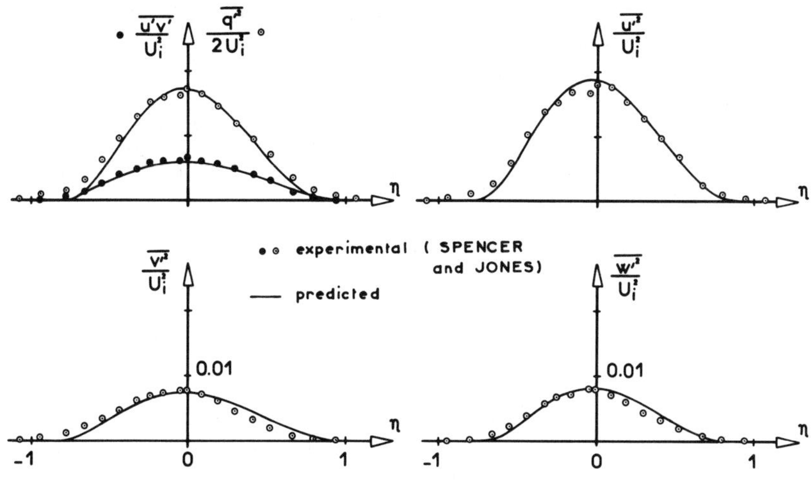

Figure 4. Distribution of the Reynolds Stresses for the Plane Two-Stream Mixing Layer, $U_e/U_i = 0.3$.

3. COMPARISON WITH EXPERIMENTS ON RELAXING SHEAR LAYERS

3.1 Plane Wake Flows

This flow configuration has been studied in detail at ONERA (Solignac, 1973) and in other research laboratories as e.g. at the CEAT in Poitiers (Tsen & Fayet, 1972) and at the Johns Hopkins University (Chevray & Kovasznay, 1969).

The experimental apparatus used at ONERA is sketched in Figure 5. It consists of a subsonic wind-tunnel of rectangular test section of 156 × 290 mm (aspect ratio ≈ 1.9). The freestream velocity is about 50 m/s. The flat plate on which the boundary layer develops is located in the horizontal symmetry plane of the tunnel. It is 500 mm long and provides at the trailing edge fully turbulent boundary layers of roughly 10 mm thickness. The Reynolds number based on the boundary layer thickness is about $4 \cdot 10^4$. The shapes of the upper and lower walls can be modified (Figure 5b,c) in order to create moderate adverse pressure gradients. These correspond to a streamwise variation of the pressure coefficient $K_p = (p(x) - p_0)/\frac{1}{2}\varrho U_0^2$ of 1 to 2 percent per boundary layer thickness at the trailing edge. Both a symmetrical (Figure 5b) and an asymmetrical (Figure 5c) arrangement have been used, the latter providing initial boundary layers which are different in shape and in thickness.

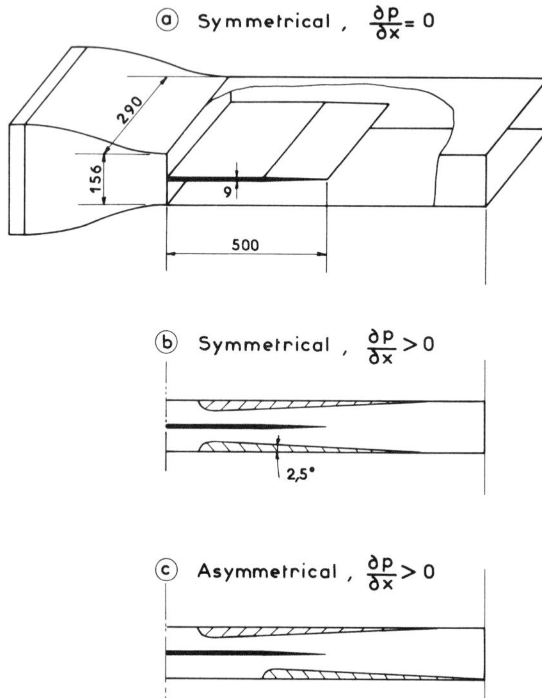

Figure 5. Plane Wake Configurations.

Figure 6 shows the streamwise variation of the minimum velocity in the wake for the symmetrical configuration without pressure gradient. The measurements of ONERA which are confined to the near region, are compared here to the data of Tsen and Fayet (1972) and Chevray and Kovasznay (1969). The experimental results viewed as a whole can be seen here to be correctly predicted by the two-equation turbulence model. The initial conditions from which the calculation starts are provided by the corresponding experimental mean velocity profiles at the trailing edge and partly extracted from the turbulence data of Klebanoff (1955), insofar as the turbulent kinetic energy profile in the boundary layer is concerned. The initial profile of the length scale is generated by using Equ. (12) and a standard expression for the mixing length profile

$$\ell = \ell_0 \tan h \left(\frac{ky}{\ell_0}\right)$$

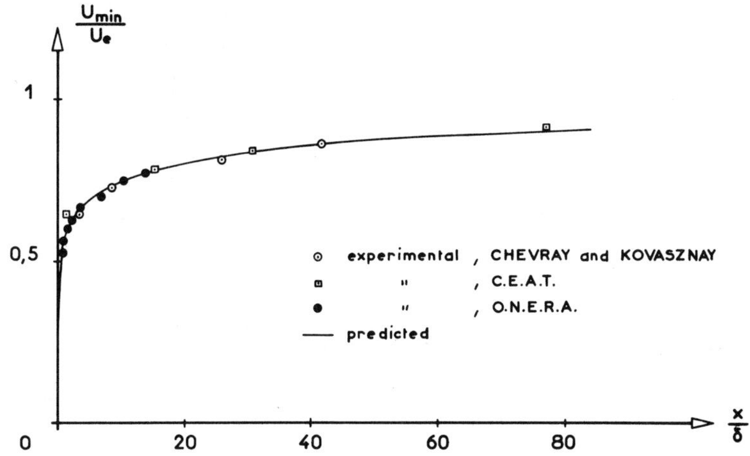

Figure 6. Minimum Velocity in the Symmetrical Plane Wake.

with ℓ_0 = 0.084 δ, k = 0.41 (the von Karman constant). δ is the boundary layer thickness and y is the distance from the wall. The constant β appearing in Equ. (12) was fixed at 0.5.

Details of the transverse profiles of the normalized streamwise velocity component are given in Figures 7 and 8 for the test series of ONERA and of Chevray and Kovasznay respectively. Reasonable agreement between the predicted and measured profiles is evidenced there.

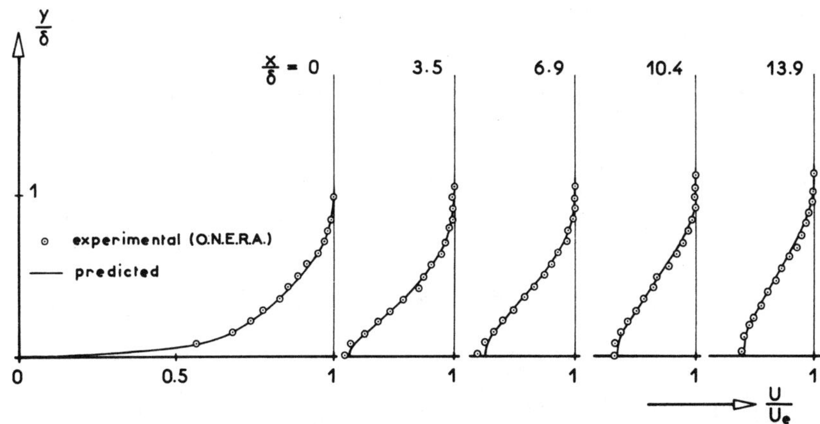

Figure 7. Mean Velocity Profiles in the Symmetrical Plane Wake (ONERA).

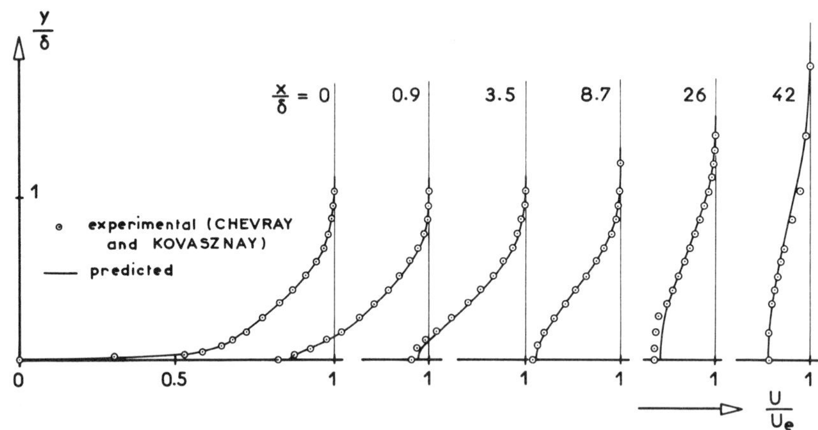

Figure 8. Mean Velocity Profiles in the Symmetrical Plane Wake (Chevray and Kovasznay).

Chevray and Kovasznay also measured the corresponding shear stress profiles by hot wire anemometry. These are compared with the calculated profiles in Figure 9. Satisfactory agreement is achieved, consistent with that of Figure 8, in spite of some minor divergence in the initial section due to the particular choice of the initial conditions for the turbulent energy and the length scale, from which the initial profile of the shear stress results.

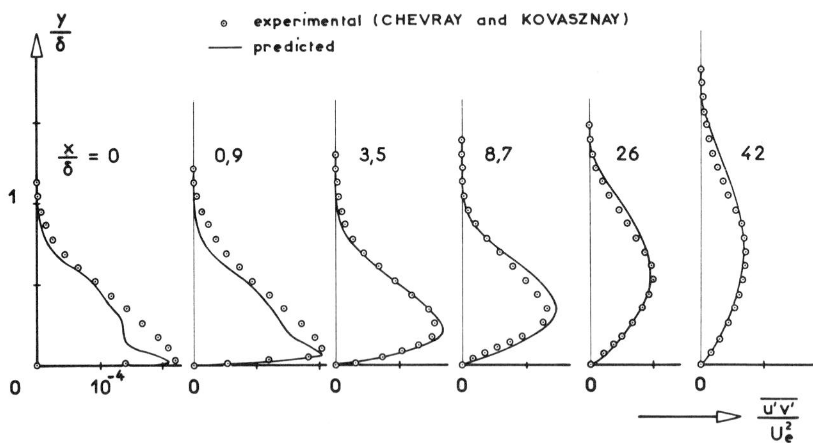

Figure 9. Turbulent Shear Stress Profiles in the Symmetrical Plane Wake.

The second example of wake flows concerns the asymmetrical configuration in the presence of an adverse pressure gradient, studied at ONERA (Figure 5c). Figure 10 shows the streamwise variation of the measured and calculated minimum velocity in the wake togehter with the corresponding pressure coefficient imposed on the outer flow. The corresponding velocity profiles are given on Figure 11. The mean velocity is normalized here by the external velocity in the initial section (U_{e_0}), so that the variation of the velocity at the boundary of the profiles reflects the action of the adverse pressure gradient imposed on the flow. The adopted model can be seen to predict correctly this asymmetric flow configuration. It can further be observed from this figure that it takes a streamwise distance of more than ten outer boundary layer thicknesses before the signature of the asymmetrical conditions at the confluence begins to dim noticeably.

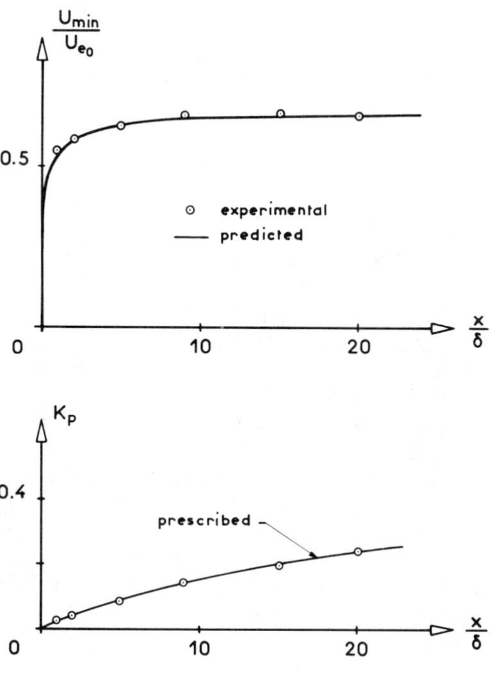

Figure 10. Minimum Velocity in the Asymmetrical Plane Wake with $\partial p/\partial x > 0$.

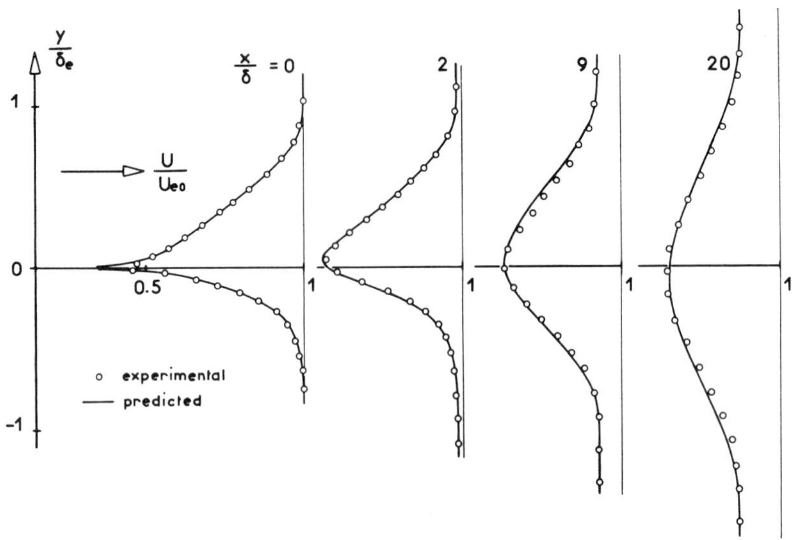

Figure 11. Mean Velocity Profiles in the Asymmetrical Plane Wake with $\partial p/\partial x > 0$.

3.2 Coaxial Jet Flows

The jet flows have been studied in a coaxial arrangement sketched on Figure 12. This can be operated either in a short nozzle version, Figure 12a, supplying the standard jet mixing configuration without initial boundary layers or in a long nozzle version, Figure 12b, providing large turbulent boundary layers at the confluence, the thickness of which can be adjusted by the length of the prolonging pipe (Leuchter, 1975). The flow velocities were 100 m/s in the primary flow and vary from 0 to 50 m/s in the secondary flow. The principal dimensions are 30 mm for the diameter of the inner pipe and 100 mm for that of the outer pipe providing conditions of an unlimited outer flow for the investigation of the near field. Typical Reynolds numbers are of the order $2 \cdot 10^5$ with respect to the flow conditions of the inner pipe. Conventional measuring techniques were employed comprising pitot and temperature probing as well as hot wire and laser doppler anemometry.

A set of typical initial conditions is given on Figure 13. It corresponds to a test case where the thickness of the inner and outer boundary layers were approximately half the inner pipe radius. Two different velocity ratios have been explored under these conditions, namely 0.25 and 0.46. The measured turbulence intensity levels in

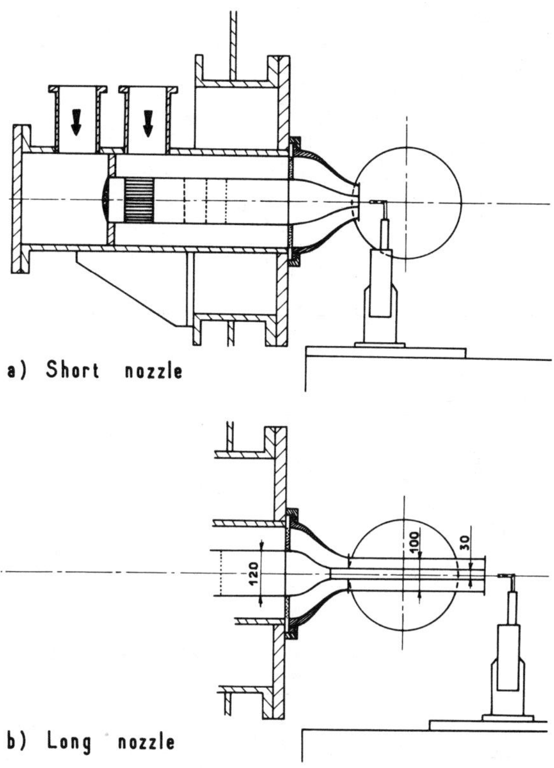

Figure 12. Jet Mixing Configurations.

the boundary layers can be seen to be of the same order as those observed in flat plate boundary layers, whereas in the regions of uniform flow outside of the boundary layers the turbulence intensity is negligibly small.

The measured and calculated mean velocity profiles developing from these initial conditions are shown on Figures 14 and 15 for the velocity ratios of 0.25 and 0.46 respectively. Again a satisfactory agreement between the prediction and the measurement is achieved. It can be noticed further that the downstream distance needed for smoothing out the initial velocity defect in the profiles is a growing function of the velocity ratio since the level of the shear stress then decreases. For the higher velocity ratio it takes more than two diameters (i.e. eight boundary layer thicknesses) before this occurs. After this distance the profiles become more and more

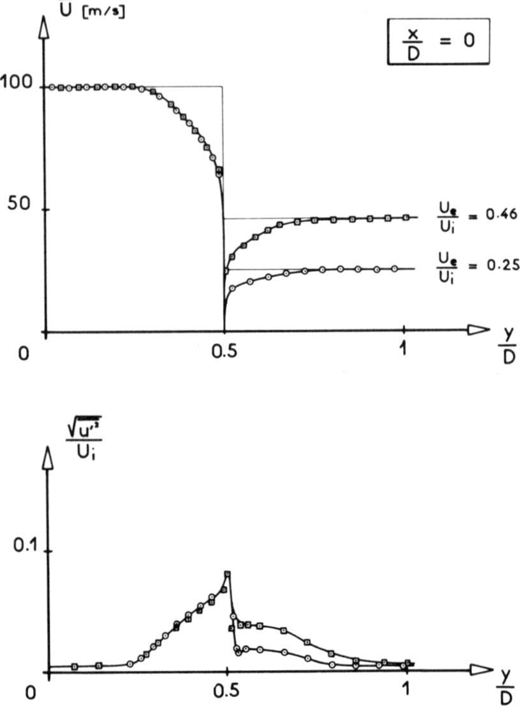

Figure 13. Initial Conditions for Coaxial Jets.

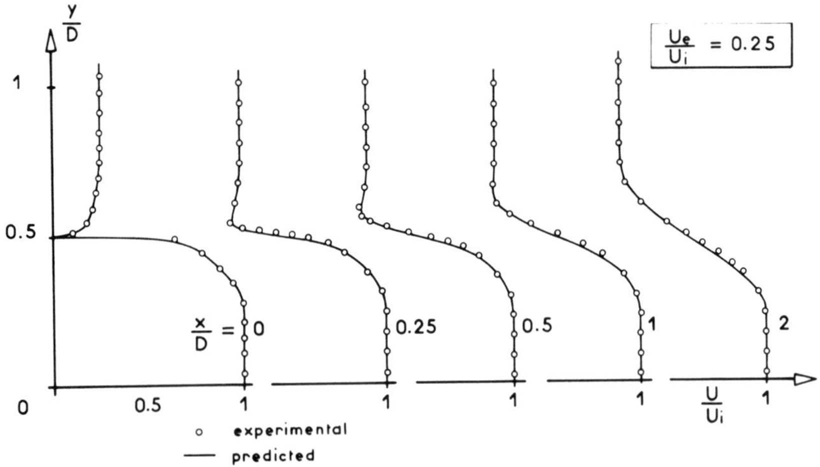

Figure 14. Mean Velocity Profiles in the Transition Region of Coaxial Jets, $U_e/U_i = 0.25$.

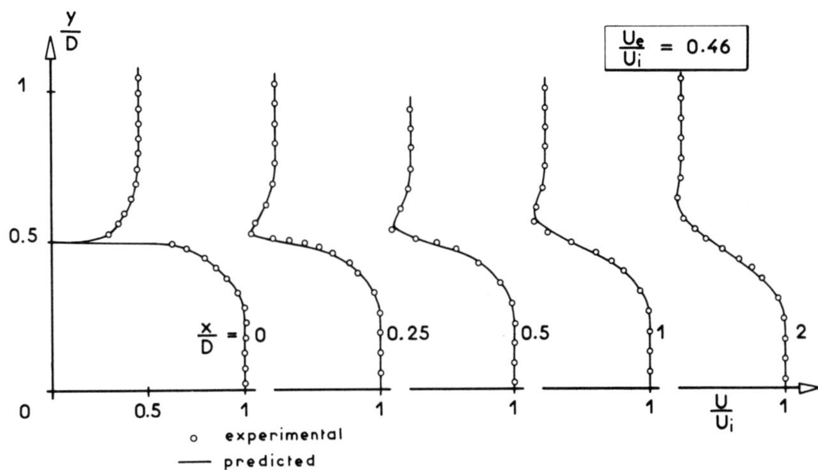

Figure 15. Mean Velocity Profiles in the Transition Region of Coaxial Jets, $U_e/U_i = 0.46$.

similar and the growth rate of the mixing layer approaches then that of the standard configuration without initial boundary layers. The only remanent effects of the boundary layers on the geometry of the mixing layer are then the upstream displacement of its virtual origin and an apparent contraction of the transverse dimensions of the jet due to the momentum defect in the initial section (Leuchter, 1975).

Figure 16 shows some measured and predicted turbulence intensity profiles for this flow configuration at the velocity ratio of 0.46. The experimental data are obtained by laser doppler anemometry. The intensity is here slightly overpredicted in the central part of the mixing layer, but both the predicted as well as the measured profiles, indicate clearly the progressive change of their shapes towards those corresponding to the asymptotical mixing layer.

As the length of the inner pipe is increased, the internal boundary layers end by joining on the center and the flow structure then develops towards that of a fully developed pipe flow. This particular jet configuration provides a strong relaxation effect for the mean and the fluctuating velocity fields and was therefore also included in the present investigation. Typical initial conditions are given in Figure 17 for this particular case, and for a stagnant outer medium. These conditions are similar to those

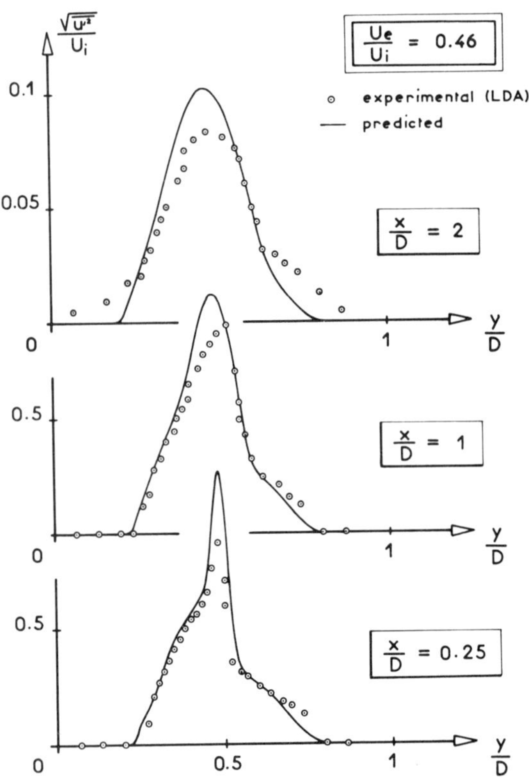

Figure 16. Turbulence Intensity Profiles in Coaxial Jets, $U_e/U_i = 0.46$.

measured by Laufer (1954) in the developed pipe flow. Experimental and calculated profiles of the turbulent intensity and the normalized mean velocity at two diameters downstream of the nozzle exit section are shown in Figure 18. In spite of a distinct overprediction of the turbulence intensity in the outer region of the mixing layer the main features of this jet flow are correctly depicted. From the shapes of the profiles it can be recognized in particular that the effects of the initial conditions are noticeably attenuated at this downstream section, so that the profiles develop further downstream in a similar way as for the mixing layers with standard initial conditions.

In this test case also streamwise space correlations of the longitudinal velocity fluctuations have been measured. It was thus

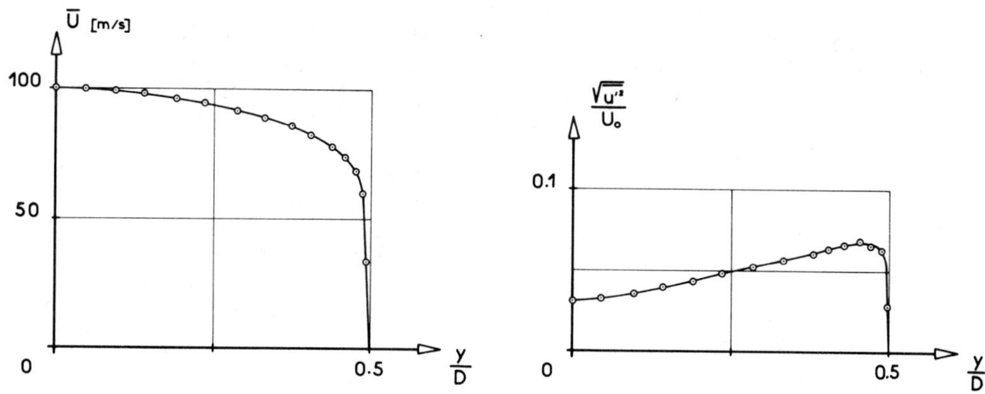

Figure 17. Jet Developing from Pipe Flow: Initial Conditions.

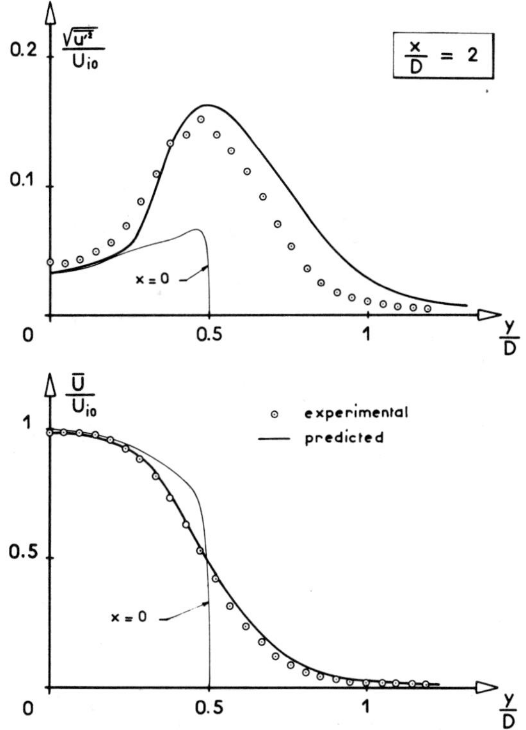

Figure 18. Predictions of Mean and Fluctuating Velocity in a Jet Developing from Pipe Flow.

possible to compare the corresponding integral length scales to the calculated length scale resulting from the prediction model. These are compared in Figure 19, where the streamwise variation of the length scales is shown for two different transverse positions in the flow, namely on the axis (Figure 19a) and half a diameter away, in the prolongation of the nozzle lip (Figure 19b). A scaling factor of 0.15 was used in order to achieve agreement between the calculated and measured length scales (instead of 0.16 in Figure 3). The main features concerning the evolution of the length scale can be seen to be correctly predicted, namely the almost linear growth in the central part of the mixing layer and the very slow variation on the axis followed by an abrupt change of the slope as the turbulence of the mixing layer reaches the axis.

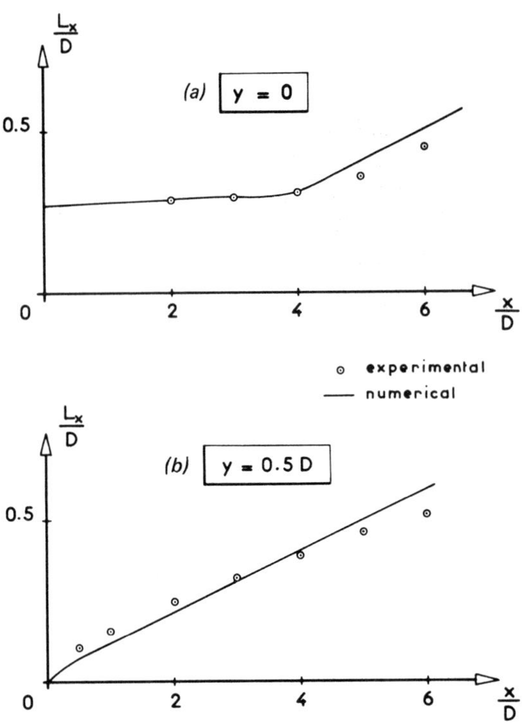

Figure 19. Integral Length Scale in a Jet Developing from Pipe Flow.

3.3 Coaxial Jet Flows with External Freestream Turbulence

This short review on our activity in the field of mixing layers and jet flows will be closed by a brief description of some recent results obtained on the effects of freestream turbulence on the mixing process. The activity dealing with these effects is still in a very early stage of development at ONERA so that the results presented here are preliminary and incomplete.

The freestream turbulence was generated by a grid which was placed into the outer stream of the coaxial jet arrangement of Figure 12a and the geometry of which is defined in Figure 20. The grid has a 10mm mesh size and was fixed in the exit plane of the nozzles, providing thus a high turbulence intensity of the order of 10% in the initial region and an integral length scale L_X of half the mesh size at two diameters downstream of the grid. L_X was there of the same

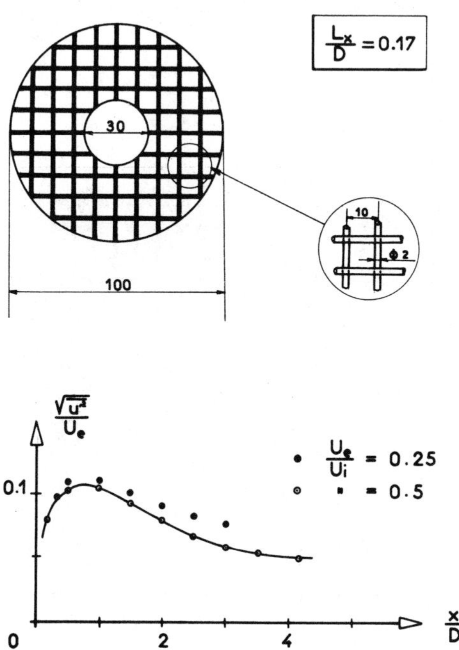

Figure 20. Turbulence Grid Configuration for Coaxial Jet.

order as the corresponding length scale in the mixing layer. Two typical values of the velocity ratio had been examined: $U_e/U_i = 0.25$ and 0.5.

Figure 21 shows that a strong interaction takes place for these particular initial conditions as can be inferred from the large increase of the mixing layer thickness in the presence of the grid turbulence. This increase appears to be much more important for the higher velocity ratio of 0.5, since in this case the freestream turbulence intensity has a higher absolute level and the turbulence intensity of the undisturbed mixing layer is at a lower level than for the velocity ratio of 0.25. It is clearly evidenced in Figure 21 that the turbulence intensity of the freestream is one of the dominant interaction parameters for this type of flow. The action of other parameters of freestream turbulence, as e.g. its length scale, could not be investigated up to now and will be examined in a later stage of this study. Figure 22 shows for the higher velocity ratio of 0.5

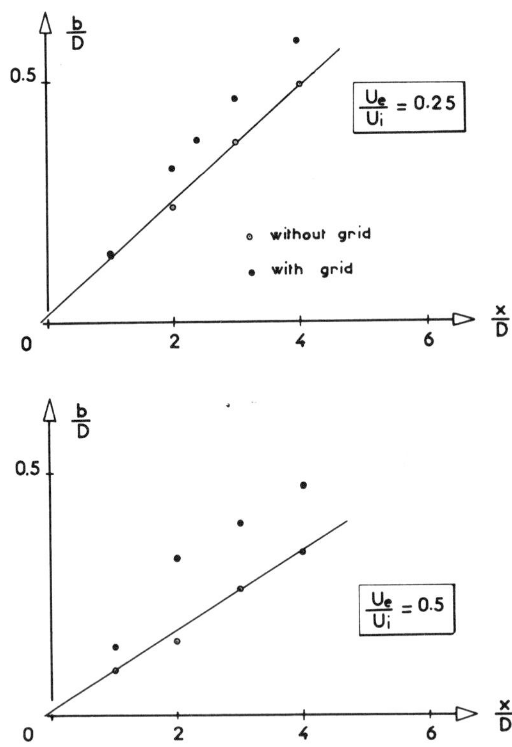

Figure 21. Effect of External Turbulence in Coaxial Jets.

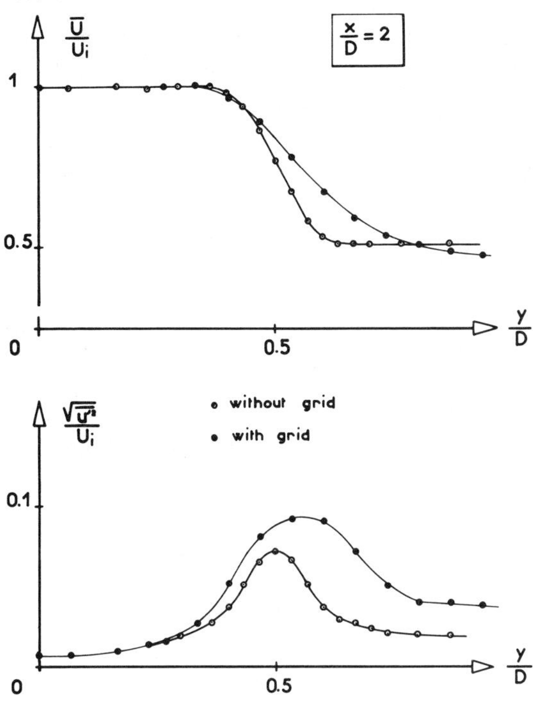

Figure 22. Effect of External Turbulence in Coaxial Jets, $U_e/U_i = 0.5$.

the modification of the mean velocity and the turbulence intensity profiles at two diameters downstream of the exit plane. It can be observed that the thickening of the mixing layer occurs simultaneously with a large increase of the turbulence intensity level in the mixing layer.

Qualitatively similar results have been obtained with the above described turbulence model, as illustrated in Figure 23. The same order of magnitude for the increase of the mixing layer thickness and of the turbulence intensity is evidenced. The numerical solution is obtained by starting from the appropriate initial conditions and by taking into account the boundary conditions deduced from the measurements. This ultimate example of relaxing shear flows demonstrates the capability of the adopted model for the qualitative prediction of turbulent flows in which strong interactions of different turbulence structures occur. Further refinements are however needed in order to achieve better quantitative agreement with the experimental data.

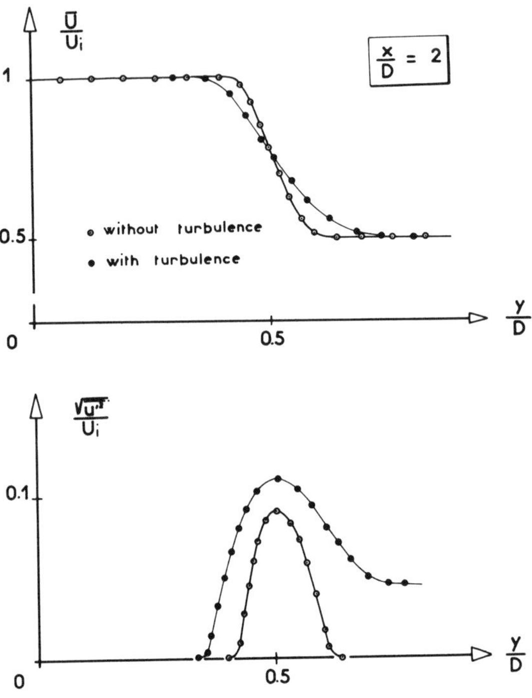

Figure 23. Effect of External Turbulence (Calculated) in Coaxial Jets, $U_e/U_i = 0.5$.

4. CONCLUSIONS

Various configurations of non-equilibrium shear flows of practical interest in turbomachinery or external aerodynamics have been examined in this paper. Comparisons between experimental and numerical results have demonstrated the capability of a simple two-parameter turbulence model for describing correctly the effects of initial boundary layers and external turbulence on the behavior of the shear layers.

The turbulence model, based on the Prandtl-Kolmogorov eddy viscosity concept, uses only two fundamental turbulence scales, namely a velocity and a length scale. The former is represented by the square root of the turbulent kinetic energy and the latter by Rotta's integral length scale. Both quantities are deduced from the numerical solution of the corresponding transport equations. It appeared to be essential for the success of the description of relaxing turbulent shear flows, like those examined in this study, that the turbulence model contains at least one balance equation for the length scale (or for any other turbulent parameter from which a length scale may be extracted) since this quantity is subjected to rapid variations in the near field of the confluence which would be too hazardous to prescribe algebraically.

Transport equations for the Reynolds stress tensor had been added to the basic two-equation model in order to get a more detailed description of the turbulent velocity field and to proceed to a more direct comparison between measured and calculated turbulence parameters (mainly $\overline{u'^2}$). The modelling of the diffusion and the pressure-strain terms of these equations need further refinements.

The turbulence model adopted in this paper appears to be a valuable tool for the engineer faced with the problem of economical predictions of realistic confluence configurations, including the effects of freestream turbulence.

REFERENCES

1. Bradshaw, P., Ferris, D. H. and Atwell, N. P. 1967. Calculation of boundary layer development using the turbulent energy equation. J. Fluid Mech. 28, 593.

2. Brown, G. L. and Roshko, A. 1971. The effect of density difference on the turbulent mixing layer. AGARD - CP - 93.

3. Chevray, R. and Kovasznay, S. G. 1969. Turbulence measurements in the wake of a thin flat plate. AIAA J. 7, 1641.

4. Davidov, B. I. 1961. On the statistical dynamics of an incompressible turbulent fluid. Dokl. Akad. Nauk SSSR, 136, 47.

5. Hanjalic, K. and Launder, B. E. 1972. A Reynolds stress model of turbulence and its application to thin shear flows. J. Fluid Mech. 52, 609.

6. Harlow, F. H. and Nakayama, P. 1968. Transport of turbulence energy decay rate. Los Alamos Sci. Lab. Rep. LA 3854.

7. Klebanoff, P. S. 1955. Characteristics of turbulence in a boundary layer with zero pressure gradient. NACA Rep. 1247.

8. Kolmogorov, A. N. 1942. Equations of turbulent motion of incompressible fluid. Izv. Akad. Nauk SSSR Seria fizichiska VI, 56.

9. Laufer, J. 1954. The structure of turbulence in fully developed pipe flow. NACA Rep. 1174.

10. Launder, B. E. 1975. Progress in the modelling of turbulent transport. VKI - Lecture Series 76, Rhode St. Genèse, Belgium.

11. Launder, B. E., Reece, G. J. and Rodi, W. 1975. Progress in the development of a Reynolds-stress turbulence closure. J. Fluid Mech. 68, 537.

12. Leuchter, O. 1973. Ueber das Verhalten der turbulenten freien Strahlgrenze mit Initialgrenzschicht. Paper no. 73-82, Annual Meeting of ÖGFT/DGLR, Innsbruck.

13. Leuchter, O. 1975. Effets de couches limites initiales sur le développement des jets turbulents. Fifth Can. Congr. of Appl. Mech. Frederichton.

14. Liepmann, H. W. and Laufer, J. 1947. Investigation of free turbulent mixing. NACA T. N. 1257.

15. Miles, J. B. and Shih, J. S. 1968. Similarity parameter for two-stream turbulent jet mixing region. AIAA J. 6, 1429.

16. Ng, K. H. and Spalding, D. B. 1972. Turbulence model for boundary layers near walls. Phys. Fluids, 15, 20.

17. Patel, R. P. 1973. An experimental study of a plane mixing layer. AIAA J. 11, 67.

18. Prandtl, L. 1945. Ueber ein neues Formelsystem für die ausgebildete Turbulenz. Nachr. Akad. Wissensch. Göttingen, Math - Phys. Klasse, 6.

19. Rodi, W. 1972. The prediction of free turbulent boundary layers by use of a two-equation model of turbulence. Ph.D. Thesis, University of London.

20. Rodi, W. and Spalding, D. B. 1970. A two-parameter model of turbulence and its application to free jets. Wärme und Stoffübertragung 3, 85.

21. Rotta, J. C. 1951. Statistische Theorie nicht homogener Turbulenz. Z. Phys. 129, 547 and 131, 51.

22. Rotta, J. C. 1971. Recent attempts to develop a generally applicable calculation method for turbulent shear flow layers. AGARD - CP - 93.

23. Rotta, J. C. 1975. Prediction of turbulent shear flows using the transport equations for turbulence energy and turbulence length scale. VKI - Lecture Series 76, Rhode St. Genèse, Belgium.

24. Sabin, C. M. 1963. An analytical and experimental study of the plane, incompressible, turbulent free shear layer with arbitrary velocity ratio and pressure gradient. AFOSR T. N. 5443.

25. Saffman, P. G. 1970. A model for inhomogeneous turbulent flow. Proc. Roy. Soc. A 317, 417.

26. Solignac, J. L. 1973. Calcul du sillage turbulent bidimensionnel en aval d'un obstacle mince. Fourth Can. Congr. of Appl. Mech., Montréal.

27. Spalding, D. B. 1969. The calculation of the length scale of turbulence in some shear flows remote from walls. Progress in Heat and Mass Transfer, 2, 255.

28. Spencer, B. W. and Jones, B. G. 1971. Statistical investigation of pressure and velocity fields in the turbulent two-stream mixing layer. AIAA Paper 71-613.

29. Tollmien, W. 1926. Berechnung turbulenter Ausbreitungsvorgänge. ZAMM 6, 468.

30. Tsen, L. F. and Fayet, J. 1972. Etude du développement d'un sillage turbulent dans un écoulement avec gradient de pression. CEAT Intern. Rept.

31. Wygnanski, I. and Fiedler, H. E. 1970. The two-dimensional mixing region. Boeing Scient. Res. Lab. Doc. D1-82-0951.

32. Yule, A. J. 1971. Two-dimensional self-preserving turbulent mixing layers at different free stream velocity ratios. University of Manchester - Repts. and Memo. no. 3683.

DISCUSSION

WILCOX: (DCW Industries)

Can you say something about how the values of the constants you fixed compare to corresponding values other researchers use for similar equations in boundary layer applications? Are they the same or do they differ much.

LEUCHTER:

I think they are essentially the same or very close. To my knowledge there is some work done by Spalding on turbulent boundary layers with this two equation model and I guess they use essentially the same constants.

BIRCH: (The Boeing Company)

I would like to repeat a remark I made yesterday. In using a two equation turbulence model to predict the effect of freestream turbulence you will find that the calculation is more sensitive to the length scale specification in the outer flow than to the turbulent intensity. Do you know what this was experimentally and what did you use?

LEUCHTER:

In the case where the external turbulence is sensitive to the development of the mixing layer we only used one length scale. This work is still going on, so we will use different mesh sizes to generate turbulence of different scales. We didn't actually vary the turbulent length scale.

MELLOR: (Princeton University)

This is the two equation model that you are talking about. Then you referred to the Reynolds stress equation. I don't understand how that all fits together with your original model.

LEUCHTER:

Yes, I introduce the Reynolds equation only to get more details on the turbulent field.

MELLOR:

After the fact of making the calculations?

LEUCHTER:

No, we didn't use the Reynolds stress actually to describe the Reynolds stress. That is we always had a viscosity model with additional transport of the Reynolds stress but passive transport. They are not coupled with mean equation.

WYGNANSKI: (Tel Aviv University)

Did you make any comparison of your coaxial jet data with previous data that was measured by Champagne and myself? I refer to this because in a coaxial jet you have the following effect: If the outer velocity is slower like the data you presented you get one type of growth but if it is the inverse case, namely, that the outer velocity is much faster then you get significantly different growth. This is so because the entrained streamliner curve in a different direction.

LEUCHTER:

No, we didn't. We only examined the case of lower outer velocity. In fact, we didn't compare the results directly with yours. I think you have a limited outer flow, so there is some interaction between the outer mixing layer and the inner mixing layer.

WYGNANSKI:

Yes, and therefore two nozzles of different diameters were used. Of course the interaction between the mixing layers occurs further downstream but in the initial four or six nozzle diameters one configuration was similar to the other.

LEUCHTER:

I didn't compare.

PART III
TURBOMACHINERY
APPLICATIONS

Measurements in Curved Flows

J. A. C. HUMPHREY AND J. H. WHITELAW
Imperial College
London, England

ABSTRACT

The application of three experimental techniques, namely, hot-wire anemometry, laser-Doppler anemometry and tracer gas analysis, to three flow configurations is described and results reported. The co-axial, swirling free jet was investigated with hot-wire anemometry, the flow in a curved duct of rectangular cross section with laser-Doppler anemometry; and the wall jets over curved surfaces with the tracer gas method.

The results obtained in the rectangular, curved duct are new and quantify the magnitude of the secondary flows. They demonstrate the upstream influence of the flow and show that the magnitude of the axial normal stresses do not exceed 15% over a substantial region. Measurements were also obtained in the same duct with the Reynolds number corresponding to laminar flow and the results compared with values obtained from a solution of the Navier-Stokes equations. The two sets of results are in general agreement. The computing time required to obtain grid-independent solutions, in the laminar flow, suggests that similar calculations in turbulent flows will not be trivial. In the particular laminar flow investigated, recirculation in the main flow direction was observed and emphasizes the need to solve the complete set of equations.

1. INTRODUCTION

This paper is intended to fulfill two purposes. The first is to demonstrate and discuss the relative advantages of three experimental techniques for the investigation of turbulent flows and the second is to present and discuss measurements of mean and fluctuating properties determined with these techniques in three flows with different forms of curvature of relevance to turbomachinery.

The experimental techniques under consideration include hot-wire and laser-Doppler anemometry. Both can lead to information of the mean and fluctuating properties of turbulent flows and each has relative advantages which are demonstrated here in the context of the swirling, coaxial jet flow of Ribeiro (1976) and the curved duct flow referred to by Humphrey, Melling and Whitelaw (1975): in the latter case, new measurements and some calculations are presented. Major advantages of laser-Doppler anemometry are its successful use in flows with recirculation and in hostile environments, see for example Durst, Melling and Whitelaw (1976), but the present comparison is made with relatively friendly flows where the turbulence intensities in the range of measurements do not give rise to the possibility of negative velocities. Secondary flows did, however, exist in the curved-duct flow and could give rise to probe-interference effects if investigated with hot-wire anemometry. The third experimental technique relates to the measurement of wall concentration of a foreign gas diffusing and convecting in wall jet flows over concave and convex surfaces. It provides practically useful information of the adiabatic-wall effectiveness at comparatively low cost: the results discussed here are due to Folayan and Whitelaw (1976).

The curved duct flow is of some relevance to turbine blade passages although the aspect ratio of the rectangular duct is unity and the initial conditions are those of a fully-developed rectangular duct flow of water. The results allow the testing of turbulence models in a flow which has strong secondary velocities arising from the pressure field and from the normal stresses. The swirling coaxial jet was also investigated to assist the development of calculation methods where a Reynolds stress closure appeared necessary. In contrast, the wall-jet flows investigated by Folayan and Whitelaw (1976) were required for film-cooling purposes and a simple eddy-viscosity hypothesis was sufficient for this engineering purpose.

The following section describes the flow configurations and instrumentation used. The description is presented in sufficient detail for present purposes and the reader should consult cited references for further information. The third section presents results, their physical interpretation and implications for calculation methods: it also provides a basis for the brief comparison of experimental techniques which is included in the discussion of the fourth section.

2. FLOW CONFIGURATIONS AND EXPERIMENTAL TECHNIQUES

2.1 Curved, Rectangular Duct

A 90 degree bend, of mean radius 92 mm, was attached to the end of the 1.8 m straight rectangular duct previously investigated by Melling (1975) and Melling and Whitelaw (1973, 1976). The cross section of the duct was 40 mm square and the bend was located in the vertical plane, with a 1.4 m length of straight duct attached to the end of the bend. Water flowed through the duct with a bulk velocity of 0.9 m/s corresponding to a Reynolds number, $U_b D_h/\nu$, of 3.6×10^4. The dimensions of the duct suggested that there would be no flow recirculation in the forward direction for this Reynolds number.

The duct and bend were manufactured from plexiglass and, together with the choice of working fluid, allowed the transmission of light. The lack of forward flow recirculation allowed the use of an integrated optical arrangement, as described by Durst and Whitelaw (1971), which did not incorporate light-frequency and which provided a fringe spacing of 1.86 μm. The light-collecting arrangement limited the number of fringes observed by the photomultiplier (EMI 9558B) and, together with the 5mW He-Ne laser, resulted in an almost continuous Doppler signal which could be conveniently processed by a frequency-tracking demodulator (DISA 55L20). The demodulated output from the tracker was presented to a digital voltmeter (SOLARTRON LM14202), after true integration (DISA 55B20), and rms meter (DISA 55D35) which resulted in values of mean velocity and normal stress, respectively.

The use of laser-Doppler anemometry is particularly appropriate to this flow configuration. The duct dimensions allow the use of an optical arrangement which results in negligible corrections for transit-time and gradient broadening effects over almost the entire duct. The forward-scattered signals, with a 5mW laser, are of high signal-to-noise ratio, and the resulting Doppler signal has a dropout of less than 10% which allows the use of electronic instrumentation, i.e. a tracker, which is convenient to use. Since the maximum turbulence frequencies present in the water flow are less than 500 Hz the dynamic capabilities of trackers are not strained and the near continuous signal ensures that the statistically required number of signal samples is rapidly achieved and without the uncertainties associated with sample biasing.

2.2 Coaxial Free Jet

In contrast, a coaxial free jet of air allows the use of frequency-tracking demodulation in the upstream region where particle concentration can be arranged to be large and where the turbulence intensity is low. The signal dropout and turbulence intensity increase with downstream distance and make the use of tracking increasingly more difficult. In such circumstances a counting arrangement may be preferred but, in any event, the precision of measurement decreases with downstream distance.

Provided the jet of air is clean, hot-wire anemometry can be used throughout the region of the flow where the turbulence intensity is less than around 25% with good precision. Indeed, particularly in the region downstream of the potential core, the precision can be better with the hot-wire than with the laser-Doppler anemometer. Certainly, in the upper region of the energy spectrum the hot wire is to be preferred and, even in terms of rms measurements, the possible error sources of laser-Doppler anemometry can render its results subject to greater uncertainty.

Ribeiro and Whitelaw (1976) described a hot-wire anemometer arrangement in which the analogue signal was digitized and processed by mini-computer. This arrangement, although expensive of equipment and of user training, is convenient and flexible and was used subsequently by Ribeiro for his examination of swirling coaxial jets. The flow emerged from two coaxial stainless steel pipes: the inner pipe was 2.83 m long, 16.13 mm inner diameter and 21.59 mm outer diameter; the outer pipe was 2.00 mm long, 44.50 mm inner diameter and 50.44 outer diameter. The two pipes were carefully aligned to ensure fully developed turbulent pipe and annular flow in the exit plane. A tangential jet swirler was incorporated in the annular flow where required.

The signal processing arrangement comprises two constant temperature anemometers (DISA 44D01) which were connected to different types of probes according to the requirements of the quantities to be measured. The signals at the output of the anemometers were linearized using function generators of the exponential type (DISA 55D10) and passed to a multiplexer via sample and hold units. A selected signal was then amplified in a programmable gain buffer amplifier and passed to an A/D converter through another sample and hold unit. Mean velocity and Reynolds stresses were determined from loops of up to 768 samples which were repeated between 200 and 400 times according to the low frequency behaviour of the signals. At the end of each loop, mean values and standard deviations for the loop were calculated in the PDP8E computer and used to update the corresponding total means. Signals proportional to u, v and w were input to the A/D interface and histograms of each quantity built up from a statistically required number of samples. The same approach was used for joint probability density distributions although here the two signals proportional to u, v and w were first held at the same instant, followed by the A/D conversion and storage.

Straight, slanting and cross-wire probes were used to obtain the measurements. The direction of mean flow was first determined and subsequent measurements obtained with the probe holder lying in the direction of the mean flow. Signals were processed with the aid of the equation

$$E_0 = S\,\overline{U}\cos\alpha\left\{1+\frac{u}{\overline{U}}+\frac{v}{\overline{U}}\tan\alpha\right\}\left[1 + \frac{\frac{h^2}{\cos^2\alpha}\frac{w^2}{\overline{U}^2}+k^2\left(-\tan\alpha-\frac{u}{\overline{U}}\tan\alpha+\frac{v^2}{\overline{U}}\right)^{\frac{1}{2}}}{\left(1+\frac{u}{\overline{U}}+\frac{v}{\overline{U}}\tan\alpha\right)^2}\right]$$

which represents the instantaneous signal in terms of the instantaneous velocity components u, v and w, the mean velocity $\overline{U}$, the angle between the normal to the wire and the mean flow direction α, the calibration constant S and the pitch and yaw factors h and k. Equations of this type, with a truncated binomial expansion to facilitate the derivation of explicit relationships between the mean velocity and higher order moments of the velocity field, have been used extensively--see for example Champagne and Sleicher (1967)--and introduce errors of increasing magnitude as the turbulence intensity increases above around 15%. An alternative procedure is to square the equation and apply the relationship between E_0^2 and the mean velocity and the correlations of the velocity components. In the present case, the former approach was preferred but the range of applicability of the method extended by least-square fitting the non-linear terms to fifth order polynomials.

2.3 Film-Cooling of Curved Surfaces

The two-dimensional film-cooling slots of Folayan and Whitelaw (1976) were arranged immediately upstream of convex surfaces of radii 76 mm and 152 mm and a concave surface of radius 152 mm. The freestream velocity, in the plane of the slot exit, was 18.5 m/s and the secondary velocity was varied. A trace of helium gas was added to the secondary flow and the wall concentration of helium measured at downstream locations by sucking samples through 0.5 mm holes in the wall and passing these through a thermal conductivity cell which had previously been calibrated. Situations with non-uniform density were readily simulated by replacing the secondary flow of air by mixtures of arcton and air in proportions designed to simulate the required density ratio.

3. RESULTS

3.1 Curved, Rectangular Duct

Measurements of the component of mean velocity in the direction of the bend axis are shown on Figure 1. Results of this type, and in similar detail, were obtained for the axial velocity components at locations upstream of the plane of the bend entrance and at locations corresponding to 0 degrees, 45 degrees, 71 degrees, 90 degrees; measurements of the radial velocity and corresponding stress components were obtained at 0 degrees and 90 degrees as well as the shear stress component $\overline{u_\theta u_R}$. In all cases, the results were converted into contour plots using a least-squares surface fitting computer program which yielded velocity contours such as those of Figures 1 to 4.

The results of Figure 1 are very similar to those obtained by Melling and Whitelaw (1976) in the absence of the bend and were obtained 2.5D upstream of the plane of the entrance to the bend. In the plane of the entrance, Figure 2, it is clear that the flow has been distorted by the downstream geometry, and the prevailing axial pressure gradient; the greater proportion of the mass flow is in the half of the duct nearer to the inner radius of the bend. The location of the maximum velocity is even closer to the inner radius at 45 degrees, Figure 3, but returned towards the center of the duct at 71 degrees, and moved towards the outer radius in the exit plane of the duct, Figure 4. From 0 degrees to 90 degrees the secondary motions deform the axial velocity profiles by forcing high speed flow in the axial direction towards the outer radius of curvature along the symmetry plane of the bend and returning it towards the inner radius along the sidewalls. This complicated flow pattern results directly from the centrifugal forces and the consequent radial pressure gradients.

It is interesting to note that, in the entrance plane of the bend, the values of normal stress are similar to those obtained in the absence of the bend, Figure 5. The asymmetry about the mid-plane is slight, by comparison with the corresponding mean-velocity contours of Figure 2; they exhibit the expected bulging towards the corners associated with secondary flows and have not been influenced significantly by the pressure gradients. In the exit plane of the bend, the normal stress contours are very asymmetric as indicated on Figure 6 but in a manner consistent with the mean-velocity contours of Figure 4.

In fully-developed square-duct flow without a bend, the maximum secondary-flow velocity is of order 1% of the maximum axial velocity. In the exit plane of the present bend, the maximum radial velocity

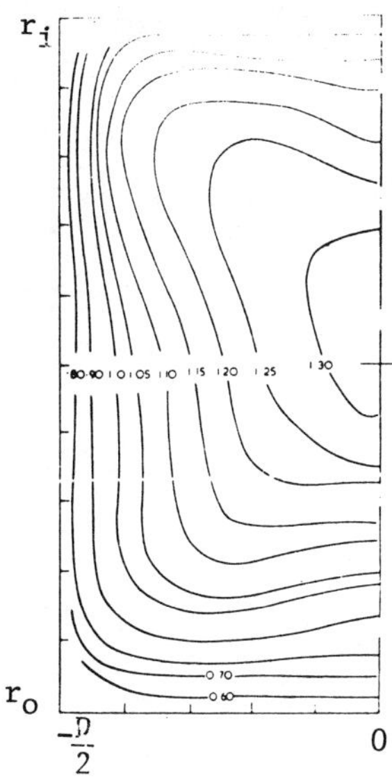

Figure 1. 0° position (entrance plane) in bend.

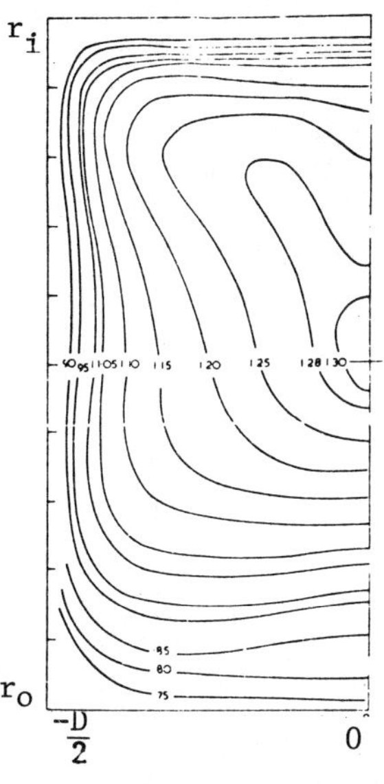

Figure 2. 45° position in bend.

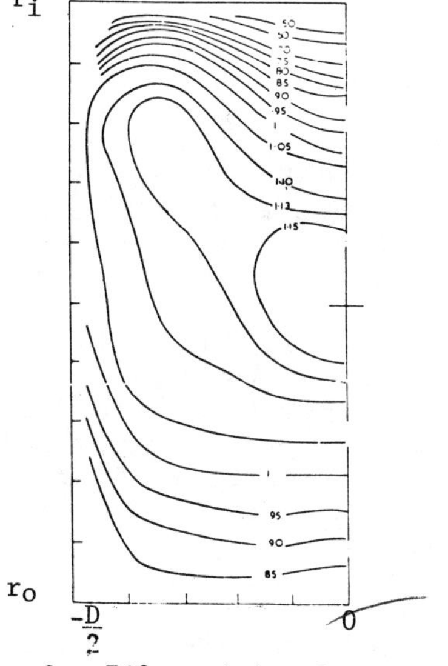

Figure 3. 71° position in bend.

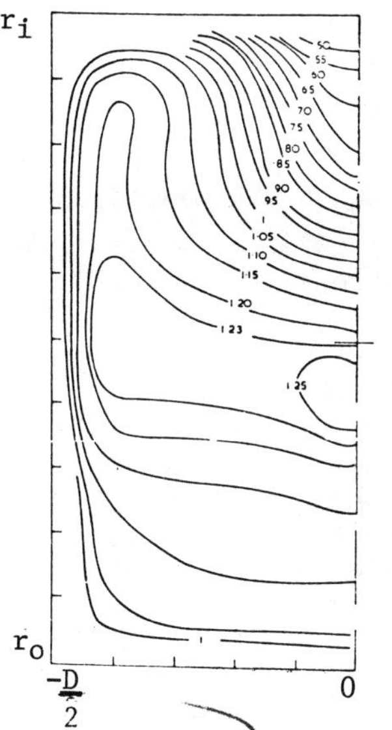

Figure 4. 90° position (exit plane) in bend.

Contour plots of normalized axial velocity component U/U_B at 0°, 45°, 71° and 90° positions in a square cross-section bend; Re = 36,000, $\overline{R}/b$ = 2.3 and U_B = .9 m/s.

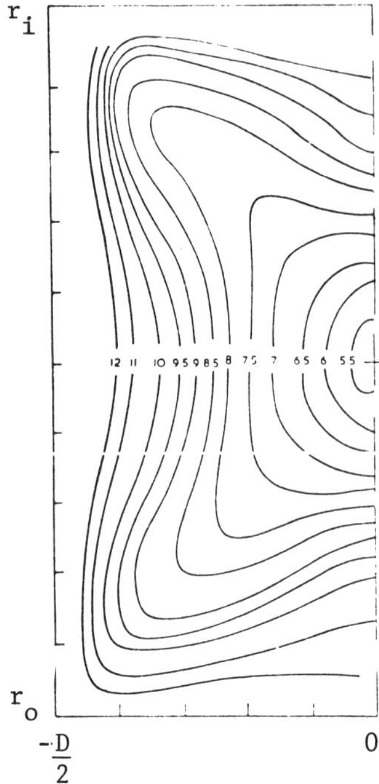

Figure 5. 0° position in bend.

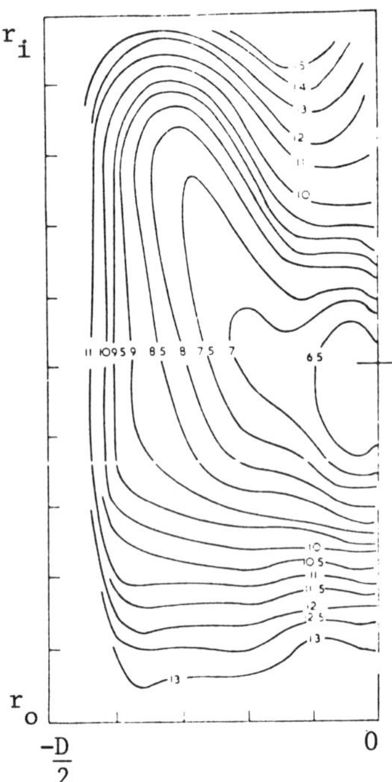

Figure 6. 90° position in bend.

Contour plots of axial component of normalized turbulence intensity $\sqrt{\overline{u^2}}/U_B \times 100$ at 0° and 90° positions in a square cross-section bend; Re = 36,000, $\overline{R}/b$ = 2.3 and U_B = .9 m/s.

was measured to be 22% of the maximum axial velocity and is undoubtedly associated with the centrifugal forces rather than the normal stresses. Calculations of a laminar flow in a duct of similar geometry, indicate secondary flow velocities up to around 13% of the bulk velocity in the exit plane: a maximum value of 47% was calculated in a plane corresponding to 27 degrees. Larger values of radial velocity can also be expected in the region immediately upstream of the exit plane of the present turbulent flow.

The measurements presented in the previous paragraph were readily obtained by a simple form of laser-Doppler anemometry. The precision and reproducibility of measurement are high and the accuracy better than could be obtained by any alternative technique. The results are relevant to the blade passages of gas turbines and suggest that flows of this type may be represented by a three-dimensional, time averaged form of the Navier-Stokes equations with a turbulence model which is rather simpler than that necessary to characterize the mean-flow properties of square-duct flow without a bend. This deduction stems from the smaller influences of the normal stresses, on the secondary flows, in the bend flow.

Nevertheless, the calculation of flows of this type can present a formidable task. The appropriate forms for the equation of continuity and momentum conservation may be written in the form:

$$\rho \left(\frac{DU_r}{Dt} - \frac{U_\theta^2}{r}\right) = -\frac{\partial P}{\partial r} + \mu \left(\nabla^2 U_r - \frac{U_r}{r^2} - \frac{2}{r^2}\frac{\partial U_\theta}{\partial \theta}\right)$$

$$\rho \left(\frac{DU_\theta}{Dt} + \frac{U_r U_\theta}{r}\right) = -\frac{1}{r}\frac{\partial P}{\partial \theta} + \mu \left(\nabla^2 U_\theta + \frac{2}{r^2}\frac{\partial U_r}{\partial \theta} - \frac{U_\theta}{r^2}\right)$$

$$\rho \frac{DU_z}{Dt} = -\frac{\partial P}{\partial z} + \mu \nabla^2 U_z$$

where

$$\nabla^2 = \frac{\partial^2}{\partial r^2} + \frac{1}{r}\frac{\partial}{\partial r} + \frac{1}{r^2}\frac{\partial^2}{\partial z^2} + \frac{\partial^2}{\partial \theta^2}$$

and

$$\frac{1}{r}\frac{\partial}{\partial r}(r\ U_r) + \frac{1}{r}\frac{\partial U_\theta}{\partial \theta} + \frac{\partial U_z}{\partial z} = 0$$

and can be solved numerically together with appropriate boundary conditions. Humphrey has solved these equations for a Reynolds number of 790 and for a geometry identical to the present bend. The calculations revealed a small region of recirculation, in the main flow direction, and this was confirmed by flow visualization with a neutrally buoyant dye. Thus for this situation, the details of the flow required the solution of the elliptic forms of the equations even though, with $8 \times 13 \times 17$ internal grid nodes in each of the entry duct, bend and exit duct, a solution required approximately 80 min of CP time on a CDC6600 computer with a convergence criterion corresponding to maximum residuals of not more than 10^{-3} in each variable.

The use of the numerical mesh referred to in the previous paragraph allowed calculations which agreed with laser-Doppler anemometer measurements to within 5% upstream of the 60 degree plane. Further downstream, the discrepancies increased to a maximum of around 15% although the trends were faithfully represented. Figure 7 compares measurements and calculations in the exit plane of the bend and indicates the extent of the agreement in the worst location. The agreement could be improved significantly by the use of a finer numerical mesh but at increased cost; the run time is given, approximately, by $0.14 \times N \times E$ where N is the number of grid nodes and

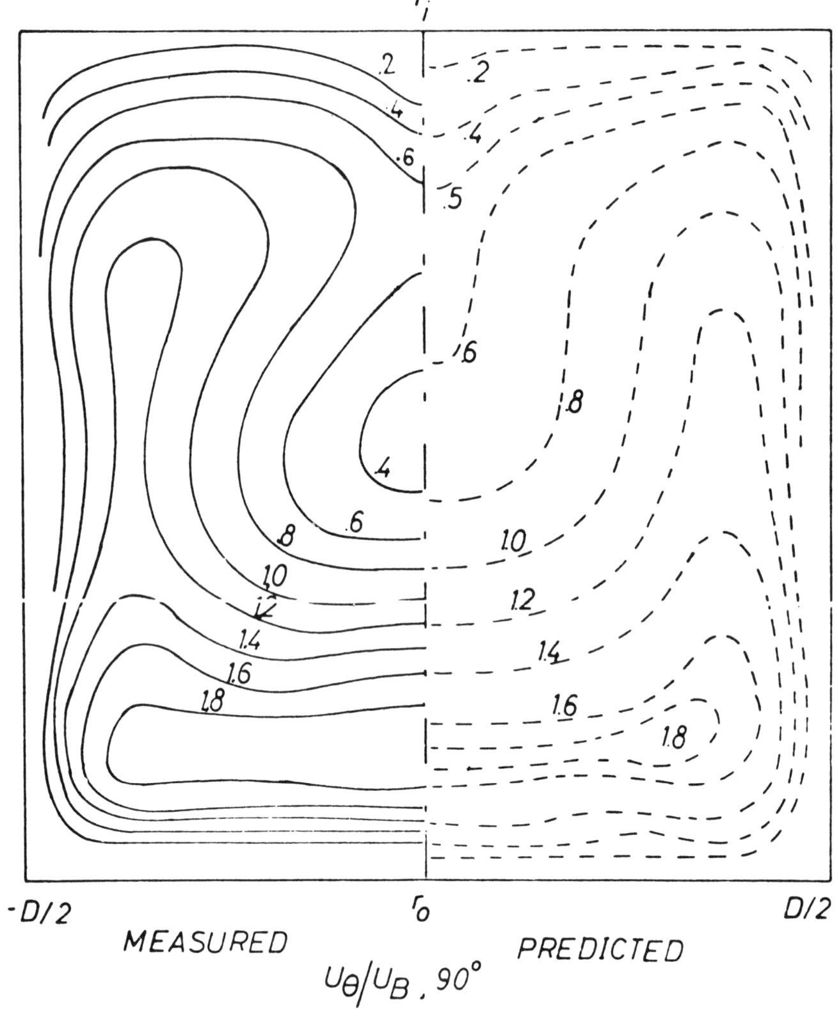

Figure 7. Comparison between experimental and calculated axial velocity components at 90° position in a 90° square cross-section bend; Re = 790, $\overline{R}/b = 2.3$, $U_B = .02$ m/s, grid is $10 \times 15 \times 19$.

E the number of equations. This also indicates the difficulties involved in obtaining solutions of equations appropriate to turbulent flow. If a two-equation turbulence model is used to represent the turbulence structure of the present turbulent flow, the computer run time would be of the order of 150 min for the numerical mesh used to obtain the results of Figure 7; and this presumes convergence in a similar number of iterations.

The previous discussion demonstrates that the calculation of the present flow is unlikely to be achieved cheaply. Of course, the numerical procedures can be improved but the cost of calculations with any three-dimensional, fully elliptic scheme will remain expensive. And yet, the use of such a scheme is necessary to correctly identify regions of forward recirculation and their consequences. The solution of partially elliptic equations (three-dimensional storage for the pressure field and two-dimensional for velocity) and three-dimensional parabolic equations can be achieved at considerably reduced cost but will not represent forward recirculation.

3.2 Co-axial Jet Flow with Swirl

The variation of the mean velocity and normal stresses along the center line are presented in Figures 8 and 9. The inverse of the center-line mean velocity is shown to be linear with distance from the origin for distances greater than around four diameters; the virtual origin appears to be located at a negative value of x, corresponding to -3.5 diameters from the inlet. The form of the distribution is similar to that obtained without swirl except that the swirling flow is accompanied by a higher spreading rate and a very much faster velocity decay for $x/D < 0.3$. This last difference is a direct result of the pressure sink in the transport equation for the axial momentum: the sudden renewal of the confinement imposed a high rate of change on the angular velocity field and the center-line pressure follows this change. The growth of the normal stresses is similar in form to that in the absence of swirl but the present values increase rather more steeply.

Figure 10 presents mean-velocity profiles at downstream distances corresponding to x/D of 1.5, 3, 5 and 6 and, together with the Reynolds stress results presented by Ribeiro (1976), allow detailed comparison with calculation methods. Comparison with measurements obtained in the absence of swirl demonstrates the considerable increase in mixing which stems from the swirl, even with the present relatively weak swirl. This is achieved by the centrifugal forces which impose a pressure gradient in the radial direction, whose effect is to increase the ratio of spread. This sets up strong gradients in the velocity field which, in turn,

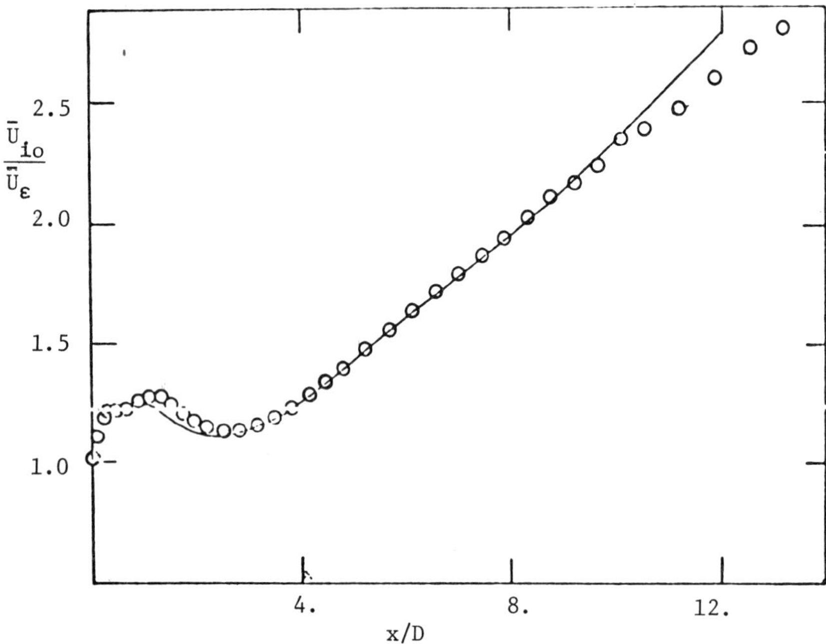

Figure 8. Center-line behaviour of the mean velocity coaxial jet with swirl. O Experiments, — Predictions.

generates turbulence. As the turbulence levels increase, so do the mixing and the rate of spread. Thus, the process is dictated initially by the centrifugal forces and gradually taken over by the increased turbulence.

Ribeiro used the calculation method of Morse (1976) in an effort to reproduce the measurements presented above. Differential equations were solved numerically with dependent variables $\overline{U}$, $\overline{rv_\theta}$, k, $\overline{u^2}$, $\overline{v_\theta^2} - \overline{v_r^2}$, $\overline{uv_r}$, $\overline{uv_\theta}$, $\overline{v_r v_\theta}$ and ε.

These conservation equations were parabolic in form. The Reynolds stress turbulence model is similar to that of Launder, Reece and Rodi (1975). The axial pressure gradient rate, immediately downstream of the pipe exit, caused a very high spreading rate over a short distance; the resulting strong deviation from orthogonality between the independent variables could not be represented by the calculation procedure and initial values had to be specified from measurements of the mean velocity and the components of the stress tensor at x/D = 1. This also obviated the need to consider the region of vortex shedding immediately downstream of the pipe wall.

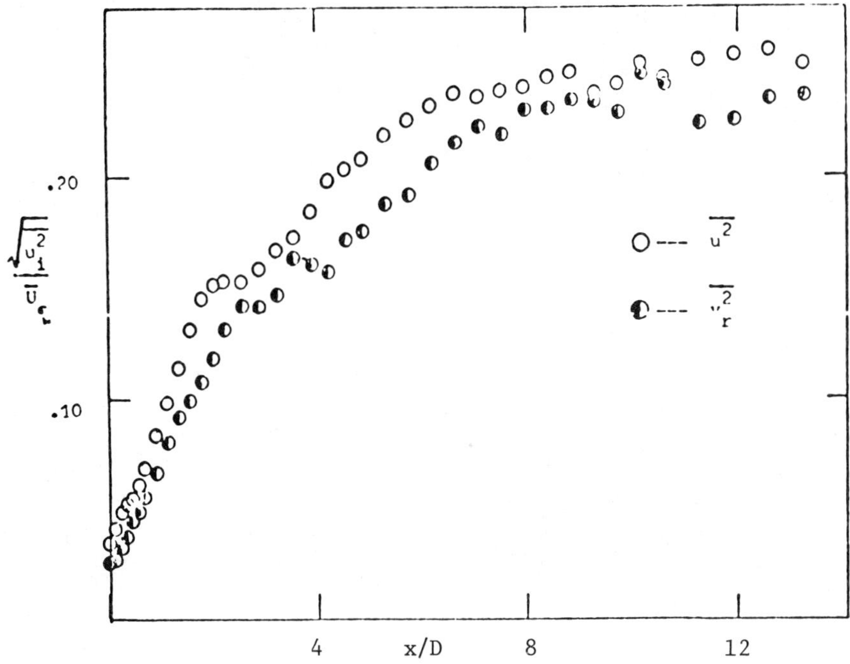

Figure 9. Center-line distribution of normal stress.

The calculated center-line variation of mean velocity is shown on Figure 8 and measured and calculated profiles of U, $\overline{u^2}$ and $\overline{uv_r}$ on Figures 10, 11 and 12 for downstream values of x/D of 1.5, 3, 5 and 6. The results show that the turbulence model has a tendency, which also prevailed in non-swirling flows, to result in high values of the shear stress; as a result, it can be anticipated that the spreading rate will be overestimated further downstream. Nevertheless, the calculations are adequate for most engineering purposes.

The region between x/D of zero and unity deserves further comment. In the region immediately downstream of the wall of the inner pipe, vortex shedding occurred with consequent increase in mixing. In the turbulence model, the production of turbulent kinetic energy was calculated in its exact form and dissipation was modelled; neither of these terms take account of periodic fluctuations. Indeed, the use of a time-averaged set of equations to predict the occurrence of a non-random oscillating velocity field will result in the underprediction of the mixing.

The region of flow close to the free boundary is also not well represented by the turbulence model since, as shown by the

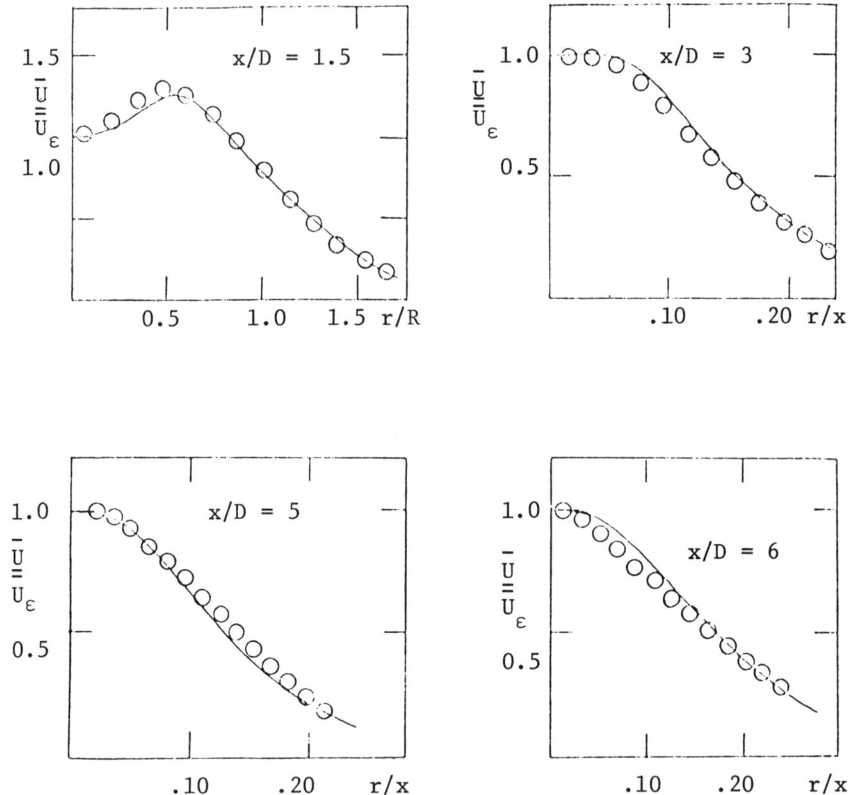

Figure 10. Radial distributions of $\overline{U}/\overline{U}_{℄}$ in a swirling coaxial jet.

probability measurements of Ribeiro, the diffusive mechanism is associated with large scale motion. In this region, the turbulent field is very asymmetric and, as a consequence, the third-order correlations influence the transport of the Reynolds stresses. In the turbulence model, and in contrast, the third-order correlations are supposed to be determined by the second-order correlations and their gradients.

3.3 Film Cooling of Curved Surfaces

The wall pressure measurements of Figure 13 suggest that the flow separates from the convex wall at longitudinal distances which depend on the radius of curvature, the slot gap and the velocity

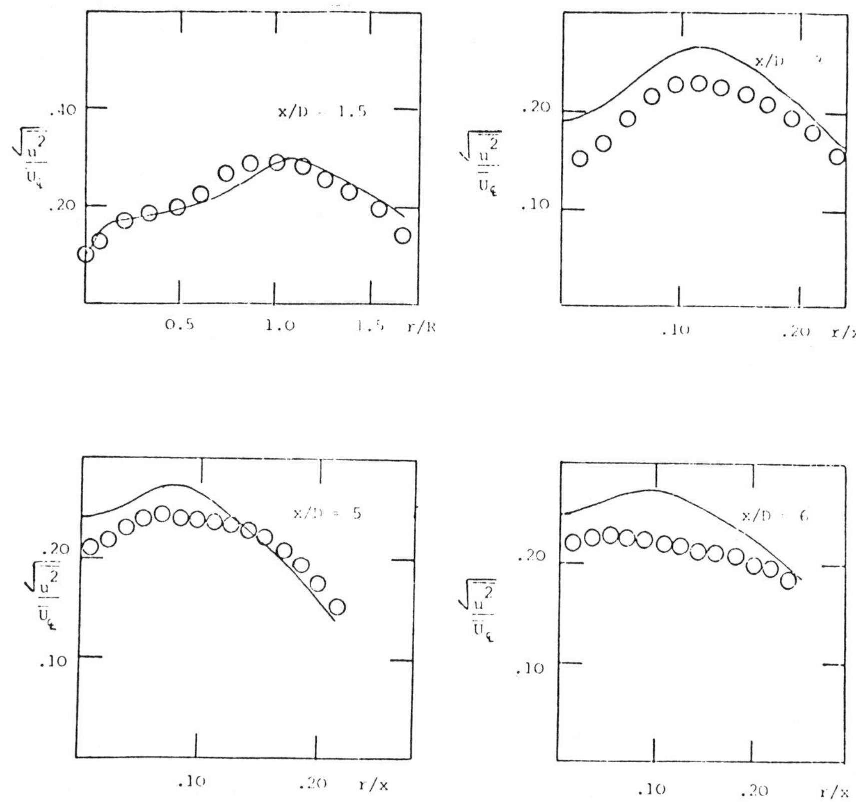

Figure 11. Radial distributions of axial normal stress in a swirling coaxial jet.

ratio. Thus, with the 76 mm radius, the film was destroyed at downstream distances less than 47 slot gaps; and this distance shortened, almost exponentially, as the velocity ratio decreases. The larger slot gap resulted in a slight increase in the length of the film and the larger radius in a larger increase. These results are in general agreement with the findings of Kind (1968) and are reflected in the effectiveness results.

Figure 14 presents measurements of the impervious-wall effectiveness obtained on flat, concave and convex surfaces and indicates that a convex surface results in improved effectiveness and a concave surface in reduced effectiveness. As suggested by the static-pressure distributions, however, the convex surface results in

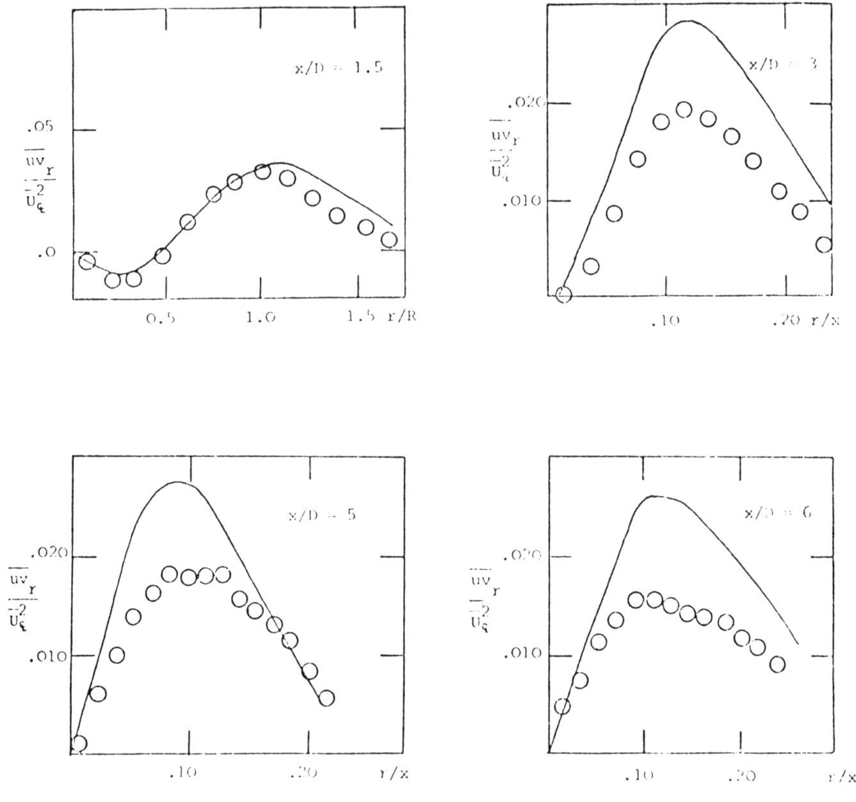

Figure 12. Radial distributions of $\overline{uv_r}/\overline{U_{\mathbb{C}}^2}$ in a swirling coaxial jet.

separation of the flow and, immediately downstream of the region where this occurs, the effectiveness falls rapidly. The effect of the flow separation is much more important than the differences in effectiveness observed in regions of attached flow. The figure shows that separation occurs at lower values of x/y_c for lower velocity ratios and for smaller radii. The influences of density ratio, for example, may be deduced by comparing Figures 14 and 15: in general, the larger density ratio results in improved effectiveness and delays the onset of separation on the convex surface.

In attempting to calculate the results of Figures 13 to 15, and thereby to provide confidence in calculations outside the range

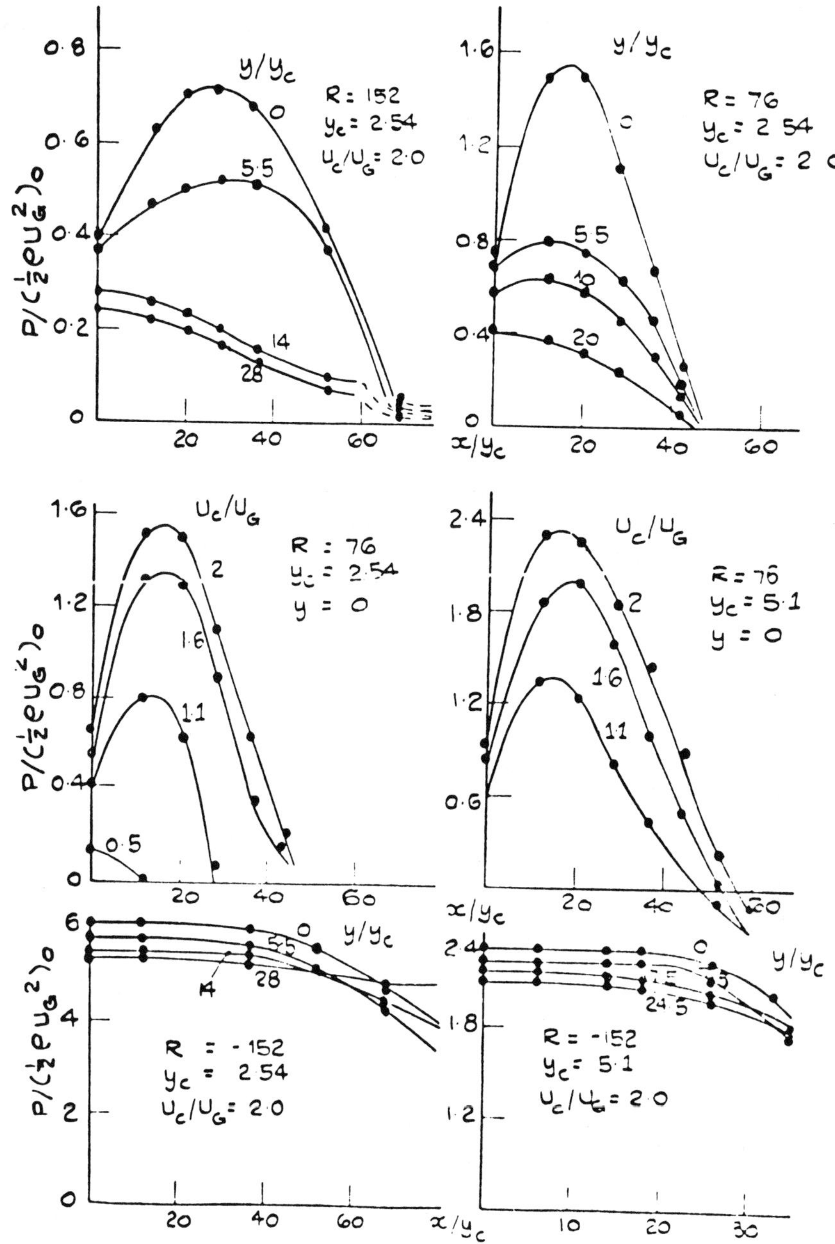

Figure 13. Static pressures on convex and concave surfaces.

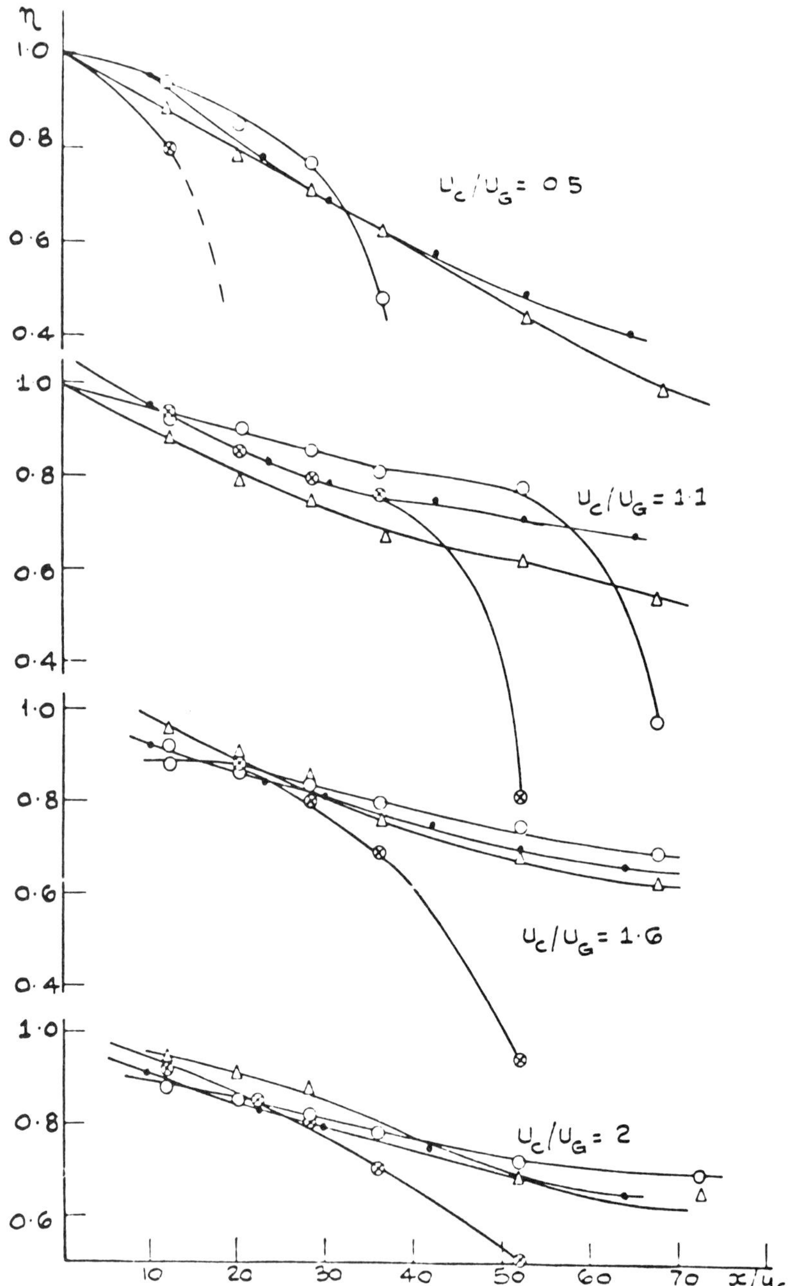

Figure 14. Impervious wall effectiveness on flat, convex and concave surfaces; ρ_C/ρ_G = 1.0.

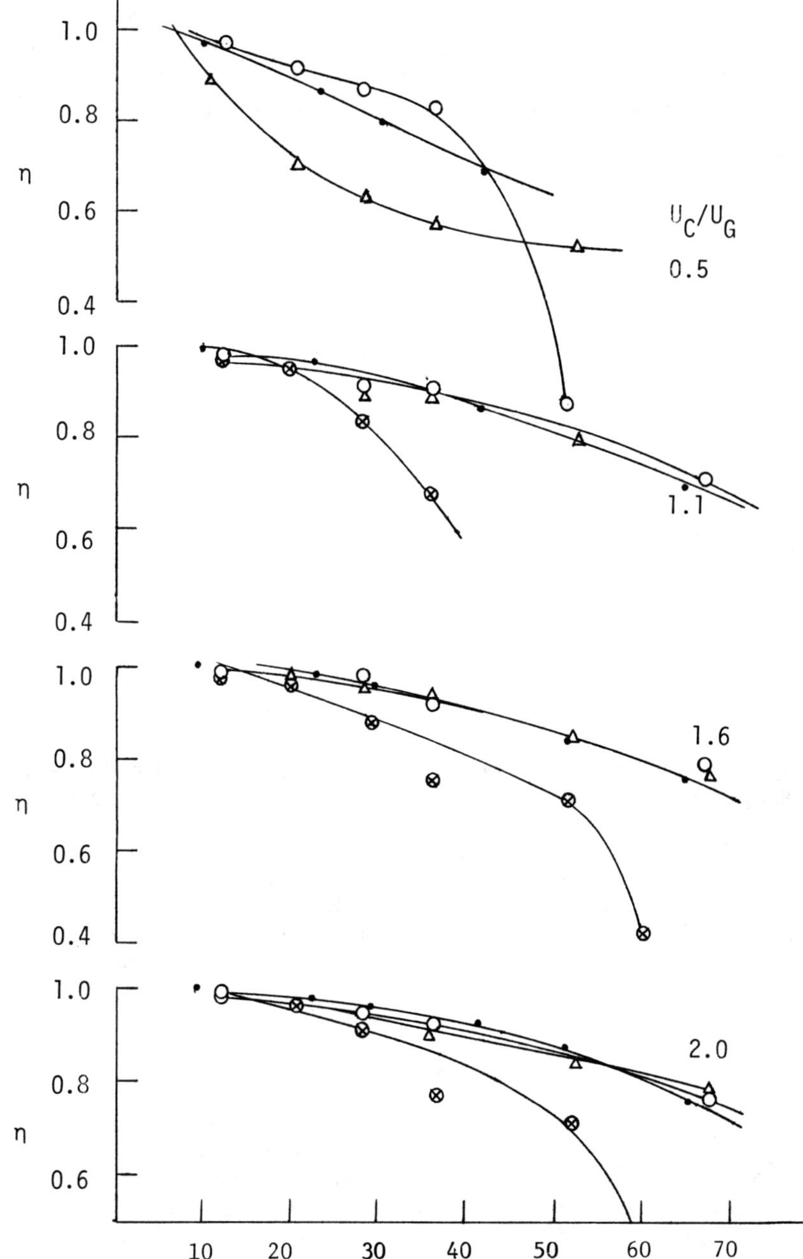

Figure 15. Impervious wall effectiveness on flat, convex and concave surfaces; ρ_C/ρ_G = 2.25.

of measurements, no attempt was made to represent the details of the turbulent flow. Time-averaged equations, appropriate to flows with boundary-layer thickness of similar magnitude to the radius of curvature, were solved with an effective viscosity hypothesis of the form:

$$\tau_{eff} = \mu_{eff} \left(\frac{\partial U}{\partial y} - A \frac{U}{R}\right) = \ell^2/\partial U/\partial y \left(\frac{\partial U}{\partial y} - A \frac{U}{R}\right)$$

and

$$\ell = \ell_0 \ (1 - \beta R_i)$$

$$\ell_0 = 0.41y \quad , \qquad 0 < 0.41y < 0.09y_G$$

$$= 0.09y_G \quad , \qquad 0.41y > 0.09y_G .$$

$$\beta = 1$$

$$A = 1$$

$$R_i = \frac{U}{R(\partial U/\partial y)}$$

An effective Prandtl number of unity was also assumed and the near-wall pressure gradient was represented in the wall functions.

Calculations and measured values of effectiveness can be compared on Figure 16. The agreement is acceptable, for engineering purposes, where calculations could be obtained. The numerical procedure failed to operate upstream of the experimentally observed region of separation and yielded increasing incorrect results as it was approached.

4. DISCUSSION

The purpose of this section is to indicate briefly the relative advantages of the experimental techniques discussed in relation to the particular investigations of the previous chapter. In general, constant temperature, hot-wire anemometry is best suited to clean flows with turbulence intensities less than around 30%. The technique is subject to the problems of probe interference and cannot resolve the direction of flow. This last difficulty has been overcome, at least in part, by the pulse-wire method of Bradbury (1969) but the method is still unsuited to flows with recirculation and to

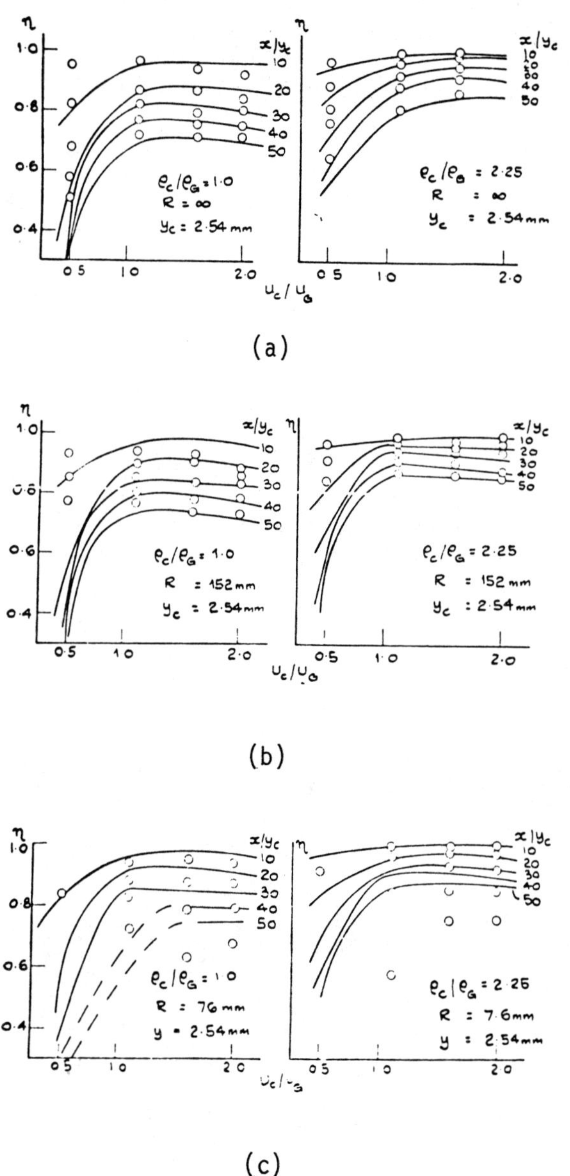

Figure 16. Comparison between measurements and calculations.

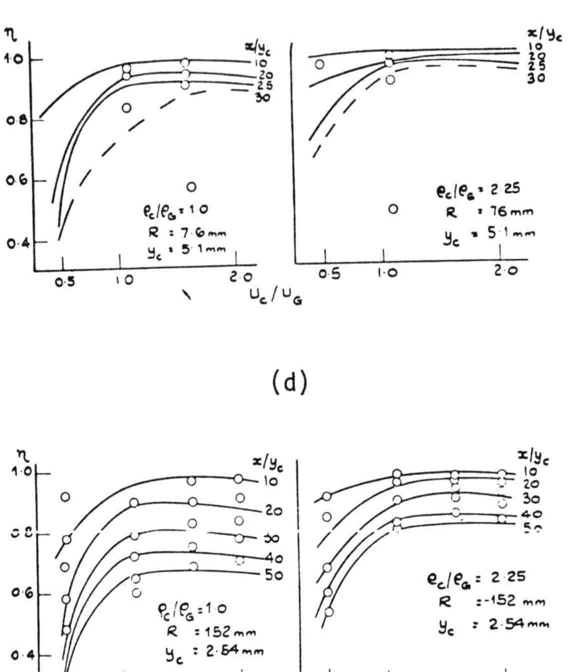

(d)

(e)

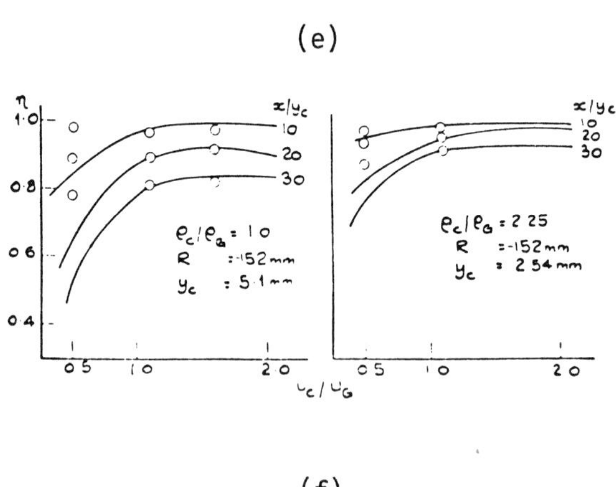

(f)

Figure 16 (cont'd.)

flows which have scalar gradients or contain particles. It is often difficult to resolve three orthogonal components of velocity due to the sensitivity of the hot wire to three components of velocity at the same time. Within these limitations, however, the hot-wire anemometry is the most accurate method available for the measurement of turbulent flow properties and is likely to be the most widely used for research into boundary-layer type flows accessible to probes.

Laser-Doppler anemometry, in contrast, can resolve the direction of velocity; is not limited in its ability to measure in highly turbulent or recirculating flows; does not interfere with the flow; can be used in dirty and hostile environments; and can resolve velocity components. It requires optical access, however, and is usually unable to provide a continuous signal. It is less accurate, in many simple boundary-layer type flows, than is hot-wire anemometry.

The previous statements suggest that hot-wire and laser-Doppler anemometry are complementary. Either technique can readily be used for the investigation of diffuser-type flows and, providing there is no recirculation, hot-wire anemometry may be preferable. In flow passages where recirculation is present or where a severe pressure gradient exists in the direction normal to the flow, laser-Doppler anemometry may be preferable. In rotating machinery, laser-Doppler anemometry has been used successfully although difficulties have been experienced in obtaining measurements close to solid surfaces. The optical technique is certainly preferable in hot and combusting flows where hot wires are unlikely to result in useful information.

The two techniques are capable of measuring the probability density distribution of velocity components and, as a result, the mean velocity and single point correlations. In this respect, the laser-Doppler anemometer has the advantage of being able to resolve the flow direction and can, therefore, result more readily in the separation of the velocity components. Correlations of two velocity components can also be obtained with both techniques as can energy spectra. In the latter case, however, the high wave numbers are likely to be better represented by hot-wire measurements than laser-Doppler measurements. In addition, since the laser anemometer seldom results in the continuous signal representing velocity information, spectrum analysis and correlation with continuous signals can cause difficulties.

In contrast to the techniques discussed in the present paragraphs, the concentration method used to obtain the results of section 3.3 is simple and provides information of engineering importance at low cost and with comparative ease.

REFERENCES

1. Champagne, F. H. and Sleicher, C. A. (1967) Turbulence measurements with inclined hot wire. Part 2 Hot-wire response equations. J. Fluid Mech. 28, 177.

2. Durst, F. and Whitelaw, J. H. (1971) Integrated optical units for laser anemometry. J. Phys. E: Sci. Instrum. 4, 804.

3. Durst, F., Melling, A. and Whitelaw, J. H. (1976) Principles and Practice of laser-Doppler anemometry. Academic Press.

4. Folayan, C. O. and Whitelaw, J. H. (1976) The effectiveness of two-dimensional film cooling over curved surfaces. ASME Paper.

5. Humphrey, J. A. C., Melling, A. and Whitelaw, J. H. (1976) Laser-Doppler anemometry for the verification of turbulence models. Proc. Conf. on Eng. Uses of Coherent Optics, 429. Cambridge U. P.

6. Kind, R. J. (1968) A calculation method for circulation control by tangential blowing around a bluff trailing edge. Aero. Quarterly 19, 205.

7. Launder, B. E., Reece, G. J. and Rodi, W. (1975) Progress in the development of a Reynolds stress turbulence closure. J. Fluid Mech. 68, 537.

8. Morse, A. P. (1976) On the calculation of asymmetric flows with a Reynolds stress closure: 1. Solution of the equations for two-dimensional parabolic flows. To be published as internal report, Mech. Eng. Dept., Imperial College.

9. Melling, A. (1975) Investigation of flow in non-circular ducts and other configurations by laser-Doppler anemometry. Ph.D. Thesis, University of London.

10. Melling, A. and Whitelaw, J. H. (1973) Measurements in turbulent water flow by laser anemometry. Proc. 3rd Symposium on Turbulence in Liquids, Rolla, 115.

11. Melling, A. and Whitelaw, J. H. (1976) Turbulent flow in a rectangular duct. Submitted to J. Fluid Mech.

12. Ribeiro, M. M. (1976) The turbulence structure of free jet flows with and without swirl. Ph.D. Thesis, Imperial College of Science and Technology, London.

13. Ribeiro, M. M. and Whitelaw, J. H. (1976) Statistical characteristics of a turbulent jet. J. Fluid Mech. 70, 1.

DISCUSSION

LAUFER: (University of Southern California)

I was rather intrigued by your results in connection with the curved channel experiment in which you showed that just ahead of the entrance to the curved region, apparently the upstream influence of the pressure field manifests itself more in the mean velocity distribution than in the Reynolds stress distribution. In the past our ideas rested on the notion that it was the other way around. Usually the Reynolds distribution is more sensitive to any kind of external influence that one imposes on the flow. I wonder if you would like to comment on that.

WHITELAW:

The measurements I showed are only an example. We have measured from something like 12 hydraulic diameters upstream and we see this trend from around 3. The influence is more on the mean than on the normal stresses. I think that perhaps the reason is that the pressure in the mean flow equations is balanced by the centrifugal effects and it operates on the mean flow equations before it comes to the correlation equations.

LUXTON: (University of Adelaide)

I wonder whether from your results on the curved bend you could say anything about the situation where the flow is not fully developed in the straight pipe and only a thin boundary layer exists on the walls before entry to the curved section?

WHITELAW:

No. Our flow was not fully developed at entry to the bend but it was nearly so. The reason I hedged on turbomachinery is that, in blade passage, the initial boundary layers are then unlike the present flow and some of the conclusions which we draw from the present result may not apply in turbomachinery. If you consider the turbomachinery situation and imagine the present geometry to be a blade passage, even though the flow is an incompressible low Mach number and low Reynolds number, the flow in the blade passage will be different because of the accelerations. First the initial boundary layer is thin and secondly the accelerations will stop any tendency towards separation in the initial region which I discussed. Having said that, on the pressure side of the blade, at least in my experience, there can be regions of recirculation in the upstream region; so to calculate the blade performance satisfactorily, there may be problems but for a different reason.

BRADSHAW: (Imperial College)

The main question is about the use of the semi-elliptic procedure for calculating this type of flow. By "semi-elliptic" I understand that you mean continued sweeps of a parabolic calculation combined with an iteration solution for the pressure field. Would you expect that if you apply such a semi-elliptic procedure to a flow in which recirculation really appeared, that (1) you would get an incorrect converged solution with no recirculation or (2) would you expect that the calculation would try to tell you something, by blowing up?

WHITELAW:

I think you could get both effects but the former is perhaps more likely.

BRADSHAW:

I wonder if I could add a second question covering something of what John Laufer said. Probably the reason that you get the same turbulence intensity on a given streamline rather than a given point in space is, as you say, a sort of "frozen stress" phenomenon; this idea has been used quite extensively for calculations of rapidly changing flows. Provided you are only interested in the bit that is rapidly changing in terms of mean flow distribution you can certainly assume a frozen stress model. But what about calculations for downstream of the bend? You have really got a model which corresponds perhaps to the flow around the front of a blade sitting on a hub. One would very much like to know what happens down towards the trailing edge. Do you have any message for us there?

WHITELAW:

First of all, I don't think that our flow is close enough to a real blade flow to be directly relevant. But I believe that we can make useful calculations, in a blade passage and near its trailing edge. I suspect that one will be able to get away with a simple type of turbulence model--a Reynolds stress model isn't necessary--with precision acceptable for engineering purposes.

NAGIB: (Illinois Institute of Technology)

You clearly stated that in the case of the swirling jet you preferred to use the hot wire instead of the laser. If the jet utilized water as the working fluid would you still prefer the former over the latter?

WHITELAW:

No, I would not.

NAGIB:

Why do you prefer to use the hot wire?

WHITELAW:

The duct with the water in it has relatively small dimensions. If I were to make the coaxial swirling free jet system with water then I would have a lot more water; it is, therefore, a lot easier to operate in air.

NAGIB:

Then I still don't quite understand why you prefer the hot wire to a laser in a flow as complex as a swirling jet with its three dimensionality, which may include recirculating flow zones.

WHITELAW:

I believe that in regions where the turbulence intensities are not too high, the hot wire is perfectly adequate for the coaxial jet flow. I don't particularly want to have to set up a very bulky water flow, just so that I can use a laser, when I can get satisfactory measurements with a hot wire system.

WYGNANSKI: (Tel Aviv University)

You have shown a pattern of large eddies moving into the corner in your rectangular channel flow. I wonder if you can comment how steady this pattern is? Did you measure these also in the curved channel flow and what is the effect of curvature on the steadiness of the pattern?

WHITELAW:

I think they are very steady. We are very cautious with this kind of flow. We use full visualization methods, with dye traces, to supplement the laser results. I believe that the secondary flows are very steady in both sets of observations.

One comment I did not make was that, in the bend, the maximum secondary flow velocities are up to 30% of the maximum axial velocity in contrast to the 7% in the straight duct.

FALCO: (Cambridge University)

I wanted to ask two questions. The first is you do not seem to have the capability in your laser system to measure reversed flow, which your calculations seem to predict. Did you have that capability?

WHITELAW:

The arrangement I showed in the slide was a straightforward two-beam system. When we have recirculation, light-frequency shifting devices may be necessary and we have them available. In laminar separated flows, however, frequency shifting is usually not necessary.

FALCO:

But the measurements you showed us and interpreted in the turbulent flow, did they have it?

WHITELAW:

No and they did not need it in the range of the present investigation.

FALCO:

My second question is the following. In your calculation of swirling jet flow, can you get the trend that is observed in single swirling jets, an overshoot in the Reynolds stress and then an undershoot, that is, the Reynolds stress is initially higher at the exit of the jet (over the first few diameters) than in the unswirled jet case, but it then decreases to lower values than found in the jet without swirl?

WHITELAW:

Figure 13, taken from the thesis of M. M. Ribeiro,* may answer your question. Further information is provided in the thesis.

EVANS: (University of Toronto)

I just want to ask, again with respect to the curved duct, whether you have compared your predictions with a simple inviscid secondary flow approach. I wonder if you might get better or, at least good engineering approximation at much less expense?

---

* M. M. Ribeiro (1976), "The Turbulence Structure of Coaxial Jet Flows with and without Swirl." Ph.D. Thesis, University of London.

WHITELAW:

We haven't. I don't think I expect to get satisfactory results, however.

EVANS:

Nowhere near the kind of accuracy you get with this system?

WHITELAW:

I think that the accuracy will be much worse.

BRADSHAW:

I think that the inviscid secondary-flow model is just one step down from the frozen-stress model; it is a neglected-stress model. You can do a little bit better if you keep the stress on a given streamline the same as it was at the start of the distortion and that, I think you will agree, will probably give reasonably good results, except possibly close to the wall where you can use the dear old log law.

SAFFMAN: (California Institute of Technology)

In the calculations which you reported did you use what is euphemistically called computer optimization or did you take the constants as given to you by the person who developed the model and then see what was calculated?

WHITELAW:

In the swirling flow we used "constants" which have been used by several authors including, for example, Pope and Whitelaw.* Perhaps I can say one more thing, these constants change from time to time but I don't believe that they change to a degree that makes me unnecessarily worried. I would like to go back to a comment that I tried to make earlier, that when you time average equations you throw away information, so we shouldn't expect perfection and we can't expect perfect universality either.

MELLOR: (Princeton University)

You said you thought the two equations wouldn't work on this particular problem, the curved flow. It seems to me that the full Reynolds stress equation, as simplified by neglecting some advective and diffusive terms, would work well.

---

* Pope, S. B. and Whitelaw, J. N., J. Fluid Mech. 73, 9, 1976.

WHITELAW:

My hunch says that we will get a long way with a straightforward K-ε model.

MELLOR:

Curvature effects--and I think Bradshaw will support me here--are strong effects. They can turn off turbulence close to the walls. The Reynolds stress equations do seem to account for this.

WHITELAW:

The proof of the pudding is in the eating and, for this flow, we haven't yet gotten indigestion.

WILCOX: (DCW Industries)

Following up on what Professor Mellor said, it is possible with the two equation turbulence model approach to account for curvature effects. However, if you just take the two equations and apply them cold they won't do a very good job. If you can make a distinction between what people usually call the turbulent kinetic energy and view the quantity more like a mixing velocity (which in the case of a boundary layer would be the fluctuations normal for the surface and not the total kinetic energy), it is possible to modify the two equations and obtain very good results.

WHITELAW:

I would like to go back to my starting point. If we try to solve or if we get into a position where we have to solve the full elliptic equations it is an expensive procedure, so we have to try with the simple assumptions and then work up from there.

BARBER: (Pratt & Whitney)

One question first. What was the ratio of the radius of pipe to the radius of curvature on the bent pipe?

WHITELAW:

The bend had a mean radius of 92mm and a 40mm × 40mm cross-section.

BARBER:

Another comment about the analogue of the bent pipe relative to the turbomachinery flow: there was a paper presented by Lee

Langston* at the Gas Turbine Meeting in New Orleans this year which made comment showing that the flow in the turbomachinery passage is dominated by the passage vortex and that this really makes the real flow situation in turbomachinery passages non-analyzable by these present secondary flow techniques, that any sort of analysis which tries to compute it as a straight secondary flow is not realistic.

WHITELAW:

I do not think I understand the point you made.

BARBER:

When you look at a turbomachinery passage or a turbine passage there is a stagnation point or a saddle point and a passage vortex which sweeps across the passage and that dominates the entire flow field. Any attempts to analyze the flow field by secondary flow analyses that do not incorporate this into the flow field will not correspond to realistic situations.

WHITELAW:

I will take your word for it.

MELLOR:

I think you are not talking about classical secondary flow calculations.

KLEBANOFF: (National Bureau of Standards)

Can you comment on how low a turbulence intensity field would be measured by a laser system?

WHITELAW:

If I lump all the spurious specimen effects and call them noise, then the noise effects with the laser system will be less than one percent, probably about 0.6%. This noise is slightly worse than that found with a hot wire.

---

* Langston, L. S., Nice, M. L., & Hooper, R. M.: "Three Dimensional Flow within a Turbine Cascade Passage," ASME paper 76-GT-50, 1976.

Small Disturbances in a Compressor with Strong Swirl

JACK L. KERREBROCK
Massachusetts Institute of Technology
Cambridge, Massachusetts (U.S.A.)

ABSTRACT

A detailed time resolved study of the flow field upstream and downstream of a high work transonic compressor rotor shows that the flow field is dominated by the downstream evolution of the viscous flow shed from the rotor blades under the influence of the strong mean swirl. The dominant periodicity in the flow changes from blade passing to 1.4 times blade passing within one chord from the blade row. A possible explanation is that the wakes evolve to a shear eigenmode of the swirling flow, as suggested by perturbation theory. Another possibility is a "propagating stall" of 16 cells, but the rotor operated near its design point. Treatments of "turbulence" in turbomachines should account for such phenomena, which originate in the strong mean swirl.

1. INTRODUCTION

The interpretation given here to "turbulence in internal flows" is rather broad; what will in fact be discussed is the structure of the unsteady flow near a high-pressure-ratio transonic rotor which is more or less typical of those near the front of high performance axial compressors. It is hoped that the discussion will serve to establish a framework for interpretation in the turbomachine context of the discussions of turbulence and turbulent mixing which constitute the body of the Workshop. To anticipate the main point, behavior of wakes behind a rotor is very different from that of turbulent wakes behind a body or wing in uniform flow, the difference being due to the strong mean swirl in the compressor annulus.

This paper will comprise mainly a report of experimental observations of the unsteady flow, but some tentative explanations for the unusual features will be given, based on a perturbation analysis of the strongly swirling flow. Both the experimental data and the theoretical ideas have been presented elsewhere in more detail. The reason for repeating them here is to relate both directly to the subject of "turbulence in internal flows."

2. UNSTEADY FLOW IN A TRANSONIC ROTOR

The observations to be reported here have been made in the transonic rotor shown in the cross section in Figure 1. It has a hub/tip radius ratio of 0.5 at the inlet and produces a pressure ratio of 1.6 at a tip tangential Mach number of 1.2. In the MIT Blowdown Compressor Facility measurements of the local static and stagnation pressures and of the three components of Mach number can be obtained with a time resolution of about one tenth of the blade passing period. This is done with a multi-sensor probe using small semiconducting diaphragms, which is traversed across the annulus during the test time of about 40 milliseconds. Since the details of both the facility and instrumentation have been fully described elsewhere (Kerrebrock et al., 1974) only the results will be discussed here.

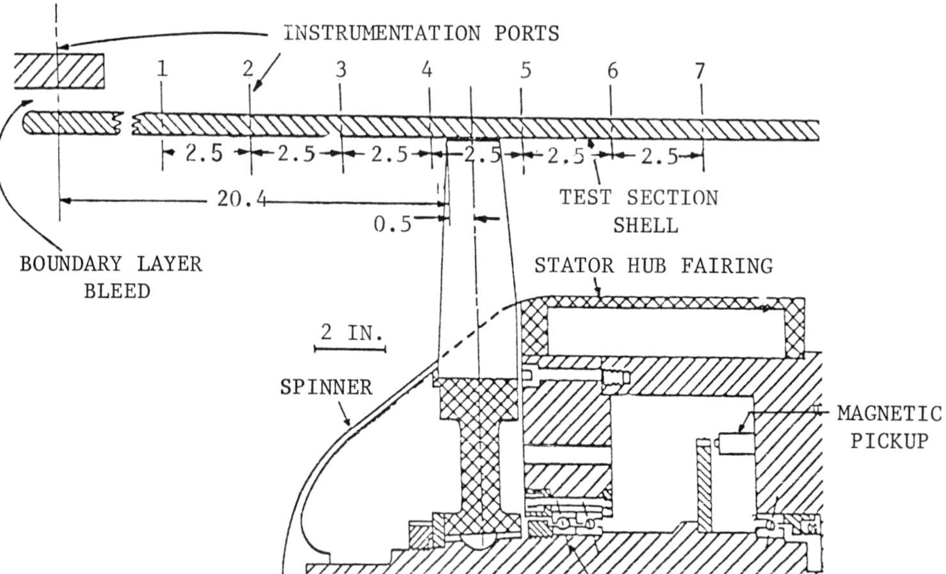

Figure 1. Scale Drawing of Transonic Rotor Installed in Blowdown Compressor.

2.1 Flow Downstream of Rotor

An example of the time-resolved output of the probe is shown in Figure 2 (Thompkins and Kerrebrock, 1975), which shows pitchwise flow angle and radial flow angle, both measured in stationary coordinates at station 5 of Figure 1 for some 5 blade passing periods, the radial probe location being at 0.87 of tip radius. The dashed vertical lines are the estimated locations of the blade trailing edges. Two points in particular should be noted: the flow is not periodic from blade to blade, and there are large radial flows on the pressure side of the passage near the blade trailing edge.

To establish that the rotor was operating at a condition of interest a stator was installed and a stage map constructed from a series of tests (Farokhi, 1976). This is shown in Figure 3. The rotor data described were taken at a back-pressure close to that for the dark point, labeled "operating point," which is approximately that for maximum efficiency at design speed and just below rotating stall. The efficiency of 0.89 is respectable if not remarkable for such a stage. Hence it is felt that the rotor flow should not be pathological. Indeed, there is evidence that rotor wakes often are not repeatable in stages near design.

By taking short time samples of the sort in Figure 2 at different radial positions of the probe the maps of entropy, pitchwise Mach number, and radial Mach number shown in Figures 4, 5 and 6 have been constructed (Thompkins and Kerrebrock, 1975). Note that all data on these figures came from one 40 millisecond test.

The entropy, shown as $(S - S_0)/C_V$, was computed from the local static pressure and temperature. The pressure is measured directly by the probe, while the temperature is inferred from the local Mach number and the stagnation temperature as computed from the Euler turbine equation with the measured tangential Mach number. It is near zero between wakes in the inner part of the flow, but shows regions of large dissipation in the wakes near the sonic radius and in general in the outer portion of the annulus. What periodicity exists is clearly at the blade passing frequency.

The radial and tangential Mach number maps similarly show periodicity at blade passing with strong radial outflow and excess tangential velocity in the wakes near the sonic radius. The radial Mach number at these points is as large as 0.6, i.e., the flow is outward at a 45 degree angle from axial.

A problem of "turbulence in internal flows" of interest to this Workshop would be to describe the evolution of this flow field as it passes downstream from the rotor. In the classical picture the

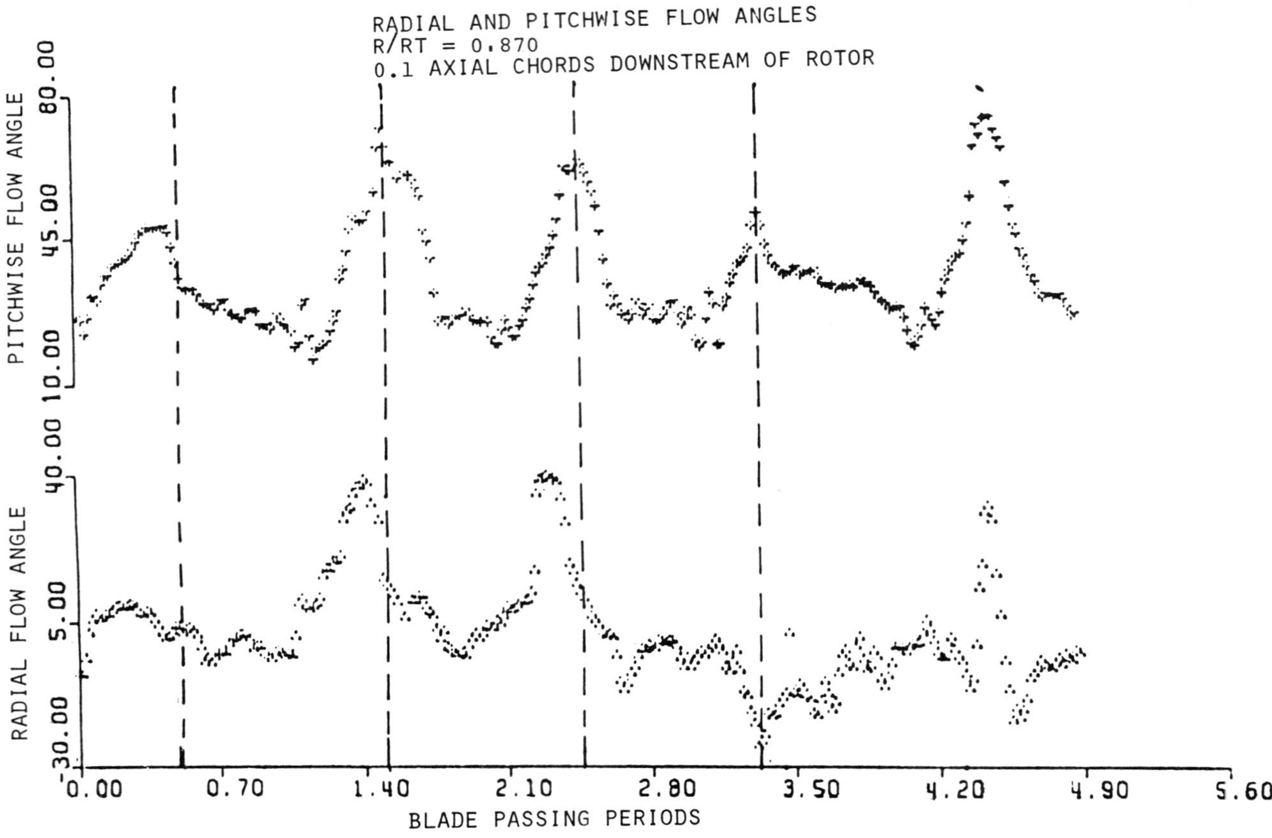

Figure 2. Time Histories of Radial and Pitchwise Mach Number at Station 5 and $r/r_T = 0.87$.

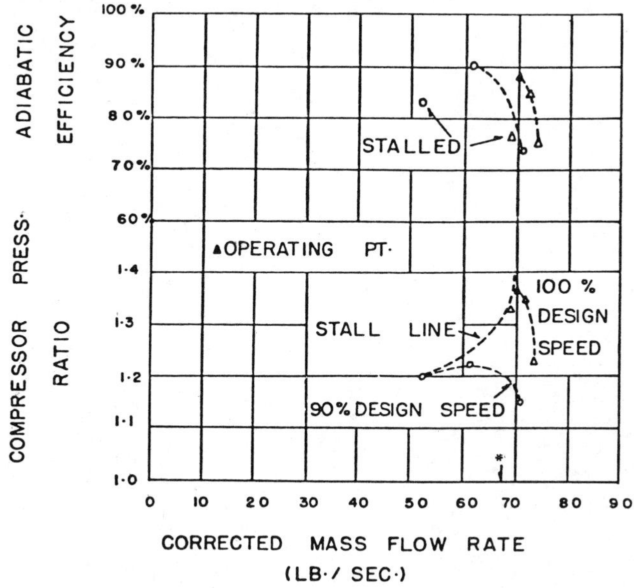

*ISOLATED ROTOR OPERATING AT 68 CORRECTED PPS.

Figure 3. Performance Map for Transonic Rotor with Stator as Determined in Blowdown Compressor Facility.

distinct wakes seen in Figures 2, 4, 5 and 6 would spread and gradually merge. In fact this is not at all what happens, as can be seen from Figures 7, 8 and 9 which show the entropy, tangential Mach number and radial Mach number at station 6, about one chord downstream of the rotor. There are still large concentrations of excess entropy, and high radial and tangential Mach numbers. But the most striking feature is that the period is now 1.4 blade passing periods. Close examination of Figures 8 and 9 shows that in the outer half of the annulus, the radial and tangential Mach numbers are about $\pi/2$ out of phase in $r\theta$. This is consistent with a cellular vortex structure such as that formed in the annulus between rotating coaxial cylinders, when the circulation decreases outward.

It must be emphasized at this point that the data in the above r versus $r\theta$ maps are temporarily, not spatially, resolved. The abscissa is time, not $r\theta$. Thus the period at station 6 is consistent with either (1) a disturbance with 16 spatial nodes in $r\theta$, rotating at the same tangential velocity which the fluid has at station 5, or (2) a disturbance with 23 nodes propagating against the rotor motion so as to give the lower frequency.

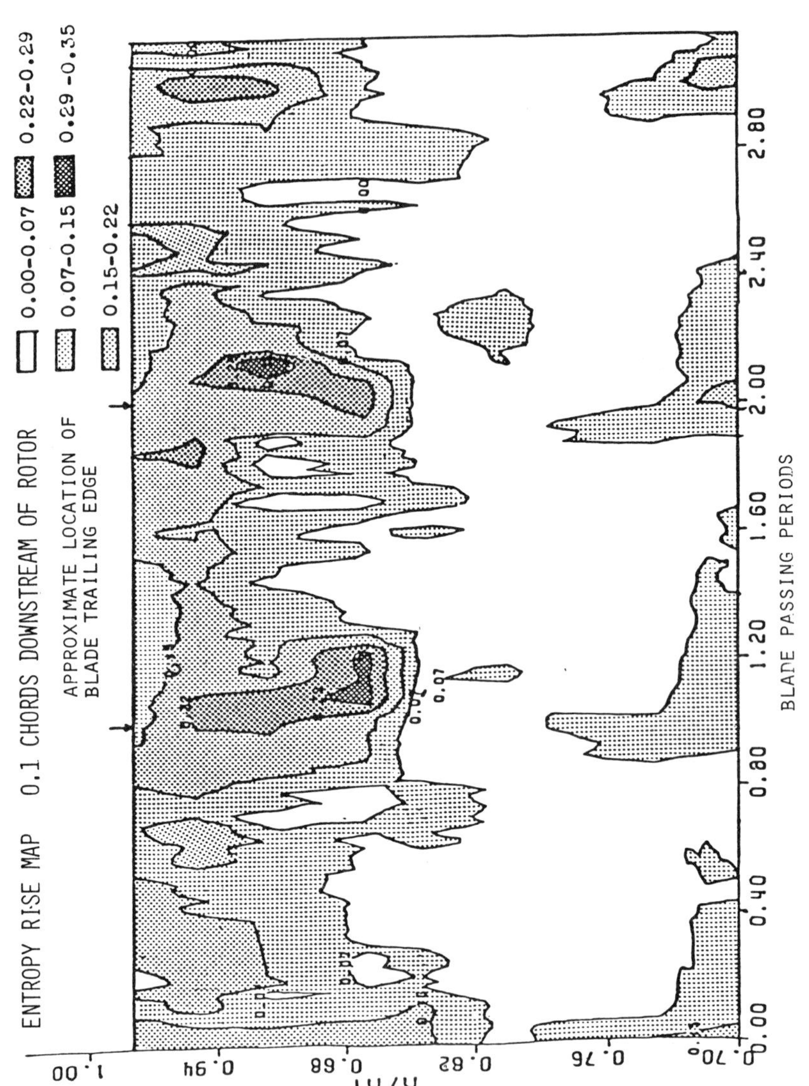

Figure 4. Entropy Relative to Upstream Flow as Determined at Station 5 as a Function of Radius and Time (Blade Passing Periods).

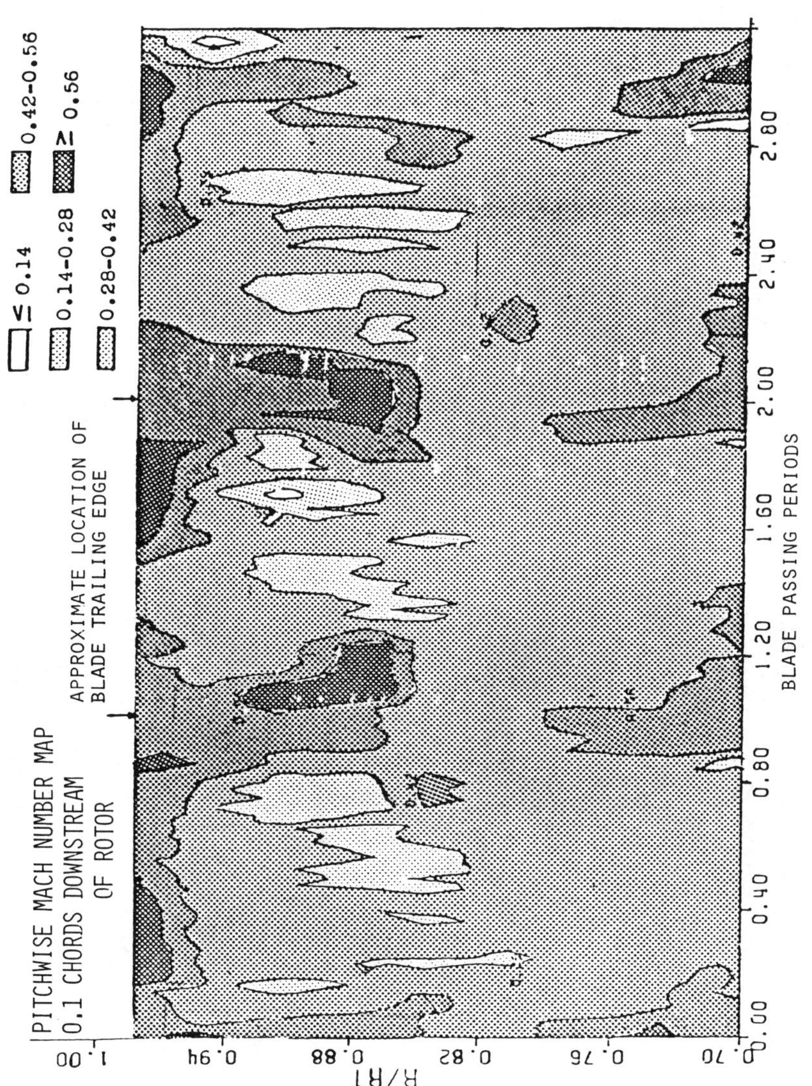

Figure 5. Pitchwise Mach Number at Station 5 as a Function of Radius and Time.

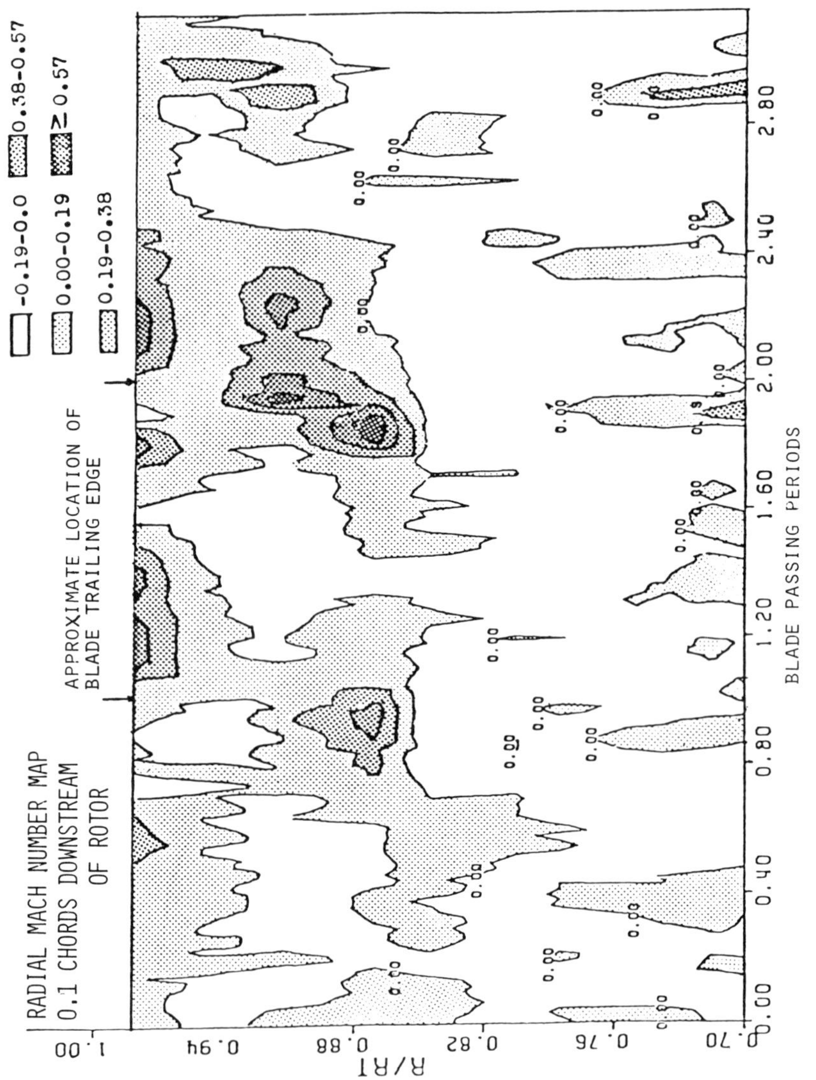

Figure 6. Radial Mach Number at Station 5 as a Function of Radius and Time.

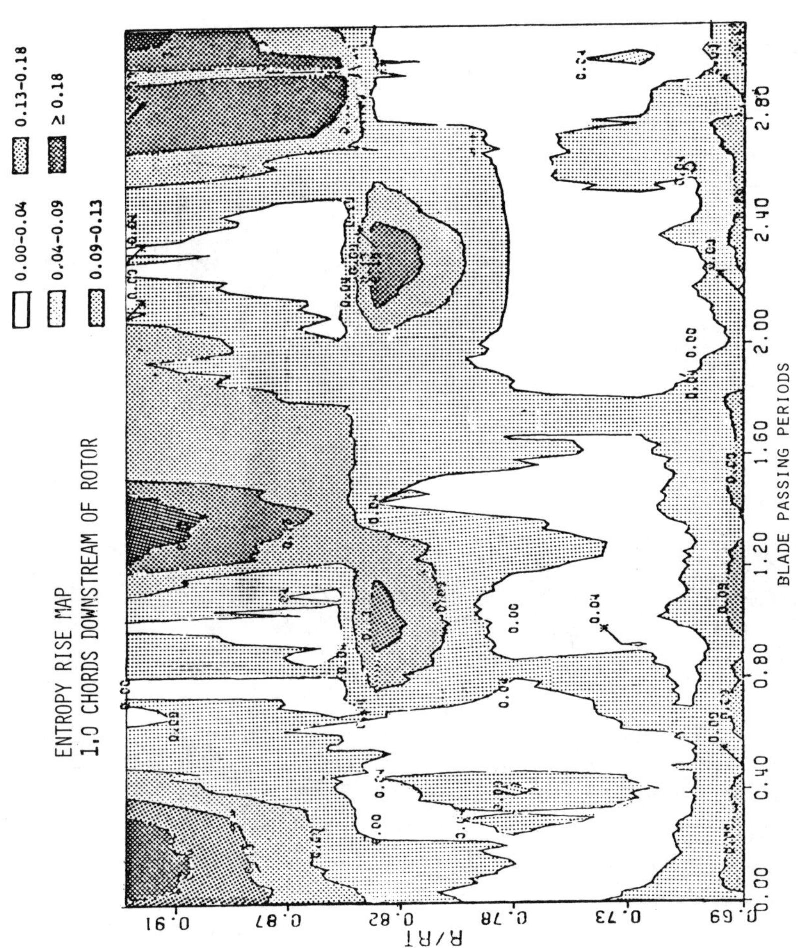

Figure 7. Entropy Relative to Upstream Flow at Station 6, One Chord Downstream of Rotor Outlet, Showing Periodicity at 1.4 Blade Passing Periods.

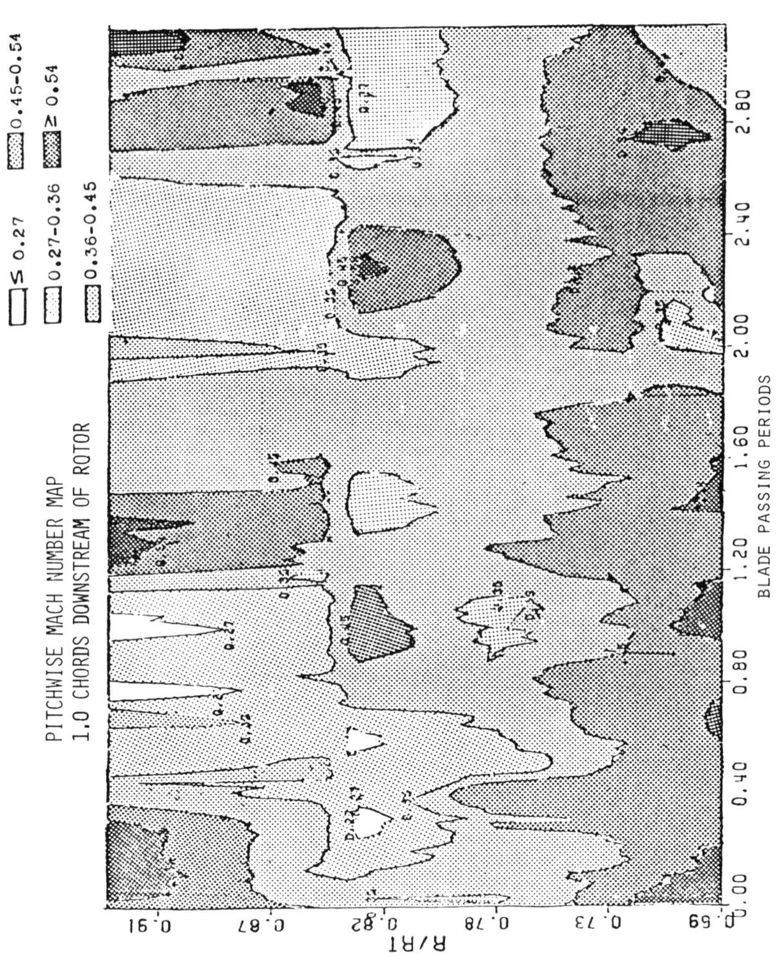

Figure 8. Tangential Mach Number at Station 6, Showing Strong Perturbation in Outer Portion of Annulus, with Periodicity at 1.4 Blade Passing Periods.

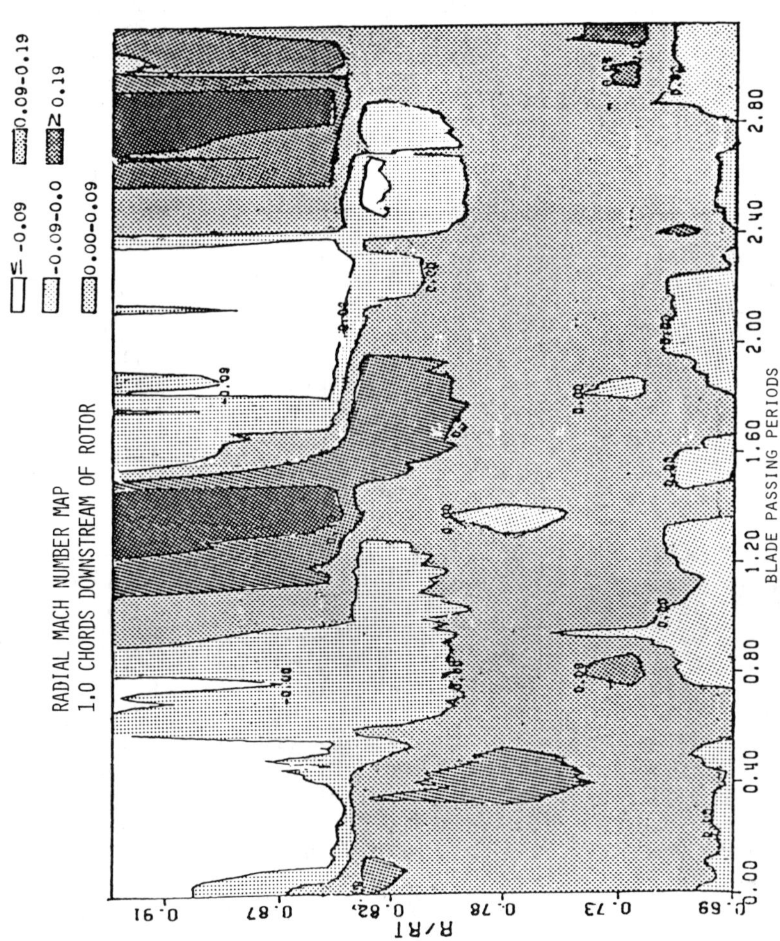

Figure 9. Radial Mach Number at Station 6, Showing Strong Perturbation in Outer Portion of Annulus, with Periodicity at 1.4 Blade Passing Periods.

2.2 Upstream Acoustic Data

Acoustic measurements carried out upstream of the rotor (Kerrebrock et al., 1974) clearly established the existence of a strong duct acoustic mode at a frequency of 16 per revolution. The harmonic analyses at the rotor face (port 4), at port 2 and at port 1 (see Figure 1) are shown in Figure 10. A typical "combination tone" structure can be seen at the two upstream stations, but the $m = 16$ and $m = 7$ harmonics dominate. It seems very likely that the $m = 16$ duct tone is excited by the downstream viscous disturbance with a period of $23/1.4 \simeq 16$ times blade passing.

This prompts the conjecture that the combination tone or "buzz saw" noise of high bypass turbofans may be excited at least in part by the viscous flow field downstream of the rotor.

Unfortunately, these upstream data do not help in choosing between the two alternative explanations for the change in periodicity which were advanced above. A visualization of the downstream flow field would resolve the question. The flow in the rotor passages has been quantitatively visualized by Epstein (1975) using the technique of gas fluorescence. This technique is presently being used to visualize the rotor outflow.

2.3 Downstream acoustic data

Acoustic measurements were also carried out downstream of the rotor (Stephens, 1974), both with a static pressure probe in the flow just inside the casing, and with a transducer flush with the casing wall. Harmonic analyses of each at stations 5, 6 and 7 are shown in Figure 11. The remarkagle point here is that the dominant tone shifts gradually from $m = 23$ (blade passing) to higher values at stations 6 and 7 in the flow, but it remains at $m = 23$ and decays much more rapidly at the wall. The latter behavior is what would be expected for a tone below cutoff. The former behavior is unexplained at present, but is most likely connected with the evolution of the viscous disturbances. (Practical difficulties with quantitative measurement of the static pressure within the flow are readily acknowledged, but what causes the frequency shift?)

3. THEORETICAL CONSIDERATIONS

A systematic perturbation theory has been developed (Kerrebrock, 1975; Thompkins & Kerrebrock, 1975) to describe the behavior of small disturbances in the strongly swirling flows characteristic of turbomachines. Only a very brief summary of the analysis and the results relevant to the above experimental observations will be given here.

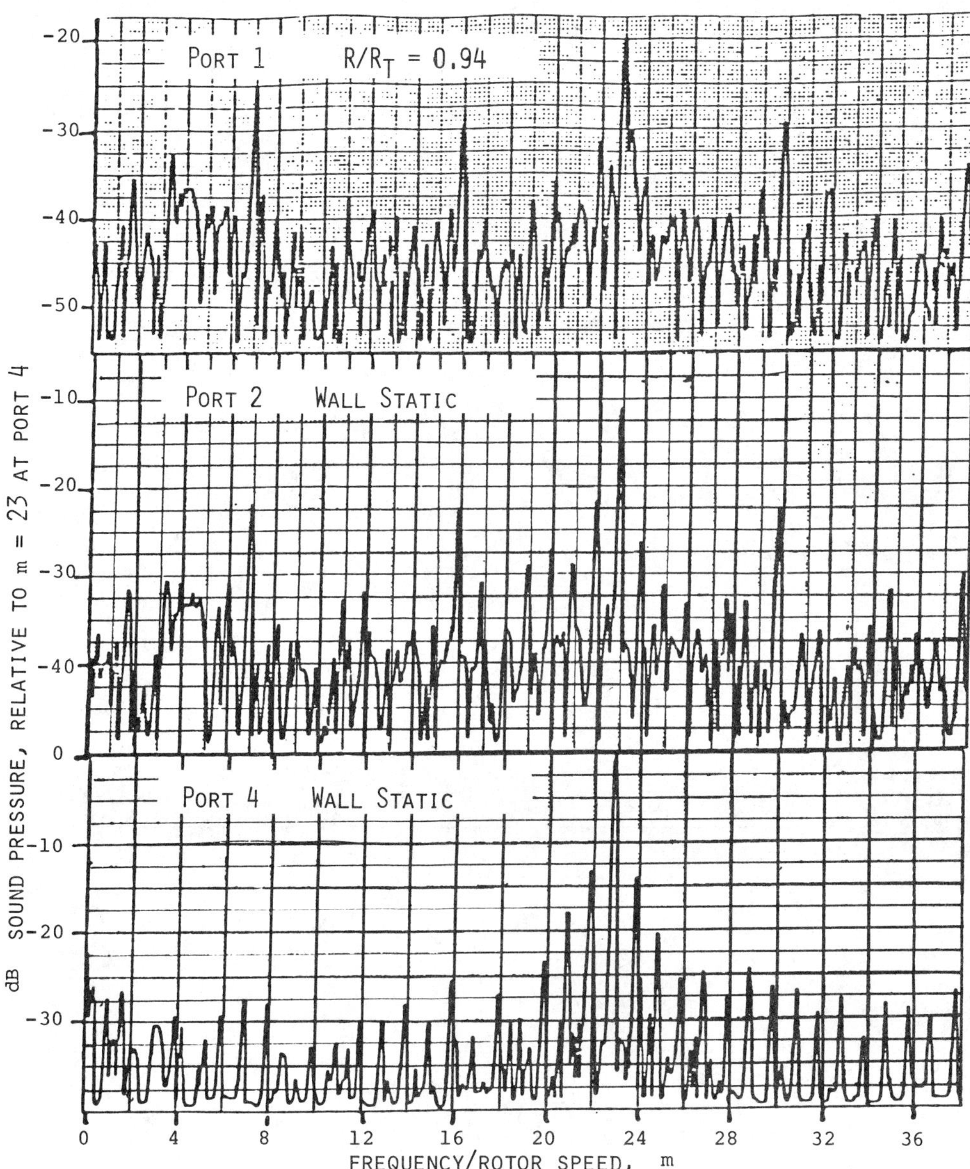

Figure 10. Harmonic Analysis of Sound Field Upstream of Rotor. Port 4 is at Rotor Face, Port 2 at 2 Chords, and Port 1 about 8 Chords Upstream. "Combination Tones" at m = 7 and m = 16 are Prominent at Upstream Points.

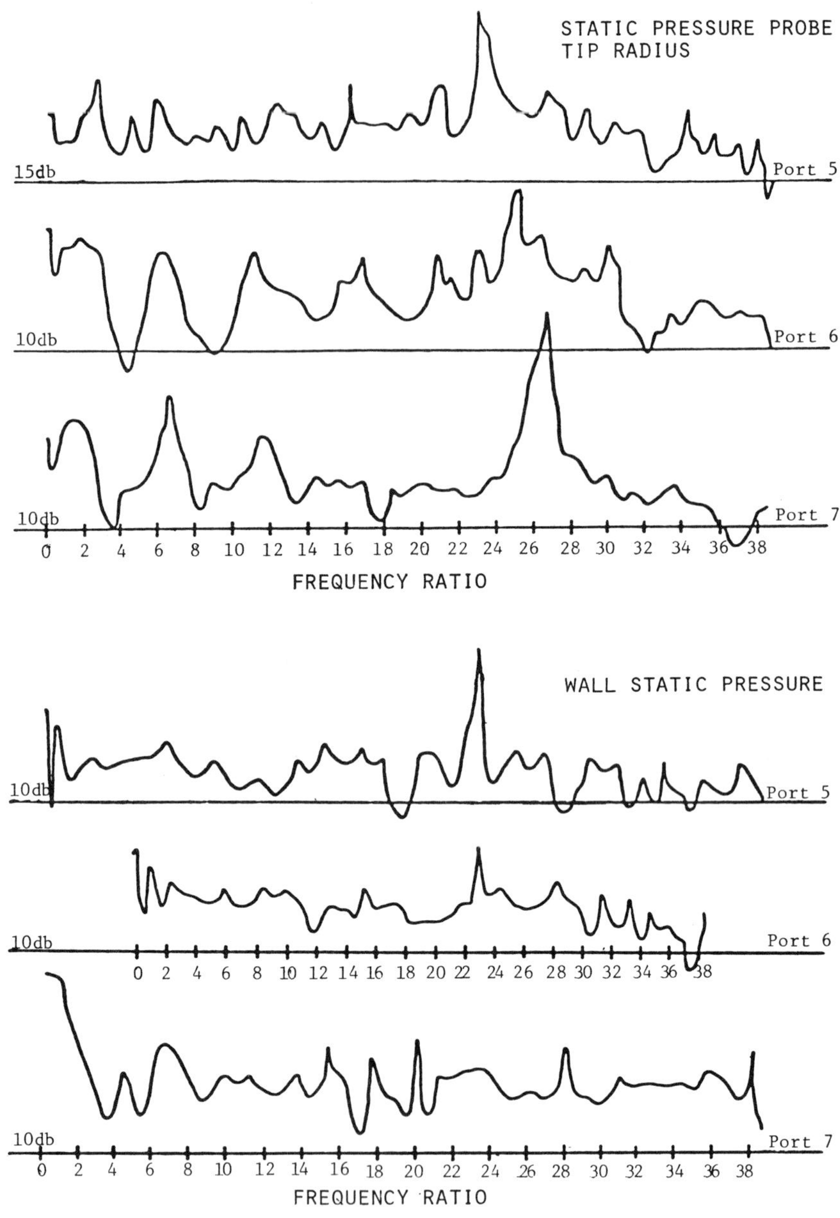

Figure 11. Harmonic Analyses of Pressure Field Downstream of Rotor at Stations 5, 6 and 7: (a) in Flow near Casing (Top), (b) from Flush Mounted Transducer (Bottom).

Assuming the geometry shown in Figure 12, where the mean tangential and axial velocities, V and W, are functions of r alone, the governing equations to first order are

$$L(u) - 2Vv/r = -\frac{1}{R}\frac{\partial p}{\partial r} + \frac{1}{R^2}\frac{dP}{dr}\rho \tag{1}$$

$$L(v) + (V' + V/r)u = -\frac{1}{Rr}\frac{\partial p}{\partial \theta} \tag{2}$$

$$L(w) + (W')u = -\frac{1}{R}\frac{\partial p}{\partial z} \tag{3}$$

$$L(\rho) + R\frac{\partial u}{\partial r} + (R' + R/r)u + \frac{R}{r}\frac{\partial v}{\partial \theta} + R\frac{\partial w}{\partial z} = 0 \tag{4}$$

$$L(s) + (S')u = 0 \tag{5}$$

where

$$L \equiv \frac{\partial}{\partial \tau} + \frac{V}{r}\frac{\partial}{\partial \theta} + W\frac{\partial}{\partial z}$$

and

$$s = c_p\left(\frac{t}{T} - \frac{\gamma - 1}{\gamma}\frac{p}{P}\right), \quad \frac{p}{P} = \frac{\rho}{R} + \frac{t}{T}$$

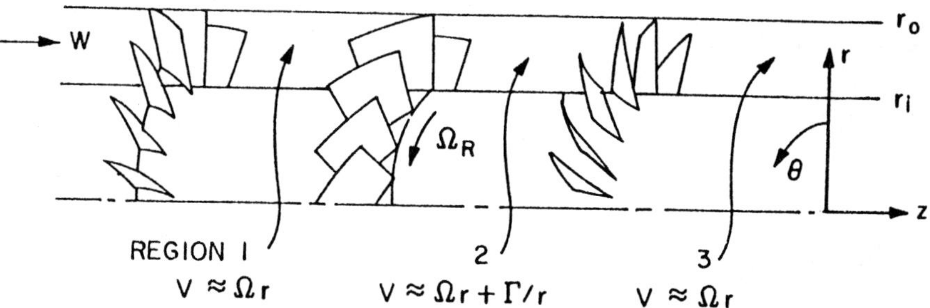

Figure 12. Schematic of Flow Geometry Used in Analysis of Behavior of Small Disturbances in Swirling Flow.

To zeroth order the mean quantities must satisfy radial equilibrium,

$$\frac{1}{R}\frac{dP}{dr} = \frac{V^2}{r} \tag{6}$$

In a nonswirling uniform flow, (1) through (5) can be reduced to (Kovasznay, 1953) $L(\vec{\omega}) = 0$, $L(s) = 0$ and $a^2\nabla^2 p - L^2(p) = 0$, so that the vorticity $\vec{\omega}$ and entropy are convected, while p satisfies a convected wave equation, and the three modes are uncoupled.

In the swirling flow, this separation is no longer valid. Taking a typical variable as

$$q(r,\theta,z,\tau) = \int q(r,m,k,\omega)e^{i(kz+m\theta-\omega\tau)}\, dk\, dm\, d\omega$$

it can be shown that the radial velocity perturbation is governed by a single, second order, ordinary differential equation with very complicated coefficients.

For the present purposes it will be sufficient to deal with a special case, where $V = \Omega r$ and $S = s = 0$. Then the density perturbation is governed by

$$r^2 \frac{d^2\rho}{dr^2} + r\frac{d\rho}{dr} - \left\{\left[\frac{\lambda^2 - 4\Omega^2}{a^2}\right]\left(1 - \frac{k^2a^2}{\lambda^2}\right)r^2 - m^2\right\}\rho = 0 \tag{7}$$

and the condition that $u = 0$ at $r = r_0, r_i$ becomes

$$\frac{r}{\rho}\frac{d\rho}{dr} = -\frac{2m\Omega}{\lambda}, \quad r = r_0, r_i \tag{8}$$

where

$$\lambda = kW + m\Omega - \omega$$

The solution is

$$\frac{\rho}{R} = Zm(\mu r) = A\, J_m(\mu r) + B\, N_m(\mu r)$$

and (8) becomes

$$\frac{\mu r}{Zm} Zm'(\mu r) = -2m\left(\frac{\Omega}{\lambda}\right) \tag{9}$$

where

$$(\mu r_0)^2 = \left(\frac{\Omega r_0}{a}\right)^2 \frac{\left[4 - \left(\frac{\lambda}{\Omega}\right)^2\right]\left[1 - M^2 + 2\frac{p}{\Omega}\left(\frac{\Omega}{\lambda}\right) + \left(\frac{p}{\Omega}\right)^2\left(\frac{\Omega}{\lambda}\right)^2\right]}{M^2} \quad (10)$$

$M = W/a$ is the axial Mach number and $p/\Omega = m(\Omega_R/\Omega - 1)$.

The eigenvalue problem posed by (9) and (10) has two distinct sets of solutions:

a) Shear waves $0 < |\lambda/\Omega| < 2$

b) Pressure waves $-\frac{p/\Omega}{1 + M} < -\lambda/\Omega < -\frac{p/\Omega}{1 - M}$

These are shown in diagrammatic form in Figure 13 for a case of $m = 1$, $\Omega r_0/W = 1$. The dotted lines are applicable to disturbances generated by a rotor, for which $\omega = m\Omega_R$ and $\Omega_R/\Omega = 3$. The dot-dash lines are for a stator, for which $\omega = m\Omega_R = 0$.

When λ/Ω is real for the pressure modes, they propagate; i.e., they are above cutoff in the acoustical sense. The rotation modifies these waves only slightly.

But the "shear waves" also propagate, are infinite in number, and are the analogue in rotating flows of vorticity or turbulence. We conclude that in strongly rotating flows all shear disturbances propagate and have associated pressure disturbances.

To illustrate this behavior, the lines of constant phase of the shear waves have been drawn for a stator and a rotor in Figure 14. We note that the shear waves come in pairs, one propagating upstream and one down, so a wake will tend to "split" or disperse as indicated, the wake crossing occurring in about one blade spacing for the example shown. This differential propagation becomes weaker as m increases.

The analysis becomes much more complex for $V \neq \Omega r$, a more general treatment is given by Kerrebrock (1975). No really satisfactory comparison of the theory with the above experimental results has as yet been realized. But a tentative calculation (Thompkins and Kerrebrock, 1975) shows that the first shear mode which fits in the outer half of the annulus, where $V \simeq \Omega r$, has $m \simeq 16$, and the structure of this mode agrees qualitatively with the observed axial, radial and tangential Mach number perturbations.

Thus, one possible explanation for the observed phenomena is that the flow evolves from the blade-periodic rotor outflow to a duct eigenmode having $m = 16$ in an axial distance of about one blade chord.

4. CONCLUDING REMARKS

It is the author's view, based on such evidence as that given above, that the flow field near a high work transonic rotor is strongly influenced by the inviscid evolution of the disturbances created by the rotor. The disturbances originate partially in irrotational effects (McCune, 1976) but primarily in the regions of high shear on the blades and casings. A classical treatment of the viscous flow may be adequate in these regions of shear dominance, but it appears the effects of mean rotation are controlling in the interblade region. Studies of "turbulence in turbomachines" should recognize this.

Eq (35):

$$\text{– – – –}\quad \frac{p}{\Omega} = m\left(\frac{\Omega R}{\Omega} - 1\right) = 1(4-1) = 3, \qquad \frac{\Omega r_o}{W} = 1$$

$$\text{— - — -}\quad \frac{p}{\Omega} = m\left(\frac{\Omega R}{\Omega} - 1\right) = 1(0-1) = -1, \qquad \frac{\Omega r_o}{W} = 1$$

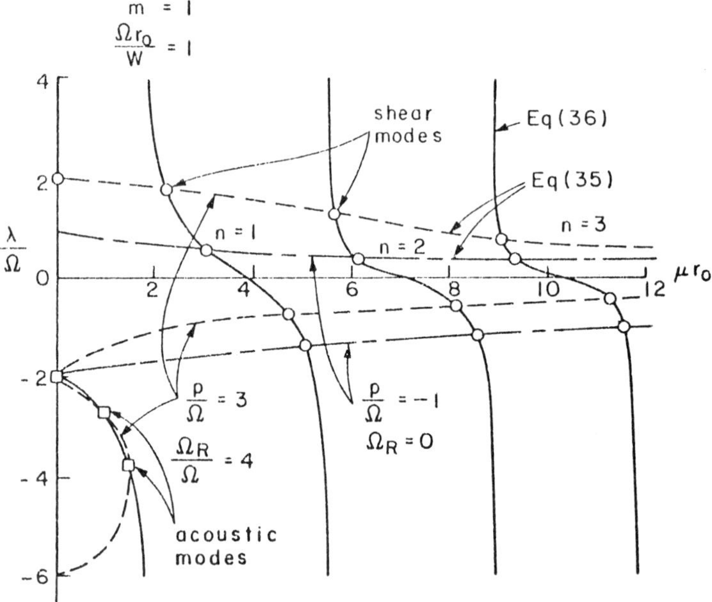

Figure 13. Diagram of Eigenvalue Problem Showing Solution for
a) Stator Shear Waves, Dash-Dot Line Intersections
b) Rotor Shear Waves, Dotted Line Intersections
c) Acoustic Rotor Modes.

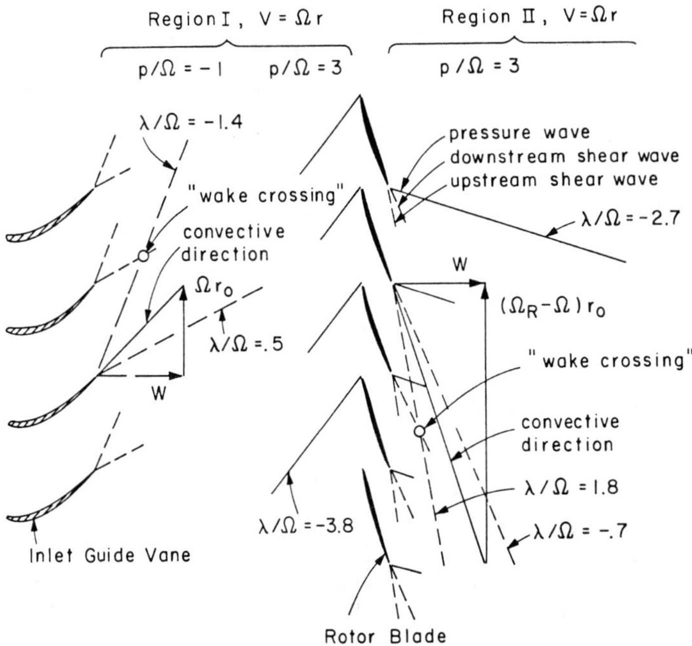

Figure 14. Wave Directions for the First Rotor and Stator Modes of Figure 13, in Regions I and II of Figure 12.

This research was supported by the NASA Lewis Research Center, by the Pratt & Whitney Aircraft Division, United Technologies Corporation, and the General Electric Company.

REFERENCES

1. Epstein, A. H. 1975 Quantitative density visualization in a transonic rotor. A.I.A.A. Paper no. 75-24. Also Ph.D. thesis, Department of Aeronautics and Astronautics, M.I.T.

2. Farokhi, S. 1976 Unsteady flow in transonic compressor stator. S.M. thesis, Department of Aeronautics and Astronautics, M.I.T.

3. Kerrebrock, J. L. et al. 1974 The MIT Blowdown Compressor Facility. J. Engr. for Power 96, 4, 394-405.

4. Kerrebrock, J. L. 1975 Small disturbances in turbomachine annuli with swirl. M.I.T. Gas Turbine Lab. Rep. no. 125.

5. Kovasznay, L. S. G. 1953 J. Aero. Sci. 20, 10.

6. McCune, J. E. 1976 Three-dimensional inviscid flow through a highly-loaded transonic compressor rotor. M.I.T. Gas Turbine Lab. Rep. no. 126.

7. Stephens III, H. E. 1974 Wake behavior downstream of a transonic compressor rotor. S.M. thesis, Department of Aeronautics and Astronautics, M.I.T. Also M.I.T. Gas Turbine Lab. Rep. no. 118.

8. Thompkins Jr., W. T. and Kerrebrock, J. L. 1975 Exit flow from a transonic compressor rotor. AGARD CPP-177, Paper no. 6. Also M.I.T. Gas Turbine Lab. Rep. no. 123.

DISCUSSION

KOVASZNAY: (Johns Hopkins University)

I was listening with fascination to your presentation and really I would like to ask a question and also make a comment. Do you have experimental means to establish with certainty that the phenomenon you observed was strictly periodic, namely phase locked with respect to the shaft?

KERREBROCK:

The answer is, yes, we do; but we have not done it yet.

KOVASZNAY:

Now the comment: When a phenomenon is strictly periodic and I think this must be the case for a single rotor with no guide wanes, no stators, there can be a conversion between frequencies (all are harmonics of the shaft frequency) only if there is a strong non-linear mechanism available. I can think of two: One of them is the merging of the wakes and the other is the merging of the shock waves. In fact both of those were observed. For the shock wave merger the relative velocities in the rotor must be supersonic. If there are small irregularities in the blades e.g., nose radii, some shocks are slightly stronger than others and will "catch up" the result if fewer shocks than blades but irregular, though it is still a strictly periodic pattern. This way it can be imagined how 23 blades can produce a 16 lobe pattern.

KERREBROCK:

Let me respond to that one before we go on. We have looked at the upstream flow field and in this compressor anyway there is no merging of shock waves. In other words the shocks from the blades are distinct as far upstream as we can trace them.

KOVASZNAY:

Again this is the same question. If you have a phase locked detection compared locked to the position of the shaft you can see the shock wave pattern clearly.

KERREBROCK:

Yes we do; the shocks are repeatable from turn to turn, from revolution to revolution, and there are 23 of them.

KOVASZNAY:

The other nonlinear phenomenon is merging of vortices. Both of these mechanisms are available and they can "demultiply" the number of periods from 23 to something else. In that way the small perturbation theory really doesn't account for these details, because a massive energy transfer is required from one given number of periods to another number of periods.

KERREBROCK:

I agree with what you say and I thank you for the comment. I think that the explanation of a propagating disturbance is most plausible at this point and having said that I want to say for the people who are not turbomachine experts, that does not allow us to say "OK this compressor is in propagating stall" because in fact it is not. If you look upstream of the rotor you find no evidence of a propagating stall and the machine is operating at its maximum efficiency point just about where you would expect a compressor to operate.

COLES: (California Institute of Technology)

Is what you see downstream of the rotor blades consistent with a multi-start thread type Taylor instability in the outer part of the compressor annulus? I don't know how many starts.

KERREBROCK:

Yes, 16.

COLES:

Not necessarily, depending on the rotational ......

KERREBROCK:

Right, it could be 16 or 23 depending on whether it is propagating relative to the rotor.

COLES:

It could be two if it is going fast enough. Does your flow visualization shed any light on this?

KERREBROCK:

It will but it hasn't. So far it has not included that part of the flow field. We are going to do flow visualization downstream of the rotor soon.

LAKSHMINARAYANA: (Pennsylvania State University)

One possible explanation of your observation is this. The profile and defect of blade wakes are all different and each blade passage seems to have differing flow. Therefore each one of these wakes propagates at different velocities. This is one of the things that we showed in the AGARD paper (AGARD CP 177, p4-1 4-1-14) where the wake centerline velocity has a different direction than the freestream even in relative flow. What might be happening is the merging of some of the wakes.

KERREBROCK:

I agree with that but I think the question then is what drives that merging and the suggestion that I am making is that in fact the flow is evolving somehow from the forced periodicity to an Eigen mode of the duct.

LAKSHMINARAYANA:

The wakes coming out of the blades are so different from one another in your pictures.

KERREBROCK:

But the question is why.

LAKSHMINARAYANA:

We measured the wakes in a subsonic rotor. We don't see this phenomena. At every station the periodicity is maintained. This is because the rotor inflow is very uniform.

KERREBROCK:

It has been true in general of compressor rotors, that wakes from successive blades are not the same and in my mind at least there has always been the question "why?" Now the answer that I am tendering is that the viscous structure of the downstream flow field in fact generates a pressure field that feeds back upstream and forces a non-blade-passing periodicity on the wake structure and on the growth of the boundary layers.

WILLMARTH: (University of Michigan)

You say "wake-splitting." I can't visualize what it means. You mean along the trailing edge of the blade the wake was going one way at one place and another way at another place?

KERREBROCK:

No. I would say, if you regard the wake as it comes off the blade as an initial condition, which the Eigen modes of the duct have to match to, then a superposition of two classes of waves is required, one propagating upstream, the other propagating downstream so that the disturbance would split as it comes off.

WILLMARTH:

So all along the radius, radially outward, the wake looks wider.

KERREBROCK:

Yes, it would break into two pieces. One propagating one way, the other one propagating the other way. And in fact in the example that I showed, those propagations are fast enough so that the one wake essentially crosses the other one in one blade chord.

WILLMARTH:

Another comment concerns your pressure measurements away from the wall, downstream of the blade. In our work using pressure measurements, we find it is really difficult to put a probe in the flow and measure fluctuating components. When you have shock waves coming over the probe, you have the strong possibility of the body of the pressure sensing device interacting with the flow field which has rapidly changing flow inclination and shocks.

KERREBROCK:

Yes, I agree with you, it is hard to measure the pressure quantitatively, but it is hard to think of things that change the frequency content of the flow field.

WILLMARTH:

Yes, but if the flow field that is hitting this probe has frequency components of 16 per revolution for example you will get pressures from it at that frequency.

KERREBROCK:

I agree, but the point that I was trying to make was that the fundamental frequency component, the dominate one, shifts from 23 to something like 27 monotonically downstream. I have not been able to come up with an explanation for that so far.

Wake Cutting Experiments†

LESLIE S. G. KOVASZNAY*
The Johns Hopkins University
Baltimore, Maryland (U.S.A.)

ABSTRACT

The interaction between a turbulent wake and a moving blade (airfoil) is in many ways central to the understanding of both the unsteady forces on the blade and the resulting acoustic emission. In order to obtain the details of a single wake cutting "event" a special experimental facility was constructed. Two-dimensional turbulent wakes were produced and, one-by-one they swept in front of a stationary (usually two-dimensional and symmetrical) airfoil. The time interval between passages was chosen long enough so that each passage could be regarded as a statistically independent experiment. The key feature of the experiment was that all flow variables (instantaneous pressure distribution on the airfoil, fluctuating velocity components in the interaction zone, acoustic far field, etc.) were measured by using time dependent ensemble averaging over a large number (500-2,000) of wake passages. The results gave instantaneous lift, drag and equivalent dipole sound source, each as a function of time. Additional experiments were performed for the "skewed" case (the plane of the airfoil and wake are not parallel, but they form a small angle), and finally some data were obtained on a cambered airfoil and on a small cascade of cambered airfoils.

---

† Research partially supported by United Aircraft Company, Pratt & Whitney Aircraft Division, East Hartford, Connecticut.

* Present Address: ONR Tokyo, U.S. Embassy, APO S.F. 96503.

1. INTRODUCTION

In fluid mechanics there is a group of problems that involves the interaction of an unsteady vortex flow with a solid body. Due to the boundary conditions at the solid surface, viscosity cannot be neglected in general and the problem becomes less amenable to theoretical treatment than either an unsteady inviscid irrotational flow or alternately a viscous flow with vorticity but steady in time. The approaching vortical flow may be either a wake (two-dimensional or axisymmetric) or it may be a trailing vortex; the solid body may be either a blunt body or it may be a lifting surface. From all of the combinations of such variables, a particular case was singled out for study where a two-dimensional wake was sweeping over a two-dimensional airfoil. Experiments were carried out and the results were reported in Refs. 1, 2 and 3. Theoretical treatment of the problem based on linearized thin airfoil theory is available in the literature (e.g. Refs. 4 and 5) and it was used as a rough guide. The primary purpose of the research was to establish experimentally the details of such an interaction, consequently no effort was made to put the "stamp of approval" on any particular theory. The problem chosen is timely on two accounts. First: in multistage rotating machinery, especially in axial compressors, the interaction between a set of wakes created by the blades of one stage are being "cut" by the blades of the subsequent stage, so the basic interaction between one wake and one blade moving relative to each other appears to be the basic phenomenon that controls both the unsteady flow around the blade as well as the sound emission. Second: from an experimental point of view the problem has become "ripe" as modern signal processing techniques have become available and they permit recovery of the relevant "deterministic" portion of the signal that describes the interaction without being obscured by large random noise contributed by turbulence in the flow.

2. STATEMENT OF THE PROBLEM

A two-dimensional airfoil is placed in a uniform parallel flow having a velocity U_∞. At a given time a two-dimensional wake having a given velocity defect distribution is carried downstream by U_∞. The wake axis is at an angle β to the undisturbed flow direction. Figure 1 shows the configuration. During the encounter of the wake and the airfoil the instantaneous flow field around the airfoil changes rapidly so both the circulation around it and the surface pressure distribution over it shows rapid transients.

Some time after the wake has left the airfoil all of the transients decay and conditions return to the original steady flow. This transient phenomenon is called wake cutting. The objective of

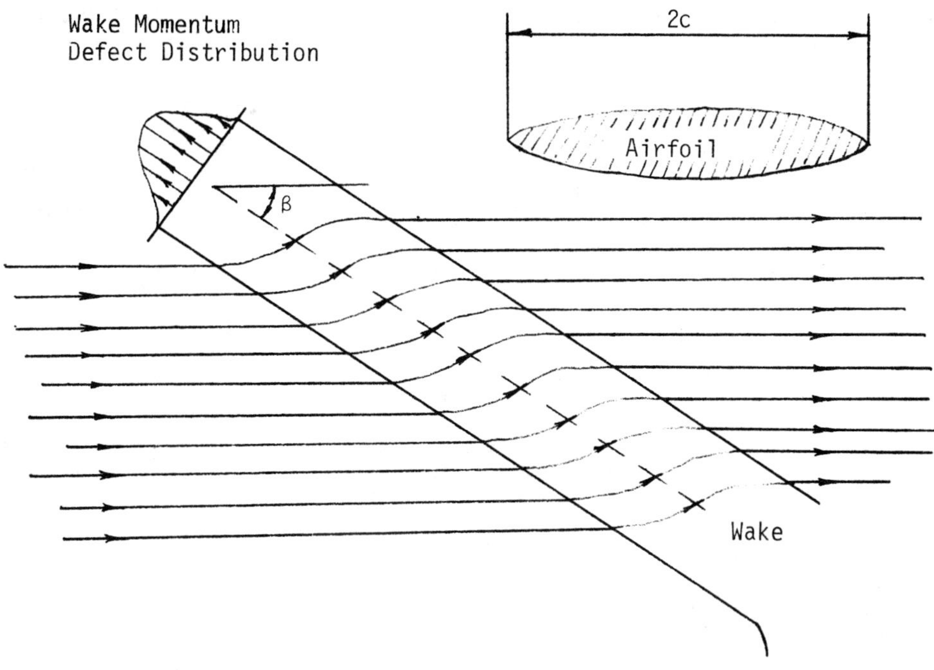

Figure 1. Wake Cutting. Wake Sweeps Upward, Instantaneous Streamlines Shown.

the research is to determine the transient behavior in terms of the given wake and airfoil parameters.

3. THEORETICAL CONSIDERATIONS

Unsteady airfoil theory has a long history. The usual assumptions are the following: linearized thin airfoil theory with the Kutta condition satisfied at all times; the approaching disturbance is specified as a "gust" namely only the velocity component, perpendicular to the undisturbed flow is specified and only within the plane parallel to the undisturbed flow and containing the stagnation line of the airfoil. The gust pattern is assumed to be a "frozen" pattern carried downstream unchanged, with the convection velocity U_∞. At this point it is customary to introduce Fourier analysis and to decompose the gust pattern into different wave number (frequency) components. One obtains similarly the transient response of the airfoil, also in terms of frequency components (see e.g. Ref. 4). In those calculations the so-called Sears function plays a rather central role. Such an approach appears to be quite adequate for obtaining the response of an airfoil to gusts.

In the case of wake cutting the wakes are relatively narrow, usually less than the airfoil half chord, c; consequently, another simplification is possible (Ref. 5). The wake width is neglected entirely and the velocity defect distribution of the wake is assumed to be a Dirac δ function weighted with the total momentum defect of the wake. On that basis the transient response of the flow around the airfoil was calculated and rather simple results were obtained. First, non-dimensional coordinates were introduced

$$x' \equiv \frac{2x}{c}; \qquad t' \equiv \frac{2U_\infty t}{c}$$

The chordwise pressure gradient becomes rather simple, namely

$$\frac{\partial p}{\partial x'} = \pm \rho U_\infty^2 \, W \, T(t') \, X(x')$$

Here the upper sign applies to the upper (suction) surface and the lower sign to the lower (pressure) side. The constant W is essentially the total integrated momentum loss of the wake. The two universal functions $T(t')$ and $X(x')$ are independent of the shape of the wake distribution (as long as it is narrow compared to the chord) and also independent of the airfoil shape or angle of attack α. The significant result of Ref. 5 is that the transient response can be expressed as the product of two factors, one depending on the time alone and the other on the chordwise position alone. In other words the chordwise pressure distribution always has the same shape, only it grows and decays proportionally in time. Similarly the time dependence is the same at every point except that it varies in magnitude according to chordwise position. The function $T(t')$ is the Fourier transform of the "Sears function" and the function $X(x')$ is given as

$$X(x') = \frac{1}{(1 + x')\sqrt{1 - x'^2}}$$

or

$$\Xi(x') = \int_{x'}^{1} X(x'')dx'' = \sqrt{\frac{1 - x'}{1 + x'}}$$

The pair of universal functions is given in Fig. 2. Both functions are singular at $t' = -1$ and $x' = -1$ respectively, but the singularity is integrable. It is a consequence of using linearized thin

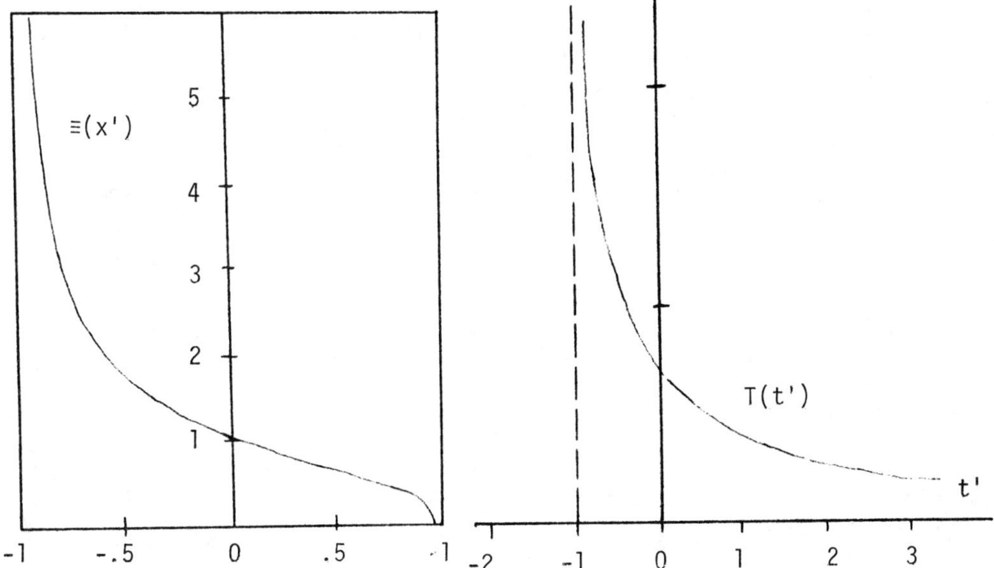

Figure 2. Universal Functions $\Xi(x')$ and $T(t')$

$$\Xi(x') = \int_{x'}^{1} X(x'')dx''$$

airfoil theory. Naturally much more elaborate calculations can be made for particular cases but the simplicity of Meyer's results (Ref. 5) was found especially attractive to serve as a simple guide to organize the experimental results.

It must be mentioned that all of the theoretical calculations of transient response in wake cutting are critically dependent on the assumption that at the trailing edge the Kutta condition remains valid even for rapid transients. In order to test this assumption experiments were carried out (Ref. 6) and the results confirmed the validity of the Kutta condition even for relatively rapid transients. Naturally it may still fail for transients so rapid that large changes occur during a time unit formed from the boundary layer thickness near the trailing edge and the freestream velocity.

4. EXPERIMENTAL FACILITY

A small wind tunnel was built especially for wake-cutting experiments (Ref. 1). It consisted of an axial fan followed by a diffuser. A settling chamber with turbulence screens was provided to reduce the freestream turbulence level, followed by a square nozzle discharging an open jet with a 30 cm × 30 cm cross section and having a nominal velocity of 38 m/sec (Fig. 3). The passing wakes were produced by a rotating "pin-wheel" consisting of two approximately 1 cm diameter cylindrical spokes cutting across the jet with a velocity typically one third of the nominal mean velocity resulting in a typical wake angle of $\beta = 22°$ (for definition see Fig. 4). The airfoils or blades fully spanned the jet and they were mounted on a steel plate that also provided safety by protecting the observer from the rotating pin-wheel. For the acoustic measurements it served as an acoustic mirror providing well defined boundary conditions.

The airfoil under test was provided with two sets of pressure taps, one set for mean pressure measurements that were connected to liquid manometers and another set that was connected to a rapid response pressure transducer. In the early phase of the experiments an F.M. operated condenser microphone was used in conjunction with a pressure switch, in the later phase a newly developed experimental electric transducer was used. It was obtained from ONERA, Paris. The far field sound pressure was measured by two calibrated

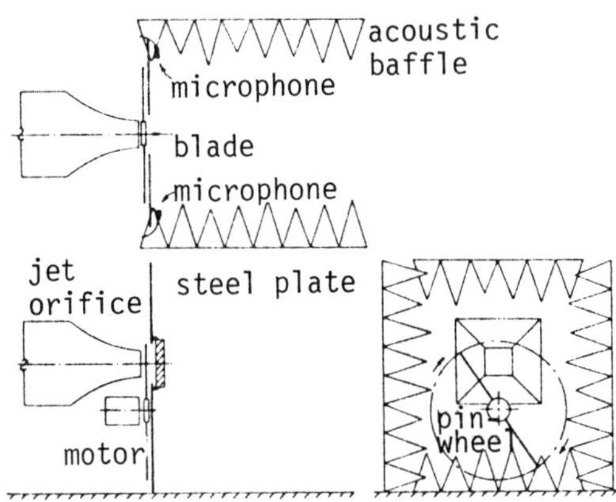

Figure 3. Experimental Facility

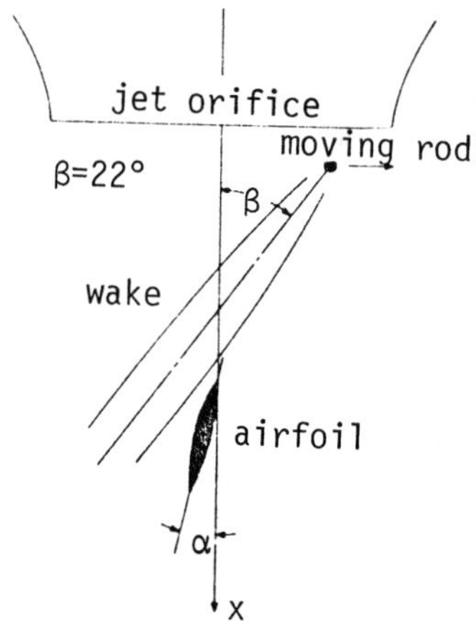

Figure 4. Definition of α and β.

condenser microphones located symmetrically along a line perpendicular to the axis of the jet and to the span of the airfoil. In order to eliminate the reflections from the laboratory walls, acoustic baffles were installed on four sides outside the jet. These consisted of wedges of rubberized hair and have proven satisfactory.

5. INSTRUMENTATION AND DATA PROCESSING

The novel feature of these experiments was the extensive utilization of conditional sampling and ensemble averaging. The wake passages were separated by long time intervals so that the transients decayed and each passage could be regarded as an independent experiment. All transient signals, such as those obtained from surface pressure transducers, hot-wire anemometer probes and far field microphones, were processed by periodic averaging. A magnetic pick-up on the shaft of the pin-wheel provided a "master pulse" giving the exact instant when the wake center passed the leading edge of the airfoil. These instants t_i, with $i = 1, 2, \ldots, N$, give the time origin for each event. The periodically sampled ensemble average of a signal $e(t)$ is defined as

$$\tilde{e}(t) = \frac{1}{N} \sum_{i=1}^{N} e(t - t_i)$$

For $N \to \infty$ the periodic average $\tilde{e}(t)$ becomes a periodic function even if $e(t)$ contained random noise. For an additive random noise that becomes statistically independent for time separations larger than one period, the r.m.s. noise level will decrease in $\tilde{e}(t)$ as $\approx N^{-\frac{1}{2}}$. For periodic averaging an analogue device (PAR Signal Eductor) was used whose output was digitized and transferred to perforated tape and further processed by computer. The choice of N is limited by two opposite considerations: with increasing N the signal-to-noise ratio improves but the effect of slow drift in the experimental conditions reduces the accuracy. Typically $500 < N < 2000$ was used and the experiments have shown good reproducibility.

6. SAMPLE EXPERIMENTAL RESULTS

First the unsteady flow field due to the wake passage was measured without the presence of the airfoil. Here the periodically averaged records have another advantage. These time records are "keyed" to the exact time origin of each wake passage (to the position of the pin-wheel shaft), so data taken at different points can be collated. Two components of the instantaneous velocity were measured by a hot-wire X probe over a lattice of points on the plane perpendicular to the wake. A total of $10 \times 20 = 200$ points were measured, so u and v gave a total of 400 records for the different x and y values. With the assumption of two-dimensional incompressible flow (the assumption of potential flow is not necessary) the instantaneous stream function $\Psi(x,y,t)$ can be reconstructed by integration. For each instant, $u(x,y)$ is integrated with respect to y and $-v(x,y)$ is integrated with respect to x. The streamlines are then obtained as Ψ = Const. (For more details see Refs. 1 and 6.) Figure 5 shows one example so obtained. The flow field shown in the perturbation flow field, namely the flow pattern seen by an observer moving with the nominal undisturbed flow or in other words it is seen in an Eulerian frame translated with velocity U_∞. In this frame the wake appears like a jet directed toward the object that produced it (the German word for wake is Nachlauf). The large vortex seen at the left edge of the jet is the remnant of the interaction between the rod and the shear layer at a time when the rod entered the jet. Whole time sequences were so obtained and the technique is straightforward; it can be applied in any two-dimensional unsteady flow at low Mach numbers.

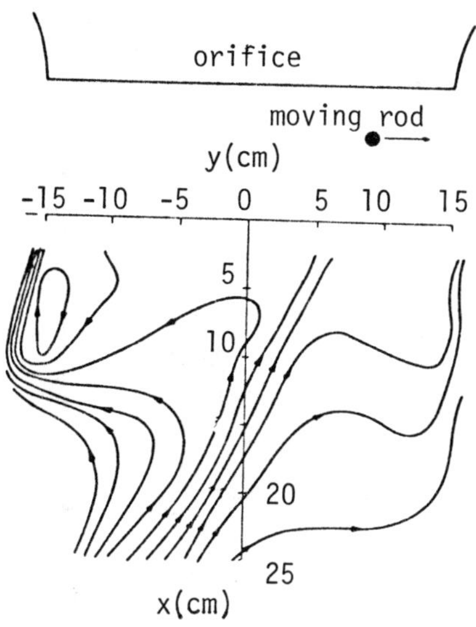

Figure 5. Instantaneous Flow Pattern as Seen from Coordinate System Moving with the Nominal Velocity U_∞.

The instantaneous pressure distribution around the airfoil was obtained at a number of chordwise stations. At each station both the static (time average) pressure was measured and the pressure fluctuation (zero time average by definition) was obtained as a periodically averaged record. The instantaneous sum of the two gave what is termed the "deterministic portion" of the pressure distribution. Figure 6 shows a sample of the pressure distribution around a symmetrical airfoil at $\alpha = 0$ for a number of instants. What is plotted here is not the pressure but $[p_+(x') - p_-(x')]$, the difference of pressure on the upper and lower surface at the same chordwise position. This quantity represents the lift per unit chord length. Let us recall here that non-dimensional time t' is equal to -1 when the wake center reaches the leading edge and t' is equal to 1 when it leaves the trailing edge. The dotted lines represent the "theoretical" distribution given by Meyer (Ref. 5). It is the function $\Xi(x')$ given in Fig. 2 except its magnitude is adjusted so that the integral (total lift) is the same as that of

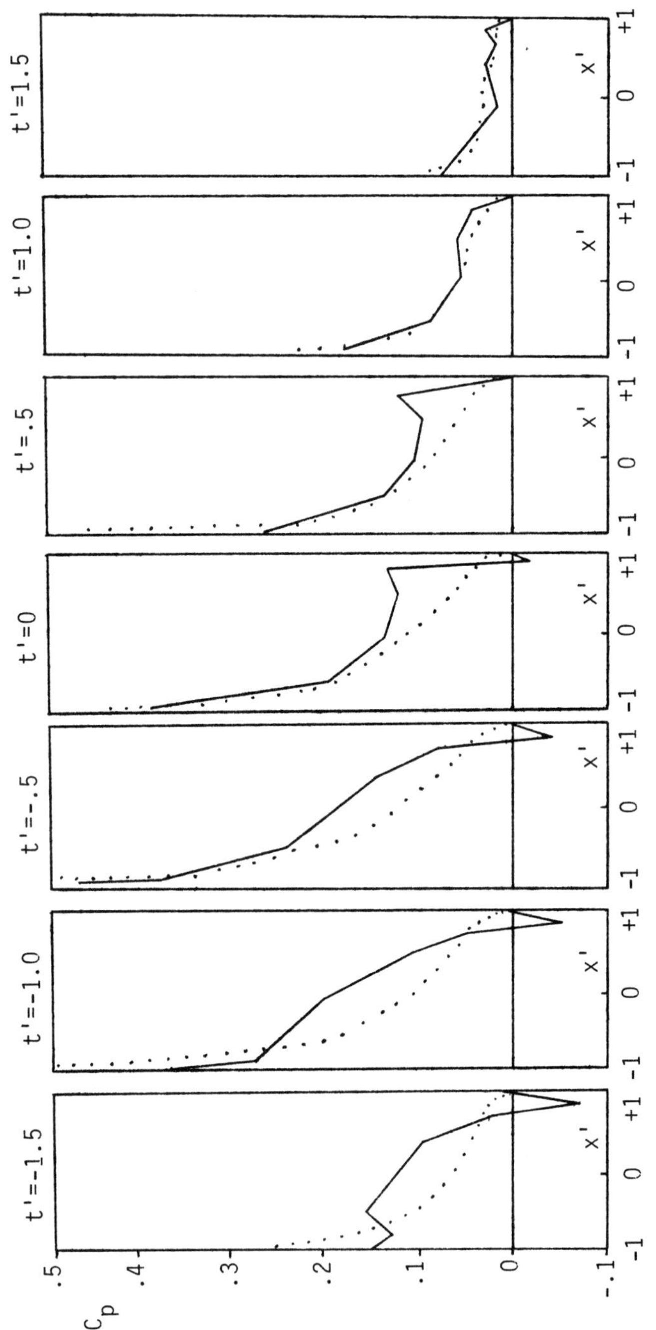

Figure 6. Pressure Distribution over Airfoil ($\alpha = 0$). $x' = \frac{2x}{c}$; $t' = \frac{2tU_\infty}{c}$; $c_p = \frac{2(p_+ - p_-)}{\rho U_\infty^2}$

the experimental distribution. The experimental distribution is in good qualitative agreement with Meyer's theory but significant departures are in evidence. The chordwise distribution is not exactly similar for all times but a certain timewise development is noticeable, namely disturbances seem to travel from leading edge to trailing edge. Probably this is due to transient readjustment of the boundary layer on the airfoil, a feature so far left out in all theories. By integration the instantaneous lift is obtained. A sample is shown in Figure 7 and it is compared with the function T(t'). The experimental curve is smooth and gentle while the theoretical curve is "spiky" at t' = -1. Nevertheless one must remember that the theoretical curve was obtained as an asymptotic case on the basis of negligible wake width. If one performs an appropriate convolution of the wake distribution function with the function T(t') a closer agreement may be obtained. The two curves have also different values for t' > 2. This is believed to be the effect of the finite size of the chord compared to the jet width (typically 1:3). Corrections can be made for this effect by including slowly variable components in the wake function to account for the large scale disturbances created in the shear layers of the jet.

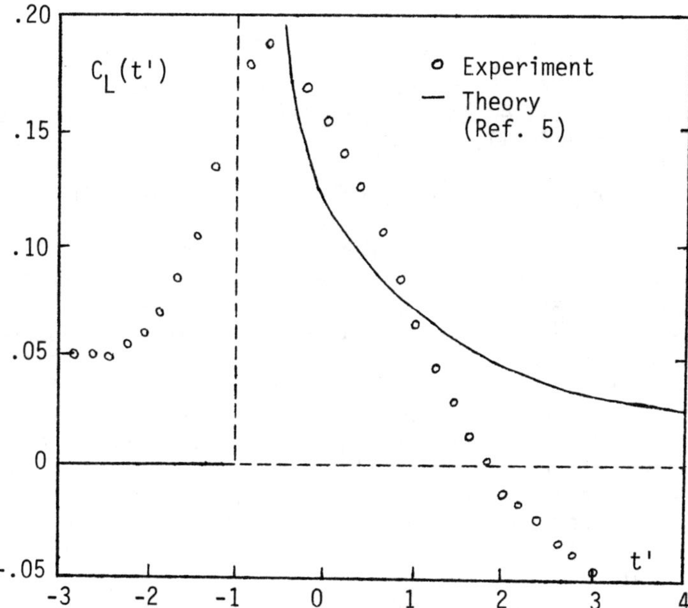

Figure 7. Transient Lift in Wake Cutting.

7. ACOUSTIC EMISSION

Wake cutting produces a detectable sound pulse. According to Lighthill's classical theory a flow without solid boundary can produce only quadrupole radiation (no lower order) but where solid boundaries are present, in general it can produce dipole radiation. The variable lift on the airfoil represents a variable body force on the surrounding fluid, correspondingly even at low Mach numbers there will be an affective acoustical dipole proportional to the time derivative of the lift. For details one must refer to Curle's paper (Ref. 7). Curle's equation can be transformed into a convenient form for the case of a plane wing and one obtains a rather simple formula for the sound pressure far from the wing (Ref. 1). Let us introduce the following quantities: A, the area of the wing; ρ, the air density; $M = U_\infty/a$, the Mach number of the undisturbed flow; R, the distance of the observer from the center of the wing; finally, the angle θ, the angle between the y axis (that is perpendicular both to x the undisturbed flow direction and z the spanwise coordinate) and the radius vector of the observer. The observed sound pressure signal then becomes the following.

$$s(R,\theta,t) = \frac{\rho U^2}{2}\,\frac{AM}{2\pi cR}\,\frac{dc_L}{dt'}\cos\theta$$

The only property of the wing that appears in that expression is dc_L/dt' the non-dimensional time derivative of the lift coefficient. The formula given is based on far field approximation. For observers that are closer there is a correction term proportional to c_L itself (see Ref. 1).

In the experiments two microphones were placed at $\theta = 0$ and $\theta = \pi$ and their sum and difference signals were both periodically averaged. The difference signal was much larger thus verifying the dominance of dipole radiation. The results were in rather good agreement with Curle's theory as Fig. 8 shows. The measured sound pressure is compared with the one calculated from the measured transient lift for four different angles of attack. Especially important is the fact that there is no adjustable constant and the agreement is good even if plotted in linear scale (and not only in decibels). It is surprising that the agreement is quite good even at $\alpha = 20°$ when the airfoil was fully stalled. One must remember however that the relationship supported by this experimental result is between the experimentally obtained body force and the radiated sound, so the validity of the linearized unsteady airfoil theory is not involved at all.

The sound signatures for the four angles of attack are shown together in Fig. 9. The curves do not differ too much. The most

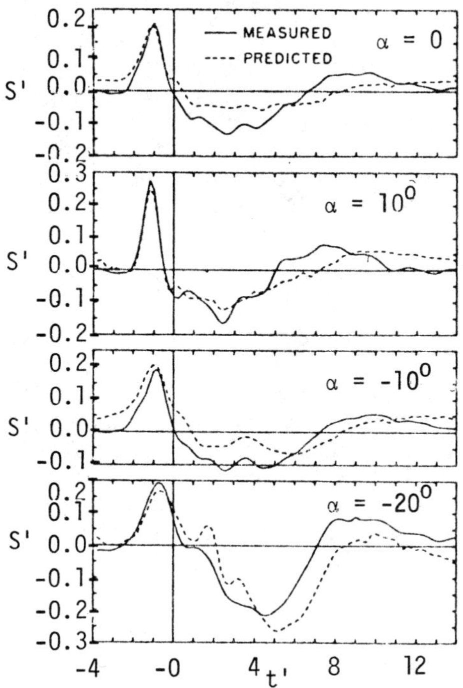

Figure 8. Measured and Predicted Radiated Sound for Four Values of α.

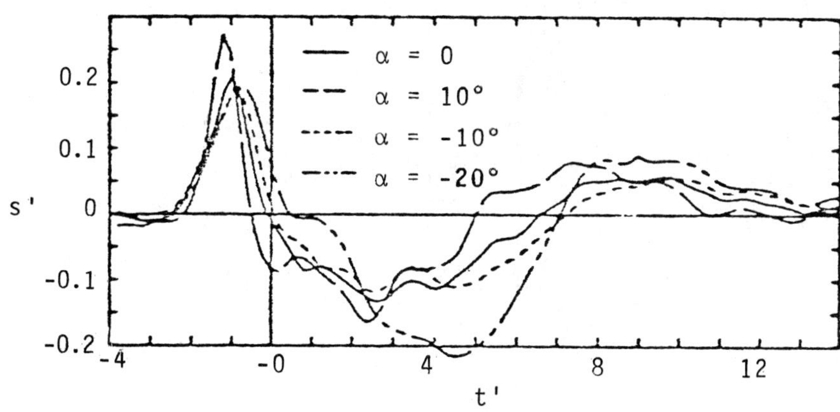

Figure 9. Comparison of Radiated Sound at Different Angles of Attack.

remarkable feature is the extra large magnitude of the first pulse in case of $\alpha = 10°$. When this case was reexamined it was concluded that the airfoil was near stall and the instantaneous angle of attack at the leading edge temporarily increased during the wake passage* and caused a separation bubble lasting only for a short time. When the flow reattached the lift suddenly increased but somewhat faster than for the non-separated case (e.g. for $\alpha = -10°$) and the larger value of dc_L/dt' caused the higher sound pulse.

8. FURTHER EXTENSIONS

After the initial success in carrying out wake cutting experiments a number of variations of the basic configuration were explored. First the "skew" case was studied. The new element was the misalignment between the plane of the wake and the plane of the airfoil by an angle θ as shown schematically in Fig. 10. The result is a large drop in the maximum lift pulse during the wake passage (Fig. 11) and an even larger drop in acoustic emission (Fig. 12). For easy comparison the value of $\cos\theta$ is also plotted. For further details see Ref. 8.

---

* See Fig. 1 for the instantaneous streamlines.

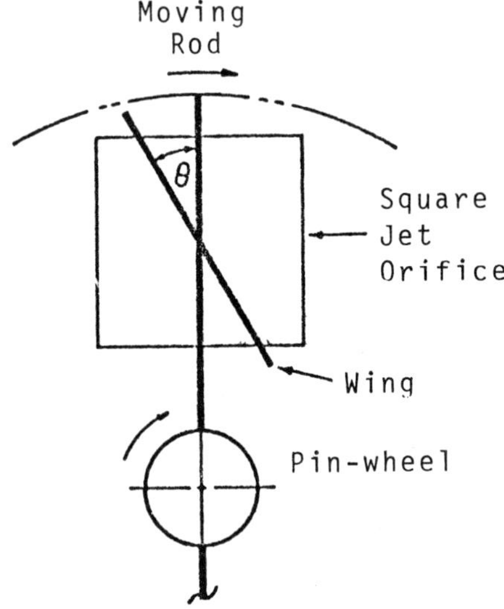

Figure 10. View of the Test Section from Downstream.

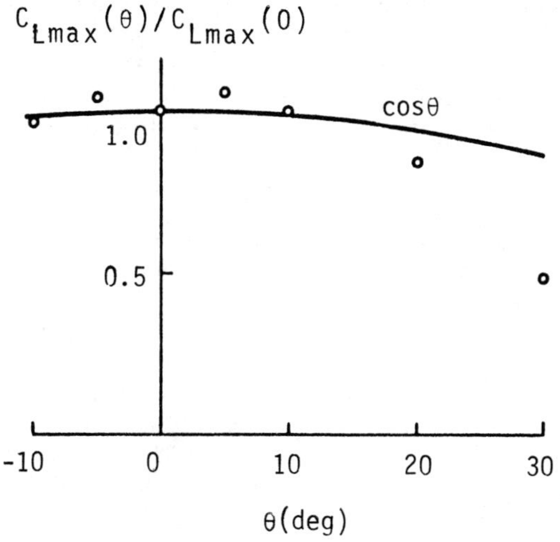

Figure 11. Maximum Value of Lift Coefficient as a Function of θ.

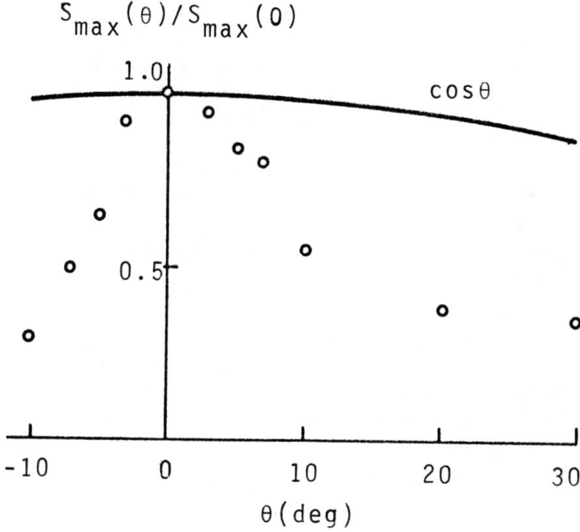

Figure 12. Maximum Value of Radiated Dipole Sound as a Function of θ.

Next, wake cutting by cambered airfoils was explored. The most interesting result then was the large difference observed in the transient response between the cases when the wake sweeps from the convex (back) side of the blade ($\beta > 0$) and from the concave (belly) side ($\beta < 0$). They may differ by as much as a factor of two and the larger response is for $\beta > 0$. For details see Ref. 2.

Finally there was a small exploratory attempt to see some gross effects in the transient pressure distribution of a highly cambered blade in a modest cascade (only three blades). The above-mentioned difference between the cases $\beta > 0$ and $\beta < 0$ were further accentuated at large angles of attack so that the ratio in the maximum transient pulses became as large as 10:1 (always the case $\beta > 0$ giving the larger pulses). In addition transient local separation bubbles were observed travelling along the boundary layer representing a large vortex perturbation. Such behavior is quite in contrast with simple unsteady airfoil theory and the unsteady perturbations and readjustment of the boundary layer seem to be involved. Such phenomena cannot be incorporated yet into the existing theoretical framework. On the other hand the relationship between unsteady lift and radiated sound appears to obey Curle's theory quite well, even in cases when the prediction of the fluctuating lift from gust theory is poor. The extension of these experiments to cascades of blades is greatly hampered by the fact that the radiated sound is strongly modified by the presence of the other blades, so only the transient surface pressure measurements are reliable.

9. CONCLUSIONS

The most important conclusion that can be drawn from the above series of experiments is the evidence of feasibility to perform systematic measurements on highly unsteady flow phenomena when reliable fast response transducers are used in combination of phase-locked periodic sampling and ensemble averaging.

As far as the transient response of an airfoil is concerned the linearized gust theory can serve as a useful guide but experiments are necessary to clarify the role of transient local separation not accounted for by perfect fluid theories. Furthermore vortex disturbances travelling along the boundary layer of both upper and lower surfaces cannot be fitted into any of the existing theories.

The validity of the simple theory of Curle as restated in Ref. 1 was well verified (Ref. 1) and it may be applied to many cases.

Quite clearly further experimentation is needed to establish the finer details in many particular applications especially in the case of thick and strongly cambered airfoils. Nevertheless the author is quite confident that the technique introduced will be utilized with profit in a large number of technologically significant unsteady flow problems.

REFERENCES

1. Fjuita, F. & Kovasznay, L. S. G. "Unsteady lift and radiated sound from a wake cutting airfoil." AIAA Journal, Vol. 12, p. 1216-21 (1974).

2. Ho, Chih-ming & Kovasznay, L. S. G. "Sound generation by single cambered blade in wake cutting." AIAA Journal, Vol. 14, p. 763-66 (1976).

3. Ho, Chih-ming & Kovasznay, L. S. G. "Wake cutting by a cascade of cambered blades." AIAA 2nd Aero Acoustic Conference Hampton, Va., March 24-25, 2975. Proceedings. p. 43-53, (1976).

4. von Karman, T. & Sears, W. R. "Airfoil theory for non-uniform motion." Jour. Ae. Sci., Vol. 5, p. 379-90, (1938).

5. Meyer, R. X. "The effect of wakes on the transient pressure and velocity distribution in turbomachines." Trans. of the ASME, Vol. 80, p. 1544-52, (1958).

6. Kovasznay, L. S. G. & Fujita, H. "Unsteady boundary layer and wake near the trailing edge of a flat plate." Proceedings of the IUTAM Symposium; Recent research on unsteady boundary layers. Editor E. A. Eichelbrenner. Les Presses de l'Université Laval, p. 805-833, (1972).

7. Curle, N. "The influence of solid boundaries upon aerodynamic sound." Proc. Roy. Soc. Ser. A, Vol. 231, p. 505-514, (1955).

8. Fujita, H. & Kovasznay, L. S. G. "Sound generation by wake cutting." AIAA Aero-Acoustic Conference, Seattle, Washington, Oct. 15-17, 1973, AIAA Paper No. 73-101.

DISCUSSION

KERREBROCK: (M.I.T.)

I don't see anything in that experiment that seems to introduce a randomness. In other words it seems to me that the whole phenomenon should be periodic.

KOVASZNAY:

The overall flow is periodic, but the turbulent wake is random.

KERREBROCK:

Is it random or is it shedding Karman vortices?

KOVASZNAY:

No, it doesn't shed Karman vortices. It is of too high a Reynolds number flow for that, so it is truly random. Actually we checked out that aspect separately and it can be done very easily with the rod. We studied that extensively. As a matter of fact, we did observe some critical r.p.m. values where the rod, every time it entered into the jet, that impact force on the rod made it vibrate and the next time around it reinforced that vibration, so there were some bad r.p.m.'s.

KERREBROCK:

If there is not a Karman vortex street there then it seems to me that the unsteadiness should be at a high frequency compared to the base phenomenon.

KOVASZNAY:

No, it isn't.

KERREBROCK:

But, I see amplitudes that are very large, which you are taking out by the signal processing.

KOVASZNAY:

Basically what you do is as follows. Looking at what I rather like to call the deterministic part of the phenomenon, the deterministic part is repeats. That is what you are extracting by periodic sampling ensemble averaging. What you destroy is the random part.

KERREBROCK:

My question, Dr. Kovasznay, is, are we sure that the deterministic result that you get from the signal processing is the same as the deterministic result that would result from a nonrandom input to such a nonlinear process?

KOVASZNAY:

Of course. By definition. The reaction is a nonlinear process, but the turbulent part, the random part, in this experiment is still confined to a rather narrow wake. Actually we determined even the profile of the turbulence intensity of the wake by the process which I just described. The point is here that you extract the deterministic part which is perturbed by the superimposed random. Now if you wish to modify even the random component and see whether you will get the same answer even this is possible, because by a second ensemble averaging one may determine statistically what the random wake is.

SAFFMAN: (California Institute of Technology)

You have measurements of the pressure as the vortex ring approaches the wall. It can be shown that for a vortex ring approaching a plane wall the total impulse communicated to the wall is zero. Did you have any way of integrating the pressure?

KOVASZNAY:

Honestly, no, because the pressure transducers used do not record DC components. This means that the long time average is unreliable and in principle you are right. In practice we know the low frequency response of these transducers, they have a cutoff at low frequency, therefore the average of the output would be zero no matter what input is, therefore this result is nonsignificant.

WILLMARTH: (University of Michigan)

Could you tell us again how you measure that fluctuating lift on the airfoil?

KOVASZNAY:

On the airfoil we had a total of 16 pressure taps to each we had a condenser microphone type of transducer connected. At each chordwise location we also measured the DC component, namely the average static pressure. To superimpose on the average static pressure the ensemble averaged fluctuating pressure (that has zero mean by value by definition) is easy so we were able to reconstruct the deterministic portion of the instantaneous pressure all around the airfoil for the same instant. For each instant this was integrated around the airfoil and the total lift was obtained.

In this case we used a doubly symmetrical airfoil which only one quarter of it instrumented and it was put into the tunnel in four ways. It had five pressure holes and these five holes were switched to the condenser microphone with a pressure switch.

HENDERSON: (Pennsylvania State University)

You mentioned experiments in your abstract with a cascade. Can you elaborate on these experiments?

KOVASZNAY:

We did a primitive cascade. We did a three airfoil cascade and just instrumented the middle airfoil.

HENDERSON:

What differences did you see between the cascade and the isolated airfoil?

KOVASZNAY:

Quite a bit. It was extremely sensitive from which direction the wake swept, it is a completely different story. If we just inserted the airfoil inverted it was a quite different response. This was the main difference. Of course the mean properties all changed when in the cascade and we had to take this into account to see clearly the changes. In addition it was clearly visible that a vortex bubble runs along the highly curved side of the airfoil in cascade. I didn't wish to prolong my talk to include the cascade case. We did the skew case too where the wake was misaligned with the leading edge of the airfoil. Therefore there is a spanwise propagating disturbance. See reference below.* We tried to compare this with the Filotas linear theory.† We did a great deal more than what I can show here. I just wanted to get the highlights on what sort of things we did.

---

* H. Fujita and L. S. G. Kovasznay. Sound Generation by Wake Cutting, Progress in Aeronautics and Astronautics, AIAA, MIT Press, Cambridge 1973, originally Paper 73-1019, Aero-Acoustics Conference, Seattle, Washington, Oct. 15-17, 1972.

† Filotas, L. T., "Theory of Airfoil Response in a Gusty Atmosphere I and II," Utias Rept. 139 and 141, Toronto, 1969.

HENDERSON:

Is the cascade data published?

KOVASZNAY:

It is in effect yes. We talked about it in this Aero-Acoustic meeting in Virginia, '75 Spring.* I also want to say a nice word for the sponsor. A great deal of this was sponsored at Johns Hopkins University by Pratt and Whitney.

---

* Ho, Chin-Ming and Leslie S. G. Kovasznay, "Wake Cutting by a Cascade of Cambered Blades," AIAA Paper 74-445, AIAA 2nd Aero-Acoustics Conference, Hampton, VA, March 24-26, 1975.

Some Turbulence and Unsteadiness Effects in Turbomachinery

R. L. EVANS
B.C. Energy Commission
Vancouver, Canada

ABSTRACT

The effects of high levels of turbulence and flow unsteadiness on turbomachine performance is investigated in three parts. The first part describes an experimental technique for separating the contributions of random turbulence and ordered unsteadiness to velocity fluctuation levels. The second part describes the effects of freestream turbulence on turbulent cascade performance, and in the final part some measurements of the unsteady boundary layer developing on a turbomachine blade are presented. The results indicate that caution should be exercised when applying steady boundary layer data to turbomachine design.

1. INTRODUCTION

In order to accurately predict boundary layer development on a turbomachine blade the freestream conditions must be known. It appears that the boundary layer develops in a region of high levels of freestream turbulence due to wakes being shed from upstream blade rows and in an unsteady pressure field due to the relative motion of the blade rows. It is likely that these two factors, high turbulence levels and periodic unsteadiness, have a significant effect on the blade boundary layer, but so far little information on the relative magnitude of the two effects has appeared in the literature.

This paper is divided into three main parts in an attempt to obtain a better understanding of the effects of turbulence and unsteadiness on turbomachine performance. The three parts present the results of:

(i) Measurements of the relative magnitude of random turbulence and periodic unsteadiness in a turbomachine.

(ii) An experimental investigation of the effects of free-stream turbulence on turbulent boundary layer development.

(iii) Measurements of the unsteady boundary layer developing on an axial flow compressor stator blade.

2. TURBULENCE AND UNSTEADINESS LEVELS IN AN AXIAL FLOW COMPRESSOR

In the past there has been a considerable effort directed toward the effects of flow unsteadiness on blade lift (see for example, Horlock, 1968), but so far there has been little experimental information on the random turbulence level and periodic unsteadiness in a turbomachine. Kiock (1973) attempted to separate periodic fluctuations in the mean flow from turbulent fluctuations by traversing a probe downstream of a stationary cascade. Whitfield et al. (1972) used a single probe rotated into each of the three coordinate directions in turn, and a phase-locked averaging procedure to produce contour plots of the periodic flow field downstream of a rotor row. The signal processing procedure used, however, lost all information of the random turbulence levels in the flow. Evans (1974a) described an ensemble-averaging procedure for separating the velocity fluctuations into ordered unsteadiness and random turbulence, and it is the results of that investigation which will be summarized in this section.

2.1 The Experimental Compressors and Coordinate System

Two low-speed single-stage compressors in the S.R.C. Turbomachinery Laboratory at Cambridge University were used for these measurements. The compressors are designated as compressor A and compressor B and are shown schematically in Figure 1. Both compressors have identical 6 in chord, 18 in span rotor blading designed for free-vortex operation at a flow coefficient C_x/U_m of 0.7, and for minimum loss at a design coefficient of 0.6. C_x denotes the axial flow velocity and U_m the rotor mid-span blade speed. Compressor A, shown in Figure 1a, consists of a single rotor row, while compressor B has a rotor row followed by a row of 1 ft chord stator blades.

The coordinate system used for the measurements is shown in Figure 1c. Axial, radial and tangential coordinates are denoted by x, r and θ respectively. The coordinate direction aligned with the mean flow velocity is the ξ coordinate and the perpendicular direction is the η coordinate. Air flow angle is given by α.

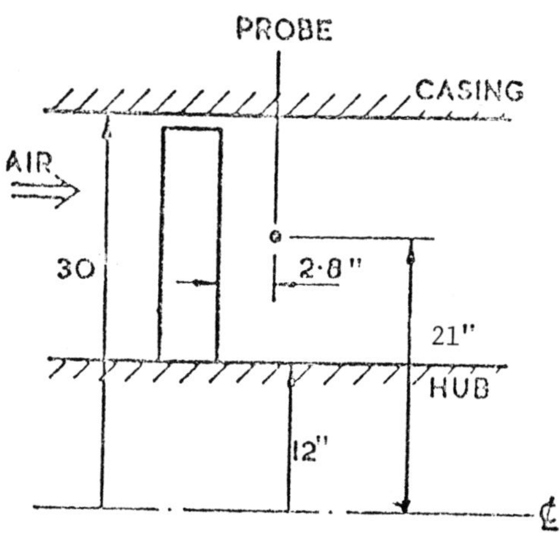

Figure 1a. Compressor 'A'

Velocities in each of the coordinate directions are denoted by C with the appropriate subscript.

2.2 Preliminary Disturbance Level Measurements

Before attempting to separate the velocity fluctuations due to turbulence from those due to periodic unsteadiness, a hot-wire X probe was used to determine the overall disturbance levels downstream of the rotor row in compressor A. The term "disturbance level" is used to indicate that the fluctuating velocity recorded by the anemometer is composed of both a random turbulent component and an ordered unsteady component. It is the value read from an rms meter connected directly to a hot-wire anemometer with no further signal processing.

The X probe was first positioned at mid-span with the plane of the probe in the x, θ plane and then rotated about the r axis into the ξ, η plane so that fluctuations C'_ξ parallel to the mean flow and C'_η perpendicular to the mean flow could be measured. The mean flow direction ξ was determined by rotating the probe about the r axis until the dc voltage from the two wires was equal. To provide a check on the C'_ξ measurements, and to measure the

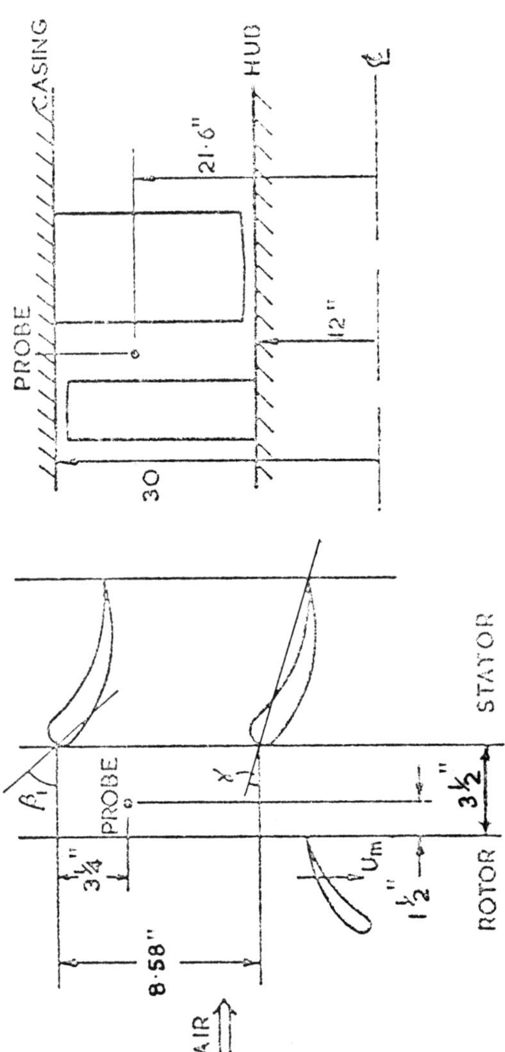

Figure 1b. Compressor 'B'

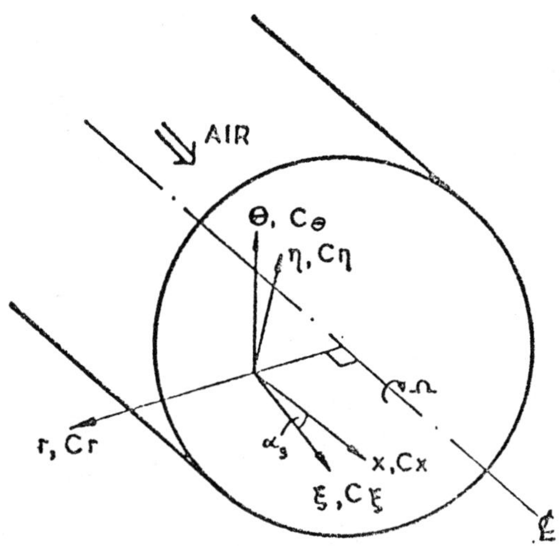

Figure 1c. Coordinate System.

fluctuations C_r' in the radial direction, the probe was then rotated about the ξ axis into the r, ξ plane.

The results of the X probe measurements are shown in Figure 2 for a range of flow coefficient from 0.45 to 0.70. The disturbance level in each of the coordinate directions approaches a minimum value near the minimum loss design flow coefficient of 0.60, while at the lower values of flow coefficient, as the rotor approaches stall, the disturbance levels approximately double their minimum value. With the X probe in the r, ξ plane the dc voltages from the two wires were equal, indicating zero mean radial velocity, although the fluctuations in the radial direction are of the same magnitude as the streamwise and tangential components.

The X probe was then replaced by a single wire probe with the wire axis aligned perpendicular to the mean flow and the anemometer signal was processed with a Fenlow spectrum analyzer. Figure 3 shows the spectra obtained at three different flow coefficients, and since only qualitative information was required the power spectral density shown is uncalibrated and simply represents a convenient scale. All the spectral measurements show a peak at blade

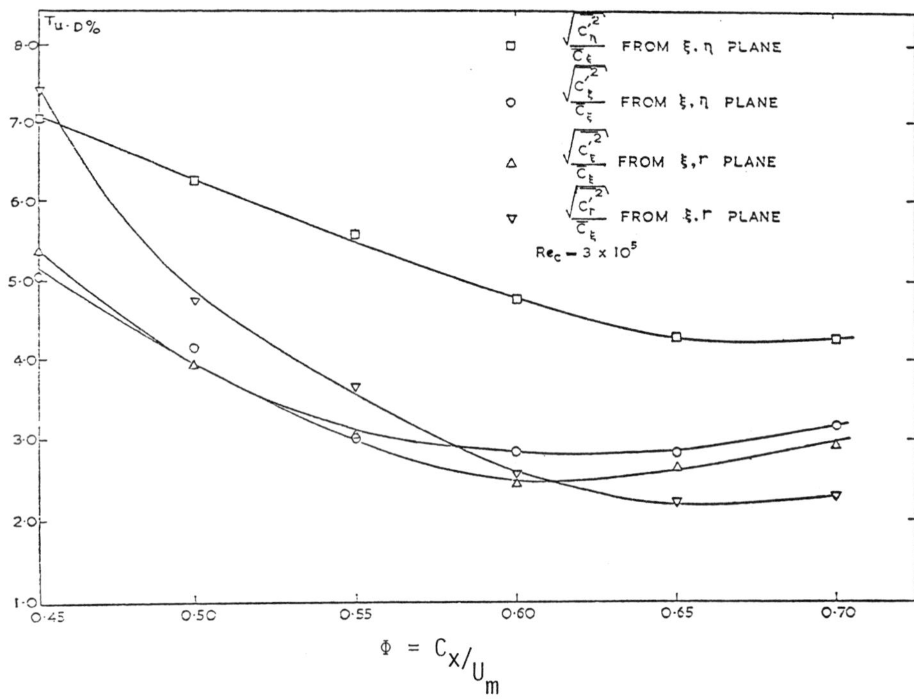

$\Phi = C_x/U_m$

Figure 2. X probe disturbance levels downstream of rotor in compressor 'A'

passing frequency and at several distinct harmonics. If plotted on a linear scale the area under the peaks would represent the fraction of the total fluctuation energy contained in the periodic oscillations, while the area of the lower band would represent the fraction of the energy contained in the turbulent fluctuations. The area corresponding to the turbulence energy is highest at the flow coefficient of 0.5 near stall, and lowest at the minimum loss flow coefficient of 0.6, while the area corresponding to the periodic fluctuations remains relatively constant.

2.3 Turbulence and Unsteadiness Measurements

The separate contributions of random turbulence and periodic unsteadiness to the velocity fluctuations were obtained using a phase-locked ensemble-averaging procedure. For these measurements a single hot-wire probe, with the wire axis perpendicular to the

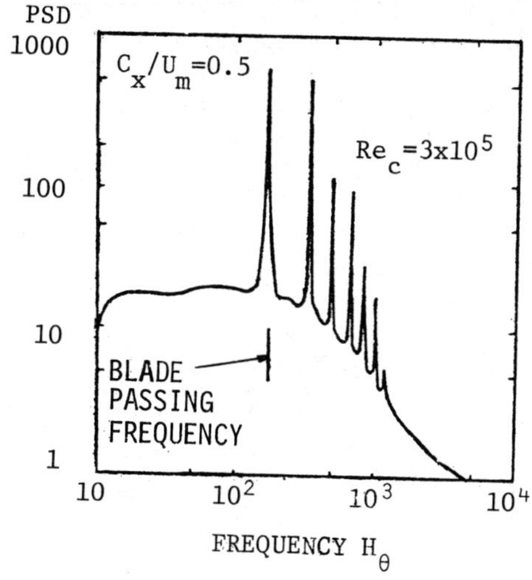

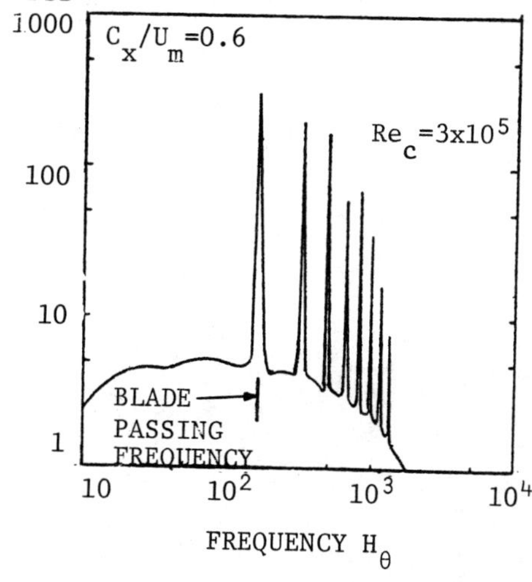

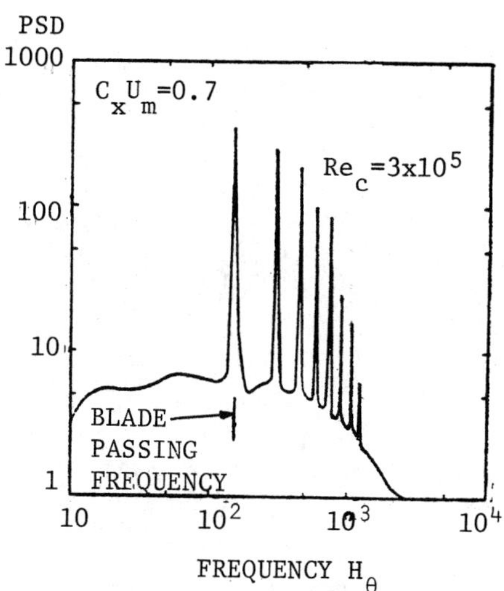

Figure 3. Power spectral density downstream of rotor.

mean flow direction, was placed between the blade rows of compressor B as indicated in Figure 1b. The linearized anemometer signal was fed directly to an rms meter and in parallel to a PDP-12 computer which, after digitizing, ensemble-averaged the signal over some 500 records. The sampling cycle of the A-D converter was initiated once per revolution by a pulse obtained from a magnetic pickup and a marker located in the rotor hub.

The ensemble averaging technique is shown schematically in Figure 4. A diagrammatic representation of many random velocity records U(t) is indicated in Figure 4a, where a typical record is the Kth record, K U(t). An ensemble average at time t_1 is the average of the instantaneous velocity $U(t_1)$ over all records, and can be given by:

$$\tilde{U}(t_1) = \frac{\lim\limits_{N\to\infty} \sum\limits_{K=1}^{N} {}^{K}U(t_1)}{N}$$

If this procedure is carried out for all points in time, the ensemble-averaged record $\tilde{U}(t)$ is obtained. A stationary random function is one for which ensemble-averages at every point in time are identical, i.e. $\tilde{U}(t_1) = \tilde{U}(t_1 + t)$ for all t (see Bendat, 1958). A time average of the Kth record is given by:

$$\overline{{}^{K}U(t)} = \lim_{T\to\infty} \frac{1}{2T} \int_{-T}^{T} {}^{K}U(t)dt$$

If a stationary random function is further restricted so that every record has the same time average, $\overline{U(t)}$, then the function is said to be ergodic and the ensemble-average is identical to the time average of any record. This ergodic hypothesis holds for homogeneous, isotropic turbulence and in that case ensemble-averages may be replaced by time averages.

The velocity-time record of a turbomachine wake passing a fixed probe is shown diagrammatically in Figure 4b. The smooth line marked $\tilde{U}(t)$ is the result of ensemble-averaging over a number of records. It can be seen immediately that the condition for the function to be stationary, namely $\tilde{U}(t_1) = \tilde{U}(t_1 + t)$, is not fulfilled. Also shown on the figure is the time-mean velocity, $\overline{U}$, taken by averaging over a long time compared with the wake passing period. The fact that the wake signal is periodic and non-stationary means that a phase locked ensemble-averaging procedure must be used.

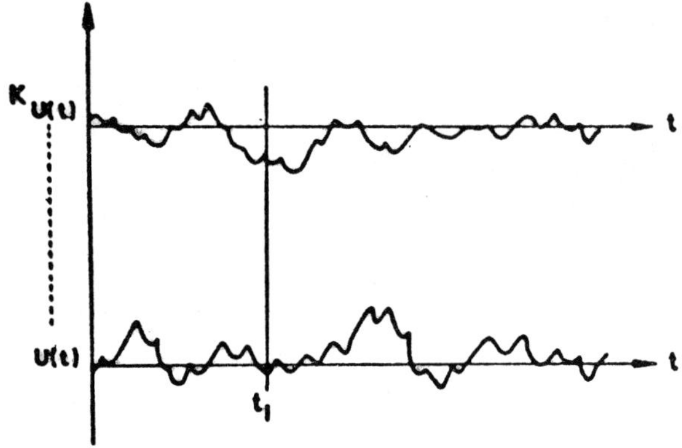

Figure 4a

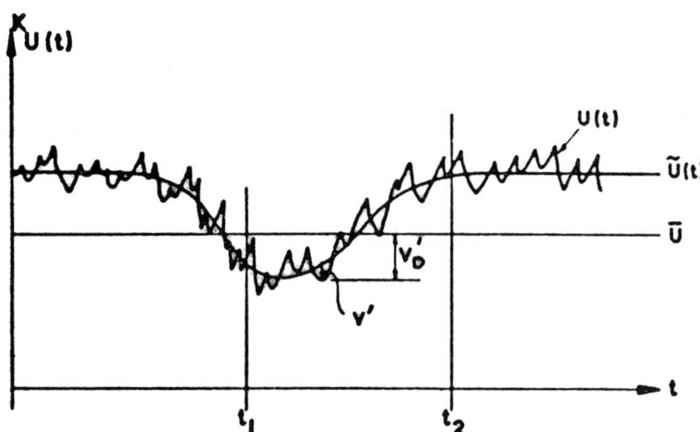

Figure 4b

Figure 4. Typical velocity records.

The phase-locking is accomplished by means of a trigger pulse obtained from a magnetic pickup on the rotor as noted previously.

Referring to Figure 4b, three different velocity fluctuations can be defined. The velocity fluctuation from the ensemble-averaged velocity, $\tilde{U}(t)$, is called the freestream turbulent fluctuation and is denoted by v'. The total fluctuation from the time-mean velocity, $\overline{U}$, is called the disturbance fluctuation and is given by:

$$V_D' = v' + (\tilde{U}(t) - \overline{U})$$

An "unsteadiness fluctuation" is defined as the difference between the ensemble-averaged velocity and the time-mean velocity. Using the usual rms definitions of these values, three different intensities may be defined:

The disturbance level: $$Tu_D = \frac{\sqrt{\overline{V_D'^2}}}{\overline{U}}$$

The freestream turbulence level: $$\hat{T}u = \frac{\sqrt{\overline{v'^2}}}{\overline{U}}$$

and the unsteadiness level: $$\tilde{T}u = \frac{\sqrt{\overline{(\tilde{U}(t) - \overline{U})^2}}}{\overline{U}}$$

Using these definitions and assuming that the turbulent fluctuations are statistically independent from the periodic wake velocity-defect fluctuations, Evans (1974a) has shown the following relation to hold:

$$Tu_D^2 = \hat{T}u^2 + \tilde{T}u^2 \tag{1}$$

A similar expression to this was first obtained by Kiock (1973) using a traversing probe downstream of a fixed cascade. In practice the disturbance level, Tu_D, is measured with an rms meter connected directly to the hot-wire anemometer. The unsteadiness level, $\tilde{T}u$, is found by taking the rms value of the difference between the ensemble-averaged velocity and the time-mean velocity, and the freestream turbulence level, $\hat{T}u$, is then found by difference from equation (1).

Figure 5 shows the results of the measurements between the blade rows of compressor B as a function of flow coefficient, C_x/U_m. The overall disturbance level, Tu_D, is seen to be a maximum at a flow coefficient of 0.50 near stall, and decreases as the flow coefficient approaches the design value of 0.60. The unsteadiness level, $\tilde{T}u$, remains relatively constant over the compressor operating range. As the rotor approaches stall the wakes become much more spread out, but the amplitude of the velocity defect is decreased, resulting in a nearly constant rms value. The free-stream turbulence level, $\hat{T}u$, obtained by difference, is a maximum near stall and decreases monotonically towards the design point.

The manner in which large changes in flow angle can occur as a rotor wake passes is illustrated in Figure 6. For a compressor, shown in Figure 6a, the relative velocity decreases from W_2 to W_2' within a rotor wake. After adding the constant blade speed U, the absolute velocity changes from C_2 to C_2', with a resultant increase in tangential velocity to C_{θ_2}' and a decrease in axial velocity to C_{x_2}'. This causes a large change in absolute flow angle within a

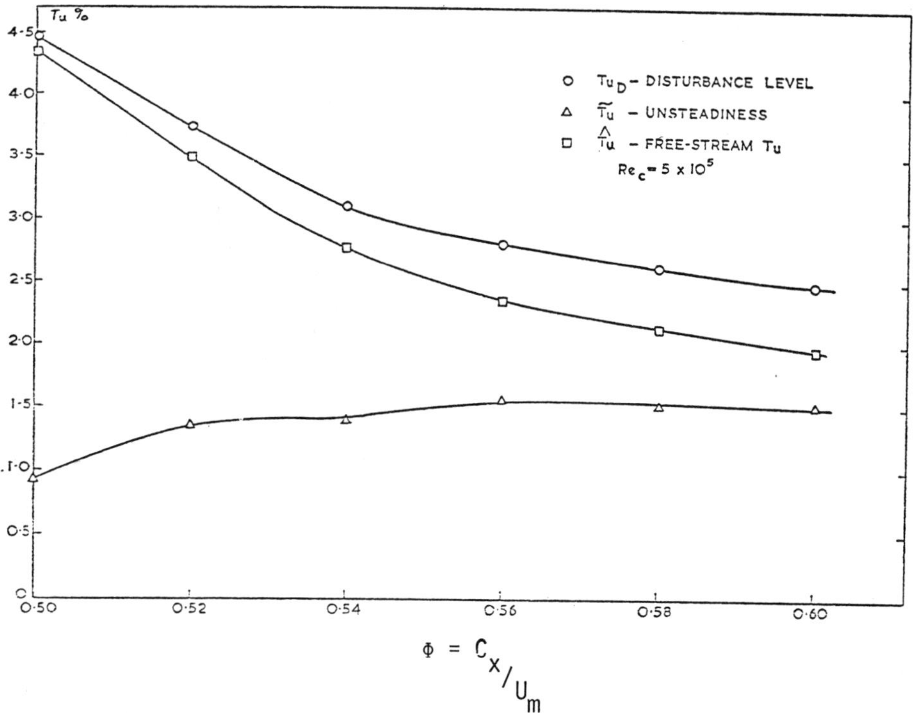

Figure 5. Turbulence and unsteadiness measurements in compressor 'B'

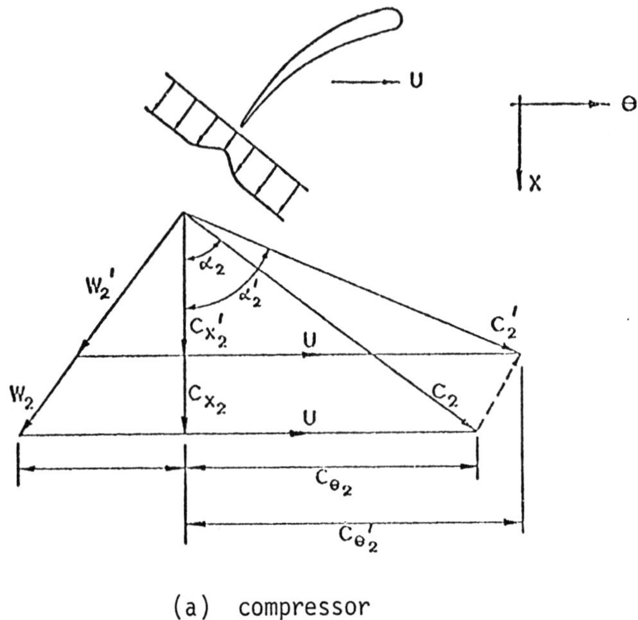

(a) compressor

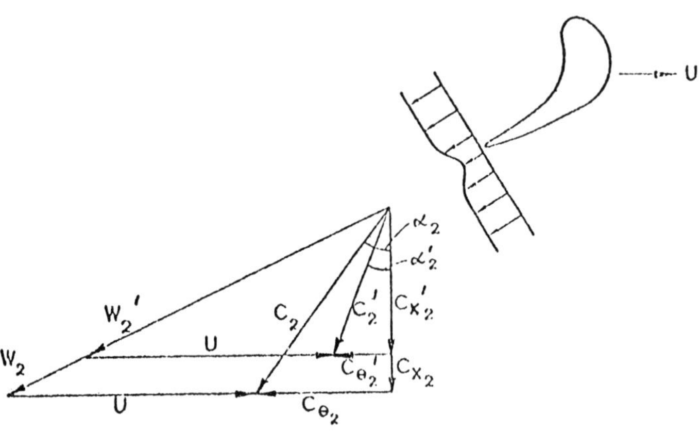

(b) turbine

Figure 6. Rotor outlet velocity triangles.

wake from α_2 to α_2'. The situation for a turbine, illustrated in Figure 6b, is similar except that both the tangential and axial velocity may decrease.

After taking ensemble-averages of both the axial and tangential velocities between the blade rows, the time resolved absolute air angle was simply found by taking the tangent of C_θ/C_X. These results are given in Figure 7, shown plotted as α_3, the air inlet angle referred to the downstream stator row, for three different flow coefficients. The stator blade inlet angle, β_1, is shown in each case to enable the stator incidence angle, $\alpha_3 - \beta_1$, to be determined. The incidence angle shows a large positive increase of between 12° and 15° every time a rotor wake passes, which might be expected to have a significant effect on the blade boundary layer.

3. FREESTREAM TURBULENCE EFFECTS ON THE TURBULENT BOUNDARY LAYER

3.1 Pipe-Wall Boundary Layer

In order to obtain an initial feeling for the effects of freestream turbulence on turbulent boundary layer development, a study of the boundary layer on the wall of an 8 in. ID pipe was initiated. The boundary layer was tripped at the pipe entrance and the freestream turbulence level adjusted with grids. Figure 8 shows dimensionless velocity profiles 5 ft downstream of the entrance at three levels of freestream turbulence. Increasing turbulence levels, Tu, can be seen to cause a significant increase in the profile fullness. There is also an increase in the profile δ_{99} thickness with increasing turbulence level, but this does not show up in the nondimensional plot.

The displacement and momentum thicknesses, δ^* and θ, are shown as a function of freestream turbulence in Figure 9. Also shown in the figure are the data of Kline et al. (1960). Both sets of data indicate an increase in the integral thicknesses with turbulence level until about Tu = 2%, and then a subsequent decrease. This type of behaviour can be explained by the two conflicting effects of increased δ_{99} thickness and increased profile fullness. At low turbulence levels the increased δ_{99} dominates, causing δ^* and θ to increase, while at higher levels the increased fullness dominates resulting in a decrease in the integral thicknesses. Further details of these measurements are given by Evans (1974b).

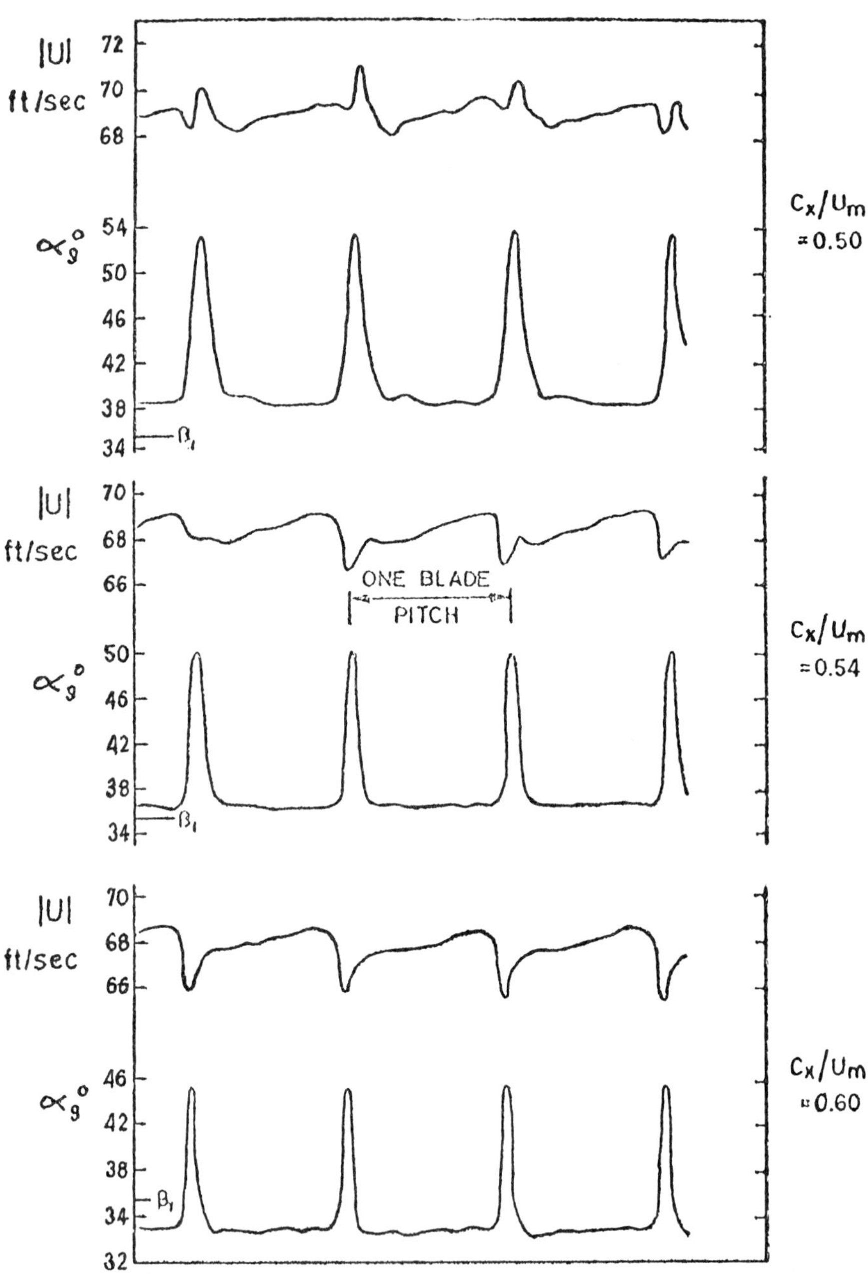

Figure 7. Stator inlet air angle and velocity vector.

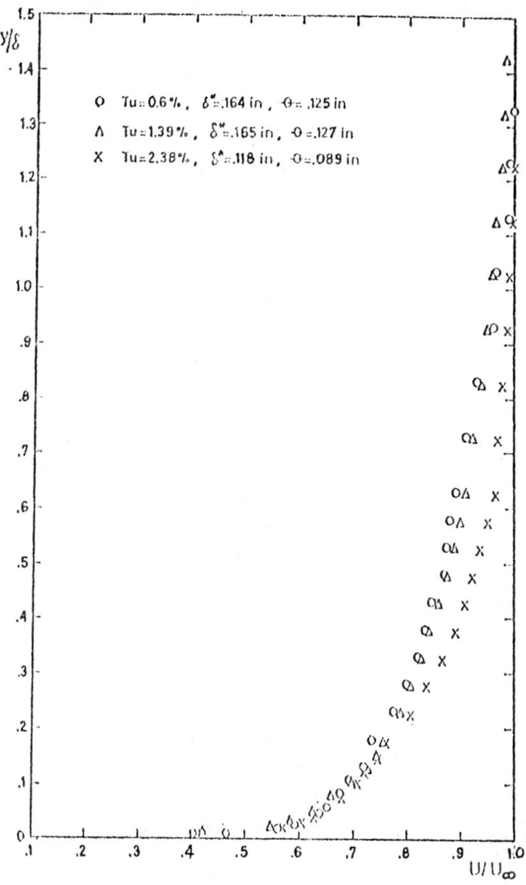

Figure 8. Dimensionless velocity profiles, X = 5 ft.

3.2 Cascade Blade Boundary Layer

A series of cascade tests was next undertaken to determine the effect of freestream turbulence on cascade performance. The cascade blades were of 1 ft chord C-4 section with the boundary layer on all blades tripped at the 10% chord position. The cascade end-walls were porous and provided with boundary layer suction to ensure an axial velocity ratio of unity. Turbulence grids were placed upstream of the cascade to vary the freestream turbulence intensity. The cascade tunnel and instrumentation have been described in greater detail by Evans (1972).

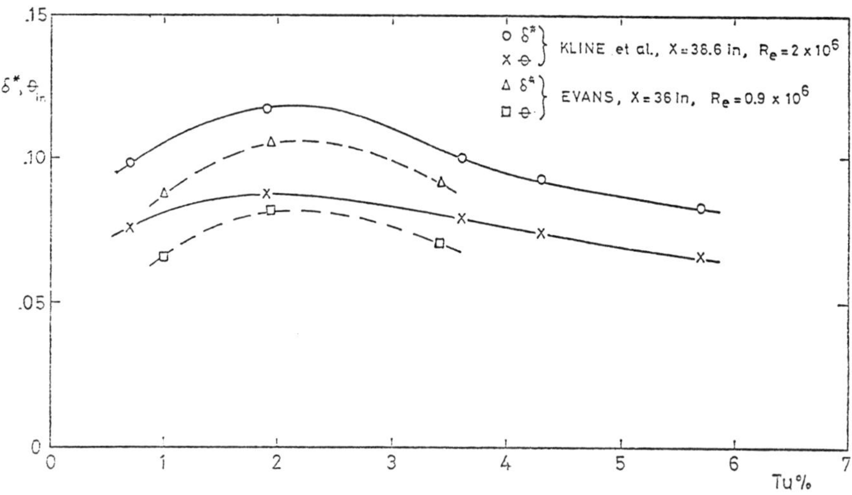

Figure 9. Boundary layer integral parameters vs. Tu.

Dimensional velocity profiles at the 70% chord position for three freestream turbulence levels are shown in Figure 10. There appears to be little increase in δ_{99} with increasing turbulence level, but there is a considerable increase in profile fullness, particularly at the highest turbulence level. The effect of the increased fullness on the displacement and momentum thickness is illustrated in Figure 11. A general decrease in both δ^* and θ is seen to take place with increasing turbulence level. The maximum effect occurs at about mid-chord, with little or no effect near the leading edge where the boundary layer is thin and near the trailing edge where separation is imminent.

Velocity profiles at the 80% chord position are plotted in semi-logarithmic coordinates in Figure 12, again for three levels of freestream turbulence. The main point of this figure is to illustrate the manner in which the outer layer wake component of the profile, and therefore the Coles wake parameter Π, decreases with increasing turbulence level. The actual form of the Coles wake function is shown in Figure 13. At the two lower turbulence levels the function is seen to be a good fit to the simple relation suggested by Coles (1956). At the highest turbulence level, however, there is seen to be a significant departure from the simple relation. This effect has already been suggested by Coles (1968) when examining many data sets for the 1968 Stanford Conference.

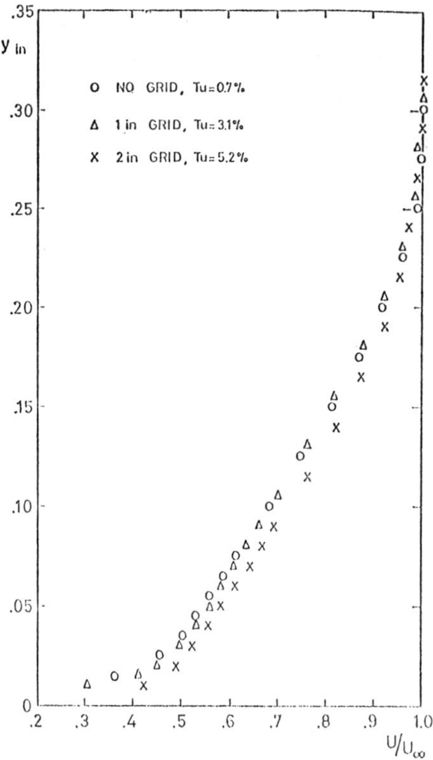

Figure 10. Velocity profiles, X/C = 0.7.

The effect of freestream turbulence on cascade performance is summarized in Figure 14. Although only three values of the turbulence level were used, the resulting trend can be explained by the previous boundary layer results. The mass averaged total pressure loss coefficient is seen to increase to a turbulence level of about 2½%, and then decreases with further increases in turbulence level. This behaviour is consistent with the momentum thickness of the pipe wall boundary layer, which showed an increase to about a 2% turbulence level, and then decreased for higher levels due to the increased profile fullness. The cascade mean deviation angle shows a monotonic decrease with increasing freestream turbulence. This result is most likely due to the separation point being moved closer to the trailing edge with increased turbulence level, because of the greater energy content of the boundary layer.

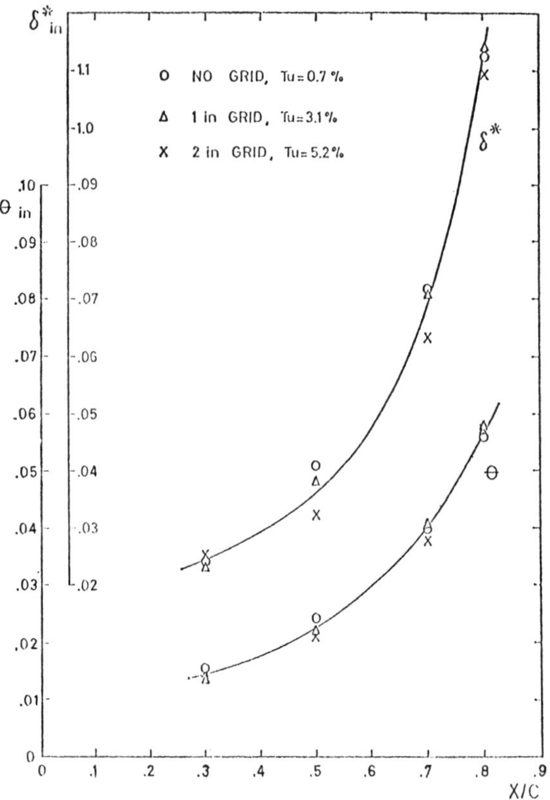

Figure 11. Boundary layer momentum and displacement thickness.

4. UNSTEADY BOUNDARY LAYER DEVELOPMENT ON A COMPRESSOR STATOR BLADE

Both time-mean and instantaneous ensemble-averaged velocity profiles were measured at mid-span on a stator blade in compressor B, which has been described in section 2.1. The hot-wire anemometer was again connected directly to the PDP-12 computer which was able to digitize and ensemble-average the signal in real time. Instantaneous and ensemble-averaged velocity records as a function of time were recorded at each station within the boundary layer. The signal was again phase-locked by means of the trigger signal generated by a pickup on one of the rotor blades. Details of the experimental setup and probe used are given by Evans (1973).

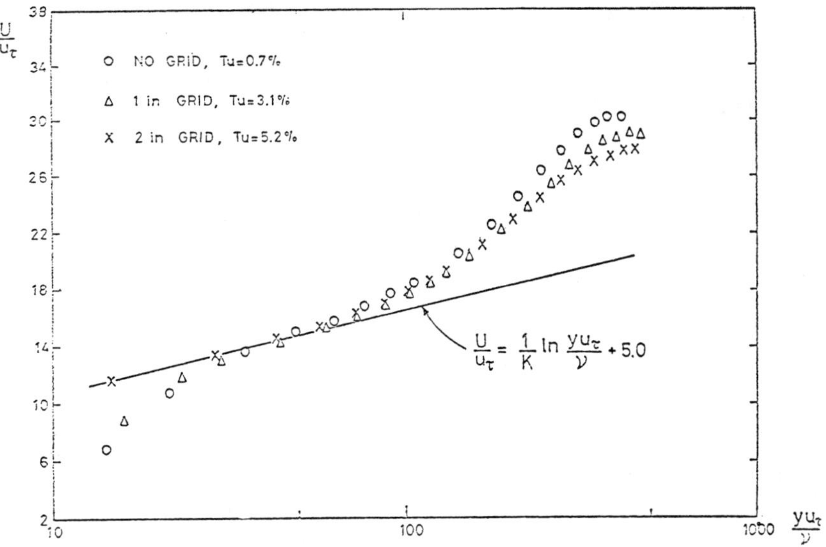

Figure 12. Semi-logarithmic profiles, X/C = 0.8.

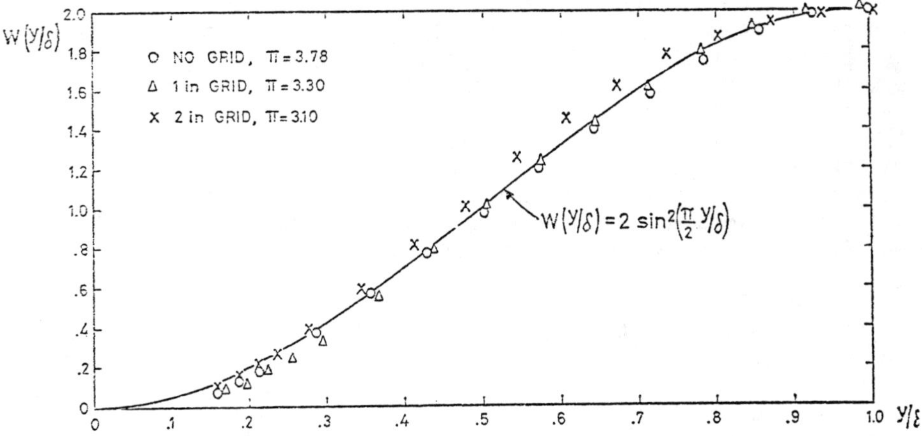

Figure 13. The Coles wake component, X/C = 0.8.

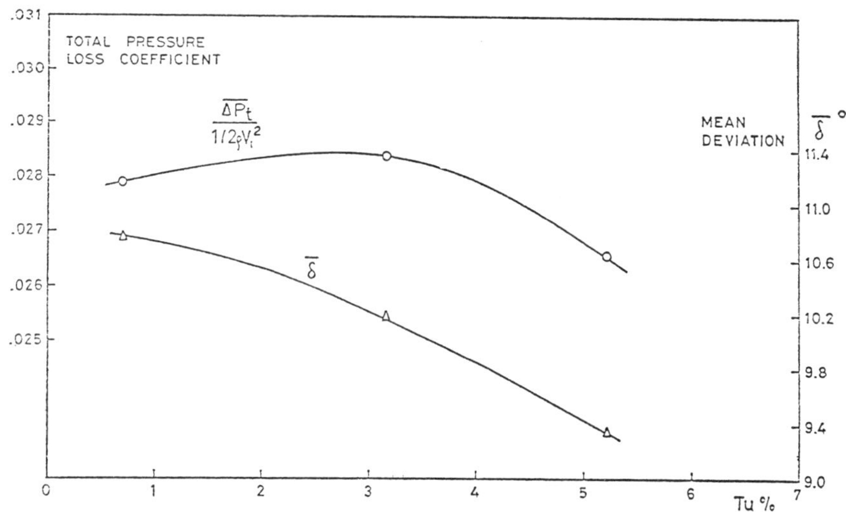

Figure 14. Cascade losses, i = +4°, $Re_C = 5 \times 10^5$, A.V.R. = 1.04

Dimensional time-mean velocity profiles at several chordwise positions are shown in Figure 15. These profiles were obtained using a DC voltmeter with an integration time much longer than the wake passing period, which was connected directly to the linearized anemometer. Also shown on the figure are Coles profiles fitted to the data. The Coles profiles appear to give quite a reasonable fit to the data except at the 80% chord position where the profile is approaching separation. On the basis of these time-mean profiles alone it would seem reasonable to assume that the boundary layer is wholly turbulent, at least from the 30% chord position.

Figure 16 shows some velocity records taken at several different positions within the boundary layer at the 30% chord position. The value of δ used to non-dimensionalize the vertical coordinate is the δ_{99} value from the time-mean profiles. In each case the top trace is the result of taking an ensemble-average over some 500 records, while the bottom trace is a typical instantaneous record. The first oscillogram is shown just outside the boundary layer at $Y/\delta = 1.14$. Rotor wakes appear as jagged peaks in the single trace and as smoothed out velocity defects in the ensemble-averaged record. Within the boundary layer the single traces indicate that

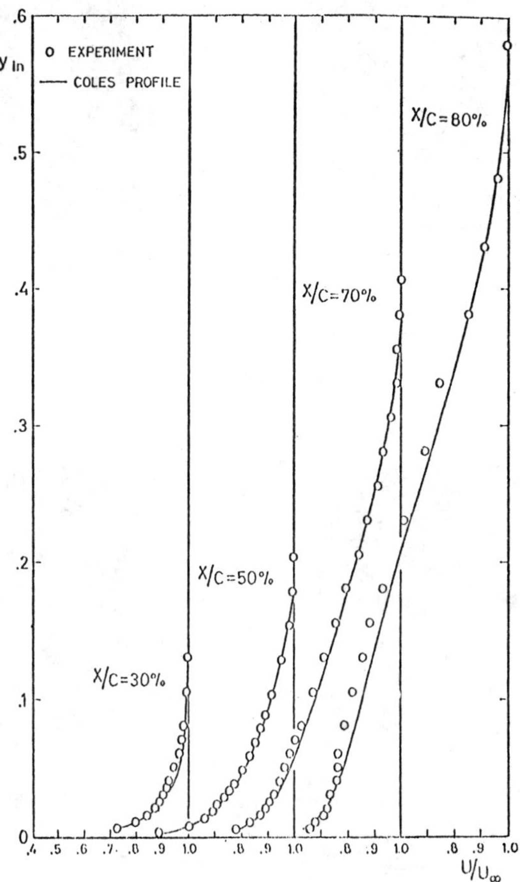

Figure 15. Dimensional mean velocity profiles, ϕ = 0.52, i = 2.5°

the flow is laminar for approximately half the time and is turbulent for the remaining half. At Y/δ = .715 the ensemble-averaged record shows a small peak appearing approximately 180° out of phase with the freestream fluctuations. Deeper within the boundary layer the phase-shifted fluctuations dominate and increase in amplitude to about Y/δ = .25, after which the restraining influence of the wall causes the oscillations to damp out. For the high frequencies involved here (240 Hz), the unsteady measurements of Karlsson (1959) indicate that any phase shift in a turbulent boundary layer is small (~35°) and limited to a very thin region near the wall. The transitional nature of the boundary layer here, as evidenced in the single trace records, appears to be the explanation for the 180° phase

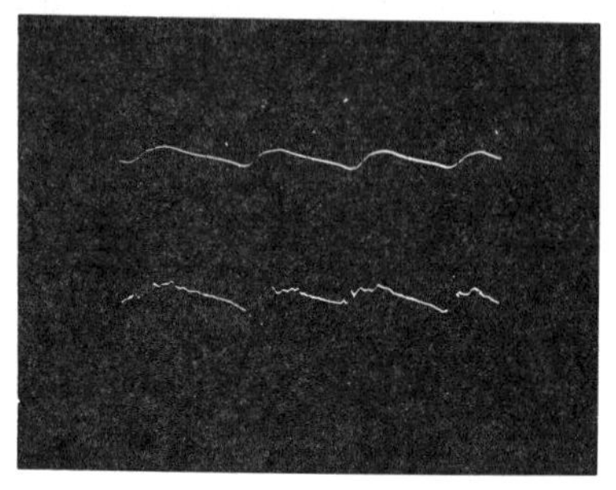

$y/\delta = 1.14$

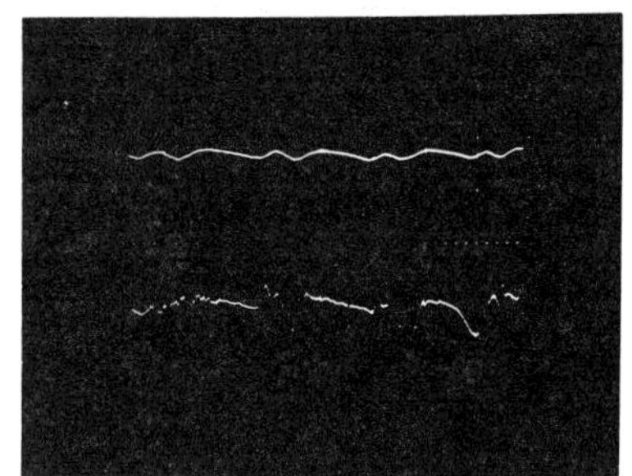

$y/\delta = .715$

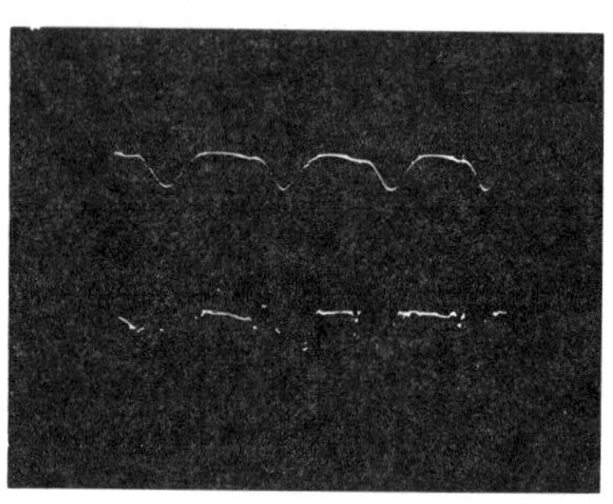

$y/\delta = .485$

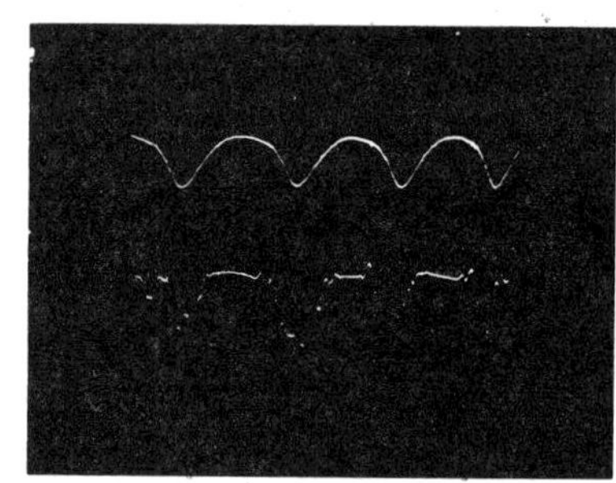

$y/\delta = .370$

VERTICAL SCALE = 8.5 ft/sec/cm

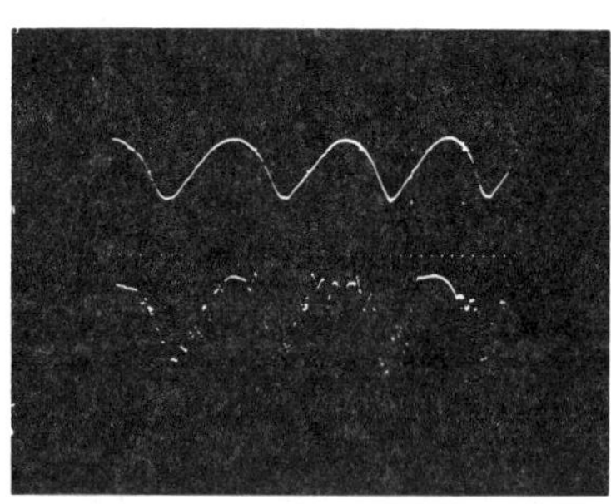

$y/\delta = .256$

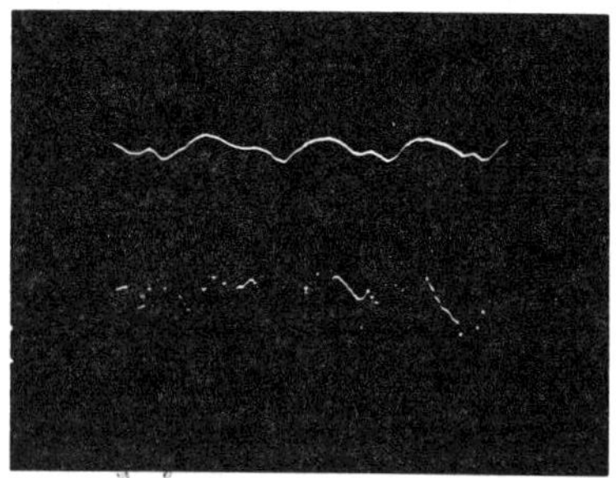

$y/\delta = .143$

Figure 16. Oscillograms of boundary layer velocity records, $\phi = 0.51$, $i = 3.5°$, $X/C = 0.3$

shift observed. This is better illustrated by the velocity profiles shown in the next figure.

Two instantaneous ensemble-averaged velocity profiles are shown in Figure 17. These profiles are obtained by taking a "slice" at a given instant in time through a series of ensemble-averaged velocity records for the whole boundary layer. For the profile labelled 0°, the slice is taken through a maximum point in the fluctuating freestream velocity record. The profile labelled 180° is then taken at a point in the freestream which is 180° out of phase with the first profile. Looking first at the profile labelled 180°, this appears to be a laminar profile with a thickness of only about 40% of the profile labelled 0°. This profile evidently corresponds to the stretches of laminar flow throughout the boundary layer which are seen in Figure 16. The profile marked 0° appears to be a turbulent profile with a thickness 2½ times that of the

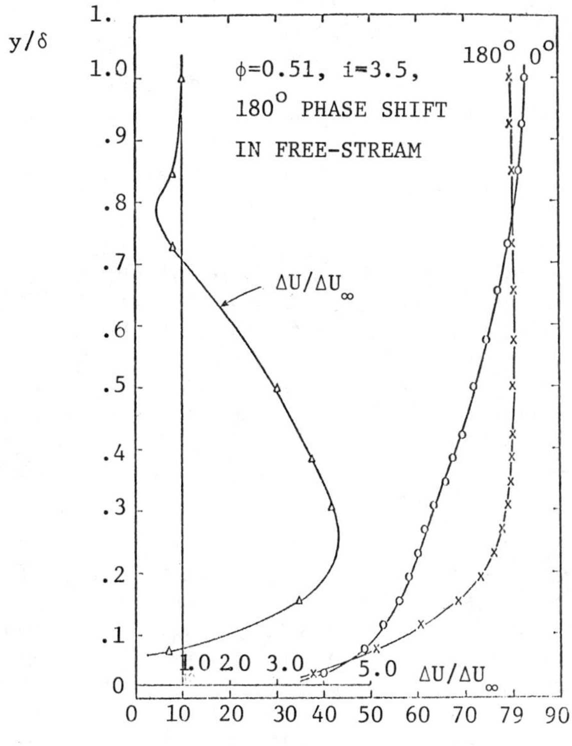

Figure 17. Instantaneous velocity profiles, X/C = 0.3

laminar profile of the other half of the cycle. The boundary layer is evidently oscillating between laminar and turbulent flow during each cycle. The increased velocity in the outer part of the turbulent profile, together with the very large increase in thickness, causes the two profiles to cross resulting in the 180° phase shift. Also shown in the figure is the ratio of fluctuation amplitude to the freestream amplitude.

A similar type of transitional boundary layer behaviour to that illustrated in Figures 16 and 17 occurred at the 50% chord position. Although the same 180° phase shift occurred due to the crossing of the profiles, there appeared to be little or no amplifications of the freestream fluctuations, and the boundary layer thickness during alternate halves of the cycle changes by about 50% rather than the 150% observed in Figure 17. Individual velocity record traces still displayed laminar regions in the outer portion of the layer, but for Y/δ less than about 0.6 the signal appeared fully turbulent. At the 50% chord position, then, there appears to be a turbulent oscillating boundary layer with a transitional shear layer superimposed on top of it.

The final two figures show the boundary layer behaviour at the 70% chord position. Examples of both ensemble-averaged and instantaneous velocity records throughout the layer are again illustrated in Figure 18. At $Y/\delta = 1.00$ the rotor wakes are clearly seen to be turbulent patches imbedded in a laminar flow. Within the boundary layer the instantaneous records indicate that the flow is fully turbulent at all times. The ensemble-averaged records show that the freestream fluctuations are damped out with no perceptible phase change. Two ensemble-averaged velocity profiles, 180° out of phase with one another, are shown in Figure 19. No crossing of the profiles occurs, and therefore no phase shift as evidenced at 30% and 50% chord. The boundary layer at this position is fully turbulent and oscillates in the manner described by Karllson (1959) for the high-frequency unsteady case.

5. CONCLUSIONS

We have seen that boundary layers on turbomachine blades develop under freestream conditions which are both highly turbulent and highly unsteady. Both of these factors have been demonstrated to have a significant effect on boundary layer development. The boundary layer on an axial-flow compressor blade is transitional over much of the blade chord, and growth is much greater than on a turbulent cascade blade. I think the single most important message which comes through here is that turbomachinery designers should exercise extreme caution when applying steady cascade data to turbomachine design.

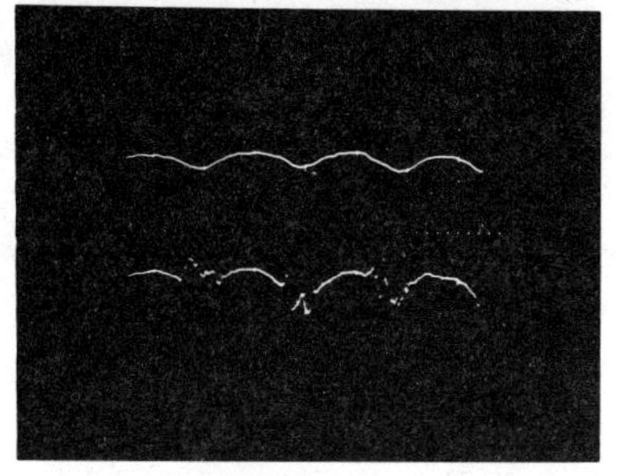

y/δ =1.00

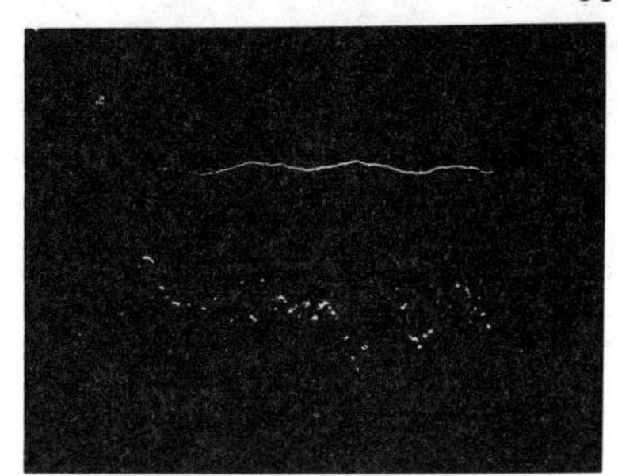

y/δ =.462

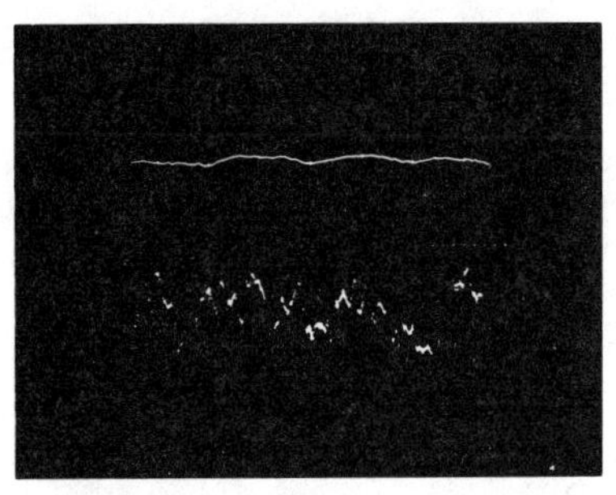

y/δ =.280

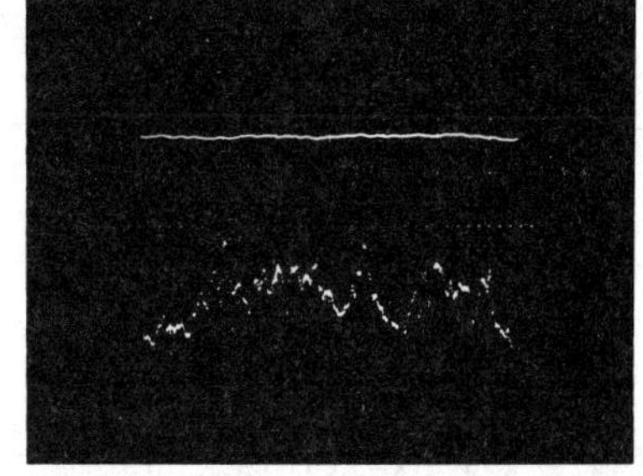

y/δ =.098

VERTICAL SCALE =11.0 ft/sec/cm

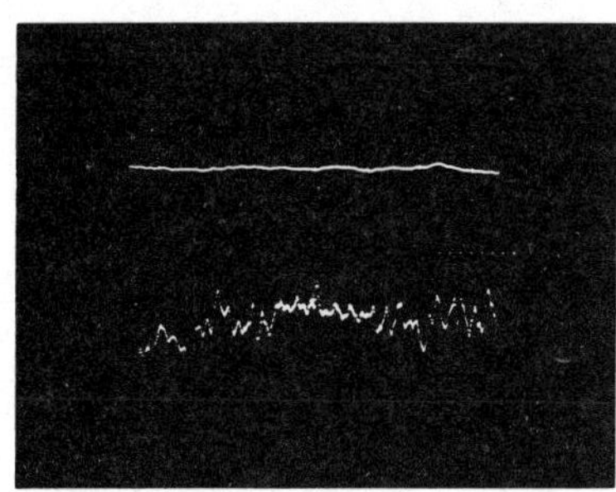

y/δ =.053

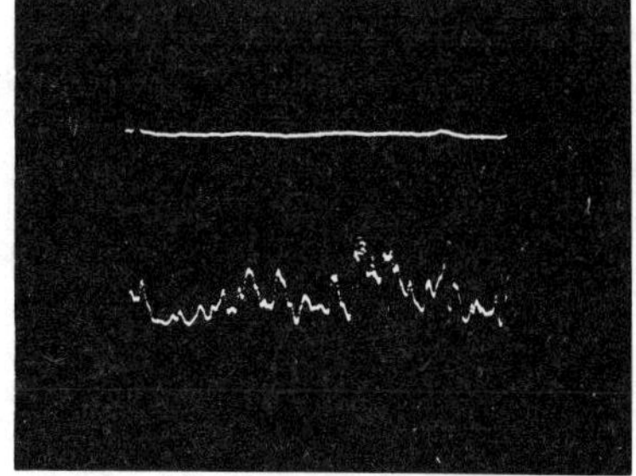

y/δ =.016

Figure 18. Oscillograms of boundary layer velocity records, ϕ = 0.51, i = 3.5°, X/C = 0.7

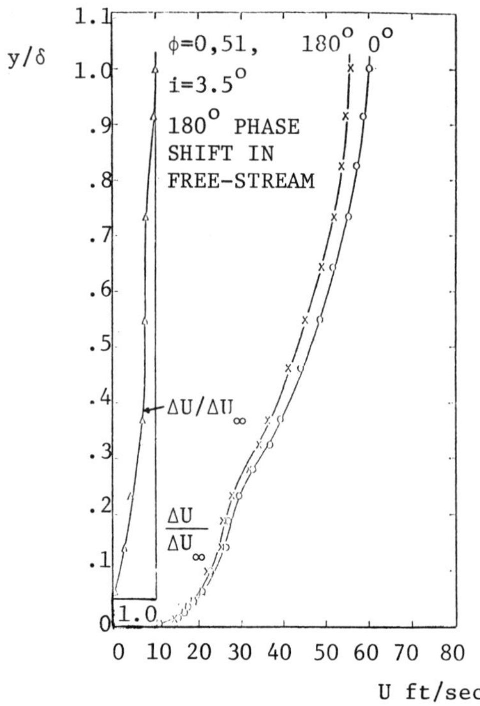

Figure 19. Instantaneous velocity profiles, X/C = 0.7

REFERENCES

1. Bendat, J. S. (1958), "Principles and Applications of Random Noise Theory," John Wiley and Sons.

2. Coles, D. (1956), "The Law of the Wake in the Turbulent Boundary Layer," JFM, 1, pp. 191-226.

3. Coles, D. (1968), "The Young Person's Guide to the Data," in Proceedings AFOSR-IFP-STANFORD Conference on Turbulent Boundary Layer Prediction, Vol. 2, Eds. D. Coles and E. Hirst.

4. Evans, R. L. (1972), "Stream Turbulence Effects on the Turbulent Boundary Layer in a Compressor Cascade," ARC Report 34 587, U.K.

5. Evans, R. L. (1973), "Turbulent Boundary Layers on Axial-Flow Compressor Blades," Ph.D. dissertation, Cambridge University.

6. Evans, R. L. (1974a), "Turbulence and Unsteadiness Measurements Downstream of a Moving Blade Row," Trans. ASME, Jnl. of Engineering for Power, 97, pp. 131-139.

7. Evans, R. L. (1974b), "Free-Stream Turbulence Effects on the Turbulent Boundary Layer," ARC CP 1282, U.K.

8. Horlock, J. H. (1968), "Unsteady Flows in Turbomachines," Paper 2674, 3rd Australian Conference on Hydraulics and Fluid Mechanics.

9. Karllson, S. K. F. (1959), "An Unsteady Turbulent Boundary Layer," JFM, 5, pp. 622-636.

10. Kiock, R. (1973), "Turbulence Downstream of Stationary and Rotating Cascades," ASME paper 73-GT-80.

11. Kline, S. J., Lisin, A. V., Waitman, B. A. (1960), "Preliminary Experimental Investigation of the Effect of Free-Stream Turbulence on Turbulent Boundary Layer Growth," NASA TN D-368.

12. Whitfield, C. E., Kelly, J. C., Barry, B. (1972), "A Three-Dimensional Analysis of Rotor Wakes," Aero. Quart., 23, pp. 285-300.

DISCUSSION

MELLOR: (Princeton University)

Some of the measurements you showed seem to go opposite to the way in which we would conceive of a decreasing flow efficient causing an increase in stall flow which would feed into turbulence; seems to me a rather major paradox.

EVANS:

My simple-minded explanation of that is I think the velocity defect wakes, the ensemble average parts, spread out but the amplitude is much decreased. It is not clear what the effect on the RMS value is, since the wakes spread out but have a decreased amplitude as the boundary layers become thicker and thicker.

MELLOR:

May I then conclude from your results, that if I was a designer and had a higher blade chord Reynolds number, then I really wouldn't need to worry about the unsteady turbulence since the major effect seems to be on transition.

EVANS:

That is my reading of it, that the unsteady effects, the effects of this massive change in incidence, every time a wake goes by, is probably more important than the details of the freestream turbulence.

KLINE: (Stanford University)

The results that you showed of ours are from Lisin and Waitman's data.* The reason that we were a little queasy about the accuracy of those data was because we got cut off by the Sputnik engendered rebudgetings and never finished in the way we would normally have done. We had hoped in fact to look at some other parameters besides intensity, including integral scale; I am still not sure that intensity is the best way to correlate freestream effects.

In regard to your 180° phase shift, I am not quite sure I understand what you are shifting. You showed two cases. One of them showed 180 about the same as zero and both were turbulent. In the other case one was laminar and one was turbulent. In the case where you had the laminar and the turbulent, how do you visualize this marching along? That is, if you draw a picture of the boundary layer, are you visualizing the flow sweeping over the blade turbulent and then at the next instant sweeping over it laminar, or are you visualizing laminar flow followed by turbulent downstream in space, or what is happening? I don't have a physical picture that goes with that curve you presented, and I am not clear on the distinction between the two cases. When were both turbulent, and when was one laminar and one turbulent? Are your results in any way related to the data that Acharya and Reynolds† got in a turbulent channel flow where they also see 180° phase shift as a function of frequency. In those data whether the oscillations are above or below a critical frequency determines whether you see a 180° phase shift; the data are all turbulent.

EVANS:

If I can start with that final point first, I don't know of that work. But if it is fully turbulent flow I don't think that phase shift has any relationship to the kind of phase shift I was talking about.

---

* See Ref. in Evans' paper.

† "Measurements and Predictions of a Fully Developed Turbulent Channel Flow with Imposed Controlled Oscillations," Report TF-8, M.E. Department, Stanford, May, 1975.

Now if I can go back and try and explain the slides. Each slide was at a given chordwise position on the blade. Something labelled zero degrees simply referred to fluctuations in the free-stream. So if my freestream ensemble average velocity looked like Figure I, I just picked a point in the freestream and arbitrarily called that zero degrees phase and then took a slice through all records to get this kind of instantaneous velocity profile. Midway in the freestream record, I took another slice and labeled that 180°. So those two profiles were 180° out of phase with respect to this freestream, if you like.

The way I visualize the boundary layer developing is that at a given point you see a laminar profile and then all of a sudden you get this switch to a turbulent profile, with the thickness increasing. If you follow it along the blade chord I think you have a laminar boundary layer with turbulence sweeping up to the front of the blade as a wake goes by and then oscillating back and forth.

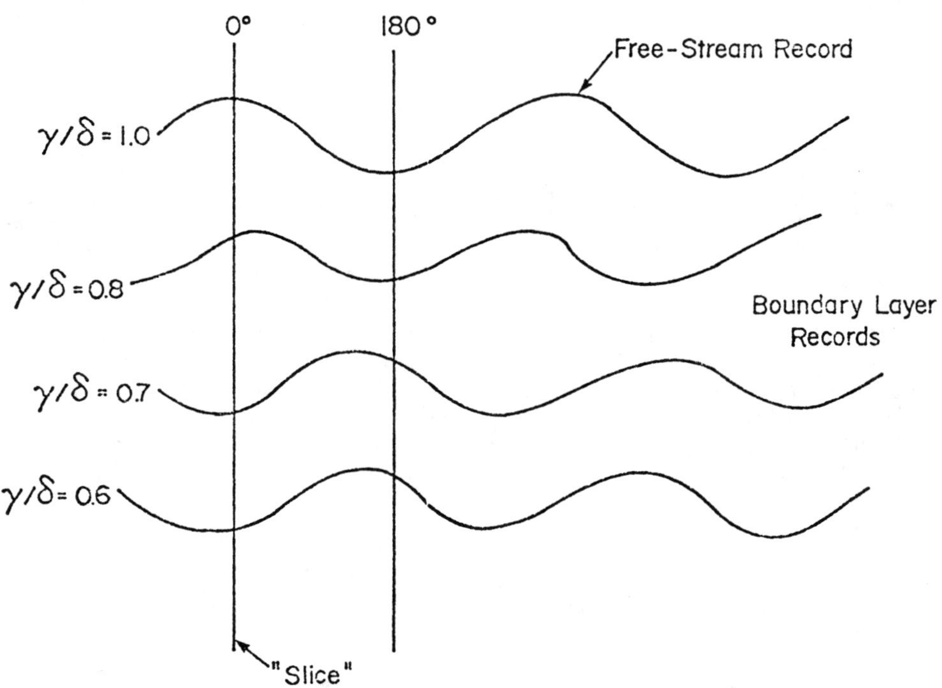

Figure I

KLINE:

So what you are saying is that if you get far enough down the blade it is turbulent all the time?

EVANS:

Yes, that is what the second slide showed.

KLINE:

So the difference is that station is farther down the blade.

EVANS:

That is down some 70% of the blade chord.

KLINE:

Somewhere in between you are seeing laminar flow, then turbulent flow, then laminar flow intermittently if you stand at that station?

EVANS:

Exactly, over quite a large chunk of the blade, at least up to 50% chord.

KERREBROCK: (M.I.T.)

I think that this is a very nice set of data. You should be congratulated on it. Naturally I am trying to correlate it with what I understand from our experiments and I must say that there are some gaps. A specific question arises when I compare your time resolved (lower) traces with the phase lock results. It seems to me that the amplitude of what you call turbulence as inferred from the real time resolved traces is really as large as, or perhaps larger than, the unsteadiness which is a result of blade passage. Is that really consistent with the picture of turbulence?

EVANS:

I guess I haven't really thought about that enough to rationalize it. I don't have information on the scale of the turbulence other than just those pictures as you suggest.

My remarks concerning the relative importance of the unsteady component of the flow were made on the basis of the ordered unsteady signal with no relation to the scale of the turbulence at all. I think the turbulence scale effect is something else that should be investigated.

KOVASZNAY: (Johns Hopkins University)

I have a modest suggestion at that point. I have tried these techniques myself. If you go just one step further in your signal processing, namely, if you take the ensemble averaged signal and subtract it from the full signal then you obtain a fluctuation; now you square it and then ensemble-average it for a second time. One obtains the time dependent turbulence level. It is really spectacular because when the probe is in the wake the level will shoot up and outside the wake, essentially in the inviscid region, it is low. Using this technique it will reveal a great deal. I suggest you do this since you have the equipment to do it so it won't be difficult.

EVANS:

I am no longer where the equipment is, unfortunately.

MILLER: (Naval Postgraduate School)

I was going to offer one more data point to your effects of turbulence on performance. Some years ago in a quick experiment we reported, in a conference I believe in Trenton Naval Air Test Station,* on the results of the replacement of some inlet guide vanes with cylinders and the effect on stage performance. As I recall there was about a 25% improvement in stage performance even when the pressure dropped across the ½ inch cylinder to replace the inlet guide vanes was charged against the pressure rise. So the intensity was out about 15% or so. So there is more to it if you go further.

LAKSHMINARAYANA: (Pennsylvania State University)

We have done exactly what Kovasznay has mentioned.† Using an ensemble averaging technique, we can derive turbulence intensity as well as correlation. I had one question: was the unsteady boundary measurement taken with a single wire coming out of the blade surface?

---

* Ref.: Effect of High Intensity Inlet Turbulence on Compressor Cascade Performance, Proc. of Environmental Effects Conference, Naval Air Station, Trenton, N.J., 1964.

† Lakshminarayana & Poncet: A method of measuring three dimensional rotating wakes behind Turbomachinery Rotors, J. Fluids Engineering, June 1976, p. 87.

EVANS:

Coming down toward the blade surface.

LAKSHMINARAYANA:

That is, you are seeing the wake that will induce three dimensional flow on the stator blade? Are you measuring the three dimensional flow?

EVANS:

This was only a single wire, so it was limited to that extent. There was very little or no mean radial flow here. But as you say, there are certainly going to be some radial fluctuations within the wake.

Visual Study of Oscillating Flow over a Stationary Airfoil

A. A. FEJER
Illinois Institute of Technology
Chicago, Illinois (U.S.A.)

ABSTRACT

Using special visualization techniques involving hydrogen bubbles the fundamental features of fluctuating flows over a stationary airfoil have been determined including the onset and nature of dynamic stall. It has been found that in periodically changing flows dynamic stall can assume a variety of forms depending on the frequency and amplitude of the unsteadiness. Furthermore, it has become apparent that a strong interplay exists between viscous and inviscid phenomena with the role of the latter being predominant.

* * * * *

In forward flight the rotor blades of helicopters move through the air with periodically changing relative velocities and angles of attack. The increase in the amplitude of these changes with flight speed may ultimately lead to abrupt changes in aerodynamic forces and moments on the retreating blade, a phenomenon referred to as dynamic stall. An experimental study seeking to identify the flow processes producing it is currently in progress at I.I.T. under U.S. Army support. As part of this study we have observed for the first time the flow field around a stationary airfoil produced by a unidirectional stream with periodically changing velocity with the aid of a hydrogen bubble visualization technique. The airfoil, an NACA 0009 profile, was set at angles of attack extending from -2° to +10°, the flow was oscillated sinusoidally with amplitudes N_A up to 70% of the mean velocity and at reduced frequencies

K* ranging from 0.15 to 3.00. The range of Reynolds numbers extended from 3,000 to 27,000.

At low frequencies of oscillation a range of "quasi-steady" flows was noted where the main features of the oscillating flow appeared at a given instant as duplicates of those in steady flow and the amplitude of oscillation was of no consequence. But when the frequency of oscillation is made to increase to a certain threshold frequency, the flow becomes amplitude dependent. At angles of attack below the angle of static stall ($\alpha < 7°$) the flow separates from the suction surface during the decelerating phase of the cycle and reattaches during the accelerating phase provided the amplitude is sufficiently large. The smallest amplitude which produces this separation-reattachment sequence depends on the frequency as shown in Figure 1, taken from Reference 1 (e.g., K = 0.6 at $\alpha = 6°$).

It can be seen that this amplitude decreases with an increase in frequency or angle of attack. At angle settings above stall ($\alpha > 7°$) there is again a "quasi-steady" range of frequencies where the airfoil remains stalled during the entire cycle and at all amplitudes. And, as seen in Figure 2, also from Reference 1, there appears to be again a threshold frequency at which the flow becomes reattached to the airfoil during the accelerating part of the cycle and separates during the decelerating part (e.g., K = 0.48 at $\alpha = 8°$).

In the amplitude dependent range of oscillations, with the airfoil set at angles below stall, the onset of separation was marked by the development of an energetic reversed flow in the boundary layer. The chordwise extent of the region of reversed flow seemed to depend on amplitude and angle of attack; at lower amplitudes it covered the rear half of the suction surface while at higher amplitudes it engulfed the entire surface. At angles above stall reversed flow was always present in the amplitude dependent regime, at least during deceleration. It first appeared near the leading edge and spread with increasing amplitudes over the entire surface.

A second feature of the oscillating flow was the generation of intense, large size vortices near the leading edge that appeared to be caused by interaction of the freestream and the reversed flow at the beginning of the accelerating phase of the cycle. Subsequently they become detached from the surface and are convected downstream.

---

* The reduced frequency K is defined in terms of the frequency f, chord c and mean velocity U as $K = \pi fc/U$.

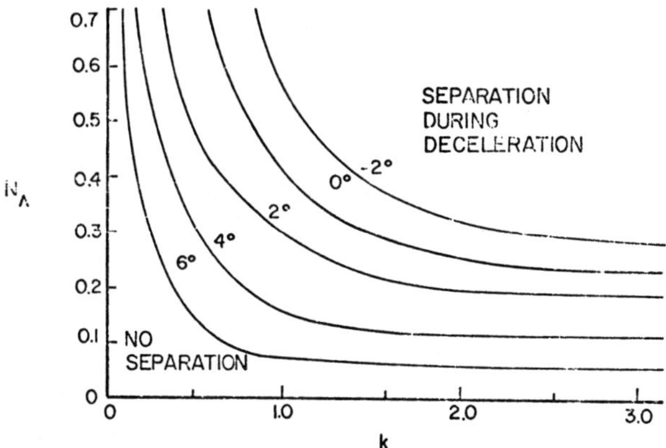

Figure 1. Duration of Separation; Boundaries Between Flows with no Separation and with Separation During Deceleration.

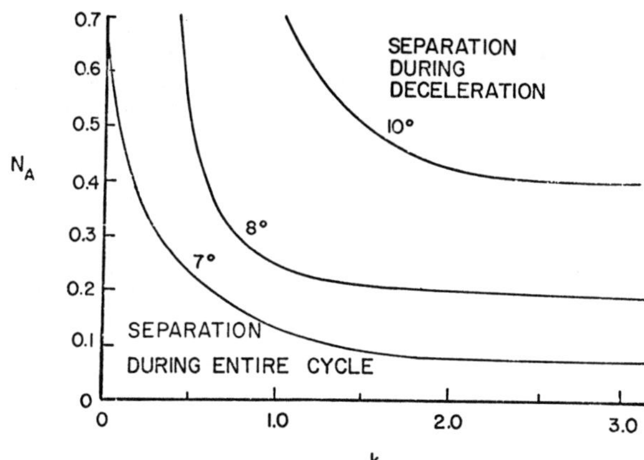

Figure 2. Duration of Separation; Boundaries Between Flows with Separation During Deceleration and Entire Cycle.

The features of the flow described above have been determined from visual observations at various combinations of angles of attack, mean velocities, amplitudes and frequencies from motion pictures. Movies made during the tests are available.

REFERENCES

1. Zolan, John J.: "Dynamic Stall on Airfoils in Oscillating Flow," M.S. Thesis, Illinois Institute of Technology, May, 1976.

DISCUSSION

MILLER: (Naval Postgraduate School)

Several years ago in a thesis* we reported essentially the same experiment in which we measured the pressure distribution and then made ensemble averages. We went to quite large angles of attack, about 15° as I recall, and then looked at both instantaneous pressure distribution ensemble averages and the time average and the ensemble average. The interesting thing was that we discovered two things. What happens is that the leading edge vortex causes a reattachment and at 15° angle of attack we regained on a time average basis about 85% of the theoretical CL. Very surprising but it explains why a helicopter flies. If it weren't for that the thing would fall out of the sky. Then we looked at the instantaneous pressure distribution from the ensemble averages and were able to detect that this reattachment occurred at about ⅓ cord and my feeling is that very sharp leading edges for example would lead to an even smaller number and it would be worthwhile to begin to look at leading edge radius effect on helicopter performance.

---

* Banning, M. R., "The Unsteady Normal Force on an Airfoil in Oscillating Flow," Engineers Thesis, Naval Postgraduate School, Department of Aeronautics, Monterey, CA, Dec. 1969.

Generation, Measurement and Suppression of Large Scale Vorticity in Internal Flows*

R. A. WIGELAND, M. AHMED, AND H. M. NAGIB
Illinois Institute of Technology
Chicago, Illinois (U.S.A.)

ABSTRACT

The main objectives are not only to develop recommended procedures for control of large scale swirling and secondary flows, but also to extract from empirical observations general concepts which design and test engineers could adapt for "tailoring and manipulating" their own special flows with different rotational characteristics. Several typical rotational flows were generated and superimposed on the flow through the test section. Careful calibration was carried out using hot-wire anemometers, and miniature vane-vorticity indicators. These flows represent the basic target conditions to be controlled by inserting various flow manipulators: honeycombs, screens and perforated plates. Comparison of the characteristics of the flow downstream of the manipulators (e.g., distribution of streamwise vorticity) to the original test flows provides a measure of the efficacy of the manipulators in suppressing large scale vorticity as well as clues to dominant mechanisms of the flow transformations.

* * * * *

For wind-tunnel experiments, a flow which has uniform mean velocity and low turbulence levels along with low acoustic disturbances is desired. Various research activities, a more recent

* Supported under ARO Grant No. DAHC04-74-G-0160. The details of the study are available in an ARO Technical Report which is based on the Master's thesis of the second author and is included here as the first reference.

one being by Loehrke and Nagib (1972), have concentrated on attaining desired levels of uniformity and turbulence intensity in the freestream of ducts and wind tunnels. In one of the flow conditions they studied, Loehrke and Nagib (1972) reported anomalous behavior which they attributed to some kind of secondary flow. Most wind tunnels and ducts exhibit various degrees of swirl and some secondary flows whether these are due to the propulsive fan or to the duct bends in presence of shear layers. As Carbonaro (1973) points out, curing these undesirable characteristics "is frequently an empirical cut-and-try study relying on ingenuity of the design engineer rather than on well established methods." At IIT in the past, such flow non-uniformities were taken care of by "overkilling" the condition, that is, by inefficient use of flow manipulators, such as screens and honeycombs, which results in a loss in the top speed of the wind tunnel and often in the gradual heating of the air in the tunnel, as discussed by Tan-atichat and Nagib (1974). Carbonaro (1973) reports that no general treatment of the problem of such rotational components has been found in the literature. With all of these statements in mind, the present systematic study on techniques for control and suppression of the mean and unsteady characteristics of swirling flows in confined streams was undertaken, with some of the initial results being presented here.

Starting with a flow facility which has uniform mean velocity and turbulence intensity levels of from 0.6% to 1.0% depending on the freestream velocity, various rotational flows had to be developed and documented so that controlled conditions exist for determining the effect of different flow manipulators on these swirling flows. Three different rotational flows have been developed so far, one which has a strong rotational core with a nearly irrotational outer part, another with angular momentum only near the walls, and the third with a swirling jet in the center of the duct which entrains the outer flow. Devices similar to those constructed for this experiment have been used by other experimenters, and references for these can be found in the report by Ahmed et al. (1976).

Construction of the first swirl condition consisted of placing two airfoils next to each other and at equal but opposite angles of attack so that a strong wing-tip trailing vortex is generated. The size and strength of this vortex can be varied by changing the angle of attack and varying the freestream velocity. The airfoils span one of the 6 inch diameter ducts identical to those forming the test section of the wind tunnel. The separation distance and hinge point for the airfoils were optimized to give the strongest and most symmetric swirl conditions, which persist throughout the length of the test section without substantial changes in their characteristics.

The second swirl generator consists of eight slots equally spaced around the periphery of one of the ducts. The slots are as tangential as possible to the inner wall of the test section. Compressed air is supplied through these slots, so that the result is a series of tangential jets providing angular momentum to the flow near the wall and having little effect on the inner portions of the axial flow. This facility has been developed to provide jets which are as uniform as possible along each slot and among the eight slots.

The third facility, called the swirling jet ejector, is a separate facility supplied by compressed air. Air is forced through an annular region containing 18 adjustable vanes which can impart angular momentum to the air. This air exits through a 1-inch diameter jet in the center of one of the circular test sections. The remaining air flow through the test section is provided by the entrainment action of the jet which draws air in from the room.

The experiments using these swirl generators are performed as shown schematically in Figure 1. With these facilities, it is necessary to have a method for measuring the swirling component of the flow. A device which is called the vane-vorticity indicator has been developed for this purpose. Vane-type vorticity indicators have been used with some success by Barlow (1972) and by Holdeman and Foss (1975) for making measurements, and by Shapiro (1974), for demonstration purposes in his movie on vorticity. Since the vane

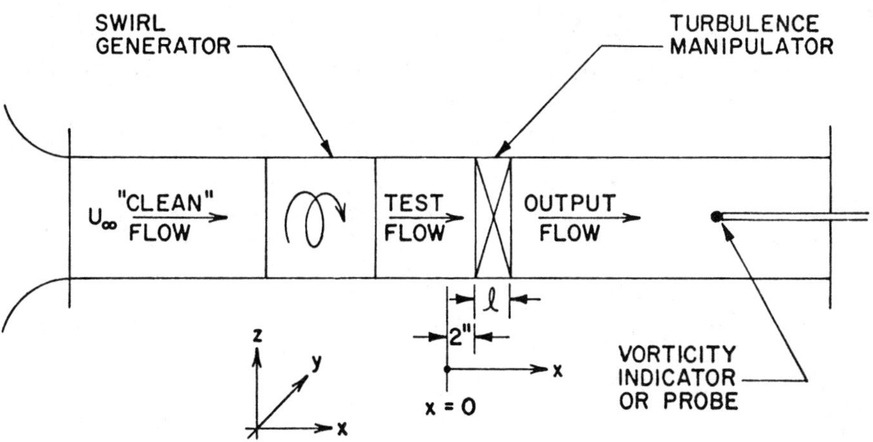

Figure 1. Schematic of Test Section Arrangement.

has a more rapid and direct evaluation of the local fluid rotation, its use in this experiment seemed promising. The miniature vanes used here have four perpendicular blades and are made from aluminum for light weight and low moment of inertia. Teflon bushings and washers were found necessary to reduce the friction and to provide a consistent and repeatable measurement of the swirl component. This vane is mounted directly upstream of a hot-wire which has a very short sensor length. When the vane rotates, the wakes from the blades pass by the sensor and can be detected in the hot-wire output signal. Measurement of the time between passage of the wakes using autocorrelation of the hot-wire signal provides a measure of the average rotational speed of the vane.

Since the airfoil swirl facility provided the most variety of clean flow conditions, it was used for the calibration of the different vanes. The vane was positioned in the center of the airfoil's trailing vortex and calibrated for vane rotational speed as a function of angle of attack of the airfoils at several freestream velocities. We found that the same calibration curve was obtained for three vanes, each of a different size and weight, so that this curve was used as the calibration standard. The independence of the curves on geometrical parameters is encouraging. However, the rotational speed of the vanes must be related to the local fluid rotational velocity.

With the aid of an x-wire probe and using the same conditions as for the calibration of the vanes, the local tangential velocity can be determined. Typical profiles of the tangential velocity as a function of the radial distance from the center of the vortex are shown in Figure 2. Once the tangential velocity is known, the local rotational speed can be calculated and used for comparison with the vane-indicator data. Using any of the vanes there are two methods which can be employed to obtain the "proper" rotational speed of the fluid for this comparison. One method is to determine the rotational speed of the solid body portion of the vortex. However, by superimposing the width of the vane on the tangential velocity profiles, we find that the vanes extend slightly beyond the region of solid body rotation. Hence, the second method uses the speed indicated at the tip of the vane as being the characteristic velocity, which is called the vane-width speed. The correct method is probably some average between the two. Taking the calibration curve for one of the vanes and plotting twice the vane speed versus the fluid rotational speed as calculated by these two methods (Figure 133 of Ahmed et al. (1976)), the agreement is quite good for airfoil's angle of attack over 5°, since the rotational friction of the vane becomes less important as the vane speed increases. Even in the lower region, a slightly different multiplication factor would render most of the remaining vane data in agreement with the x-wire data. This shows that the ratio of the vane rotational speed to the fluid

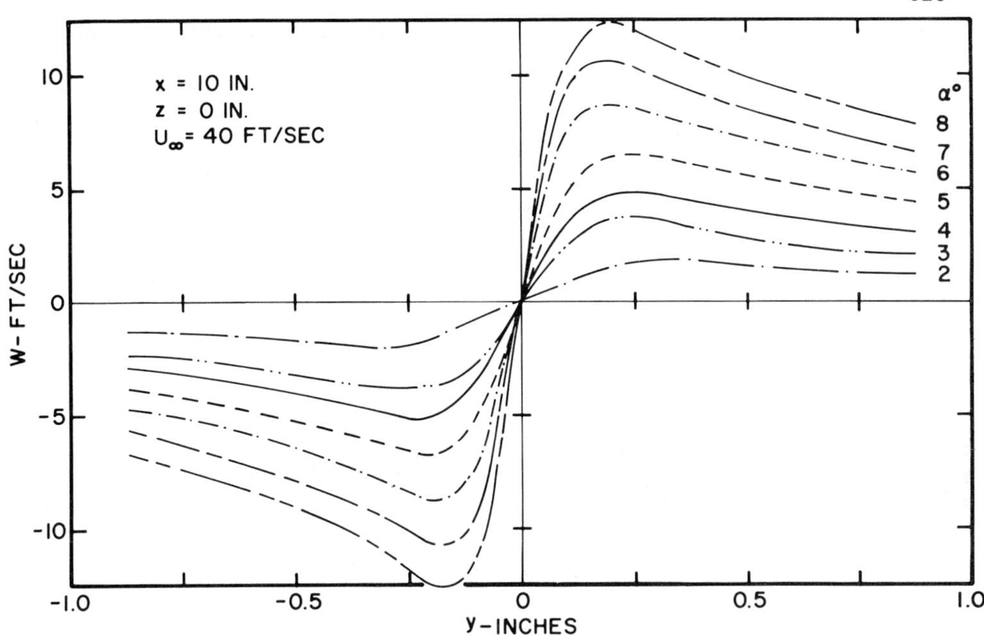

Figure 2. Tangential Velocity Profiles Across Vortex of Airfoil Swirl Generator for Various Angles of Attack.

rotational speed, denoted by η, is somewhat independent of the fluid rotational velocity. Comparing the x-wire data to the vane data for various freestream velocities (Figure 134 of Ahmed et al. (1976)), a similar independence of the ratio η is observed.

Another important test is for the effect of the vorticity gradient, $\partial\Omega_x/\partial r$. In Figure 3 the streamwise vorticity is plotted as a function of radial distance from the center of the vortex. The x-wire tangential velocity data were converted to vorticity by assuming axial symmetry and using the relation

$$\Omega_x = \frac{W}{r} + \frac{\partial W}{\partial r} ,$$

where W is the tangential velocity. The vane vorticity indicator data were interpreted using the calibration curves. The corrected vane data agree quite well with the calculated x-wire data until the level of vorticity drops too low for the vane to rotate due to friction. The interpretation of the vane data was carried out using both methods for determining the fluid rotational speed, as was discussed for Figure 2. In summary, it appears that the ratio

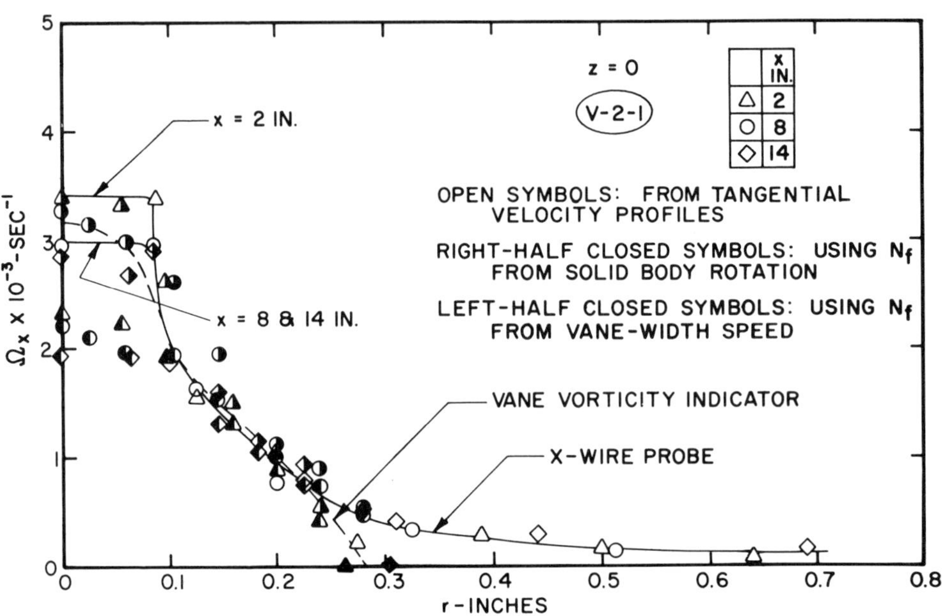

Figure 3. Comparison of Radial Distribution of Local Streamwise Vorticity Obtained from X-Wire and Vane-Vorticity Indicator Measurements in Test Flow Condition "V-2-1."

η is independent of the fluid rotational speed, the freestream velocity, and the spatial gradient of vorticity for the range of these parameters tested. The vane does then provide a good picture of the streamwise vorticity of the flow.

This device is then used to determine the test flow conditions generated by the swirl generators. In each case, changing of the parameters involved was done until flow conditions with reasonable swirls were obtained. These flow conditions are used to test the effect of the different flow manipulators. One flow condition from the tangential jet swirl generator, one from the swirling jet ejector, and seven from the airfoil swirl generator with different strengths and different sizes were utilized.

The flow manipulators used were various screens, honeycombs, and perforated plates. The screen has a mesh of 0.0357 inch with a solidity of 0.35, i.e., made of wire 0.007 inch in diameter. The honeycombs were #3 which is 1-inch long, with a 0.257 inch mesh and 0.014 solidity; #2 which is 2-inch long, with a mesh of 0.135 inch and a solidity of 0.008; and #1, a 2-inch long honeycomb, 0.068

inch mesh and 0.03 solidity. The perforated plates were both 0.063 inch thick, #2 with 0.14 inch holes, 0.188 inch mesh and 0.49 solidity, and #3 with 0.25 inch holes, 0.313 inch mesh and 0.42 solidity. Figure 4 displays the pressure drop coefficient of these manipulators as a function of freestream velocity. The importance of the pressure drop coefficients will be brought out in the following discussion on manipulator performance.

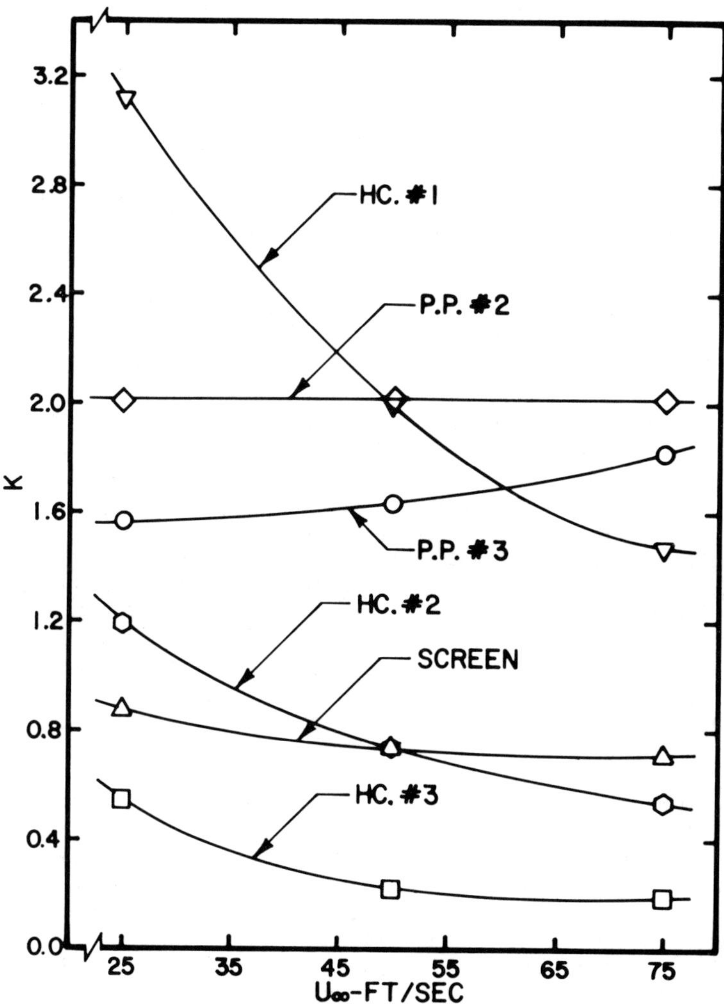

Figure 4. Pressure Drop Coefficient of Manipulators Versus Freestream Velocity.

For each of the test flow conditions, manipulators were placed a certain distance downstream of the swirl generators, and their effect on the swirl flow condition was determined by the vane indicator. Figure 5 is a sample of some typical data taken in two of the test flow conditions. The top figure is in the airfoil swirl flow condition "V-2-1," while the bottom one is in the swirling jet ejector, i.e., flow condition "SJ-1." In the top figure, the vane rotation defining the flow condition is the top curve, designated as no manipulator, and the other curves demonstrate the effects of the different manipulators as indicated. It can be seen that the screen has the least effect, then the perforated plates, which have more effect and then the honeycomb, which eliminates the measurable vorticity completely at some small distance downstream. The drop in vane rotation with downstream distance immediately after the manipulator is viewed as a homogenization of the pieces of the swirl condition that emerge from each cell of the perforated plate or the honeycomb. Only in the manipulator itself can the angular momentum be absorbed; afterwards the flow from each cell must rejoin. Notice that even the screen can take some torque from the flow, and does not show the decrease in the rotational speed immediately downstream due to its small mesh, low solidity, and short length. In the lower figure, the screen and perforated plate have approximately the same effect while the honeycomb again eliminates the swirl. We conjecture that a different mechanism is working in this case. Further elaboration can be given based on Figure 6. The peak for P.P. #3 in the top figure is much flatter than it is in the lower figure, while the profiles downstream of the screen do not exhibit much difference between them. The major difference between these two conditions is the size of the swirling part of flow; i.e., the swirl in the lower figure is about 3 times larger. When the size of the swirl is compared to the characteristic mesh size of the manipulator, there is not much change for the screen between the two cases but there is for the perforated plate. It becomes apparent that a scaling between the swirl size and the manipulator characteristic length is important, and for the best effect, the proper ratio must be chosen.

Another important result can be demonstrated by Figure 7. In the top part of the figure the effect of a given manipulator on all of the swirl conditions is summarized. The swirl reduction ratio presented here is one minus the ratio of vane rotational speed in the flow downstream of the manipulator to that in the flow at the same position when there is no manipulator present. As outlined in the figure, a range is defined which bounds the reduction in swirl of the conditions employed. Using the dashed marks to indicate this range, and taking similar results from the other manipulators, a composite map of the effect of the different manipulators can be drawn, as is shown in the lower part of the figure. An important thing to notice in this composite plot is that the same

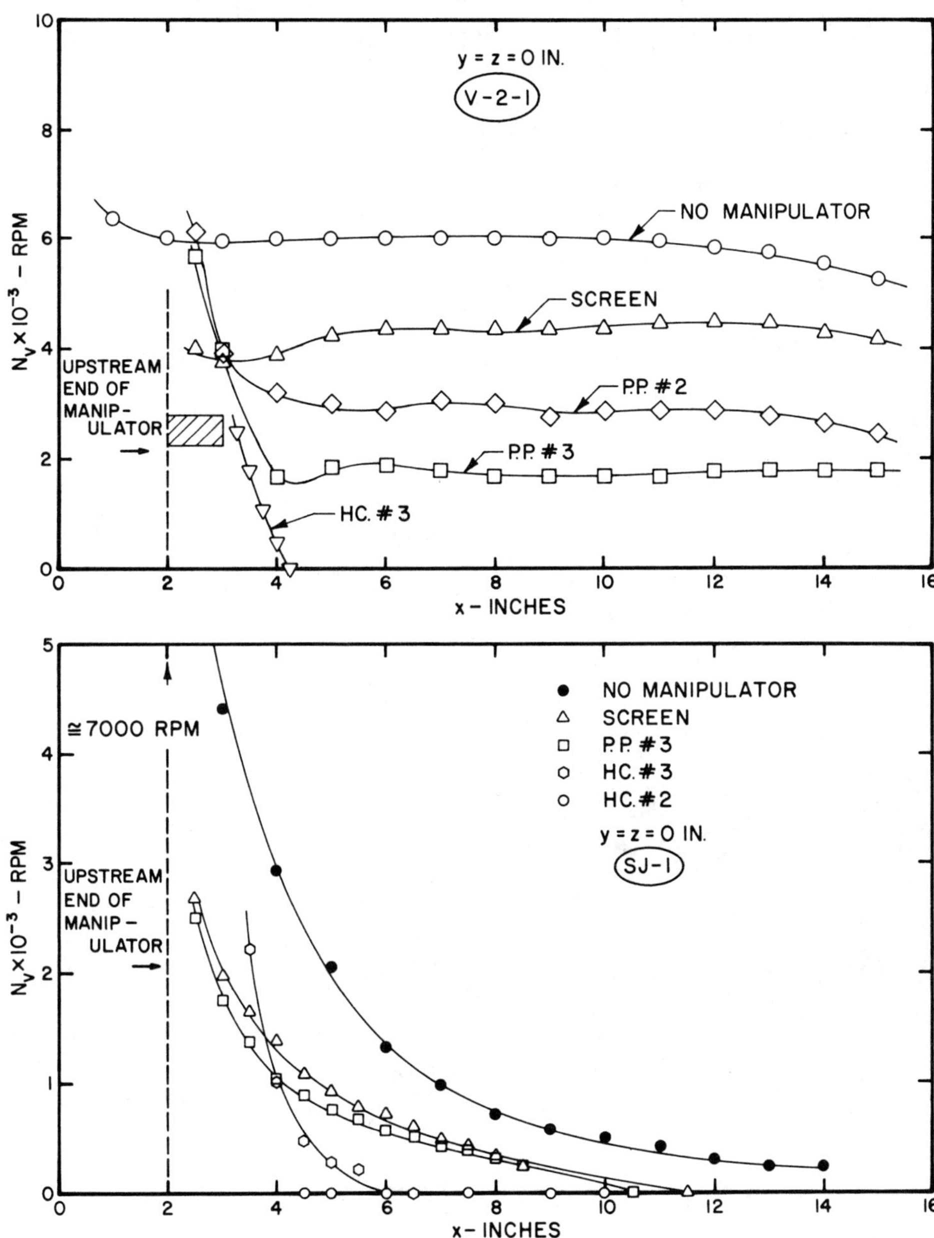

Figure 5. Effect of Various Manipulators on Downstream Development of Test Flow Conditions "V-2-1" and "SJ-1."

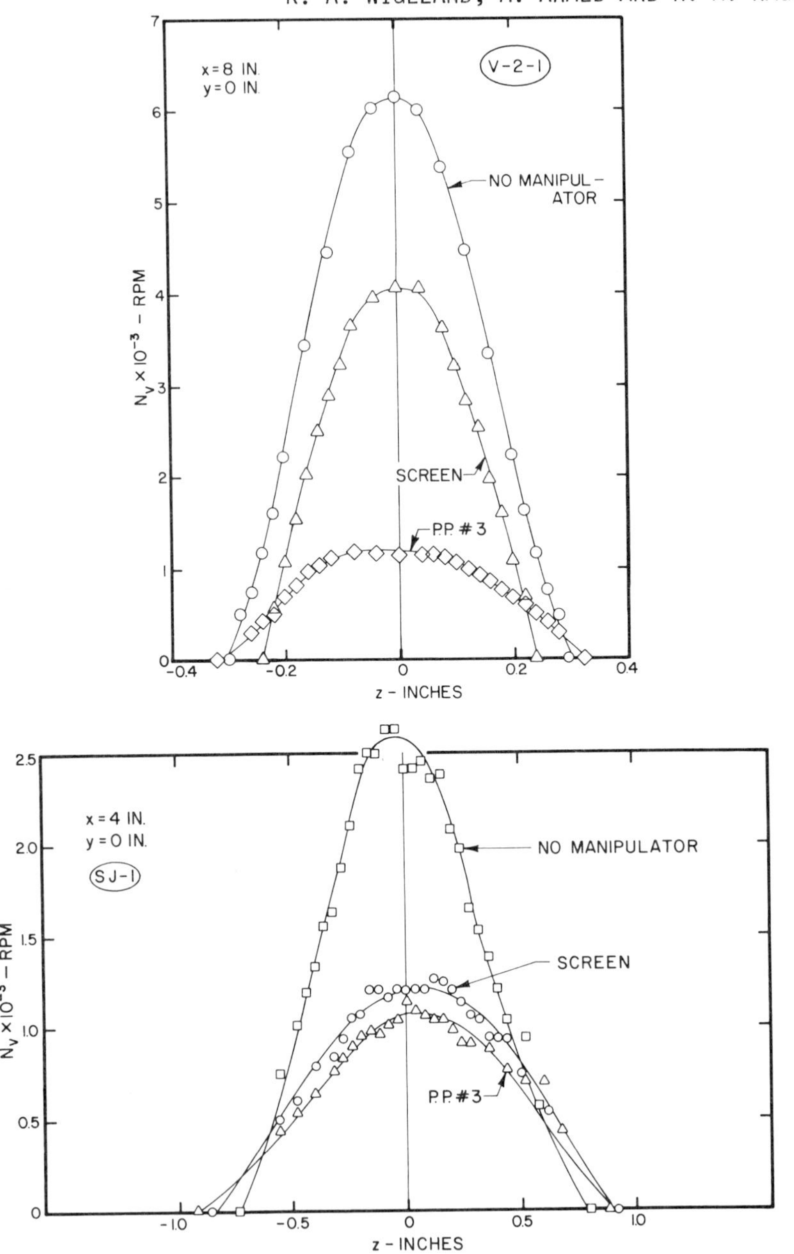

Figure 6. Effect of Screen and P.P. #3 on Radial Distribution of Vorticity Indicator Rotation in Test Flow Conditions "V-2-1" and "SJ-1."

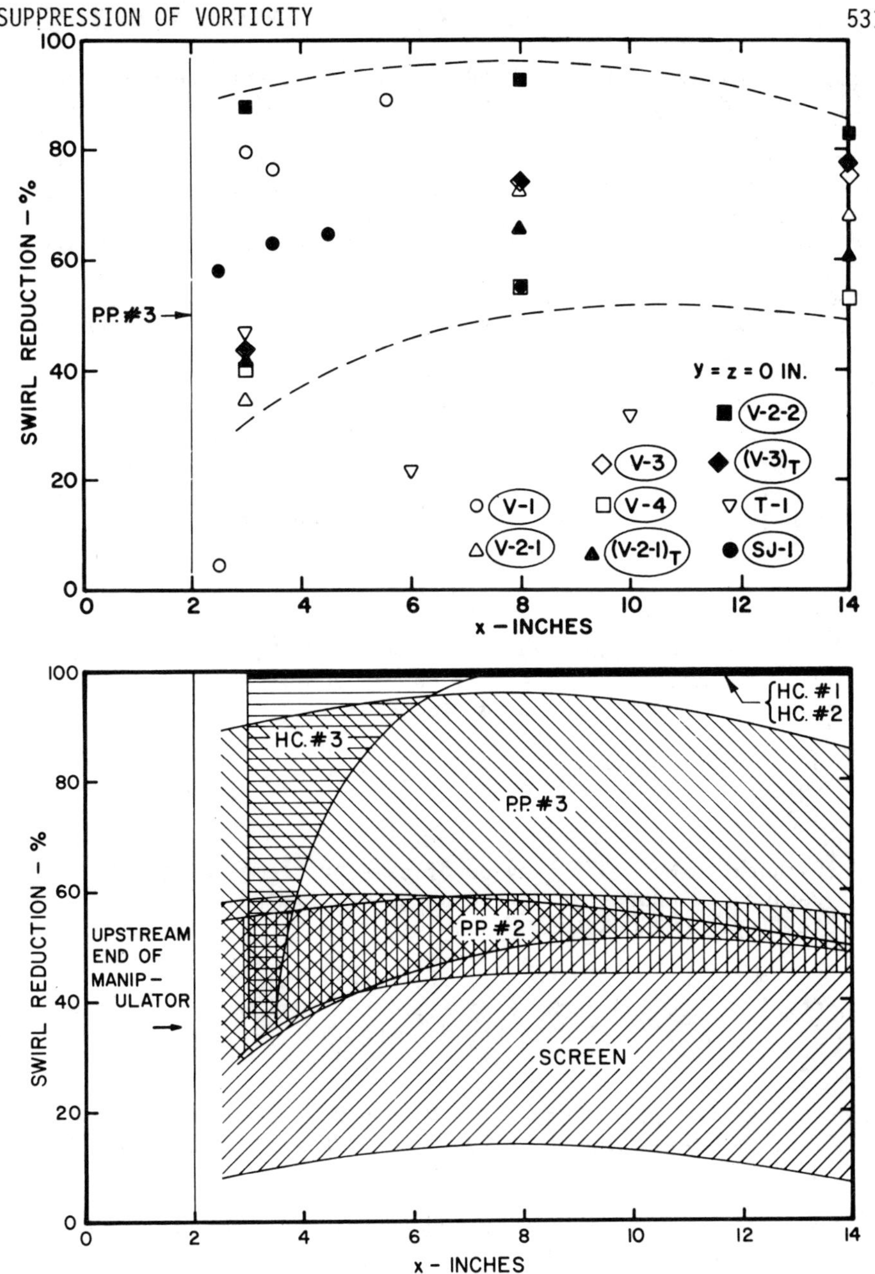

Figure 7. Swirl Reduction in Different Test Flow Conditions for One Flow Manipulator and Composite of Reduction Range for Several Manipulators.

reduction in swirl can be achieved using any one of several manipulators, as indicated by the overlapping regions. This makes it possible for the design engineer to select a certain manipulator with a known level of performance which may have other desirable characteristics for his case, where another manipulator with the same swirl reduction ratio may not have them. Another, probably more important result is that the selection can be optimized with respect to the total pressure drop, an important parameter in wind tunnel design. In particular, note that honeycomb #3, which eliminates the swirl completely, also has the lowest pressure drop coefficient, as presented in Figure 4. Given a certain swirl condition, it should be possible to choose the best manipulator for the job based on the other factors involved. It is perhaps better to work on the swirl reduction first when modifying a given flow, and then to operate on the turbulence, rather than the other way around. It may also be possible to achieve better results with combinations of manipulators as far as both of these items are concerned.

REFERENCES

1. Ahmed, M., Wigeland, R. A. and Nagib, H. M. 1976 Generation and management of swirling flows in confined streams. I.I.T. Fluids and Heat Transfer Report R76-2.

2. Barlow, J. B. 1972 Measurement of wing wake vorticity for several spanwise load distributions. University of Maryland Report.

3. Carbonaro, M. 1973 Review of some problems related to the design and operation of low-speed wind tunnels for V/STOL testing. AGARD Report No. 601.

4. Holdeman, J. D. and Foss, J. F. 1975 The initiation, development, and decay of the secondary flow in a bounded jet. Journal of Fluids Engineering, Vol. 97, Series 1, No. 3, p. 342-352.

5. Loehrke, R. I. and Nagib, H. M. 1972 Experiments on management of free-stream turbulence. AGARD Report No. 598; AD749891.

6. Shapiro, A. 1974 Illustrated Experiments in Fluid Mechanics. MIT Press.

7. Tan-atichat, J. and Nagib, H. M. 1974 Measurements near bluff bodies in turbulent boundary layers intended to simulate atmospheric surface layers. I.I.T. Fluids and Heat Transfer Report R74-2 or U.S. Air Force Office of Scientific Research 74-0964; AD 782090.

DISCUSSION

SAFFMAN: (California Institute of Technology)

Do you know if this vortex is laminar or turbulent. How can you tell?

NAGIB:

The ones I presented here today are all laminar as can be seen from hot wire measurements. In addition, by introducing a grid upstream of the airfoil, for the same angles of attack, we have looked at the same conditions with a different level of controlled background turbulence.

SAFFMAN:

Because of the rigidity of rotating flows with high rotational speeds, it is usually dangerous to believe that any measurement of the trailing vortex which you get by sinking a probe into the flow is reliable. The probe may disturb the flow a great deal more than you think. How can you check up on this?

NAGIB:

We tried a number of things. One of them was through visualization by adding some smoke. We were quite aware of the problem and checked for anything like vortex breakdown or similar instabilities. We looked at the flow conditions very carefully in order to generate strong and stable vortices. I remember there was one curve that showed some high RPM--in the range 6,000-8,000--and some lower ones. We had no problems with the higher ones, but at the very low rotation rates weak vortices were sensitive to the probe. There was one slide where the vane was held in someone's fingers that showed you the physical size of the vane and how small it is. The vortex was usually much larger than the vane.

CORRSIN: (The Johns Hopkins University)

I would like to remark that H. B. Squire was doing things like that a long time ago and there might be some experiences in Imperial College related to your work.

Another thing that I wanted to remark on is the peculiar kind of nonlinearity, to which these are subject. People who use vane anemometers in meteorology have finally discovered that they get large errors in the mean when fluctuations are present. Because of the inertia things tend to slow down more gradually than they speed up. In a gust they respond very quickly and when the wind dies down, they reduce in speed very slowly and there are errors in the mean.

NAGIB:

In regard to your first comment we are aware of many of the references but we could not find any details. For example, there is one report from the University of Maryland by Barlow, which we included in our list of references. However, there were no details of careful calibration of the instrument which include tests for spatial resolution and so forth; but maybe all of the information is buried in some report which we never found.

CORRSIN:

There was a Ph.D. thesis from the University of Bombay. (Published reference: papers by Sundaram Ramachandran in the Quarterly Journal of the Royal Meteorological Society, Volume 95, No. 403, page 163, 1969 and Volume 96, No. 407, p. 115, 1970).

NAGIB:

In regard to your second comment, the swirling flows we used were all steady state conditions. If you have low frequency turbulence fluctuations or flow unsteadiness the nonlinearity you mentioned becomes very important. We did not suffer from it because our deviations from the mean were very small and because of the small moment of inertia of the vanes. However, I agree with you that these nonlinearities are present in the vane-vorticity meters as in cup anemometers.

BRADSHAW: (Imperial College)

I am afraid that Imperial College has no information as far as the vorticity meter is concerned.

Denis Bryer of the National Physical Laboratory used a very similar rotating-vane device in delta wing trailing vortices about ten years ago, but did not write it up because the correspondence between vorticity and r.p.m. did not seem to be consistent.

PART IV
PANEL DISCUSSION AND SUMMARY

Panel Discussion

PANELISTS

J. Laufer (University of Southern California, L.A.) - Chairman

D. Coles (California Institute of Technology)
S. J. Kline (Stanford University)
E. Reshotko (Case Western Reserve University)
W. W. Willmarth (University of Michigan)

S. N. B. MURTHY: (Purdue University)

Welcome to the Panel Discussion, an important part of this Workshop.

The principal objectives here are to discuss the subject of the Workshop in general terms but, of course, in depth and to point out the basic advances and outstanding questions as far as we can see them.

I suppose I could add here parts of my initial letter to Professor John Laufer who has agreed to chair the Panel.

"I believe the speakers and participants at the Workshop will discuss in various ways the following:

a) measurements in curved and rotating flows and in wakes;

b) modeling of curved and rotating flows and wake flows;

c) structure of turbulence as it is affected by various interactions in confined flows; and

d) specific measurements in cascades and turbomachinery.

"In putting together the program, I had, in fact, in mind discussion of such topics from the point of view of establishing:

i) What can we calculate with some rationale?

ii) What is the possibility of synthesizing the approach based on examining the large scales of turbulence in terms of coherent structures and the turbulent stress-energy-statistical viewpoint?

iii) What kind of detailed measurements are being made and are required in complex flows?

iv) Are there any lessons in the foregoing for people engaged in the design of diffusers and turbomachinery?

"I also thought that some other questions of broader import may also be raised during the panel discussion, for example:

a) What are some of the implications of advances in data processing?

b) Should we emphasize measurements of pressure-velocity spacewise correlations in boundary free flows?

c) To what extent are scalar quantities useful in experiments as indicators?

"Having said the foregoing, I would like to leave the entire business of the panel discussion in your hands. Your contributions to the field are universally respected and I am sure you are in a unique position to extract the best out of the panel and the audience. From the point of view of those sponsoring the Workshop, I hope, a program of research will evolve during the Workshop."

Now, Professor Laufer.

J. LAUFER: (University of Southern California, L.A.)

Incidentally, if anybody would like to take my position I would be more than happy to let them have it. I did promise that I would do my best to steer the discussion in a more or less unified direction. Since the interest of the people present here covers wide areas, to keep this discussion in some sort of focus will be difficult indeed. I would like to remind everybody that the primary purpose of our meeting here is to bring together the practitioners of the turbulence problem with those individuals whose primary interest is applications and in particular applications which happen to have extremely difficult geometry.

Now I discussed with the members of the panel what would be the best procedure here in conducting such a discussion and I think all of us agreed to try to keep the discussion as open as possible, to keep it unstructured and put only the following boundary condition on the discussion. We have approximately one hour and a half. Let's divide it into two periods. In the first period, let us try and concentrate on problems having to do with the philosophy of experiments as applied to the problems under discussion, and in the second half let us concentrate then on the question of modeling: on the theoretical approach to this problem.

In order to start the discussion let me just present you with the following observation as far as the experiments are concerned: When I listened to the various presentations having to do with experiments I noted that they could be divided into two distinct groups. One set of experiments considered relatively simple geometries such as mixing layers, turbulent boundary layers and superimposed on such simple flows were certain types of perturbations, like freestream turbulence level and outer vorticity layer; thus it considered a slightly more complicated situation than the simple "basic turbulence problem" itself.

The second group of experiments tried to simulate in the laboratory the more complicated "real life" problems connected with turbomachinery; the most interesting example was Kerrebrock's experiment.

The two approaches are obviously different in nature and I would like to present you with this question: What are the advantages and disadvantages in following these two different philosophies in formulating the experiments themselves?

So I invite the members of the panel or the audience in considering this question.

H. W. LIEPMANN: (California Institute of Technology)

With a complex and technical, important, problem like turbulence, one has to distinguish clearly in one's mind whether one attempts to contribute to the understanding of the physics or whether one attempts to solve (or fix) urgent, important technical problems. Both directions of work are important but they do differ of course in the methods used. In the one, the problem is simplified as much as possible without losing the essential ingredients and results of general validity are wanted. In the other, a very complex set of conditions has to be accepted but specific results are usually sufficient.

The main difficulty, in my opinion, with much of the work in turbulence is the lack of distinction between these two avenues. A lack of understanding does not necessarily foreclose the possibility of important contributions to industrial problems (e.g., the lack of a fundamental theory of phase transition has not affected the steam turbine). Conversely, the successful application of a turbulence model to complex technological problems may not contribute to an understanding of turbulent shear flows.

Both approaches are necessary and much fruitless discussion and argument stems from a lack of keeping the distinction between the two in mind. Indeed a meeting like the present one has, in my opinion, succeeded if it exposes the fundamentalists to the modellers and both of them to the industrial problems. For example, Prof. Kerrebrock's talk dealing with the complexity of turbulent flow in real, rotating machinery emphasized for me the need of more fundamental experiments with turbulence in rotating systems and in other systems where the usual diffusive transport is wavelike, e.g., in stratified flow and liquid Helium II. Maybe an even more "radical" talk dealing with industrial problems would have been useful in a further exposure of fundamentalists to technology and vice versa.

E. RESHOTKO: (Case Western Reserve University)

On this panel I seem to be low man on the totem pole experimentally, so maybe that is why I am talking first. Now, I may wish to object to the division of problem areas into experimental and theoretical. I support what Prof. Liepmann has just said. The problem classes that he defined are just as valid theoretically as experimentally and that we really should consider these approaches simultaneously, first in discovering new phenomena and then once a phenomenon has been defined, to refine our knowledge of it to the point where it becomes of engineering utility. I don't know if I can pressure the chairman to change the orientation of the discussion but I feel it is really not easy to discuss experiment as divorced from theory or vice versa.

The question in any meeting of this kind is: What can we hope to accomplish toward solving the general problems motivating our work? The title of the symposium was "Turbulence in Internal Flows." The presented work dealt both with steady turbulent phenomena in boundary layers and other types of flow field elements as well as with some attempts at the unsteady turbulent problem. What are the applications in a propulsion sense? What kind of information do we really want?

As far as turbomachinery goes we already have, and for many years, have had, turbomachinery elements that have of the order of 90% stage efficiency for compressors or certainly between 85% and 90%. The likelihood of improving stage efficiency is not very great. We may be able to improve the performance in the sense of pressure ratio per stage, to have fewer stages for a given machine. This looks hopeful except that it will lead more toward transonic and supersonic staging. I believe that while transonic work was mentioned, most of the activity that we heard about today was nearly incompressible. As staging becomes more transonic, and perhaps even supersonic, we will require different kinds of capabilities both in terms of our experiments as well as in our calculations.

The other point that was evident in the presentations, primarily today and highlighted in Prof. Kerrebrock's talk, was the significant influence of unsteady phenomena in turbomachinery. Professor Kerrebrock described particularly the interactions between flow phenomena on the blading and modes of oscillation within the configuration, in his case the duct modes. I think we have to ask ourselves, in what sense are these important? In terms of performance, perhaps, they are not all that important because of only the very slight improvements possible. But they are certainly important, for example, in terms of acoustics. We have to start asking ourselves, to what extent must we really consider these phenomena? How much effort are they worth? It would seem that this is something that we might think about as a result of this meeting. It certainly represents a departure from the general trend of the work displayed in the presentations.

J. LAUFER:

I think Eli Reshotko justly emphasized the importance of unsteady flows and Fejer's very interesting movies certainly dramatized the difficulties associated with, for instance, unsteady separation. I would just like to mention that by now our experimental techniques have advanced far enough that we could and should in fact re-examine this problem that in the past we avoided because of experimental difficulties. Today with the use of conditional sampling techniques, phasing, etc., we should be seriously thinking of attacking these problems.

S. J. KLINE: (Stanford University)

I guess in summarizing this meeting we have to say that the state of the art is turbulent. What I mainly want to talk about is based on the 1968 meeting, but before that I want to say something about Prof. Reshotko's remarks as applied to diffusers. There are some diffuser questions that still need work. In the first place, no present method of calculation will predict some important design optima. We are still optimizing off-data correlations, which means, when we get into extensions, we are still in trouble. One case where large turbomachinery manufacturers are currently asking for information is turbine after-diffusers that have struts. In particular, annular configurations in large machines are sometimes troublesome because mechanical limitations squeeze the design; the diffuser has to be short. This area looks like a one or two step extension from some of the wake calculations we are now able to do. More work is needed, and there are other such examples.

In terms of the 1968 conference, I am afraid that I will have to violate your ground rules also and make it more turbulent. What I want to talk about is the interaction between theory and experiments. I don't see how to avoid that. I wanted to ask a question, "What, if anything, did we learn from the 1968 conference that can be applied?" (Professor Coles was also deeply involved in that conference, but I assume, knowing Don Coles, that if he disagrees with me he will say so.) I am talking about the results in the two volumes entitled "Computation of Turbulent Boundary Layers--1968 AFOSR-IFP-Stanford Conference." What we tried to do in those volumes was compare the existing theories on turbulent boundary layers with good, standardized data. I think, however, that it is worth reflecting that we scooped up something like 75 years worth of data. We also need to remember the obvious fact that most of that data, not all of it, but a very large share of it, was really motivated by problems of aeronautical engineering. In order to do wing flows people have done a lot of work on smooth surface incompressible turbulent boundary layers. That was the class of flows to which we restricted ourselves. We did so because that was the class of flows for which we believed there were enough data to do a good job. Hence one does not get into the difficulty--that Prof. Liepmann very properly just pointed to--of basing a calculation method on one set of data. This lesson certainly comes out of the results of the conference very clearly. Up to the time that those volumes were completed, papers in the published literature typically checked two or three sets of data. However, if you look closely, you will find that there are no two or three sets of data in those volumes that can tell whether a given method is any good or not. So that, in fact, the 16 mandatory data sets in those volumes represent something like a minimum rather than a maximum for verification of a calculation method. I take "verification" here to denote the idea that if you extrapolate to another flow, there is a high probability of obtaining reasonable results.

Several people asked us about a year or two ago, "Should we repeat such a conference in 1978--10 years after the first one?" There are two difficulties. First, I am not sure I want to do that much work again. Leaving that aside, there is a much more serious difficulty. If you question what data sets would you put together, what classes of flows would you try to treat, then the extensions run off into at least four directions that haven't been adequately classified.

First, are other effects on the boundary layers; we saw evidence on one of these in this conference, that is, the effects of freestream turbulence. The last time I put this together with some help from a number of leading researchers, many of whom are here, we got up to, as I recall, 14 classes of effects that significantly alter what goes on in a turbulent boundary layer including coriolis forces, centrifugal forces, curvature, blowing or sucking, roughness, and so on. Most of these had more effect than freestream turbulence, not less. So that is one kind of extension, none of these 14 or so effects were seriously taken into account in the 1968 conference.

A second class of extension which is wholly different is recirculating flows where you have in effect a branched flow field.

A third case is more complex geometries than airfoil--like shapes that strain the flow field differently and alter turbulence production and/or decay; we saw some in the results that Mr. Bradshaw showed us yesterday and two years ago.

Finally, people are mentioning unsteady flows and there may still be some other classes. I have not yet even got to things like compressibility which are obviously not trivial. Hence there is a vast number of classes of flows, but we do not seem to have enough data on any one of these classes to be sure that the data are reliable, to know how to parameterize well, to verify calculation schemes in the sense I used a moment ago.

I have suggested this before in public, but I would like to repeat myself. I would really like to see some of the funding agencies encourage the formation of a committee to examine the types of questions I am raising, systematically, not with one or two people but by correspondence in this country and with our able colleagues abroad, some of whom are here. Such a committee should concern itself with which classes are more important in both senses that Prof. Liepmann already raised, that is, both scientifically and technologically. Which classes need more good data sets and how many? How many data sets ought we to have before we do another conference of the 1968 type in order to make sense out of the more advanced models that now seem to be approaching the limits of what people can do using current concepts? We don't seem to be getting many new ideas. We are getting, rather, polishing of existing concepts of things and attempts to extend them a little. I think that if we don't organize, cooperate and plan more effectively, that in 1988 we are still going to be in a position where we are not going to be able to collect enough sets of data that are believable, that are standardizable, to do the sort of thing that we did in the 1968 conference for further classes of problems. There are too many kinds of problems and too few investigators for effective progress of this sort to be likely through pure chance.

Finally, I wanted to raise one other question about the nature of recirculating flows. We have been doing quite a bit of work on fully-separated flows which are a kind of branched flow field. In such flow fields, the use of the parabolic equations on simple extensions of what are basically parabolic methods worry me considerably. I have serious doubts that they can succeed. This comment goes to the theorist, and hence I will hold details for the moment.

W. WILLMARTH: (University of Michigan)

I came here hoping to learn more about internal flows and what the experimental problems are in internal flows, and when Hans Liepmann said, why don't you have the people with the problems bring the problem to us and we can look at it, well, this is just the way I still feel. If someone in industry had a problem and brought it to me, and I've had this happen, the first thing I would do would be to suggest using visual methods (I thought Steve Kline was going to say this) and to learn more about the problem. It appears to me that flow visualization of a new phenomena would be the first cut at the problem, so you would then know where to put your probes, what kind of probes were needed and what you were measuring. I can appreciate that experiments on rotating flow machinery with flow visualization can be very difficult.

I really don't understand what all the significant problems are. There seem to be so many. Is there any listing?

D. COLES: (California Institute of Technology)

I think that organizers of mixed meetings like this one run a certain risk. It is very like the risk that is run by the zoo keepers in the Washington zoo every spring, when they open the door between the boy Panda and the girl Panda, and nothing happens. Well, something happened with me. My education was considerably advanced by the presentation of Prof. Kerrebrock.

We had a private discussion on a point which I want to raise. Maybe it will die, but I will raise it anyway. This is our primitive state of understanding of effects of rotation. There is one book on this subject, by Harvey Greenspan, who is linearized. I personally have had experience with two problems in this area. One, long ago, was circular Couette flow. The other, more recently, was spinup and spindown of fluid in a cylindrical container in the nonlinear and turbulent regimes. I have never really understood effects of rotations, although I know all the good things. On the one hand, the linearized equations in rotating coordinates are hyperbolic. On the other hand, the motion during spinup and spindown is complicated by the existence of inertial modes, which represent in some sense the same kind of thing. On another hand, there are inviscid and viscous instabilities, one of which I think is involved in Kerrebrock's study. I have never really been able to sort this out. I keep telling myself, all right, it is just gyroscopic rigidity and gyroscope and everything goes fine. It doesn't work out that way. I am still trying to put together some kind of primitive understanding of general and special effects of rotation in the case of internal flows. I think there is some delinquency on the part of the academic community. You would expect new developments, and synthesis of ideas that have been advanced in various forms in this area, to come from the academic community. So far they have really not done so. I simply point out what I see as a blank spot in the picture. Maybe I am mistaken about the degree of understanding that industrial people have on this subject.

A. A. FEJER: (Illinois Institute of Technology)

I would like to comment on Professor Reshotko's statement that everything is well understood as far as the axial compressor is concerned. In the past we have designed engines for a narrow band of operating conditions. In other words the system was arranged in such a way that the engine was always operating close to its design point. With increasing fuel prices the global efficiency of the system is becoming increasingly important and more and more emphasis must be placed on broadening the operating range of engines. In order to do this without costly and time consuming trial-and-error type development procedures that have been commonly used in the past we must be able to understand what is going on in engines at off-design conditions. In addition, in particular in the case of airplane engines, we will be raising operating temperatures to the highest values that the metallurgists will permit us to use. This will require higher pressure ratios and smaller frontal areas. Other new areas may also be opening up that should be anticipated. One must also consider the aspect of engine noise which is not yet fully understood. So I think there are plenty of problem areas that can benefit from the modern experimental techniques that Dr. Laufer was referring to in his opening remarks and lead to a better understanding of the flow processes in gas turbines.

D. HUFFMAN: (Indianapolis Center for Advanced Research)

I would like to amplify Dr. Fejer's remarks by saying that the major problem area with modern turbine engines which have compressors operating at transonic blade speeds is aerodynamic vibrations either induced by turbulence or self-excited flutter. There have been some very spectacular failures as a result of this mechanism, i.e., the RB 211 and the JT9D, both of which required extensive redesign of the compression system. Both of these failures involved unsteady aerodynamics which to date are still not well understood. Experimental techniques are still under development in this area and a good deal of work remains to be done. Unsteady aerodynamics in conjunction with high temperature environments are probably the most important problem areas in modern aircraft turbine engines.

G. SOVRAN: (General Motors Research Laboratories)

As one of the few representatives of an industry in the audience, I would like to address some comments to Prof. Liepmann. His separation of problems into scientific and technological is a separation that itself happens to be a crucial difficulty that industry often faces. A typical definition of an industrial problem is, "It doesn't work, fix it!" Professor Liepmann would like to think of a situation where the industry would come to basic researchers and ask for assistance, but without suggesting what needs to be done. The question is, to who would you go? You have to know something reasonably detailed about the problem and have a fairly good idea of the basic difficulty and what needs to be learned before you can even consider who the proper individuals might be to do the required basic research. In internal flows, one of the difficulties is that the fluid mechanics are so complicated that it is extremely difficult to identify the real problem that needs to be worked on. I will illustrate the difficulty with the procedure that Prof. Liepmann thinks is desirable by using a current problem from my industry. The emissions from engines are too high and the fuel economy is too low, fix it! To whom do you go, and what specific research do you ask them to do?

H. W. LIEPMANN:

I don't think you understood me correctly. Significant problems in industry usually involve a host of external constraints including financial conditions and government regulations which are foreign to, and not well understood in, universities. Furthermore, very often a quick, specific solution is expected and needed. In contrast, university research tends toward slow progress in developing understanding and in finding general solutions in contrast to very specific ones. The time constant in university research is, after all, largely set up by the maturity of research students. Direct help from a university team for an "industrial fix" is available if the team has already worked in the appropriate field for some time to lay the groundwork. To this end, exposure of university research to important industrial problems is very important indeed, because it frequently results in a change in research emphasis, and certainly readies the university group for the next time a similar industrial problem is posed. It is, therefore, crucial to keep a continuous dialogue between industry and universities; but the panic button approach is very rarely successful. Hence, industry must be prepared to preach to the universities, often without immediate response, and the university researchers have to be prepared to listen, often with the frustrating feeling of helplessness. The occasional clicking of a successful exchange makes this very worthwhile.

G. SOVRAN:

Apparently I did not say the right things. I am agreeing with you in principle, but we may not have the same interpretation of what is involved in "just tell us the problem." In the example I used about internal combustion engine emissions and efficiency, if I just present it to this group in the manner that I did, it won't define anything for you. None of you is going to rush off and do research for me because you would have no idea of what is really involved. The industry has first to have sufficient understanding of its problem so that it can identify the expertise required, and to at least point the people who have it in the right direction. What is it that I really need to know about the fluid mechanics inside an engine cylinder in order to improve either emissions or fuel economy, and hopefully both? I don't know, and therefore I have great difficulty posing relevant problems to any of you gentlemen.

This is also a difficulty with most internal flows encountered in the industrial situation. Their degree of complexity is so much greater than for most of those we have been discussing at this meeting. There is no such thing as a two-dimensional flow or low free-stream turbulence. The flows are highly three-dimensional, and it is often very difficult even to conceive the kind of information that you want to know. I am treading on somewhat shaky ground here with so many of you people having an aeronautical background but, relatively speaking, the configurations and flows you have dealt with outside of engines are very simple compared with the kind of things you run into inside engines and the normal industrial flow system.

W. WILMARTH:

And furthermore you have got to find a problem you can work with that is simple enough, so that a student can help you and/or do his thesis work.

P. M. BEVILAQUA: (Rockwell International)

The organizers of a workshop like this one have a very difficult job, which really only begins when they open the door between the boy Panda and the girl Panda. I would like to step into the adjoining cage by providing Prof. Liepmann with a short list of internal flow problems that will be of continuing interest to us.

(1) Extra strain rates have a significant, and in some cases, even dominant effect on the rate of confined jet mixing. The effect of streamwise curavture is generally recognized, but there is also a strong influence of lateral straining. We have observed that changes in duct cross section, as in the transition from square to rectangular sections, have a large influence on the entrainment of a jet passed through the distortion. The lateral strain rates change the structure of the turbulence in a way not predicted by current turbulence models. We have tinkered with some of these models, but in the spirit of Prof. Liepmann's request, I won't go into that.

(2) In very long ducts the question of turbulent memory becomes important. How long does it take for the turbulence in a ducted jet to forget that it originated in a jet and recognize that it is now part of a pipe flow? As another example, how long does it take the turbulence in the initial boundary layers to recognize that they have merged and become part of a pipe flow? While this information could be used to develop a sophisticated transport model, I would like to make the observation that for reasons of cost and complexity, such methods are slow to find engineering application. A simple answer to this question in terms of eddy scale, or mixing length, would be very useful.

(3) Lastly, we have a particular interest in predicting the separation of turbulent boundary layers and turbulent wall jets in adverse pressure gradient of an internal flow. As you may know, the presence of the opposite wall seems to have a stabilizing effect on internal separation. Something analogous to the Stratford criterion for external flows would be useful to have.

These questions would seem to be sufficiently general to be of interest for a long time to come.

E. RESHOTKO:

This remark is directed as well to the representatives of the funding agencies here. The academic community cannot wait for industry to confront it with its problems. Because the academic community works at a rather slow rate, it is incumbent upon the academic community and the sponsors of its work to some extent. I do have a list of some things along the lines considered at the workshop that I think ought to be undertaken in the next several years.

For one thing I think we will be dealing more and more with flows in which there are interactions. I am going to use the word interaction rather broadly for a moment. In steady flows the interactions are the ones just stated. In a confined flow there is no inviscid freestream, so that the flow away from the boundary is very much influenced by what is going on along the boundaries. We have to have procedures that will allow for the interaction between the flow away from the wall with the boundary layers such that there is mutual interaction. While the development of such procedures is ongoing for external flows on wings and bodies, they have to be initiated in internal flows for a great combination of flow regimes. We have to consider cases where the outer flow might be subsonic or supersonic or transonic and within a boundary layer flow, of course, we are dealing with the low speeds up to sonic, and maybe higher depending on how the outer and inner flow fields are identified. When it comes to transonic or supersonic outer flows we have to be able to trace waves, particularly shock waves, toward the wall so that we know where the impingements are and what their tendency is toward separating the boundary layer, because again this is an element of interaction that will have a great influence on the overall flow field and on performance.

When it comes to subsonic external flows there is not the economy of the hyperbolic calculation procedures that one has for supersonic flows so we now have some massive elliptic interaction schemes. We are going to have to learn how to simplify these procedures so that they enter into the realm of the possible. Jack Kerrebrock mentioned a three-dimensional inviscid calculation. I guess that is the one that Dave Oliver has been working on. That is a very impressive calculation of the inviscid flow through a three-dimensional flow passage. And yet, if the boundary layer development on those passages is significant, then there has to be a way to alter that inviscid flow in order to get a better view of the entire flow field.

When it comes to unsteady interactions I think our eyes were opened up today to the real unsteadiness of wakes and the nonreproducibility of wakes in turbomachinery, particularly behind rotor rows. These nonreproducibilities may be due to vortices generated from the hub-blade corners or from the scraping of the housing boundary layers by the blades, or else, very simply, that the outer portions of the boundary layers developing on the blades themselves are intermittent and each wake is being caught at a different part of its intermittency history. For whatever reason these nonreproducibilities exist the next row of blades has to accommodate the consequent variety of conditions and a design according to the time-average of all of these wakes may be inadequate. Perhaps there is some improvement possible by considering specifically the unsteady elements of these blade flows.

We now come to the interaction between these wake structures and the duct modes. Are these important? The evidence that we saw today suggests that they are important at distances on the order of one chord length downstream of a rotor. But in a piece of turbomachinery, as also mentioned, the spacing between stators and rotor, is much less than one chord length. Is there really opportunity for the duct modes to develop at each stage? Possibly not. Are the duct modes important just behind the last stage? More generally, what is the importance of the duct modes in turbomachinery? Are they important because of their influence on noise problems in turbomachinery or also because of their influence on the flow field?

The aforementioned are examples of interaction phenomena. They will require more work on three-dimensional boundary layers about which we heard very little today and yesterday. Professor Saffman mentioned his interest in extending his techniques into three-dimensional boundary layers. Peter Bradshaw, I understand, has some three-dimensional procedures in his bag of tricks and I also have been working on three-dimensional steady turbulent boundary layers with the idea of incorporating them into interaction procedures.

The development of flow calculation procedures takes many, many years before one gets anything that is even semi-reliable. We cannot wait for industry to identify every little problem. We have to start out ahead of time because of the rate at which we work and perhaps the standards to which we subject ourselves; since it will take some time before we can obtain results that are useful to the industry. These are the elements of anticipation.

J. LAUFER:

We have heard from a number of individuals who have in the past been working in turbulence. They indicated that indeed they have learned something at this meeting and have gotten some ideas for future research. I think it would be very interesting to hear now from those individuals who are actually worrying about the practical aspects of the problems. I wonder if I could put Jack Kerrebrock on the spot here and find out from him, has he really gained any useful information from the turbulence practitioners and whether he has any suggestions what other problems one should look into?

J. KERREBROCK: (M.I.T.)

The answer to the question is, yes, of course, I have learned a great deal!

I want to respond to two or three of the remarks here. First the last one Eli Reshotko made, about the importance of the evolution of the flow behind one blade row before it comes to the next one. I would just point out that the rotor-stator spacing in the big new fan engines is many chords. Those fans are pushing an awful lot of us across the country and 1% on fan efficiency is 1% on fuel consumption and there are thousands of dollars lost because of the inefficiency in the fan. So there is a motivation for you.

To Prof. Coles, welcome on board and I am sure that if you and some people like you get interested in this question of rotating flows in turbomachines we will make more progress than we have in the last few years.

Lastly, I would like to come back to the comment that Eli Reshotko made and differ with my good friend and colleague for the moment. I think that the position he took was that since we already have a polytropic efficiency of 90% in good compressors, we were not going to make big gains there. I don't think that is right, for two reasons. One is that if you look at the evolution of modern compressors, over the last 10 years, you will find that as we raised the tip speed in order to get higher pressure ratio per stage, the polytropic efficiency fell. We have lost about 5% or more in the last 10 years so that the modern stage is less efficient than the one built in 1955. I submit that there is no fundamental reason for that. If you calculate the shock losses in that modern transonic stage you will find that they are very small as limited by fundamental considerations like the pressure ratio across the shocks which are necessary, given the Mach number at which the blading operates. So there is something going on which we don't understand. Let me just push that point a little bit further and say that if you calculate the fundamental loss that a turbomachine has to sustain because of the Reynolds number limited shear stresses on all of the internal surfaces of it, you will come up with an efficiency that is closer to 97% or 98% than to the less-than 90% levels that we find now. I don't think that we really understand where that six to eight percent has gone but it is there for the having if we understand it in the same sense that I think some people in the air frame industry understand the flow over a wing or a body. If we can gain a few points of efficiency it is really worthwhile in terms of fuel consumption and dollars.

E. RESHOTKO:

I don't disagree with you at all. I don't know where or how we are going to find that extra 7% for current compressors. I do believe that because people have been looking for it now for more than 20 years, and I assume will continue to look, that maybe we will find a percent or two. But you are absolutely right that as we push the pressure ratio per stage upward, we will have to fight very hard to maintain even the current levels of polytropic efficiency; and that is certainly worth doing because the drops are very severe once you go into transonic and supersonic staging. Maybe I am a skeptic here, but I don't know if we will ever really recover that extra 7% that you are looking for.

J. LAUFER:

I would like to give the opportunity to the audience at this stage to consider the problems connected with modeling, with theoretical approaches and I would therefore like to invite discussion on this topic.

R. E. FALCO: (Cambridge University, England)

I was somewhat impressed that people who do modeling work haven't brought to our attention any new modeling problems due to the new input into turbulence, that is, rotation or other forcing effects. In my talk I indicated that if you put rotation on a turbulent flow that you might in fact see a very different picture. After proceeding downstream to axial positions where the turbulent intensities were lower than non-swirled jet values, the observed microscales were larger, which means that dissipation is going down. Stan Corrsin made a comment on the classical understanding that dissipation is Reynolds number independent in turbulent flow and reiterated that it is not a well understood phenomena. I think that it is very possible that in the case of swirling flow, at least, that it may not be true. Would anyone like to comment on that, particularly people involved with the modeling problem? Maybe we have to learn a little more about the physics of dissipation in turbulence under a strong rotation before we will be able to properly model it.

J. LAUFER:

I think that before we go into much specifics, could we rather address the general problem of modeling, how we go about it, what are future improvements and so on?

P. SAFFMAN: (California Institute of Technology)

I would like to say something very general about the philosophical attitude of the theoretician. I think that the theoreticians are now far too conservative. The interaction between theory and experiment is now one way and is one way by default. That is, the experimentors are doing the experiment and then the theoreticians are running along afterwards to see how successfully they can "postdict" all the data and it has become clear that they can postdict anything that you like.

Now that we have the models and we have the computers the theoreticians can be more positive. They can calculate flows now and they can try and investigate phenomena theoretically which the experimentor has not yet got around to measuring. That is, we can actually make genuine predictions of flows that have not yet been studied and I think that the theoreticians should now start doing this. It is something that we intend to do as far as possible and I hope that the experimentors will then cooperate in the way that theoreticians have been cooperating in the past. That is, that when the theoretician comes up with a prediction the experimentor will seriously consider how he can do an experiment on the phenomenon which the theoretician has been calculating. I hope also that the experimentor will also take care to make sure that he doesn't postdict because I know of cases where incorrect theories have been successfully checked experimentally!

S. KLINE:

Yes, I agree strongly with Prof. Saffman. I would wholeheartedly endorse that. It was one of the things that I was going to say, but I would like to go one step farther. If the predictions can be made in such a way that it makes possible the use of strong inference rather than induction it would be extremely valuable. What I mean by strong inference is the following. Set a hypothesis pair that exclude each other, say, A1 and A2. If you can <u>disprove</u> A1, then you are left with A2. Whereas if you merely do induction, which is all we have been doing so far in the turbulence and turbulent boundary layer problems, then the logic does not close. It therefore lends itself to endless arguments, and we have some of those going. Let me give a ridiculously simple example. If we make the hypothesis: "Every animal that has two legs is human," we would shortly discover that there are two-legged duck in the lake just outside the building, and that would get rid of this ridiculous hypothesis. We would not have to go on arguing about it endlessly; negative inference settles the question with finality.

J. T. C. LIU: (Brown University)

In connection with Prof. Saffman's remarks, I should like to say that ultimately the prediction schemes for turbulent shear flows must include the mechanisms of the large-scale structure interacting with the much smaller scale random turbulence (and with the mean flow), such as those situations observed in turbulent free shear flows by Brown and Roshko (1974) and in turbulent boundary layers by Kim, Kline and Reynolds (1971). In certain turbulent shear flows where such organized large-scale structures do occur, it might be fruitful to sort out the large-scale structure from under the overall average and look at the interactions rather than sweep everything under the overall average. To this end, nonlinear hydrodynamic stability theory can be of much service towards the understanding of the large-scale structure problem. Of course, one ought not to look at such problems as one in hydrodynamic stability to the exclusion of the small-scale turbulence, nor from the present prediction point of view ignoring entirely what now observationally appears to be the omnipresence of the large-scale organized structure, but rather from a healthy blend of the two viewpoints. Recently, Dr. L. Merkine and I have made an attempt towards the understanding of the interactions between the two different scales of fluctuating motions in a turbulent shear flow (Liu and Merkine, 1976). It would be extremely helpful if experimentalists and theoreticians were to interact along the way rather than to compare notes after each had done his own thing!

References

1. Brown, G. L. and Roshko, A. 1974 On density effects and large structure in turbulent mixing layers. J. Fluid Mech. 64, 775.

2. Kim, H. T., Kline, S. J. and Reynolds, W. C. 1971 The production of turbulence near a smooth wall in a turbulent boundary layer. J. Fluid Mech. 50, 133.

3. Liu, J. T. C. and Merkine, L. 1976 On the interaction between large-scale structure and fine-grained turbulence in a free shear flow. Part I. Proc. Roy. Soc. Lond. A (to appear).

D. COLES:

In response to Prof. Saffman and others, I submit that a modeler who likes is welcome to calculate the unsteady mean flow past a smooth circular cylinder in a uniform stream at a Reynolds number of 140,000. When you are finished, I will show you the data.

I. WYGNANSKI: (Tel Aviv University)

I would like to pick up on Prof. Saffman's suggestion. I would like those people who do models and try to make predictions that they will also make a parametric variation of the constants that they put in their models so that we will be able to see and differentiate, what's the effect of changing slightly some of those constants on the general development of the flow. Very often we are faced with something which fits one particular data set because the manner in which the constants have been adjusted.

J. KERREBROCK:

I would like to speak in favor of people doing both theory and experiment and therefore being internally honest.

D. E. ABBOTT: (Purdue University)

Jack Kerrebrock has a very important point and I couldn't agree more. However, in addition to the turbulence modeling discussed up to now, there is an additional area that needs to be emphasized and no one has picked up on as yet. We have heard from something like three to five predictors of wake-type flows and some tested full elliptic equations while others have solved parabolic equations. The intent of most of these investigators (I believe Dave Huffman mentioned this) is to calculate a turbulent flow as it interacts with some downstream condition. For example, Jim Whitelaw commented on the first day that at a bend in a duct, he measured an upstream influence of some 3 to 5 or 5 to 6 duct diameters. Certainly for such cases of strong upstream interaction, parabolic equation models are incorrect for any turbulence model. I would like to suggest this as an additional area where good experiments are needed to direct the proper formulation of the mathematical equation, a step that may be almost as challenging as turbulence modeling in the long run.

S. KLINE:

That is hardly new. The experimentors told us this in 1911. If you look at the data* for subcritical flow over a circular cylinder normal to the stream, you see separation well before 90°; the

---

* See for example, Schlichting, H. "Boundary Layer Theory," 6th Ed., McGraw-Hill, 1968, p. 23.

point of maximum thickness. You cannot get that result out of the parabolic equations. You have to put in the actual pressure distribution to make the boundary layer (parabolic) calculation come out right. If you read carefully what Schlichting says at various places, you will see that is how he solves the problem. He does put in data, but he does not comment that this is not *a priori* predictive; it only postdicts as Prof. Saffman aptly put it. The total problem is really elliptic--not parabolic. Most of my own research right now is on separated flows, and we are now able to predict several classes quite well, but we have to include elliptic effects to do so. If there is a single general conclusion from our work thus far, it is that if you get the interaction right between the separated zone and the other flow in the sense of putting in the displacement thickness (or the blockage or whatever you want to call the separated zone), then everything will fall out well. Moreover, what boundary layer method you use is not nearly as critical as some of the early discussion suggests so long as it's a decent one. If you get the interaction, the elliptic effects right, a simple method will work well; if you don't then nothing works. I am referring here to turbulent cases. It is different, of course, for the laminar cases; you can put them into the computer and do them exactly. For the turbulent cases we are using approximate models. This implies that we are using some kind of intermediate modeling and there is nothing to say that we can't use zonal models. Indeed, we get good results with zonal models when elliptic effects are included. But it does worry me to see people using parabolic methods for branched flow fields, as I noted above. The problems do show up in the numerical work. You saw this in George Mellor's presentation. He had to do something about the flow off that corner in order to get the reattachment right. If you had a flow where you didn't have Doug Abbott's data, how would you close that problem in a numerical scheme? Maybe there is a way to do this, but the problem concerns me. Maybe George can comment on this. I do think that we need to look more carefully at how to do those interactions. That is the point I am really making.

G. MELLOR: (Princeton University)

Sure I would like to comment on that. The corner problem I feel is strictly a resolution problem. It is not a physical problem. The point of my discussion on that particular point was that, if you get smaller grid scales the problem vanishes. It is not an independent case at all.

S. KLINE:

I don't understand. You said that you had to put some condition which would amount to a Kutta condition at that edge. How would you have known which one to choose if you had not had Doug Abbott's data? Resolution or no resolution, you are going to have a grid size.

G. MELLOR:

I saw from the data that the flow did not go around the corner at acute angles. I think probably you would guess at that without seeing the data at all. There are only two things that it can do. The flow can go straight off the surface or it can all of a sudden have an acute angle. I think it is pretty reasonable to rule out the second possibility.

S. KLINE:

It is quite clear when you get the sharp corner, that is the easiest possible case. If you take a case where you have a separation off a gradual surface, in fact you don't know where that streamline goes and then you really have a matching problem.

G. MELLOR:

The problem is not there any more. The singularity goes away and that case ought to be calculated quite easily.

J. WHITELAW: (Imperial College, London)

I would like to make one and a half points. The half point is to reinforce what Jack Kerrebrock said about interactions between experiments and calculations. I believe that if you perform experiments in conjunction with calculations you are more likely to improve knowledge of turbulent flow and the ability to calculate flow than by doing one without the other.

Now to come to my major point. I don't believe that models *a priori* are applicable to all flows. I said this before in this meeting and I say it again. We take equations that we believe to be exact and time-average them. As a consequence, we have thrown away information and need to build hierarchies of turbulence models which come to some point of diminishing return. Who is going to believe that we can find universal constants, or that we can find problems that will be exactly represented by this approximate set of equations? No one. Given that we have equations that are approximate, then a lot of the difficulties can be avoided by simple experiments which guide the calculation; the calculations can then be used as a means to interpolate and extrapolate, interpolate with confidence and extrapolate with considerable care. So I am in favor of an interaction between calculations and measurements. I have no great faith in calculations without some checking: but considerable checking has already been done and, in a fairly wide range of flow configurations, we can calculate and reasonably expect the results to be correct to a precision satisfaction for many purposes.

E. RESHOTKO:

Just one short remark. The cylinder problem that Steve Kline mentioned is an example of an interaction problem, such as I was referring to earlier, where you have to deal with the boundary layer and the external flow together for the overall flow to be properly predicted. I guess the use of the experimental pressure distribution simply replaced the interactive calculation of the external flow.

J. LAUFER:

I go along with Dr. Whitelaw's comments. There is no question that there has to be an interaction. On the other hand if the so-called modeling calculations are indeed predictive in the true sense of the word, they should give us somewhat more confidence than we have in them at the present time. I am leaving this meeting with a rather uncomfortable feeling that modeling techniques can predict the experiments extremely well, *a posteriori* and I am sure that you don't mean to imply that. What are the limitations? What are the areas of application for a particular model? I don't think at the present time this is clearly indicated.

J. WHITELAW:

I was trying to say that I don't believe in many cases, *a priori*, that a calculation method based on turbulence models is going to give me good results until some experimental testing has been done. I say again that we operate with inexact equations and undoubtedly inexact assumptions, many of which we have no means to check directly. We can't check dissipation equations exactly and directly. So we have to have an interaction between the experiment and the calculation.

P. BRADSHAW: (Imperial College, London)

I would like to second what Jim Whitelaw has said, I am quite sure that the constants are not constant. I also doubt, as Jim does, whether experiments are going to answer all the questions which are posed by calculation methods simply because a lot of the terms in equations like the dissipation equation, or even the pressure strain term in the Reynolds stress equation, are currently unmeasurable. I think that even to get calculation methods of the current levels to operate over a reasonably wide range of flows (excluding some of the most awkward cases) we are going to have to rely on computer simulations of the time-dependent turbulence to extract these unmeasurable quantities. I really see this as the big hope of perhaps the next 10 years. When you look at the results that the large eddy simulators are getting they are starting to look like rather crummy experiments. There is a lot of scatter, there aren't very many points on the profiles, and so on, but they are improving. In particular, when the subgrid scale modeling, which is the main difficulty at the moment, has been sorted out a little bit better, I believe that we will be able to extract data for "unmeasurable" quantities from these simulations, plug the quantitative data into our turbulence models and both get a better physical understanding of what our models represent and extend the range of validity of those models. I think that in the foresseable future, which means the working lifetime of more-or-less everybody here, we are not going to have another set of equations with constant coefficients which describe turbulence. Navier and Stokes have given us one set of equations with constant coefficients and I don't believe that there is another set.

P. SAFFMAN:

I hope that I have misunderstood what Mr. Bradshaw has said. He wants to calculate random solutions of the Navier-Stokes equations so that he can calculate things like the pressure-strain correlations which he will then feed into the model equations and then solve them. But if you can solve the Navier-Stokes equations then you don't need these models.

P. BRADSHAW:

I should explain myself a little bit better. What I am suggesting is a <u>small</u> number of expensive Navier-Stokes simulations (analogous to basic experiments) for some carefully chosen flows, including some of the complex flows, which we have been playing with over the past few years and which we take to be initially well behaved turbulent flows and kick them around with extra strain rate interactions and so on. Given a few of those Navier-Stokes simulations (a number equal in order of magnitude to the number of good data sets we've got), we can very much extend the validity and reliability of our models. Clearly if computing ever becomes cheap enough for us to do Navier-Stokes solutions for each and every engineering configuration then the modellers are right out of business for good and, in effect, experimental turbulence research will have gone the way of experimental stress analysis at least for a generation or so.

P. SAFFMAN:

Can I just answer that and say that in connection with the pressure-strain correlation, which seems to play a fundamental role for the people who model averages of the Navier-Stokes equations, term by term, the pressure is not a local quantity. It is determined by nonlocal effects, by the whole flow field. As Dr. Liepmann keeps saying, you should perhaps think of it as a Lagrangian multiplier for the constraint enforced by the requirement that the fluid is incompressible. I think that to believe that you will find a local modeling of terms involving pressure is to be very optimistic indeed.

S. KLINE:

I just want to comment, there are a couple of cases that have been run which are precisely what Mr. Bradshaw has been describing. These are the recent work by Clark and Shaanan, who are students of J. H. Ferziger and W. C. Reynolds at Stanford; they do tell us a little bit. Most of these are subgrid calculations, but they also include one "honest" calculation which is a complete numerical solution of the Navier-Stokes equation for a low Reynolds number flow. What emerges when you look at the structure of those computations is that you can only do the complete "honest" calculation for relatively low Reynolds numbers and relatively simple cases. As everybody known, when you go to higher Reynolds numbers, the ratio of scales that you have to cover in your computer is proportional to the Reynolds number. Therefore you get into monumental time and cost problems at higher Re. If you stick with lower Reynolds numbers and with simpler cases, you can run a complete calculation for simple turbulent cases. Reynolds, Ferziger and their students are doing the kind of sorting out that Peter Bradshaw is talking about, and that will help us, though only for certain very specific cases, just as he says.

P. BRADSHAW:

I agree one cannot solve exactly all the way down to the small eddies; you have got to have a subgrid scale model. When you have got a subgrid scale model there shouldn't be any limit on Reynolds number with the possible exception of dealing with near wall flows.

J. LAUFER:

I would like to proceed to cover one more area and address my question to the modeling people. So far, at least, most of the presentations here yesterday and today used the average Reynolds equations as a starting point for their modeling. In view of some of the experimental observations of the past few years, I wonder, is it at all possible at the present time to exploit the "double structure" nature of turbulence? That is to say, is it feasible to formulate the problem in a fashion that takes into account the two rather distinct length scales observed in turbulent flows?

E. RESHOTKO:

Let me, in that context, raise the question: How can we incorporate our knowledge of the large-scale coherent structures into flow field calculation?

It is evident that the large-scale structures have always been there. I guess most of us are convinced of this now although for many problems they did not appear explicitly in the modeling because the modeling was either not sophisticated enough or detailed consideration was not required.

We find that the large-scale structures show themselves most prominently in free shear flows as well as in the outer portions of boundary layers, where even our present day modeling techniques have indicated that the mixing lengths or characteristic eddy lengths are of the order of the thickness of the free shear layer or of the boundary layer. So in a sense the large structures have in a time-averaged manner already been modeled. We have to ask ourselves the question in dealing with the large-scale structures, where do they make a difference? In what problems must we really know something specific that is readily identifiable with large-scale structures? We have a few examples: One very clear example has to do with the mixedness between two streams and the relationship to combustion. It is important that we have some idea of the contact surface between a fuel and an oxidant that are being mixed in a free shear flow prior to reaction. A knowledge of the dynamics of large-scale structures should be helpful here. I believe that in the experiments of Roshko and co-workers they have made especial attempts to measure mixedness in order to help in the evaluation of such combustion theories.

If we are interested in considering the unsteady aspects of turbulent boundary layers and the wakes produced by such boundary layers then we will have to do some more specific modeling of the large scale structures. But I emphasize that we are interested in the time dependent or unsteady aspects rather than the time averaged evaluation of the large scale structures.

P. M. BEVILAQUA:

I would like to express an opinion, based on Prof. Wygnanski's work presented at this meeting. I suspect that the contribution of large structures to our modeling may be in what we call now the Kármán constants, the unknown constants, that is as the large structures were changed in Prof. Wygnanski's experiment we always saw the same generally self-preserving structure but the spreading rates were different. Perhaps that might be the approach to take to include the large structures in the modeling.

W. WILLMARTH: (University of Michigan)

No one has said much about the small scale turbulent structure. I don't think we know much about turbulence at high Reynolds numbers especially near transonic speeds. In our laboratory we have made very small X wires to use in an ordinary flat plate boundary layer. We find that below $y^+ = 500$ there are very small scale turbulent structures which are intermittent and are not measured properly by ordinary length hot wires. The hot wires that we used had a length of $L^+ = 4$ (i.e. four viscous lengths). From what I have seen in the turbomachinery results, you must have a lot of small scale turbulence structures in these internal flows that must be dissipating energy but this I am sure has not yet been measured properly.

S. N. B. MURTHY:

It remains for me to thank the members of the Panel and each of you among the participants for a thoughtful discussion. I shall try to produce a set of recommendations based on your statements, no doubt realizing that some of them may indeed be tentative in a subject as complex as turbulence in flows of rather complex geometry.

Summary Report

S. N. B. MURTHY
Purdue University
West Lafayette, Indiana (U.S.A.)

For the purposes of this summary report including recommendations, the various topics discussed at the Workshop may be conveniently grouped under the following.

1. Turbomachinery flows
2. Analysis and
3. Fundamental problems.

1. TURBOMACHINERY FLOWS

The sponsors of the present Workshop on internal flows have been interested in general in jet propulsion engines. One aspect of turbulence in jet propulsion engine flows is transport and chemical reaction (Ref. 1). A second aspect is the influence of complexities due to geometry, curvilinear motion and rotation. The current workshop is concerned with the status, problems and prospects for advances in the latter category of problems. The central feature of these problems is, in the words of P. Bradshaw, the existence of strong interaction and deformation in shear flows and the influence of rotation. In jet propulsion technology, such flows arise in several contexts in connecting ducts and elbows, diffusers, nozzles and turbomachinery.

The engineering interest in such problems arises ultimately because of (a) momentum loss, (b) heat transfer, (c) entrainment of fluid from one flow into another, (d) dynamical coupling between flow and mechanical structure, (e) passage of distortion through the machine and (f) noise. While none of those processes was discussed specifically with respect to turbomachinery, the contributions made at the workshop have important implications to the effects of turbulence in those processes.

Among practical flows of interest in turbomachinery some aspects of the following flows have been discussed in the current workshop.

1. Duct flows
2. Flow manipulators
3. Coaxial jet with swirl
4. Injection of fluid into a stream
5. Flow at blade-hub junction
6. Cascade flows with freestream turbulence
7. Nonstationary rotor-generated flows in ducts
8. Unsteady boundary layers over blade surfaces
9. Interacting wakes.

They may be regrouped as follows from the point of view of the nature of complexities.

a. Flow with deformation
    i. Flow undergoing deformation in a curved duct
    ii. Flow at the junction of an airfoil with a wing
b. Flow with interaction
    i. Boundary layer flow with freestream turbulence
    ii. Unsteadiness in ducted rotor-generated flows
    iii. Interacting wakes
    iv. Mixing of a jet with a freestream
c. Flow with rotation
    i. Coaxial jet with swirl
    ii. Flow manipulators for swirling flows

From the point of view of design of turbomachinery, the main problems discussed in this workshop are the following.

1. Influence of freestream turbulence on the development of blade boundary layers under steady and nonsteady state conditions.
2. Influence of freestream turbulence and initial conditions on the development of wakes downstream of stators and rotors
3. Implications of the presence of organized unsteadiness simultaneously with turbulence
4. Implications of the simultaneous existence of secondary flows of the first and second kind
5. Modification of a given level of nonuniformity introduced by turbulence.

Each of those broad topics requires further investigation.

It was pointed out in the workshop that a general survey on the problems of turbulence in turbomachinery does not eixst. It is of course too early to attempt such a survey, principally for two reasons: (1) it is not yet clear what the precise effects of turbulence are in a flow as complex as that in a turbomachine and (2) experimental data are only beginning to be acquired that can separate the effects of turbulence from those of other features of the flow. The contributions of Bradshaw, Evans, Fejer, Kerrebrock, Kovasznay and Whitelaw illustrate the latter clearly.

2. ANALYSIS

A number of well-defined (and idealized, in the sense that they emphasize one feature of a flow more than others in a given flow situation) flows can now be calculated with varying degrees of empiricism. Measurements have also been undertaken on some of those flows. The following flows were discussed in some detail at the workshop.

1. Flow in a curved duct with a forward recirculation zone under laminar flow conditions (Whitelaw)
2. Development of turbulent mixing layers including the effects of initial boundary layers and freestream turbulence (Saffman and Leuchter)
3. Turbulent flows subjected to deformation and rotation (Saffman and So)
4. Fully separated flow (Mellor and Simpson)
5. Turbulent wake (Ng and Huffman)
6. Transitional flows (Wilcox).

The four basic turbulence models employed in the calculation of those flows are (1) effective viscosity model, (2) nonisotropic mixing length model, (3) Reynolds stress model and (4) phenomenological model.

In the case of fully separated flow at a step, Mellor introduces additional approximations for vorticity at the corner and in the viscous sublayer. In the general problem of predicting separated flow, Simpson shows that turbulent normal stresses are important near and downstream of separation.

The calculation of turbulent wakes (Ng and Huffman) is based on the "interaction hypothesis" of Bradshaw (1972) according to which two shear stress fields can be superimposed so long as the turbulence structure in each field can be assumed to be unaffected by the presence of the neighboring field. The applicability of weak interaction hypothesis depends upon the establishment of limits for such weak interaction.

The phenomenological model of turbulence has been undergoing steady development by Saffman since 1970. There is no term-by-term modelling of Navier-Stokes equations in this. Without altering the nature of the equations one has to deal with finally, Saffman attempts to model certain physical processes associated with the presence of turbulence. The resulting two-equation model (for e and ω) has been employed in predicting a number of flows. In the paper presented at this workshop, Saffman applies this model (a) to a flow including scalar diffusion with a new similarity variable representing the scalar quantity and (b) to simple flows with deformation and rotation utilizing the concepts of an equilibrium Reynolds stress and the approach of the actual Reynolds stress to the equilibrium value through a relaxation diffusional process.

Predictive methods are necessary in engineering practice. It is important not only to be logical and consistent but to have sound physical bases for models. One can object to time-averaged equations and show their limitations. However, in the context of such equations, one has to be selective about terms or processes or relations among them for each class of flows, if only to reduce empiricism. The question therefore is probably not "should we do further research in this subject" but "who should investigate modelling development." At the same time, it is clear that investigations are required at various levels in understanding the implications of superposability, similarity and relaxation with respect to certain turbulent processes, the method of including (and eliminating) certain stresses in various situations and the method of accounting for additional strains as in rotating flows.

3. FUNDAMENTAL PROBLEMS

Turbulence has remained a basic challenge in fluid mechanics and therefore in any group concerned with turbulent flows fundamental questions will arise. Some of the more interesting questions raised pertain to the following.

1. Effect of initial conditions on the structure of turbulent mixing layers (Wygnanski)
2. Effect of freestream turbulence on transitional processes in free shear layers (Bradshaw)
3. Superposability of turbulence and accommodation in structure in a developing flow (Bradshaw)
4. Eruption and streak formation as a consequence of interaction between concentrated vorticity and a viscous layer (Walker and Abbott)

5. Possibility of investigating characteristics of local (space and time) and instantaneous flow phenomena with computer experiments (Corrsin).

Computer experiments on variously idealized flows will certainly be performed extensively in the future. The success of such experiments will depend on the same factors that physical experiments depend on: clear definition and sufficient analysis to elucidate the basic processes. Computer experiments are probably the only practical means of establishing the anisotropy of pressure-velocity correlations. Similarly, in many flows undergoing deformation, computer experiments will be valuable whether one is dealing with moment (conservation) or process equations, for example in the interaction of large scale vorticity with turbulence. However, there is no complete substitute for physical observation in physics of fluids.

Each of the five areas mentioned earlier involve various controversies and important basic ideas that need further investigation.

All of the experimental results presented at this workshop pertain to laboratory flows. Some of the results were obtained with flow visualization (Falco and Bradshaw). Quantitative measurements had been obtained using either imbedded probes or nonobtrusive measurement technique, for example laser-Doppler velocimetry. It may be pointed out that these techniques are in general complementary although in special situations one or the other technique may be entirely inapplicable.

The workshop occasionally discussed data processing techniques, especially various types of conditional sampling. That technique and image processing, although sometimes misunderstood and misapplied, do have important applications in experimental research in turbulence.

REFERENCE

1. Murthy, S. N. B., Turbulent Mixing in Nonreactive and Reactive Flows, Plenum Press, New York, 1975.

INDEX

boundary layer	2, 37, 132
effect of external turbulent shear flow	101
effect of free stream turbulence	497
separation	311
unsteady--	502
structure	3, 132
"wake region"	103
burst	132, 134
dependence on Reynolds number	137
mechanism	131, 151
calculation methods	557, 559, 561
cascade	
of blades	297, 307, 482, 499
of vortices	130
cellular motion	11
sheared	11
wind tunnel version	12
complex flow	35, 543
compressor	439, 485, 515, 540, 545
with strong swirl	439
computer	
as a laboratory	12
correlation	
pressure-strain	27, 28, 29, 30, 31
cylinder	
flow past	554, 555, 558
wake	186
diffusers	1, 35
distortion	1, 522
eddy	3
microscale	114
scale	125, 129
"typical"	114, 116

film cooling 411, 420

flow
  curved 250, 407, 409, 412, 437
  fully developed 60, 61, 64
  secondary 2, 52, 53, 59, 60, 437, 521
  sudden expansion 251

free stream turbulence 36, 245
  and transition 42
  in jets 41, 395, 403
  near a wall 41
  scale effect 42, 58

helicopter
  rotor blade 517, 520

hot-wire anemometer 169, 433
  calibration 186
  constant temperature 169
  nonlinear theory 169

interaction 35, 36, 64, 101, 105, 110, 151, 383, 550
  approach 283, 385
  hypothesis 284
  measurement 64

interface 61, 63, 105

jet 41, 84, 548
  coaxial 388, 403, 409, 417
  effect of swirl 113, 125, 417, 434

laser-Doppler anemometry 429, 434

Law
  of the wall 38

length scale 37, 58, 110, 244, 250, 277
  equation 250, 279
  master length scale 250

mixing layer 45, 56, 67, 371
  compressible 196, 230
  developing region 89
  effect of density gradients 96
  effect of initial conditions 67, 86, 91
  effect of streamwise diffusion 194
  modeling 194, 371
  passive scalar diffusion 204
  shadowgraphs 97, 99
  similarity solutions 191, 227
  spreading rate 68, 81, 87, 231

oscillations
  duct mode 541

period
  quiescent 136, 163

phenomena
  frozen stress 432

pressure field
  calculation 21
  in rotors 461, 462

reentry
  vehicle 236

relaxation stress model 210
  homogeneous turbulence 219
  fluid in solid body rotation 221, 229, 230

sampling
  temperature-conditioned 39

second moment model 249
  for fully separated flow 249

separation 2, 229, 249
  behind step 260, 276, 556
  intermittent 340
  prediction 319, 324, 548

shear flow 11
  homogeneous 11
  wave model 17

smoke injection 43, 59, 63, 533

stall 341
  dynamic 517
  propagating 459

structure
  large scale 45
  small scale 562
  three dimensional 48, 56
  two-dimensional 47, 61
  vortex-ring 116, 125

superposition 39, 63, 65, 283

swirling flow 113, 347, 417, 521, 545, 551, 552
  generators 522
  modeling 349, 552
  stability 362, 545

Taylor instability 459

trailing edge 306

transition 3, 233, 511
  in blades 502, 511, 512
  low turbulence 244, 246, 247
  prediction 233

transonic rotor 439
  acoustic data 450
  unsteady flow 440

turbomachines 35, 372, 431, 485, 540, 545, 551, 562, 563

turbulence
  and centrifugal instability 127
  in internal flow 1
  in turbomachinery 1
  memory 32, 33, 548
  production 3
  stability theory 554
  structure 36, 115, 120, 127, 554 562
  theory 3

two equation model 191, 227
  curved flow 436
  mixing layer 194, 196
  scalar diffusion 204

unsteadiness
  ordered 36, 37, 485
  random 480, 481, 485, 514

velocity scale
  for swirling flow 347

viscosity
  tensor turbulent 227, 369

visualization 39, 114, 460
  volume technique 113

vortex pairing 43, 45, 46, 47, 48, 62

vorticity
  indicator 523, 533, 534
  suppression 522

wake 49, 75, 342, 372, 383, 439, 461, 516
  acoustic emission 474
  asymmetric 297
  cutting 463
  initial conditions 307
  merging 458, 460
  modeling 203, 283, 306, 383, 555
  of rotor 441
  unsteadiness 549

wall layer 131
  modeling 215
  structure 131